Springer Texts in
Electrical Engineering

Consulting Editor: John B. Thomas

Springer Texts in Electrical Engineering

Donald L. Snyder   Michael I. Miller

Random Point Processes in Time and Space

Second Edition

With 76 Illustrations

Springer-Verlag
New York Berlin Heidelberg London
Paris Tokyo Hong Kong Barcelona

Donald L. Snyder
Electronic Systems and Signals Research Laboratory
Department of Electrical Engineering
Washington University
St. Louis, MO 63130
USA

Michael I. Miller
Electronic Systems and Signals Research Laboratory
Department of Electrical Engineering
Washington University
St. Louis, MO 63130
USA

Library of Congress Cataloging-in-Publication Data
Snyder, Donald L. (Donald Lee), 1943–
    Random point processes in time and space. — 2nd ed. / Donald L.
  Snyder, Michael I. Miller.
      p.   cm. — (Springer texts in electrical engineering)
    Rev. ed. of: Random point processes. 1975.
    Includes bibliographical references and indexes.
    ISBN-13: 978-1-4612-7821-4   e-ISBN-13:978-1-4612-3166-0
    DOI: 10.1007/978-1-4612-3166-0
    1. Point processes.   I. Miller, Michael I.   II. Snyder, Donald L.
  (Donald Lee), 1943–   Random point processes.   III. Title.
  IV. Series.
  QA274.42.S69   1991
  519.2—dc20   91-10891
                                               CIP

First edition published by John Wiley & Sons, 1975 as *Random Point Processes*.

Printed on acid-free paper.

© 1991 Springer-Verlag New York, Inc.
Softcover reprint of the hardcover 2nd edition 1991

Camera-ready copy prepared by the authors.

9 8 7 6 5 4 3 2 1
ISBN-13: 978-1-4612-7821-4

PREFACE

This book is a revision of *Random Point Processes* written by D. L. Snyder and published by John Wiley and Sons in 1975. More emphasis is given to point processes on multidimensional spaces, especially to processes in two dimensions. This reflects the tremendous increase that has taken place in the use of point-process models for the description of data from which images of objects of interest are formed in a wide variety of scientific and engineering disciplines. A new chapter, *Translated Poisson Processes*, has been added, and several of the chapters of the first edition have been modified to accommodate this new material. Some parts of the first edition have been deleted to make room. Chapter 7 of the first edition, which was about general marked point-processes, has been eliminated, but much of the material appears elsewhere in the new text. With some reluctance, we concluded it necessary to eliminate the topic of hypothesis testing for point-process models.

Much of the material of the first edition was motivated by the use of point-process models in applications at the Biomedical Computer Laboratory of Washington University, as is evident from the following excerpt from the Preface to the first edition.

"It was Jerome R. Cox, Jr., founder and [1974] director of Washington University's Biomedical Computer Laboratory, who first interested me [D.L.S.] in the aspects of random point processes that form the theme of this book. At the time I joined BCL in 1969, there was already an active project dealing with the computer processing of radioactive-tracer data obtained daily in clinical and research laboratories in the Mallinckrodt Institute of Radiology of the Washington University School of Medicine. There was then, as now, a sizable group of enthusiastic participants in this BCL effort: Carol S. Cobel, Rexford L. Hill, Kenneth B. Larson, Joanne Markham, Nizar A. Mullani, and Maxine L. Rockoff were all members of this group. BCL's effort was conducted in collaboration with E. James Potchen and Michael M. Ter-Pogossian, directors of the Divisions of Nuclear Medicine and Radiation Sciences, respectively, and John O. Eichling, Michael E. Phelps, and Roger H. Secher-Walker. My thoughts on point processes have been influenced to some extent by all these people who became my colleagues in 1969. But it was Jerome R. Cox, Jr. who suggested my using the statistical estimation and decision theory I had previously applied only to communication and control problems for the analysis of radioactive-tracer data."

The changes we have introduced in this revision are in large part an outgrowth of our research over the past seven years on the development of methods for quantitative image-formation. Our research at the BCL has been a major stimulus for the newly developed models for positron-emission tomography and electron-microscopic autoradiography, all presented in Ch. 3. We also include new work in astronomical imaging of faint objects. We are indebted to Dr. Lewis J. Thomas, Jr., Director of the BCL, for his encouragement and support in the biomedical imaging projects. We are also indebted to our coworkers who participated with us in this work, especially Timothy J. Holmes, John M. Ollinger, David G. Politte, Badrinath Roysam, and Timothy J. Schulz, who contributed much to our understanding of the problems and the formulation of the new models we describe.

All the applications in which we have been involved are examples of nonparametric density estimation, which provides the major motivation for our new results on constrained estimation techniques. For these applications, the use of unconstrained maximum-likelihood estimation fails because the estimates are not consistent in the statistical sense; they do not converge, with increasing amounts of data, towards the quantity being estimated. Regularization of the estimates is, therefore, absolutely essential. There has been an explosion of methods for such regularization in nonparametric density estimation, including Grenander's method of sieves, penalized likelihood approaches, and Rissanen's description length complexity method. In the new Ch. 3, we focus on the use of Grenander's sieves and on Good's and Tapia and Thompson's work in penalized likelihood estimation. The application of these approaches for regularization is developed for point process estimation.

In Ch. 6, we also present new results for estimating hazard functions for self-exciting point-process models. This has been motivated by the recently developed models of neural discharge. We wish to express our appreciation to Dr. Charles E. Molnar for encouraging us to examine this problem. For solving the nonparametric problem of estimating a hazard function, we show how a Bayesian approach to selecting the model-order, from a family of hazard functions of varying orders of complexity, induces Rissanen's method of penalized likelihood estimation based on minimum complexity. We thank Kevin Mark for his help in Ch. 6 in describing his work on this problem.

Section 1.3 in Chapter 1 contains a detailed preview of this book. The strategy we have used starts in Chapter 2 with a detailed development of the Poisson process in time and space. Then, in subsequent chapters, a hierarchy of increasingly more complex point-process models is derived by making a series of modifications of the properties of the Poisson process. Each modification is introduced by indicating its usefulness in a real application. Thus, the approach is to start from the specific and work towards generality, with practical applications motivating the extensions along the way.

We are grateful to many students and colleagues who read the manuscript of this edition and made helpful suggestions for its improvement. Dr. Costas Georghiades of Texas A&M University, Dr. Joseph L. Hibey of Old Dominion University, Dr. Timothy J. Holmes of the Rensselaer Polytechnic Institute, and Mr. Abed Hammoud of Washington University all receive our special thanks for their help.

We owe thanks to the National Institutes of Health, through grant RR001380 from the Division of Research Resources and grant DC00333 from the Institute on Deafness and Other Communication Disorders, and the National Science Foundation, through grants ECS-8215181, ECS-8407910, ECE-8552518, and MIP-8722463 for support during the preparation of this text.

St. Louis, Missouri
January, 1991

Donald L. Snyder
Michael I. Miller

Contents

CHAPTER ONE

POINT AND COUNTING PROCESSES: INTRODUCTION AND PRELIMINARIES

1.1 Introduction

A random point process is a mathematical model for a physical phenomenon characterized by highly localized events distributed randomly in a continuum. Each event is represented in the model by an idealized point to be conceived of as identifying the position of the event. If X denotes the continuum space, then a realization of a random point process on X is a set of points in X. The number and variety of phenomena for which this type of stochastic process provides a reasonable mathematical model is surprisingly large. Here are ten examples indicating some of the forms X may have.

Example 1.1.1 *Electron Emissions* ───────────────────────

Suppose that the heating element in a video tube is energized at time t_0 and that electron emissions from the cathode are observed for the subsequent T seconds. Here, the space X is an interval $\{t : t_0 \leq t \leq t_0 + T\}$ of the real line $\mathcal{R}^1$. Each element of X represents a time during the observation period. A realization of the random point process on X is a sequence of instants $\{t_1, t_2, \cdots, t_N\}$, where $t_0 \leq t_1 < t_2 < \cdots < t_N \leq t_0 + T$. Here, t_i is the emission time of the ith electron, and N is the number emitted during the observation interval. ∎

Example 1.1.2 *Electrical Activity in Nerves* ───────────────

A microelectrode inserted into a nerve fiber can detect electrical activity in the form of randomly occurring spike discharges. An example observed in the auditory nerve of a cat is shown in Fig. 1.1. Suppose that the electrode is inserted at time t_0 and that discharges are observed for the following T seconds. The space X is that of Ex. 1.1.1. Here, t_i is the occurrence time of the ith discharge, and N is the number of discharges during the observation interval. ∎

Figure 1.1. Spontaneous electrical activity recorded in the auditory nerve of a cat.

Example 1.1.3 *Radioactivity* ————————————————————

Gamma or annihilation photons generated in radioactive decays are emitted at a rate that depends on the quantity of radioactive material present. Emissions in a T-second interval starting at time t_0 lead to the space X of Ex. 1.1.1. The times t_i are photon emission times, and N is the number of photons produced in the interval $[t_0, t_0 + T]$. ∎

Example 1.1.4 *Photoelectron Emissions* ——————————————

An optical field incident on a detector such as a photomultiplier results in the emission of photoelectrons intermixed with extraneous, internally generated thermoelectrons. Emissions in a T-second interval starting at time t_0 lead to the space X of Ex. 1.1.1. The times t_i are electron emission times. An example of a pulse process observed at the output of a photomultiplier is shown in Fig. 1.2. ∎

Figure 1.2. Photomultiplier pulses observed by F. Davidson of Johns Hopkins Univ.

Example 1.1.5 *Lightning Discharges* —————————————————

Lightning discharges occurring in the earth's atmosphere produce interference in a radio receiver. This type of interference is called *atmospheric noise* and is significant at frequencies below about thirty kilohertz. Examples of atmospheric noise observed in Malta are shown in Fig. 1.3. Suppose that the receiver is energized at time t_0, and discharges are observed for the following T seconds. The space X is that of Ex. 1.1.1. Each t_i is the instant of a discharge. ∎

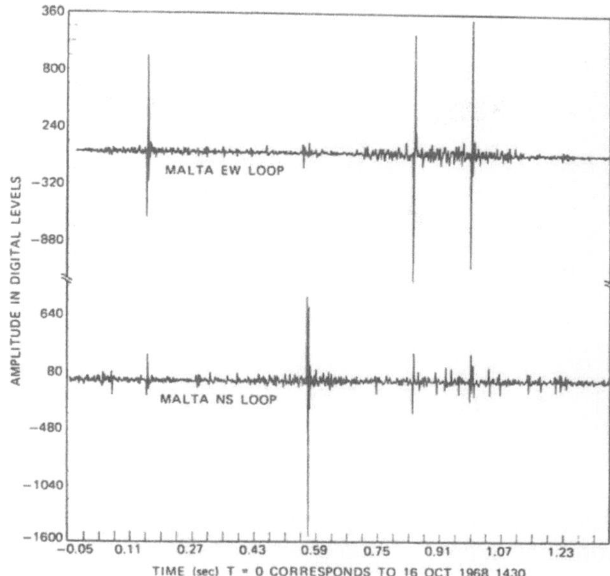

Figure 1.3. Atmospheric noise recorded in Malta from antennas having East-West and North-South orientations. From J. Evans [5].

Example 1.1.6 *Astronomy* ——————————————————

Suppose that stars in some part of the universe are observed. Here, the space X can be taken as a region in a three-dimensional Euclidean space $\mathcal{R}^3$. A realization of the random point process on X is the set of positions $\{r_1, r_2, \cdots, r_N\}$, where r_i is the position of the center of the ith star, and N is the number of stars in the region. ∎

Example 1.1.7 *Nuclear Medicine* ————————————————

A technique used in nuclear medicine to visualize biochemical phenomena occurring in internal organs is to introduce radioactively-labeled, biochemically-active substances. Radioactive emissions are then observed with external radiation detectors, with the number of emissions in a region depending on the concentration of the tracer in that region. The positions of the emissions form a random point process on X in which the realization is the set $\{r_1, r_2, \cdots, r_N\}$, where r_i is the position of the ith emission, and N is the number of emissions. An example showing radioactive water in a human brain is in Fig. 1.4. This image was formed using the point-process estimation methods developed in Ch. 3. ∎

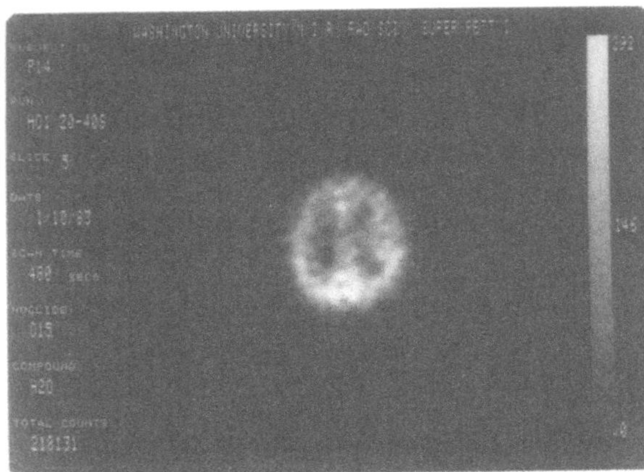

Figure 1.4. Radioactive water in the human brain.

Example 1.1.8 *Electron Microscopy* ———————————————

In electron microscopy, a flood of electrons illuminates a microscopic object, such as a thin section of body tissues. The electrons are detected and their positions recorded. The space X can be taken as a region of a two-dimensional Euclidean space $\mathcal{R}^2$ corresponding to the detector's surface. A realization of the stochastic process on X is the set $\{r_1, r_2, \cdots, r_N\}$ of positions of the electron impacts in the detector, and N is the number detected. In electron microscopic autoradiography, the thin section is also covered with a photographic emulsion. Minute quantities of a radioactive substance in the tissue section expose silver grains in the emulsion. When the emulsion is developed, these grains appear superimposed on the organelle structures in the electron micrograph, as seen in Fig. 1.5. Then, $\{r_1, r_2, \cdots, r_N\}$ are the positions of the developed silver grains, which are used to estimate the concentration of radiotracer in the various organelles, as described in Ch. 3. ∎

Example 1.1.9 *Dynamic Studies in Nuclear Medicine* ———————————

A technique used in nuclear medicine to determine parameters relating to dynamic physiochemical phenomena such as blood flow-rate, metabolic exchange-rates, and lung ventilation is to introduce a radioactive tracer into the organ or region of interest and then to monitor the time course of radioactivity as the tracer is removed from the field of view of the detector as a result of the phenomenon of interest. The space X can be taken as the Cartesian product of the time interval $[t_0, t_0 + T]$ corresponding to the observation interval and a region of a two-dimensional Euclidean space $\mathcal{R}^2$ corresponding to the surface of a scintillation crystal used to detect the radioactivity. A realization of the random point process on X is the set $\{(t_1, r_1), (t_2, r_2), \cdots, (t_N, r_N)\}$, where t_i and r_i are the time and location of the ith photon, and N is the number of detections. ∎

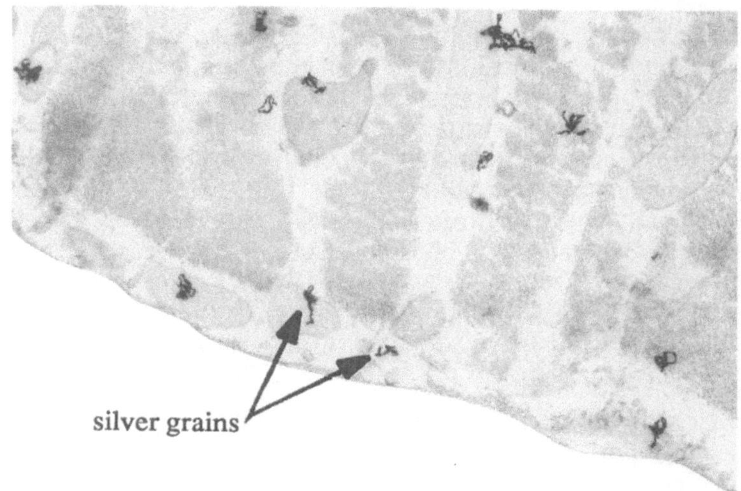

Figure 1.5. Electron Microscopic Autoradiograph of Heart Tissue.

Example 1.1.10 *Seismic Events* ────────────────────────────────

Seismic events such as earthquakes can be observed on a worldwide basis with an array of seismometers. Suppose that seismic events are observed for T seconds starting at time t_0 and that the epicenter of each such event in the earth is determined from the observations. Here, the space X can be taken as the Cartesian product of the time interval $[t_0, t_0 + T]$ and a region of three-dimensional Euclidean space corresponding to the earth. A realization of the random point process on X is the set $\{(t_1, r_1), (t_2, r_2), \cdots, (t_N, r_N)\}$ of times and locations of the seismic events. ∎

An observer of point phenomena is often not directly interested in the point realizations themselves but, rather, in what can be inferred from them about some underlying mechanism that influences where the points occur. For instance, in Ex. 1.1.2, an observer of spike discharges in an auditory nerve may be interested in the transformation that takes place between an acoustic pressure signal applied at the outer ear and the electrical activity measured in the nerve. In Ex. 1.1.3, an observer of radioactivity may want to know the half-life of the isotope being monitored. In Ex. 1.1.4, photoelectron emissions in optical communication and radar systems are influenced by analog or digital signals that modulate the optical field incident on the photodetector in the receiver; there is a desire to recover these modulating signals from the photodetector's output. In Ex. 1.1.8, an electron microscopist wants structural information about the object illuminated with electrons. Finally, in Ex. 1.1.9, a radiologist using radioactive tracers in dynamic-function studies wants to determine physiological parameters from the observed time-course of the radioactivity.

Thus, underlying mechanisms of interest are often associated with observed point phenomena. As a result, there is a need for a systematic theory of information processing for such observations. The formulation of this theory requires mathematical models for point phenomena; a wide variety of models is developed in the text. These are then used with statistical estimation theory to make inferences about underlying mechanisms that influence observed point phenomena.

The space X in the above examples is seen to represent time (Ex's. 1.1.1-1.1.5), a geometrical space (Ex's. 1.1.6-1.1.8), or a combination of these (Ex's. 1.1.9-1.1.10). In a general setting of the theory, the space X can be quite abstract. However, this generality will not be needed for our purposes, and we will usually take X to be either a semiinfinite real line representing time, a subset of Euclidean space representing a spatial region, or a combination of these.

There are two important features of a random point process for which statistical characterizations are normally sought. The first is interpoint spacings, with most emphasis placed on nearest-neighbor spacing. Thus, for instance, the statistics of intervals between spike discharges in Ex. 1.1.2 or photoelectron emissions in Ex. 1.1.5 are desired. These are in general termed *location statistics*; they are termed *time statistics* when X is a time interval. Location statistics are characterized for a wide variety of models.

The second feature of a point process for which statistical characterizations are sought is the number of points falling in subregions of the space where points occur. For example, with reference to Ex. 1.1.3, the statistics of the total number of photons emitted during $[t_0, t_0 + T]$ may be desired, or with reference to Ex. 1.1.10, the statistics of the number of seismic events in the years 1940 to 1970 in New Zealand may be desired (see D. Vere-Jones [20] for an interesting discussion of this topic). These are termed *counting statistics*. Counting statistics are characterized for a wide variety of models. A particularly useful concept in discussing counting statistics is that of a counting process. This is defined in the next section.

1.2 Counting Processes

In practice, it is often of interest to count the numbers of points in subsets of the space X on which a point process is defined. A *counting process* is introduced for this purpose and can be associated with every point process. The idea is as follows. We assume that a realization, call it ω, of a random point process on X is a denumerable point set of X. This means that ω can be enumerated as $\omega = \{x_1, x_2, \cdots\}$, where each x_i denotes the coordinate of a point in X. Now, let A be a subset of X, and denote by $N(A:\omega)$ the number of points in ω that lie in A. Formally,

$$N(A:\omega) = \sum_i I_A(x_i), \tag{1.1}$$

where each x_i is the coordinate of a point in ω, and where $I_A(x)$ denotes the set characteristic-function of A,

$$I_A(x) = \begin{cases} 1, & x \in A, \\ 0, & x \notin A. \end{cases} \tag{1.2}$$

When viewed as a function of A and ω, $N(A:\omega)$ defines a nonnegative, integer-valued random process on X. A process constructed in this way is called a *counting process*. It is customary to suppress the dependency of $N(A:\omega)$ on ω and write the number of points in A simply as $N(A)$. We adhere to this notation in the remainder of the book.

As an example, suppose that $X = \{t:t \geq t_0\}$, corresponding to a point process in time for times after and including t_0. If A is an interval of time for $t \geq t_0$, then $N(A)$ is the number of events having epochs in the interval. For the set $A = \{t:s \leq t < u\}$, where $t_0 \leq s$, we introduce the more compact notation

$$N(\{t:s \leq t < u\}) \stackrel{\Delta}{=} N(s,u). \tag{1.3}$$

Thus, $N(s,u)$ is the number of points in the interval from s to u possibly including a point at s but not at u. We further introduce a stochastic process $\{N(t):t \geq t_0\}$ according to the definition

$$N(t) \stackrel{\Delta}{=} N(t_0) + N(t_0,t), \tag{1.4}$$

where $N(t_0)$ is a nonnegative integer that most often will simply be set to zero. Viewed as a function of t, $\{N(t):t \geq t_0\}$ has a unit jump at the epoch of each point, as shown for a typical sample function in Fig. 1.6.

Suppose that $A_1, A_2, \cdots$ are disjoint sets in X. Then

$$I_{\bigcup_k A_k}(X) = \sum_k I_{A_k}(X),$$

and we have from (1.1) that

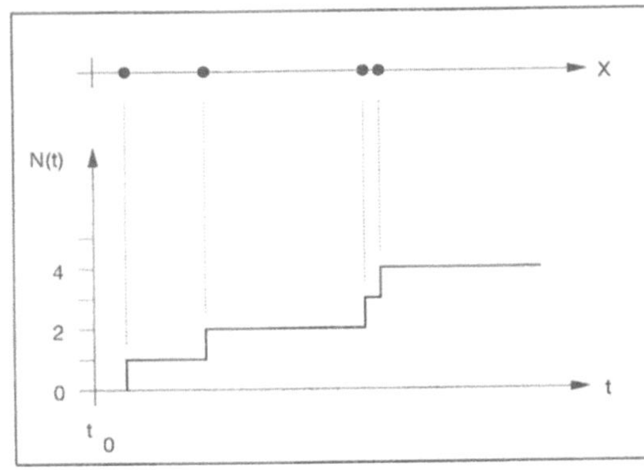

Figure 1.6. A point process and its associated counting process.

$$N\left(\bigcup_k A_k\right) = \sum_i \sum_k I_{A_k}(X_i) = \sum_k N(A_k). \tag{1.5}$$

Thus, $N(A)$ is an additive set function on X; the number of points in a union of disjoint sets in X is the sum of the number in each set. As an example, suppose that $X = \{t : t \geq t_0\}$. If $s \in [t_0, t)$, then $[t_0, t) = [t_0, s) \cup [s, t)$ and $[t_0, s) \cap [s, t) = \emptyset$. Consequently, by the additivity property of N, we have

$$N(s, t) = N(t) - N(s). \tag{1.6}$$

Thus, $N(s, t)$ is the random variable denoting the increment of the counting process on the interval $[s, t)$.

1.3 Organization of the Book

The organization of the text, with chapter dependencies, is indicated in Fig. 1.7. The *Poisson process* in time is the basic building block. We begin with its definition and an investigation of its properties in Ch. 2. Qualitative conditions that a point process must have to be a Poisson process are given. These facilitate making judgements about the appropriateness of a Poisson model in particular applications. The point-location and counting statistics for Poisson processes are developed, and on the basis of these, procedures are outlined for numerically generating a Poisson process for use in computer simulation or Monte Carlo studies of situations too complex for direct analysis. The important practical problem of estimating parameters that influence the intensity with which points occur in a Poisson process is discussed. Major emphasis in the chapter is placed on unconstrained maximum-likelihood estimation.

Translated Poisson-processes are developed in Ch. 3. These point processes result by translating each point of a Poisson process by some random amount to another location. They are of interest as models for instruments that measure point locations with only a finite precision. Other sources of instrumentation error also occur and are included in these

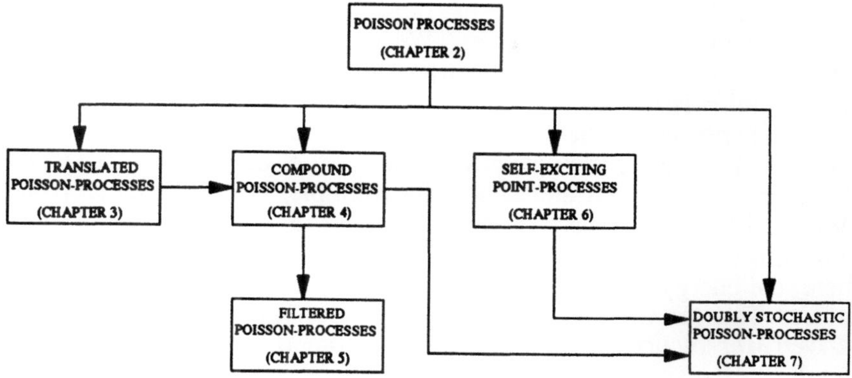

Figure 1.7. Organization of the book.

models. Extraneous points unrelated to those being measured can be included, and some points can be missed altogether. The model developed
in Ch. 3 includes translations, insertions, and deletions of points of a
Poisson process. Parameter estimation for translated Poisson-processes is
addressed using unconstrained maximum-likelihood estimation. Constrained estimation for translated Poisson processes is developed in this
chapter for the problem of estimating the intensity function of a Poisson
process in time or space because unconstrained estimates are very rough.
The concept of Grenander's sieves is introduced and related to translated
Poisson processes. The use of sieves and penalties for introducing constraints that regularize maximum-likelihood estimation is developed. The
use of Markov random-fields is also explained for incorporating
constraints. Numerical procedures are presented for producing constrained maximum-likelihood estimates based on the expectation-
maximization algorithm. A brief discussion of special computer
architectures for producing constrained maximum-likelihood estimates is
given.

Compound Poisson-processes are studied in Ch. 4. These are obtained from the Poisson process by associating an auxiliary random variable, called a *mark*, with each point occurrence. Successive marks are
selected independently from a common distribution. A compound Poisson
process might be used, for instance, as a first approximation model for the
lightning discharges of Ex. 1.1.5. Here, the Poisson process would model

the occurrence times of the discharges, and the marks would indicate the energy of each discharge. In this way, the amplitude as well as the occurrence time of the pulses in Fig. 1.3 can be modeled. A compound Poisson process is shown to be decomposable into a linear superposition of time-space Poisson processes. Decompositions of this type have a traditional role in the theory of random processes, particularly for independent-increment processes. We use the decomposition to specify procedures for estimating unknown parameters that influence an observed, compound Poisson-process.

Point processes often have a direct influence on systems or devices of interest. The classic example is shot noise, which occurs with the electron emissions of Ex. 1.1.1 and which, for example, limits the signal detectability in radio and radar receivers. Another example is lightning-induced atmospheric noise, which also limits signal detectability in radio receivers; here, the energy marks described above play a role in characterizing the noise. Thus, in Ch. 5, we study *filtered Poisson processes* and *Poisson-driven Markov processes*. These processes are the response of linear and nonlinear dynamical systems to the marked Poisson processes of Ch. 4. There are two major sections in the chapter. The response of linear systems is considered first. Here, the response is a superposition of the individual responses to each marked point. Statistics for such processes are developed in detail, including the historic Campbell's theorem for power spectra. Applications to the modeling of shot noise and photoelectron noise are given. The second major section of the chapter deals with the modeling of responses that are not a superposition of separate responses to each point. The main mathematical tool is the theory of stochastic differential and integral equations, which we develop in some detail but not with the generality or rigor that may be found in more mathematical texts, such as the one by R. Lipster and A. Shiryayev [10]. The use of stochastic differential and integral equations for describing the response of linear and nonlinear systems driven by a continuous random process, rather than a point process, is by now nearly standard in the engineering and applied sciences literature; see, for example, the text by E. Wong and B. Hajek [21]. Our treatment in this chapter for the response to a point process parallels these developments. While a familiarity with these alternative discussions is helpful, it is not needed for understanding Ch. 5.

A distinguishing property of a temporal Poisson process is that it evolves in time without aftereffects. This means that the occurrence times and number of points realized before some time have no influence on the occurrence times and number of points subsequent to that time. While this property is often met to a reasonable degree for many point phenomena encountered in practice, it is not a good hypothesis universally. An example where it is not a good hypothesis is provided by the nerve discharges described in Ex. 1.1.2. It is observed experimentally that subsequent to each discharge there is a brief refractory period during which additional discharges do not occur. A similar effect can be observed in scintillation detectors used in the dynamic studies in nuclear medicine described in Ex. 1.1.9. The restriction that points occur without aftereffects is removed in

Ch. 6 where we define and study *self-exciting point processes*. For these processes, the future evolution of points can be influenced by all past occurrences of points. In many applications, the evolution does not depend on the entire past but only a particular feature of the past such as the most recent occurrence time or the total number of past occurrences. The most notable of these processes because of their wide use are *renewal processes* and *Markov birth-processes*. Self-exciting point processes with limited memory are characterized in Ch. 6.

Doubly stochastic Poisson processes are obtained as a generalization of the Poisson process when the intensity with which points occur is allowed to be a random process; a doubly stochastic Poisson process is conditionally a Poisson process given the stochastic intensity process. Doubly stochastic Poisson processes are reasonable models for the photoelectron emissions of Ex. 1.1.5 when the incident optical field is incoherent Gaussian light as emitted, for example, by a light-emitting diode or an astronomical source. The photoelectron generation rate in an optical communication system employing coherent laser light as a carrier is stochastic because of intentional modulation at the transmitter by information-bearing signals and because of unintentional modulation by fading processes in the atmospheric propagation. In Ch. 7, we develop algorithms for processing a sequence of observed point occurrences in order to estimate random signals that influence the underlying and unobservable intensity with which the observed points occur. Two approaches are considered. In the first, the estimate is constrained to be a linear function of the observed point process. A realization of the estimator as a Kalman-Bucy filter is given. The linearity constraint is removed in the second approach, which parallels the estimation theory initiated by R. Stratonovich [19] and H. Kushner [9] for observations of a continuous random process rather than a point process.

Major sections are numbered in each chapter; for instance, Sec. 3.4 is the fourth section of Ch. 3. Examples and problems are numbered by sections; for instance, Ex. 3.4.2 and Problem 3.4.2 are the second example and problem of Sec. 3.4. References and problems appear at the end of each chapter. Indices to cited authors and to examples are at the end of the text. The symbols ❑ and ∎ mark the end of theorem proofs and of examples, respectively.

1.4 Mathematical Preliminaries

In this section, we survey certain elements of the theory of probability and stochastic processes that are used frequently in the chapters that follow. The survey is intended mainly as a reference and not as a source material seen for the first time. This section also serves to establish the notation to be used subsequently. More detailed treatments of the material summarized here are readily available; in particular, we mention the elementary texts by A. Bharucha-Reid [2], A. Papoulis [14], and E. Parzen

[15,16] and the more advanced texts by J. Doob [4], I. Gikman and A. Skorokhod [7], M. Loeve [11], J. Neveu [13], and E. Wong and B. Hajek [21].

1.4.1 Probability Spaces

A basic concept in the theory of probability and stochastic processes is that of a *probability space*. A probability space is comprised of three quantities: a *sample space* Ω, a collection $\mathcal{F}$ of certain subsets of Ω called a *field of events*, and a *probability measure* μ. These quantities are defined as follows.

Sample Space. An element ω of Ω is a possible outcome of a random experiment. In the simplest situations, Ω consists of a denumerable collection of elements $\omega_1, \omega_2, \cdots$. Examples include the random experiment of tossing a die, for which Ω consists of six elements with ω_i corresponding to the *i*th face showing on the toss, and the random experiment of tossing a coin until a head shows, for which Ω consists of a denumerably infinite number of elements with ω_i corresponding to a head on the *i*th toss. Nondenumerable sample spaces are common. Examples arise with the random experiment of breaking a unit stick, for which $\Omega = \{\omega : 0 \le \omega \le 1\}$, and with the random experiment of failures of a light bulb placed in service at time t_0, for which $\Omega = \{\omega : \omega \ge t_0\}$. Nondenumerable sample spaces predominate in later chapters. More complicated sample spaces are possible. These include function spaces in which, for instance, each ω corresponds to a continuous function of t for $0 \le t \le 1$. A situation of more interest here occurs when each ω corresponds to a denumerable point set in some Euclidean space. While we will not do so, it is possible to develop a theory of stochastic point processes starting with such a space; see J. Moyal [12], and P. Fishman and D. Snyder [6].

Sigma Field of Events. A subset of Ω is called an *event*. An example is the subset $A = \{\omega : 0.25 \le \omega \le 0.75\}$ of the sample space $\Omega = \{\omega : 0 \le \omega \le 1\}$ associated with the random experiment of breaking a unit stick; in words, A is the event that the break occurs in the central half of the stick. The sigma field of events $\mathcal{F}$ is a particular collection of events. For denumerable sample spaces, $\mathcal{F}$ may consist of all possible subsets of Ω, but this is not necessary. For nondenumerable sample spaces, $\mathcal{F}$ cannot consist of all possible subsets of Ω but, rather, only of certain subsets called *measurable events*. What is required of $\mathcal{F}$ are the following three properties:

1. $\mathcal{F}$ is not empty;

2. if an event A is in $\mathcal{F}$, so is its complement $A^c = \{\omega : \omega \in \Omega, \omega \notin A\}$;

3. if A_1, A_2, $\cdots$ are events in $\mathcal{F}$, so is their union $\cup_i A_i = \{\omega : \omega \in A_i \text{ for some } i\}$.

It follows immediately from these properties that Ω and its complement $\Omega^c = \varnothing$ are sets in $\mathcal{F}$; in this context, Ω and $\varnothing$ are called the *sure* and *null* events, respectively. Moreover, if $A_1, A_2, \cdots$ are events in $\mathcal{F}$, so is their intersection $\cap_i A_i = \{\omega : \omega \in A_i \text{ for all } i\}$.

Probability Measure. The probability measure $\mu(\cdot)$ is a function that maps each measurable set in $\mathcal{F}$ into the unit interval $[0,1]$. This function is required to have the following properties:

1. $0 \leq \mu(A) \leq 1$ for every measurable set $A \in \mathcal{F}$;
2. $\mu(\Omega) = 1$;
3. if $A_1, A_2, \cdots$ are disjoint sets in $\mathcal{F}$, then $\mu(\cup_i A_i) = \sum_i \mu(A_i)$.

So far as the theory is concerned, these are the only restrictions on μ; any function on $\mathcal{F}$ satisfying (1)-(3) is an admissible probability measure. However, it is only by a judicious selection of μ that the mathematical model can be made useful in predicting occurrences in an actual physical application. An event A is said to occur *with probability one* or *almost surely* if $\mu(A) = 1$.

A probability space, denoted by the triplet $(\Omega, \mathcal{F}, \mu)$, consists of a sample space Ω, a sigma field $\mathcal{F}$ of subsets of Ω, and a probability measure μ.

1.4.2 Random Variables, Distributions, Densities, Expectations

Certain real-valued functions defined on a probability space are called *random variables*. To specify such functions, let $(\Omega, \mathcal{F}, \mu)$ be a probability space, and denote the real line by $\mathcal{R}$. A function $x(\omega)$ with domain Ω (that is $\omega \in \Omega$) and range $\mathcal{R}$ (that is, $x(\omega) \in \mathcal{R}$) is a random variable if for every $X \in \mathcal{R}$, the set $\{\omega : x(\omega) \leq X\}$ is in the sigma field of events $\mathcal{F}$. The restriction to functions of this type, called *measurable functions*, ensures that the probability $\mu(\{\omega : x(\omega) \leq X\})$ is well defined for all $X \in \mathcal{R}$. Moreover, if B is any subset of $\mathcal{R}$ obtainable by forming countable unions, intersections, and complements of subsets of $\mathcal{R}$ of the form $\{r : r \leq X\}$, then $\mu(\{\omega : x(\omega) \leq X\})$ is well defined. The function of X defined by $P_x(X) = \mu(\{\omega : x(\omega) \leq X\})$ maps the real line into the unit interval $[0,1]$ and is called the *probability distribution function* for the random variable $x(\omega)$. We note that $P_x(X)$ is a nondecreasing function of X with $P_x(-\infty) = 0$ and $P_x(+\infty) = 1$.

A particular form for $P_x(X)$ is often postulated in applying the theory. In these cases, the probability space only plays a subsidiary role, and the explicit dependence of $x(\omega)$ on ω is normally suppressed. Thus, throughout most of the book, we simply refer to a random variable x having a given probability distribution function $P_x(X)$. We also use the notation $\Pr(B)$ for the probability a condition B on the random variable holds; that is, $\Pr(B) = \mu(\{\omega : B \text{ holds}\})$. Thus, $\Pr(x \leq X)$ is another way we write the probability distribution function $P_x(X)$ of x. It is readily verified that $\Pr(X_1 < x \leq X_2) = P_x(X_2) - P_x(X_1)$ for $X_1 < X_2$.

A random variable $x(\omega)$ is called *discrete* if it has only a countable number of possible values in R. Suppose the set of values is $\{X_1, X_2, \cdots\}$, where $X_i \in \mathcal{R}$ for each i. Then

$$P_x(X) = \sum_i I_{B(X)}(X_i)\mu(\{\omega : x(\omega) = X_i\}), \tag{1.7}$$

where $B(X) = \{r : r \in \mathcal{R}, r \leq X\}$ is the set of real numbers less than or equal to X, and where

$$I_{B(X)}(X_i) = \begin{cases} 1, & X_i \in B(X) \\ 0, & X_i \notin B(X). \end{cases}$$

It follows from (1.7) that the probability distribution function for a discrete random variable is piecewise constant with jump discontinuities at X_1, X_2, $\cdots$; the amplitude of the discontinuity at X_i is $\mu(\{\omega : x(\omega) = X_i\}) = \Pr(x = X_i)$.

A random variable $x(\omega)$ is called *continuous* if there exists a function $p_x(X)$, called the *probability density function* for $x(\omega)$, such that

$$\Pr(x \in B) = \int_B p_x(X)\,dX$$

holds for any subset B of R for which $\mu(\{\omega : x(\omega) \in B\})$ is defined. The following relations between the probability distribution and density functions for $x(\omega)$ apply whenever the density exists:

$$P_x(X) = \int_{-\infty}^X p_x(\alpha)\,d\alpha,$$

and

$$p_x(X) = \frac{dP_x(X)}{dX}.$$

Both discrete and continuous random variables will be encountered in our studies of random point processes. The most prominent discrete variable, which arises repeatedly, is the number of points falling in some specified set; the values this random variable may take are nonnegative integers. The time interval between successive points of a point process evolving in time will always be a continuous random variable in the point processes we study.

The extension to more than one random variable defined on the same probability space is straightforward. Thus, if $x_1(\omega)$, $x_2(\omega)$, $\cdots$, and $x_n(\omega)$ are n random variables defined on $(\Omega, \mathcal{F}, P)$, the column vector $x(\omega) = [x_1(\omega) \quad x_2(\omega) \quad \cdots \quad x_n(\omega)]'$ defines a mapping from Ω to $\mathcal{R}^n$, where $\mathcal{R}^n$ denotes n-dimensional Euclidean space. This mapping is such that the set $\{\omega : x_1(\omega) \leq X_1, x_2(\omega) \leq X_2, \cdots, x_n(\omega) \leq X_n\}$ is in $\mathcal{F}$ for every choice of $X = [X_1 \quad X_2 \quad \cdots \quad X_n]' \in \mathcal{R}^n$. Consequently, the probability of this set is well defined, and we define

$$P_{x_1, x_2, \cdots, x_n}(X_1, X_2, \cdots, X_n)$$

$$= \mu(\{\omega : x_1(\omega) \leq X_1, x_2(\omega) \leq X_2, \cdots, x_n(\omega) \leq X_n\}). \tag{1.8}$$

This function is called the *joint probability distribution* for $x_1(\omega)$, $x_2(\omega)$, $\cdots$, and $x_n(\omega)$. For conciseness in notation, we also use $P_x(X)$ and even $P_x^{(n)}(X)$ for this function, the latter especially when there is possible ambiguity in the number n of variables involved. The random variables $x_1(\omega)$, $x_2(\omega)$, $\cdots$, and $x_n(\omega)$ are said to be *statistically independent* if their joint distribution function is the product of their individual distribution functions; that is, if

$$P_x(X) = \prod_{i=1}^{n} P_{x_i}(X_i)$$

for all $X \in \mathcal{R}^n$. If there is a function $p_x(X) = p_{x_1, x_2, \cdots, x_n}(X_1, X_2, \cdots, X_n)$ such that $\Pr(x \in B) = \int_B p_x(X) dX$ holds for an arbitrary set $B \in \mathcal{R}^n$ for which the probability $\mu(\{\omega : x(\omega) \in B\})$ is defined, then the variables $x_1(\omega)$, $x_2(\omega)$, $\cdots$, and $x_n(\omega)$ are jointly continuous, and the function $p_x(X)$ is called the *joint probability density* for $x_1(\omega)$, $x_2(\omega)$, $\cdots$, and $x_n(\omega)$.

Suppose that $g(\cdot)$ is a function mapping $\mathcal{R}^n$ into $\mathcal{R}^m$; a point X in $\mathcal{R}^n$ is mapped into a point $Y = g(X)$ in $\mathcal{R}^m$. If $x(\omega)$ is an $\mathcal{R}^n$-valued random variable on $(\Omega, \mathcal{F}, \mu)$, then $y(\omega) = g(x(\omega))$ is a composite function mapping Ω into $\mathcal{R}^m$, but $y(\omega)$ in general may not be a measurable random variable; $y(\omega)$ is an $\mathcal{R}^m$-valued random variable on $(\Omega, \mathcal{F}, \mu)$, defined by $x(\omega)$ and $g(\cdot)$, provided $g(\cdot)$ is such that the set $\{\omega: y_1(\omega) \le Y_1, \cdots, y_m(\omega) \le Y_m\}$ is in $\mathcal{F}$ for all $Y \in \mathcal{R}^m$. Continuous functions mapping $\mathcal{R}^n$ into $\mathcal{R}^m$ satisfy this condition. Suppose, now, that $g(\cdot)$ is a one-to-one, invertible mapping of $\mathcal{R}^n$ into $\mathcal{R}^m$ such that both $g(X)$ and its inverse function $g^{-1}(Y)$ have continuous partial derivatives with respect to the elements of X and Y, respectively. Then, $g(\cdot)$ is a continuous function, and $y(\omega) = g(x(\omega))$ is an $\mathcal{R}^n$-valued random variable. Furthermore, if $x(\omega)$ is a continuous random variable with joint probability density $p_x(X)$, then $y(\omega)$ is also a continuous random variable with joint density

$$p_y(Y) = \left| \det\left(\frac{\partial g^{-1}(Y)}{\partial Y} \right) \right| p_x(g^{-1}(Y)), \tag{1.9}$$

where $\det(M)$ denotes the determinant of the matrix M, and $\partial g^{-1}(Y)/\partial Y$ denotes the Jacobian matrix of $g^{-1}(Y)$; the i,j-element is $\partial g_j^{-1}(Y)/\partial Y_i$. As an application of (1.9), suppose that $x_1(\omega)$, $x_2(\omega)$, $\cdots$, and $x_n(\omega)$ are statistically independent, normally distributed random variables, with $x_i(\omega)$ having mean zero and variance σ_i^2. Then,

$$p_x(X) = (2\pi)^{-n/2} [\det(\Lambda)]^{-1/2} \exp\left(-\frac{1}{2} X'\Lambda^{-1}X \right), \tag{1.10}$$

where

$$\Lambda = \mathrm{diag}(\sigma_i^2) = \begin{bmatrix} \sigma_1^2 & 0 & \cdot & \cdot & 0 \\ 0 & \sigma_2^2 & \cdot & \cdot & 0 \\ \cdot & \cdot & \cdot & \cdot & \cdot \\ 0 & \cdot & \cdot & \cdot & \sigma_n^2 \end{bmatrix}.$$

Let $Y = g(X) = MX + m$, where M is a nonsingular matrix. Then, $g^{-1}(Y) = M^{-1}(Y - m)$, and the probability density function for $y(\omega) = g(x(\omega))$ is given by

$$p_y(Y) = \frac{1}{\sqrt{(2\pi)^n \det(\Sigma)}} \exp\left[-\frac{1}{2}(Y-m)'\Sigma^{-1}(Y-m)\right], \qquad (1.11)$$

where $\Sigma = M\Lambda M'$. Random variables having this density are called *jointly normal* or *jointly Gaussian*; we sometimes abbreviate $p_y(Y)$ in (1.11) by $N(m, \Sigma)$.

The *expectation* or *mean value* of a random variable (r.v.) $x(\omega)$ is denoted by $E(x)$ and defined by

$$E(x) = \int_{\mathcal{R}^n} X dP_x(X)$$

$$= \begin{cases} \sum_i X_i \Pr(x = X_i), & x \quad \text{a discrete r.v.} \\ \int_{\mathcal{R}^n} X p_x(X) d^n X, & x \quad \text{a continuous r.v.} \end{cases} \qquad (1.12)$$

Suppose that $y(\omega) = g(x(\omega))$ is a random variable defined by $x(\omega)$ and the function $g(\cdot)$, which maps $\mathcal{R}^n$ into $\mathcal{R}^m$. Then (1.12) also defines the expectation of $y(\omega)$ according to

$$E(y) = \int_{\mathcal{R}^m} Y dP_y(Y), \qquad (1.13)$$

where $P_y(Y)$ is the probability distribution function for $y(\omega)$. This expression for the expectation of $y(\omega)$ is often inconvenient because of the need to determine $P_y(Y)$ before the integral (1.13) can be evaluated. This complication can be avoided because the expectation can be alternatively determined according to

$$E(y) = \int_{\mathcal{R}^n} g(X) dP_x(X)$$

$$= \begin{cases} \sum_i g(X_i) \Pr(x = X_i), & x \quad \text{a discrete r.v.} \\ \int_{\mathcal{R}^n} g(X) p_x(X) d^n X, & x \quad \text{a continuous r.v.} \end{cases} \qquad (1.14)$$

Special cases of (1.14) occur so frequently in later chapters that it is convenient to introduce additional terminology for them. Suppose, first, that $g(x(\omega)) = e_i'\{x(\omega)x'(\omega) - E(x)E(x')\}e_j$, where e_i is an n-dimensional unit vector with a one in row i and zeros elsewhere. The expectation of

$g(x(\omega))$ is called the *covariance* of $x_i(\omega)$ and $x_j(\omega)$ when $i \neq j$ and the *variance* of $x_i(\omega)$ when $i = j$. We denote $E[g(x(\omega))]$ by λ_{ij}. The matrix $\Lambda = (\lambda_{ij})$ is termed the *covariance matrix* of $x(\omega)$. These second moments occur naturally in estimation problems where the estimator is constrained to be a linear transformation of the observable data. As an example, suppose that $x_1(\omega)$, $x_2(\omega)$, $\cdots$, $x_{n-1}(\omega)$ model observed data, and on the basis of an observation of these quantities it is desired to estimate $x_n(\omega)$, which cannot be observed directly. It is a straightforward calculation to verify that if the estimate is constrained to have the linear form

$$\hat{x}_n = a + b'x^{(n-1)}(\omega) \tag{1.15}$$

for some choice of a and b, where $x^{(n-1)} = [x_1 \quad x_2 \quad \cdot \quad \cdot \quad \cdot \quad x_{n-1}]'$, then the selections of a and b minimizing $E[(x_n - \hat{x}_n)^2]$ are given by

$$a = E(x_n) - b'E[x^{(n-1)}]$$

and

$$b = \begin{bmatrix} \lambda_{11} & \lambda_{12} & \cdot & \cdot & \cdot & \lambda_{1,n-1} \\ \lambda_{21} & \lambda_{22} & \cdot & \cdot & \cdot & \lambda_{2,n-1} \\ \cdot & \cdot & \cdot & \cdot & \cdot & \cdot \\ \cdot & \cdot & \cdot & \cdot & \cdot & \cdot \\ \lambda_{n-1,1} & \lambda_{n-1,2} & \cdot & \cdot & \cdot & \lambda_{n-1,n-1} \end{bmatrix}^{-1} \begin{bmatrix} \lambda_{1n} \\ \lambda_{2n} \\ \cdot \\ \cdot \\ \lambda_{n-1,n} \end{bmatrix}.$$

In later chapters, we study versions of this linearly constrained estimation problem in which the data $x_1, x_2, \cdots, x_{n-1}$ are derived from observations of a point process, and x_n is an underlying variable influencing the point process.

Complex valued random variables will occasionally occur in our subsequent studies. The real and imaginary parts of these are random variables defined on the same probability space as above. A frequently occurring example in later chapters is associated with the complex exponential function $g(x) = \exp[j < v, X >]$, where $v \in \mathcal{R}^n, X \in \mathcal{R}^n$, and

$$< v, X > \overset{\Delta}{=} \sum_{i=1}^{n} v_i X_i.$$

As $\text{Re}[g(X)] = \cos\langle v, X\rangle$ and $\text{Im}[g(X)] = \sin\langle v, X\rangle$ are continuous mappings from $\mathcal{R}^n$ to $[-1, 1] \subset \mathcal{R}$, the random variables $y_r(\omega) = \cos\langle v, x(\omega)\rangle$ and $y_i(\omega) = \sin\langle v, x(\omega)\rangle$ are well defined. Thus, we can define the complex random variable $y(\omega)$ by

$$y(\omega) = y_r(\omega) + jy_i(\omega) = e^{j\,<v,x(\omega)>}.$$

The expectation of $y(\omega)$ plays an important role in later chapters; we denote it by $M_x(jv)$. Thus,

$$M_x(jv) = E(e^{j\,<v,x>})$$

$$= \int_{\mathcal{R}^n} e^{j\,<v,X>} dP_x(X) \tag{1.16}$$

$$= \begin{cases} \displaystyle\sum_i e^{j\,<v,X_i>} \Pr(x = X_i), & x \quad \text{a discrete r.v.} \\[2em] \displaystyle\int_{\mathcal{R}^n} e^{j\,<v,X>} p_x(X) d^n X, & x \quad \text{a continuous r.v.} \end{cases}$$

The function $M_x(jv)$ is called the *characteristic function* of x. Moments of x can be determined from the characteristic function of x. For example, suppose that x is a continuous random variable with values in $\mathcal{R}^1$. Then, from (1.16), the mth derivative of $M_x(jv)$ with respect to jv is

$$M_x^{(m)}(jv) = \frac{d^m}{d(jv)^m} M_x(jv) = \int_{-\infty}^{\infty} X^m e^{jvX} p_x(X) dX.$$

Hence, $M_x^{(m)}(j0) = E(X^m)$, which is called the *moment-generating property* of the characteristic function.

There is a one-to-one correspondence between the characteristic function and probability distribution function because they form a reciprocal Fourier-transform pair. For instance, if x is a continuous random variable, then

Table 1.1. Some Discrete Probability Laws and Their Characteristic Functions

Discete Prob. Law	Probability $\Pr(x=n)$	C. F. $M_x(jv)$	Mean $E(x)$	Variance $E(x^2)-E^2(x)$
Poisson $\lambda>0$ $n=0,1,2,\ldots$	$\dfrac{\lambda^n}{n!}e^{-\lambda}$	$\exp[\lambda(e^{jv}-1)]$	λ	λ
Negative Binomial $m=1,2,\ldots$ $n=0,1,\ldots$ $0\le p\le 1$ $q=1-p$	$\dfrac{(m+n-1)!p^mq^n}{(m-1)!n!}$	$\left(\dfrac{p}{1-qe^{jv}}\right)^m$	$mp^{-1}q$	$mp^{-2}q$
Bose -Einstein $\lambda>0$ $n=0,1,\ldots$	$\dfrac{\lambda^n}{(1+\lambda)^{n+1}}$	$\dfrac{1}{1+\lambda(1-e^{jv})}$	λ	λ

$$p_x(X)=\frac{1}{(2\pi)^n}\int_{\mathcal{R}^n} e^{-j<v,X>}M_x(jv)\,d^nv. \qquad (1.17)$$

Characteristic functions for some distributions encountered in this book are given in Tables 1.1 and 1.2 for discrete and continuous random variables, respectively.

As an example using Table 1.2, let $g=<v,y>$, where $y=\{y_1, y_2, \cdots, y_n\}$ are jointly normal random variables with density (1.11). As g is a linear combination of normal random variables with mean-value vector m and covariance matrix Σ, it is normally distributed, and its mean and variance are readily found to be $<v,m>=v'm$ and $v'\Sigma v$, respectively. Using this and the characteristic function in Table 1.2 for a Gaussian random variable, we conclude that the joint characteristic function for y is given by

Table 1.2. Some Continuous Probability Laws and Their Characteristic Functions

Cont. Prob. Law	Probability Density $p_x(X)$	C. F. $M_x(jv)$	Mean $E(x)$	Variance $E(x^2) - E^2(x)$
Exponential $\lambda > 0$ $X \geq 0$	$\lambda e^{-\lambda X}$	$\dfrac{\lambda}{\lambda - jv}$	$\dfrac{1}{\lambda}$	$\dfrac{1}{\lambda^2}$
Gamma $\alpha > 0$ $\beta > 0$ $X \geq 0$	$\dfrac{\alpha^{\beta+1} X^\beta e^{-\alpha X}}{\Gamma(\beta+1)}$	$\left(\dfrac{\alpha}{\alpha - jv}\right)^{\beta+1}$	$\dfrac{\beta+1}{\alpha}$	$\dfrac{\beta+1}{\alpha^2}$
Uniform $\alpha < X < \beta$	$\dfrac{1}{\beta - \alpha}$	$\dfrac{e^{jv\beta} - e^{jv\alpha}}{jv(\beta - \alpha)}$	$\dfrac{\alpha+\beta}{2}$	$\dfrac{(\beta-\alpha)^2}{12}$
Normal or Gaussian $-\infty < m < \infty$ $\sigma^2 > 0$ $-\infty < X < \infty$	$\dfrac{\exp[-(X-m)^2/2\sigma^2]}{\sqrt{2\pi\sigma^2}}$	$\exp\left[-\dfrac{1}{2}\sigma^2 v^2 + jvm\right]$	m	σ^2

$$M_{y_1, y_2, \ldots, y_n}(jv_1, jv_2, \ldots, jv_n) = E[e^{j\langle v, y\rangle}]$$

$$= \exp\left(-\frac{1}{2}v'\Sigma v + jv'm\right). \qquad (1.18)$$

Joint random variables having this characteristic function are jointly normal with density (1.11).

In later chapters, we will often be interested in information that observed data provides about random quantities that cannot be measured directly. The concepts of *conditional probability* and *conditional expectation* play a central role in this issue. We now give an heuristic overview of these concepts; for a precise treatment, we refer to J. Doob [4, p. 15].

If $x(\omega)$ and $y(\omega)$ are discrete random variables defined on a common probability space and having values $X_1, X_2, \cdots$ in $\mathcal{R}^n$ and $Y_1, Y_2, \cdots$ in $\mathcal{R}^m$, respectively, then the conditional probability that $x(\omega) = X_i$ given that $y(\omega) = Y_j$ is defined for all Y_j such that $\Pr(y = Y_j) > 0$ by

$$\Pr(x = X_i \mid y = Y_j) = \frac{\Pr(x = X_i, y = Y_j)}{\Pr(y = Y_j)}.$$

The conditional probability distribution function for $x(\omega)$ given that $y(\omega) = Y_j$ is then defined to be

$$P_{x|y}(X \mid Y_j) = \sum_i I_{B(X)}(X_i)\Pr(x = X_i \mid y = Y_j),$$

provided $\Pr(y = Y_j) > 0$, which is the counterpart of (1.7) given that $y(\omega) = Y_j$. The conditional expectation of $x(\omega)$ given that $y(\omega) = Y_j$ is defined for all Y_j such that $\Pr(y = Y_j) > 0$ by

$$E(x \mid y = Y_j) = \int_{\mathcal{R}^n} X dP_{x|y}(X \mid Y_j) = \sum_i X_i P_{x|y}(X_i \mid Y_j). \qquad (1.19)$$

The situation for continuous random variables is analogous. If $x(\omega)$ and $y(\omega)$ are jointly continuous random variables defined on a common probability space and having values in $\mathcal{R}^n$ and $\mathcal{R}^m$, respectively, then the conditional probability density for $x(\omega)$, given that $y(\omega) = Y$, is defined for all Y such that $p_y(Y) > 0$ by

$$p_{x|y}(X \mid Y) = \frac{p_{x,y}(X, Y)}{p_y(Y)}, \qquad (1.20)$$

and the conditional probability distribution function is[1]

$$P_{x|y}(X \mid Y) = \Pr(x \leq X \mid y = Y) = \int_{\mathcal{R}^n} I_{B(X)}(\xi) p_{x|y}(\xi \mid Y) d^n\xi. \qquad (1.21)$$

---

[1] Inequalities between vectors are to be interpreted as applying for all the corresponding components of the vectors.

The event $\{\omega: y(\omega) = Y\}$ associated with these expressions is to be interpreted in the sense of the limit of the event $\{\omega: y_i - \Delta Y_i < y_i(\omega) \leq Y_i; i = 1, 2, \cdots, m\}$ as $\max_i |\Delta Y_i|$ tends to zero.   The conditional expectation of $x(\omega)$ given that $y(\omega) = Y$ is defined for all Y such that $p_y(Y) > 0$ by

$$E(x \mid y = Y) = \int_{\mathcal{R}^r} X dP_{x \mid y}(X \mid Y) = \int_{\mathcal{R}^r} X p_{x \mid y}(X \mid Y) d^n X. \qquad (1.22)$$

The conditional expectation of $x(\omega)$ given $y(\omega) = Y$ defines an $\mathcal{R}^r$-valued function of Y, $g(Y) = E(x \mid y = Y)$.   Hence, $g(y(\omega)) = E(x \mid y(\omega))$ is an $\mathcal{R}^r$-valued random variable. A property of conditional expectation that will be useful in later chapters is that the expectation of this random variable is the expectation of $x(\omega)$; that is

$$E(x) = E[E(x \mid y(\omega))] = \int_{\mathcal{R}^r} E(x \mid y = Y) dP_y(Y). \qquad (1.23)$$

This *iterated expectation* formula is frequently useful for evaluating the expectation of expressions involving more than one random variable. As an application motivated by Ex. 1.1.5, suppose that the number n of lightning strokes in a day is a discrete random variable with given probabilities $\Pr(n = N), N = 0, 1, \cdots$ and that the energy released in the ith stroke z_i is a continuous random variable. Assume that the energies released in successive strokes are statistically independent, and all are distributed as a random variable z having a probability density function $p_z(Z)$. The total energy released in a day is

$$x = \sum_{i=1}^{n} z_i.$$

The expected energy released in a day can easily be evaluated using the iterated expectation formula $E(x) = E[E(x \mid n)]$. We see that

$$E(x \mid n = N) = E(z_1 + z_2 + \ldots + z_N) = N E(z).$$

Hence, if the energy in a stroke is independent of the number of strokes in a day, we have $E(x) = E[nE(z)] = E(n)E(z)$. The total expected energy released per day is then the product of the expected number of strokes per day and the average energy released per stroke.

There is another property of conditional expectation that will be useful in later chapters where we encounter estimation problems having a minimum mean-square error fidelity criterion. Consider the estimation problem associated with (1.15). The variables $x_1(\omega)$, $x_2(\omega)$, $\cdots$, and $x_{n-1}(\omega)$ model observed data, and on the basis of these data, it is desired to estimate $x_n(\omega)$, which cannot be observed directly. Suppose that the estimate is not constrained to have the linear form (1.15). Rather, what we seek is the function $\hat{x}_n(\omega) = g(x_1(\omega), x_2(\omega), ..., x_{n-1}(\omega))$ of the data that minimizes the mean square-error $E[(x_n - \hat{x}_n)^2]$. By iterating the expectation, we note after some manipulation that

$$E[(x_n - \hat{x}_n)^2] = E\{E(x_n^2 \mid x_1, x_2, ..., x_{n-1})$$

$$+ [\hat{x}_n - E(x_n \mid x_1, x_2, ..., x_{n-1})]^2$$

$$- E^2(x_n \mid x_1, x_2, ..., x_{n-1})\}.$$

Only the second term on the right side depends on the selection of $\hat{x}_n$, and for any selection made, the contribution of this term to the mean square-error is nonnegative. Consequently, the choice

$$\hat{x}_n(\omega) = E[x_n(\omega) \mid x_1(\omega), x_2(\omega), ..., x_{n-1}(\omega)] \tag{1.24}$$

minimizes the mean square-error over all other functions of the observables. The conditional mean, therefore, has an important property for optimization problems: it defines the structure of minimum mean square-error estimators. In later chapters, we study versions of this estimation problem in which x_1, x_2, $\cdots$, and x_{n-1} are derived from observations of a point process, and x_n is an underlying random quantity influencing where the points occur.

1.4.3 Stochastic Processes

A stochastic process $\{x(t,\omega): t \in T\}$ is a family of random variables all defined on the same probability space $(\Omega, \mathcal{F}, \mu)$ and indexed by a parameter t that takes values in a parameter set T called the *index set* of the process. The indexing parameter t represents time, space, or a combination of these in most of our considerations. The process $\{x(t,\omega): t \in T\}$ is called a *discrete-parameter process* if the index set T is a countable set $\{t_1, t_2, \cdots \}$, and it is called a *continuous-parameter process* if T is a subset of $\mathcal{R}^n$ for some n. Many of the processes we consider in subsequent chapters are continuous-parameter processes in which T is a subset of $\mathcal{R}$ corresponding to a time interval; for example, T could be times after some initial time, $T = \{t : t \geq t_0\}$. Other processes we consider are continuous-parameter processes in which T is a subset of $\mathcal{R}^2$ corresponding to a two-dimensional image; for example, T could be locations in a unit square centered at the origin. For fixed $t \in T$, $x(t, \omega)$ as a function of ω is an $\mathcal{R}^n$-valued random variable. For each fixed $\omega \in \Omega$, $x(t, \omega)$ as a function of t is an $\mathcal{R}^n$-valued time function called a *sample function, realization, or path* of the process. The set of all these functions obtained as ω varies over Ω is called the *ensemble* or *sample-function space* of the process.

Example 1.4.1. *Counting Process* ──────────────────────────────
Let each $\omega \in \Omega$ be a finitely denumerable point set in the interval $[0,1]$. Then, each ω can be enumerated in the form $\{t_1(\omega), t_2(\omega), \cdots, t_n(\omega)\}$, where $0 \leq t_1(\omega) \leq t_2(\omega) \leq \cdots \leq t_n(\omega) < 1$ are the coordinates of the points and $n < \infty$. Define $\{N(t, \omega); 0 \leq t < 1\}$ by

$$
N(t, \omega) = \begin{cases} 0, & 0 \leq t \leq t_1(\omega) \\ i - 1, & t_{i-1}(\omega) < t \leq t_i(\omega), \quad 2 \leq i \leq n \\ n, & t_n < t < 1. \end{cases}
$$

Then, for each fixed $t \in [0,1]$, $N(t, \omega)$ is an integer-valued, discrete random variable. For each fixed $\omega \in \Omega$, $N(0, \omega) = 0$ and $N(t, \omega)$ is a piecewise-constant, left-continuous function of t with unit positive jumps at $t_1(\omega)$, $t_2(\omega)$, $\cdots$, and $t_n(\omega)$. The sample-function space consists of all step functions of this type that are initially zero and have a finite number of unit jumps. Defined in this way, $\{N(t, \omega): 0 \leq t \leq 1\}$ is a counting process. ∎

A stochastic process $\{x(t, \omega): t \in T\}$ is said to be *completely characterized statistically* if the joint distribution function

$$
P_{x(t_1), x(t_2), \cdots, x(t_k)}(X_1, X_2, \cdots, X_k) = \Pr(x(t_1) \leq X_1, x(t_2) \leq X_2, \cdots, x(t_k) \leq X_k)
$$

for the random variables $x(t_1, \omega)$, $x(t_2, \omega)$, $\cdots$, and $x(t_k, \omega)$ is known for any finite collections $\{t_1, t_2, \cdots, t_k\}$ and $\{X_1, X_2, \cdots, X_k\}$, where $t_i \in T$ and $X_i \in \mathcal{R}^n$ for $1 \leq i \leq k$. The following are remarks on the complete statistical characterization of some particular stochastic processes we shall encounter in later chapters.

Markov Processes. Let the index set T of the process $\{x(t,\omega): t \in T\}$ be a countable set $\{t_1, t_2, \cdots\}$ of ordered time instants, $t_1 < t_2 < \cdots$. This process is by definition a discrete-time Markov process if for any $k \geq 2$, $t_k \in T$, there holds

$$\Pr(x(t_k) \leq X_k | x(t_1) \leq X_1, x(t_2) \leq X_2, \cdots, x(t_{k-1}) \leq X_{k-1})$$

$$= \Pr(x(t_k) \leq X_k | x(t_{k-1}) \leq X_{k-1}). \tag{1.25}$$

This condition can be interpreted qualitatively in two useful ways. The first is that for a process to be a Markov process, it must be amnesic to the extent that given the "past" $\{x(t_1), x(t_2), \cdots, x(t_{k-1})\}$, the "present" $x(t_k)$ depends on only the most recent past $x(t_{k-1})$. For the second interpretation, note that if (1.25) holds, so does

$$\Pr(x(t_1) \leq X_1, ..., x(t_{j-1}) \leq X_{j-1}, x(t_{j+1}) \leq X_{j+1}, \cdots, x(t_k) \leq X_k | x(t_j) \leq X_j)$$

$$= \Pr(x(t_1) \leq X_1, \cdots, x(t_{j-1}) \leq X_{j-1} | x(t_j) \leq X_j) \tag{1.26}$$

$$\times \Pr(x(t_{j+1}) \leq X_{j+1}, \cdots, x(t_k) \leq X_k | x(t_j) \leq X_j).$$

Thus, a Markov process in discrete time has the property that its "past" $\{x(t_1), \cdots, x(t_{j-1})\}$ and "future" $\{x(t_{j+1}), \cdots, x(t_k)\}$ are independent given the "present" $x(t_j)$.

A discrete-time Markov process can be completely characterized statistically by only two quantities, the distribution for $x(t_1)$ and, for $k \geq 2$, the conditional distribution for $x(t_k)$ given $x(t_{k-1})$. This follows by first noting that in general

$$\Pr(x(t_1) \leq X_1, \cdots, x(t_k) \leq X_k)$$

$$= \Pr(x(t_1) \leq X_1) \prod_{i=2}^{k} \Pr(x(t_i) \leq X_i | x(t_1) \leq X_1, \cdots, x(t_{i-1}) \leq X_{i-1}).$$

Thus, if $\{x(t,\omega): t \in T\}$ is a Markov process, (1.25) implies that

$$\Pr(x(t_1) \leq X_1, \cdots, x(t_k) \leq X_k)$$

$$= \Pr(x(t_1) \leq X_1) \prod_{i=2}^{k} \Pr(x(t_i) \leq X_i | x(t_{i-1}) \leq X_{i-1}). \tag{1.27}$$

The quantity $\Pr(x(t_i) \leq X_i \mid x(t_{i-1}) \leq X_{i-1})$ is called the *transition distribution function* and is the principal quantity characterizing the time evolution of the process.

We may also employ the joint probability density

$$p_{x(t_1),x(t_2),\cdots,x(t_k)}(X_1,X_2,\cdots,X_k)$$

of $\{x(t_1,\omega), x(t_2,\omega), \cdots, x(t_k,\omega)\}$ when it exists. For a Markov process,

$$p_{x(t_1),x(t_2),\cdots,x(t_k)}(X_1,X_2,\cdots,X_k)$$

$$= p_{x(t_1)}(X_1) \prod_{i=2}^{k} p_{x(t_i)\mid x(t_{i-1})}(X_i \mid X_{i-1}). \tag{1.28}$$

The conditional probability density function $p_{x(t_i)\mid x(t_{i-1})}(X_i \mid X_{i-1})$ is called the *transition density function*.

Continuous time Markov processes have similar properties and characterizations.

Markov Random-Fields. Markov random-fields extend Markov processes to multidimensional spaces by generalizing the concepts of time ordering and past dependence that are fundamental in the definition of Markov processes. This generalization becomes important when dealing with multidimensional Poisson-processes and, in particular, with the applications in Ch. 3 to quantum-limited imaging. The earliest application of Markov random-fields is contained in the well known works of E. Ising, and later L. Onsager, on the classic Ising random field models for characterizing magnetic domains. See K. Huang [8] for a complete treatment of these results, and refer to R. Dobrushin [3] and D. Ruelle [18] for a complete mathematical treatment of random fields. J. Besag [1] discusses a broad variety of Markov random-fields and their applications.

The generalization of a temporal Markov process to a multidimensional process is not straightforward because the notions of past and future, which are natural with a one-dimensional time coordinate, do not have counterparts in higher-dimensional spaces. Instead the notion of *neighbors* is used. Let $\{x(t): t \in T\}$ be a random process with a discrete index set $T = \{t_1, t_2, \cdots, t_k\}$; here, the elements of T can be vectors from a multidimensional space, such as the two-dimensional plane $\mathcal{R}^2$. Assume that the random variables $x(t_1), x(t_2), \cdots, x(t_k)$ are continuous, so their joint-probability density $p_{x(t_1),x(t_2),\cdots,x(t_k)}(X_1,X_2,\cdots,X_k)$ exists. We will also assume for each $i, 1 \leq i \leq k$, that the joint density of the $k-1$ random variables $\{x(t_j); j \neq i, 1 \leq j \leq k\}$ is strictly greater than zero,

$$p_{x(t_1)\cdots,x(t_{i-1}),x(t_{i+1}),\cdots,x(t_k)}(X_1,\cdots,X_{i-1},X_{i+1},\cdots,X_k) > 0. \qquad (1.29)$$

This *positivity condition* insures the existence of the conditional densities

$$p_{x(t_i)|x(t_1),\cdots,x(t_{i-1}),x(t_{i+1}),\cdots,x(t_k)}(X_i|X_1,\cdots,X_{i-1},X_{i+1},\cdots,X_k)$$

for $1 \le i \le k$, which appear in the following discussion.

Markov random-fields are influenced by their neighborhoods, where index or site t_j is said to be a *neighbor* of t_i if the above conditional density for $x(t_i)$ is a function of X_j. Denoting $N(t_i)$ as the set of all the neighbors of t_i, then $\{x(t): t \in T\}$ is said to be a Markov random-field with neighborhoods $\{N(t_i): t_i \in T\}$ if

$$p_{x(t_i)|x(t_1),\cdots,x(t_{i-1}),x(t_{i+1}),\cdots,x(t_k)}(X_i|X_1,\cdots,X_{i-1},X_{i+1},\cdots,X_k)$$

$$= p_{x(t_i)|\{x(t_j); t_j \in N_i\}}(X_i| \{X_j; \text{all } j \text{ such that } t_j \in N_i\}).$$

The neighbor relationship is a symmetric one because if $t_j \in N_i$, then $t_i \in N_j$.

A random process $\{x(t): t \in T\}$ with a discrete index set $T = \{t_1, t_2, \cdots, t_k\}$ and a set of neighborhoods $\{N_1, N_2, \cdots, N_k\}$ and conditional densities is called a *Markov random-field*. We now give three examples of these processes.

The simplest example of a Markov random-field is one in which for each i, the neighbors of t_i are all the other indices $\{t_j; j \ne i, 1 \le j \le k\}$. For this choice, $N_1 = \{t_2, t_3, \cdots, t_k\}$, $N_k = \{t_1, t_2, \cdots, t_{k-1}\}$, and $N_i = \{t_1, \cdots, t_{i-1}, t_{i+1}, \cdots, t_k\}$ for $2 \le i \le k-1$. This example with no restrictions on the statistical dependency between the variables indicates that Markov random-fields can be quite general. Gaussian random-processes, which are defined below, represent one interesting example of Markov random-fields that have this neighborhood structure.

Temporal Markov random-processes of the type described in the previous section are Markov random-fields with a nearest-neighbor structure. To see this, suppose that $\{x(t): t \in T\}$ is a Markov random-process with a discrete index set $T = \{t_1, t_2, \cdots, t_k\}$; here, each t_i is a real number representing time, and the joint density of $\{x(t_1), x(t_2), \cdots, x(t_k)\}$ satisfies (1.28). We also assume that the positivity condition is satisfied. Applying the definition of conditional probability densities (1.20), we have that

$$p_{x(t_k)|x(t_1),\cdots,x(t_{k-1})}(X_k \mid X_1,\cdots,X_{k-1})$$

$$= \frac{p_{x(t_1),x(t_2),\cdots,x(t_k)}(X_1,X_2,\cdots,X_k)}{\int_{-\infty}^{\infty} p_{x(t_1),x(t_2),\cdots,x(t_k)}(X_1,X_2,\cdots,X_k)\,dX_k}. \tag{1.30}$$

Substitution of (1.28) into the right-hand side of this equation yields

$$p_{x(t_k)|x(t_1),\cdots,x(t_{k-1})}(X_k \mid X_1,\cdots,X_{k-1}) = p_{x(t_k)|x(t_{k-1})}(X_k \mid X_{k-1}),$$

which implies that t_{k-1} is a neighbor of t_k; since it is the only neighbor, $N(t_k) = \{t_{k-1}\}$. Similarly,

$$p_{x(t_1)|x(t_2),\cdots,x(t_k)}(X_1 \mid X_2,\cdots,X_k) = p_{x(t_1)|x(t_2)}(X_1 \mid X_2), \tag{1.31}$$

which implies that t_2 is a neighbor of t_1, and $N(t_1) = \{t_2\}$. For $2 \le i \le k-1$, this same procedure yields

$$p_{x(t_i)|x(t_1),\cdots,x(t_{i-1}),x(t_{i+1}),\cdots,x(t_k)}(X_i \mid X_1,\cdots,X_{i-1},X_{i+1},\cdots,X_k)$$

$$= \frac{p_{x(t_i)|x(t_{i-1})}(X_i \mid X_{i-1})\,p_{x(t_{i+1})|x(t_i)}(X_{i+1} \mid X_i)}{\int p_{x(t_i)|x(t_{i-1})}(X_i \mid X_{i-1})\,p_{x(t_{i+1})|x(t_i)}(X_{i+1} \mid X_i)\,dX_i}, \tag{1.32}$$

which implies that both t_{i-1} and t_{i+1} are neighbors of t_i, and $N(t_i) = \{t_{i-1},t_{i+1}\}$ for $2 \le i \le k-1$. Thus, Markov random-processes are an example of a Markov random-fields with neighborhoods $N(t_1) = \{t_2\}$, $N(t_k) = \{t_{k-1}\}$, and $N(t_i) = \{t_{i-1},t_{i+1}\}$ for $2 \le i \le k-1$. The converse can also be shown; namely, a Markov random-field with a discrete, real-valued index set and these neighborhoods is a Markov random-process.

Another simple example of a Markov random-field can be constructed as follows. Define the index set to be of the form $T = \{t_{ij}; t_{ij} = (i\Delta x, j\Delta y), \ 0 \le i \le N_x - 1, 0 \le j \le N_j - 1\}$ corresponding to the lattice points of a discrete pixelization of a finite rectangular-region of the plane. Suppose that the neighbors of an interior point t_{ij} are the lattice points lying to the North, East, South, and West, so the neighborhood of t_{ij} is $N(t_{ij}) = \{t_{i,j+1}, t_{i+1,j}, t_{i,j-1}, t_{i-1,j}\}$. The neighbors of boundary points of the region are similarly defined but only include lattice points within the region. This set of neighborhoods along with a specified conditional density for the process at each lattice point,

$$p_{x(t_{ij})|\{x(t_{mn}), t_{mn} \in N_{ij}\}}(X_{ij}| \{X_{mn}, \text{ all } m, n \text{ such that } t_{mn} \in N_{ij}\}), \tag{1.33}$$

defines a two-dimensional Markov random-field with nearest-neighbor dependency.

Being able to deduce the joint density for the random variables $\{x(t_i); t_i \in T\}$ from given conditional densities and neighborhoods is important for completing the statistical characterization of a Markov random-field. For a Markov random-process, the joint density is readily obtained from given conditional densities by using (1.28). The analogous result for a Markov random-field with given conditional densities and cliques is not nearly as straightforward, but it is likewise possible to deduce the structure of the joint density from the given conditional densities. J. Besag [1] gives one way to do this by observing first that any joint density can be written in the following form:

$$p_{x(t_1), x(t_2), \cdots, x(t_k)}(X_1, X_2, \cdots, X_k)$$

$$= p_{x(t_k)|x(t_1), \cdots, x(t_{k-1})}(X_k \mid X_1, \cdots, X_{k-1})$$

$$\times p_{x(t_1), x(t_2), \cdots, x(t_{k-1})}(X_1, X_2, \cdots, X_{k-1}). \tag{1.34}$$

The first factor on the right can be replaced by the given conditional density of $x(t_k)$ given $\{x(t_j); t_j \in N_k\}$, but there is no straightforward way to continue in this manner with the second factor because

$$p_{x(t_{k-1})|x(t_1), \cdots, x(t_{k-2})}(X_{k-1} \mid X_1, \cdots, X_{k-2})$$

is not easily expressed in terms of the given conditional densities and neighborhoods. To circumvent this problem, J. Besag [1] introduces a fixed value for $x(t_k)$, and expresses the second factor so as to obtain

$$p_{x(t_1), x(t_2), \cdots, x(t_k)}(X_1, X_2, \cdots, X_k)$$

$$= \frac{p_{x(t_k)|x(t_1), \cdots, x(t_{k-1})}(X_k \mid X_1, \cdots, X_{k-1})}{p_{x(t_k)|x(t_1), \cdots, x(t_{k-1})}(Y_k \mid X_1, \cdots, X_{k-1})}$$

$$\times p_{x(t_1), x(t_2), \cdots, x(t_k)}(X_1, X_2, \cdots, Y_k), \tag{1.35}$$

where Y_k is some arbitrary but fixed value that $x(t_k)$ may have. Repeating this for $x(t_{k-1})$ in the last factor on the right-hand side yields

$$p_{x(t_1),\cdots,x(t_{k-1}),x(t_k)}(X_1,\cdots,X_{k-1},Y_k)$$

$$= \frac{p_{x(t_{k-1})|x(t_1),\cdots,x(t_{k-2}),x(t_k)}(X_{k-1}\mid X_1,\cdots,X_{k-2},Y_k)}{p_{x(t_{k-1})|x(t_1),\cdots,x(t_{k-2}),x(t_k)}(Y_{k-1}\mid X_1,\cdots,X_{k-2},Y_k)}$$

$$\times p_{x(t_1),\cdots,x(t_{k-2}),x(t_{k-1}),x(t_k)}(X_1,\cdots,X_{k-2},Y_{k-1},Y_k). \tag{1.36}$$

Continuing this process eventually yields the following expression for the joint density:

$$\frac{p_{x(t_1),\cdots,x(t_{k-1}),x(t_k)}(X_1,\cdots,X_{k-1},X_k)}{p_{x(t_1),\cdots,x(t_{k-1}),x(t_k)}(Y_1,\cdots,Y_{k-1},Y_k)} \tag{1.37}$$

$$= \prod_{i=1}^{k} \frac{p_{x(t_i)|x(t_1),\cdots,x(t_{i-1}),x(t_{i+1}),\cdots,x(t_k)}(X_i\mid X_1,\cdots,X_{i-1},Y_{i+1},\cdots Y_k)}{p_{x(t_i)|x(t_1),\cdots,x(t_{i-1}),x(t_{i+1}),\cdots,x(t_k)}(Y_i\mid X_1,\cdots,X_{i-1},Y_{i+1},\cdots,Y_k)}.$$

The denominator on the left-hand side can be viewed as the requisite normalization factor so that the left-hand side is unity when integrated over all values of $X = \{X_1, X_2, \cdots, X_k\}$. The conditional densities on the right-hand side can be replaced by those given in the specification of the Markov random-field. While in principle this indicates how to determine the joint density from given conditional densities in a Markov random-field, there are difficulties in its use to obtain explicit results.

Independent and Orthogonal Increment Processes. A temporal random process $\{x(t,\omega): t \in T\}$, where T is a time interval, is called an *independent increment process* if the k random variables $x(t_1,\omega)$, $x(t_2,\omega)-x(t_1,\omega), \cdots, x(t_k,\omega)-x(t_{k-1},\omega)$ are statistically independent for any finite collection for times $t_1 < t_2 < \cdots, t_k$, where $t_i \in T$ for $1 \le i \le k$. If these variables are pairwise uncorrelated, $\{x(t,\omega): t \ge t \in T\}$ is called a process with *uncorrelated increments* or, if in addition the increment variables have zero expectation, *orthogonal increments*. We will say that a stochastic process $\{x(t,\omega): t \in T\}$ on a higher dimensional space T is an independent increment process if the random variables $\{x(t_i,\omega); t_i \in A_i, i = 1, 2, \cdots, k\}$ are statistically independent for any collection of disjoint sets $\{A_i, i = 1, 2, \cdots, k\}$ in T and any k. The most important examples of such a process will be the Poisson process on multidimensional spaces $T = \mathcal{R}^n$, especially for $n = 2$ and 3, and the Poisson process in time and space $T = [t_0, \infty) \times \mathcal{R}^n$.

A stochastic process $\{x(t,\omega); t \in T\}$ with independent increments is completely characterized statistically by its first-order transition distribution (that is, the probability distribution for $x(t,\omega)$ and by its increment distribution (that is, the distribution for $x(t,\omega)\text{-}x(s,\omega), s < t$). This can be verified as follows. Let $t_1 < t_2 < \cdots, t_k$ be an ordered set of times in T, and denote the joint characteristic function for $x(t_1,\omega), x(t_2,\omega), \cdots, x(t_k,\omega)$ by

$$M_{x(t_1),x(t_2),\cdots,x(t_k)}(jv_1, jv_2, \cdots, jv_k) = E\left\{\exp\left[j \sum_{i=1}^{k} v_i x(t_i)\right]\right\}.$$

Now,

$$\sum_{i=1}^{k} v_i x(t_i) = \left(\sum_{m=1}^{k} v_m\right) x(t_1) + \sum_{i=2}^{k}\left(\sum_{m=i}^{k} v_m\right)(x(t_i) - x(t_{i-1})).$$

Hence, if $\{x(t,\omega): t \in T\}$ has independent increments,

$$M_{x(t_1),x(t_2),\cdots,x(t_k)}(jv_1, jv_2, \cdots, jv_k)$$

$$= M_{x(t_1)}\left(j \sum_{m=1}^{k} v_m\right) \prod_{i=2}^{k} M_{x(t_i)-x(t_{i-1})}\left(j \sum_{m=i}^{k} v_m\right). \qquad (1.38)$$

The joint characteristic function is, therefore, determined by the characteristic functions for $x(t_1,\omega)$ and $x(t_j,\omega)\text{-}x(t_{j-1},\omega)$. As characteristic functions and distributions have a one-to-one relationship through Fourier transformation, it follows that the joint distribution for $x(t_1,\omega), x(t_2,\omega), \cdots, x(t_k,\omega)$ is determined by the first-order and increment distributions for the process.

Independent increment processes are basic building blocks for a wide variety of stochastic processes used as models for physical phenomena. In later chapters, we will encounter two particularly important independent increment processes, the *Poisson process* and the *Wiener process*. A fundamental role is played by these processes because all other independent increment processes are a linear combination of them. The Poisson counting process is defined and studied in detail in Ch. 2; it is an independent increment process with discontinuous sample functions. The Wiener process appears in Ch's. 5 and 7. It is an independent increment process with continuous sample functions. By definition, $\{x(t,\omega), t \geq t_0\}$ is a Wiener process if

a. $\{x(t,\omega): t \geq t_0\}$ has independent increments;
b. $x(t_0,\omega) = 0$ with probability one;

c.   the increment $x(t,\omega) - x(s,\omega)$, where $t_0 \le s \le t$, is normally distributed with mean zero and variance $(t-s)\alpha$, where $\alpha > 0$.

According to property (c), the distribution of the increment depends only on the difference $t-s$ and not t and s separately. A process with this property is said to possess *stationary increments*. $x(t,\omega)$ is normally distributed with zero mean and a variance $(t-t_0)\alpha$ that increases linearly with time. The constant α determines how rapid this increase is. The quantity α^{-1} is a characteristic time for the process, which accumulates one (r.m.s.) unit of drift in $t-t_0 = \alpha^{-1}$ units of time. A Wiener process is said to be *normalized* or *standardized* if $\alpha = 1$. An $\mathcal{R}^n$-valued, standard Wiener process is a vector of n independent, standard Wiener processes. By using the expression in Table 1.2 for the characteristic function of a normal random variable, and by using (1.38), we see for a Wiener process that

$$M_{x(t_1),x(t_2),\cdots,x(t_k)}(jv_1, jv_2, \cdots, jv_k)$$

$$= \prod_{i=1}^{k} \exp\left[-\frac{1}{2}\alpha(v_i + v_{i+1} + \cdots + v_k)^2 (t_i - t_{i-1}) \right]$$

$$= \exp\left(-\frac{1}{2}v'\Lambda v \right), \tag{1.39}$$

where $v = [v_1 \quad v_2 \quad \cdots \quad v_k]'$ and Λ is a $k \times k$ matrix with i,j-element λ_{ij} given by

$$\lambda_{ij} = \begin{cases} \alpha(t_i - t_0), & i \le j \\ \alpha(t_j - t_0), & i > j \end{cases} = \alpha \min(t_i - t_0, t_j - t_0).$$

Hence, from (1.11) and (1.18)

$$p_{x(t_1),x(t_2),\cdots,x(t_k)}(X_1, X_2, \cdots, X_k) = \frac{1}{\sqrt{(2\pi)^k \det(\Lambda)}} \exp\left(-\frac{1}{2}X'\Lambda^{-1}X \right), \tag{1.40}$$

where $X = [X_1 \quad X_2 \quad \cdot \quad \cdot \quad \cdot \quad X_k]'$. Thus, we conclude that for a Wiener process, the random variables $x(t_1,\omega)$, $x(t_2,\omega)$, $\cdots$, $x(t_k,\omega)$ are jointly normal with mean zero and covariance matrix Λ. While the sample functions of a Wiener process are continuous with probability one, they are nondifferentiable. For a proof of this assertion as well as further properties of the Wiener process, we refer to J. Doob [4, Ch. 3].

Gaussian Processes. A stochastic process $\{x(t,\omega): t \in T\}$ is called a *Gaussian* or *normal* process if for any finite set $\{\{t_1, t_2, \cdots, t_k\}$, the random variables $x(t_1,\omega)$, $x(t_2,\omega)$, $\cdots$, $x(t_n,\omega)$ are jointly normal $N(\mu, \Sigma)$, where the i*th* element of the mean vector μ is $m_x(t_i) = E[x(t_i)]$, and where the i,j*th* element of the covariance matrix Σ is

$$K_x(t_i, t_j) = E(x(t_i)x(t_j)) - E(x(t_i))E(x(t_j)). \tag{1.41}$$

The functions $m_x(t_i)$ and $K_x(t_i, t_j)$ are called the *mean-value function* and *covariance function* of the process. It is seen that these first and second moments completely characterize a Gaussian process statistically because knowing them is equivalent to knowing the joint distributions. From (1.40), a Wiener process is a Gaussian process with mean-value function $m_x(t) = 0$ and covariance function $K_x(t, u) = \alpha\min(t, u)$. Gaussian processes are also an example of Markov random-fields in which the neighborhood for t_i contains all the indices t_j for which $j \neq i$ and $K(t_i, t_j) \neq 0$.

Homogeneous and Wide-Sense Stationary Processes. A stochastic process $\{x(t,\omega): t \in T\}$ is said to be *homogeneous* if its mean-value function $m(t) = E[x(t)]$ is a constant independent of t and its covariance function $K(t, u) = E[x(t)x(u)] - E[x(t)]E[x(u)]$ is a function only of $t - u$ and not t and u separately. If the index set T corresponds to time, then the process is also called *wide-sense stationary* under these conditions. For an homogeneous process, $K(t, u) = K(t - u, 0)$. It is common practice to write $K(t - u, 0)$ as $K(\tau)$, a function of a single variable $\tau = t - u$. The *power-density spectrum* for an homogeneous process is the Fourier transform of its covariance function,

$$S(f) = \int_{-\infty}^{\infty} K(\tau) e^{-j2\pi<f,\tau>} d\tau, \tag{1.42}$$

where $<f, \tau> = \sum_i f_i \tau_i$. Conversely,

$$K(\tau) = \int_{-\infty}^{\infty} S(f) e^{j2\pi<f,\tau>} df. \tag{1.43}$$

While the power-density spectrum introduces nothing new statistically, it has important physical interpretations. Its total area $K(0) = \int_{-\infty}^{\infty} S(f) df$ is the variance of $x(t)$ for any t. By virtue of the relation

$$K(0) = \lim_{T \to \infty} \frac{1}{2T} \int_{-T}^{T} x^2(t)\, dt,$$

which holds for an ergodic process, this variance is often interpreted as the total power in $x(\cdot)$. The integral $\int_F S(f)\, df$ is interpreted as the power in $x(\cdot)$ associated with components in the frequency set F. This may be motivated using the relationship that exists between the power-density spectra of processes at the input and output of a linear shift-invariant filter. Thus, if

$$y(t) = \int_{-\infty}^{\infty} h(t - u) x(u)\, du, \tag{1.44}$$

then it is known that

$$S_y(f) = |H(f)|^2 S_x(f), \tag{1.45}$$

where

$$H(f) = \int_{-\infty}^{\infty} h(t) e^{-j2\pi ft}\, dt \tag{1.46}$$

is the frequency response function of the filter [14, 21]. Setting $H(f) = 1$ for $f \in F$ and $H(f) = 0$ otherwise shows that

$$K_y(0) = \int_{-\infty}^{\infty} S_y(f)\, df = \int_F S_x(f)\, df.$$

Hence, the total power in $y(\cdot)$ is the power in those components of $x(\cdot)$ in the frequency set F. It is intuitively clear that the power in $y(\cdot)$, namely

$$\int_{-\infty}^{\infty} |H(f)|^2 S_x(f)\, df,$$

will differ by only a small amount from

$$N_0 \int_{-\infty}^{\infty} |H(f)|^2\, df$$

if $S_x(f) \approx N_0$ over all frequencies where $H(f)$ has significant nonzero values. In this case, it is common to assume that $x(\cdot)$ is a *white-noise process*. The power-density spectrum for a white-noise process is a constant for all frequencies; equivalently, its covariance function is an impulse function. While white-noise processes do not exist physically, they provide a convenient approximation and manipulative tool. We do not use the notions of white noise and power-density spectra in most of the text; the main exception is in Ch. 5, where we discuss shot noise models.

E. Wong and B. Hajek [21, Ch. 3] may be consulted for further discussion of homogeneous processes.

1.4.4 Sequences and Convergence Concepts

A sequence of random variables $x_1(\omega), x_2(\omega), \cdots$ is a particular form of stochastic process in which the index set is the nonnegative integers $\{1,2, \cdots\}$. A recurring problem in the analytical investigation of stochastic processes is the resolution of what transpires in the limit as the indexing parameter tends to infinity. For instance, this is the central issue in the study of the continuity, differentiability, and integrability of processes. We now briefly review convergence concepts for sequences of random variables. A thorough treatment of this topic is given by M. Loeve [11,Ch. 2] and I. Gikman and A. Skorokhod [7, Ch. 2].

Convergence Everywhere. Let the sequence of random variables $x_1(\omega), x_2(\omega), \cdots$ and the random variable $x(\omega)$ all be defined on the same probability space $(\Omega, \mathcal{F}, \mu)$. This sequence is said to converge *everywhere* to $x(\omega)$ if for every $\omega \in \Omega$, there holds

$$\lim_{n \to \infty} x_n(\omega) = x(\omega). \tag{1.47}$$

The probability measure μ plays no role in this form of convergence. The following forms do reflect μ in various ways. These alternatives are of greater importance in the theory of probability and stochastic processes.

Convergence Almost Everywhere. The sequence of random variables $x_1(\omega), x_2(\omega), \cdots$ is said to converge *almost everywhere, almost surely*, or *with probability one* to $x(\omega)$ if the set of ω for which (1.47) fails to hold has zero probability measure. Thus, the sequence may not converge for every $\omega \in \Omega$, but the event that it does not occurs with zero probability. We occasionally use the notation

$$x_n \xrightarrow{a.s.} x$$

to indicate that $x_n(\omega)$ converges almost surely to $x(\omega)$.

Convergence in Probability. A sequence $x_1(\omega), x_2(\omega), \cdots$ is said to *converge in probability* to $x(\omega)$ if for every $\varepsilon > 0$,

$$\lim_{n \to \infty} \Pr(|x_n - x| > \varepsilon) = 0. \tag{1.48}$$

If the sequence $x_1(\omega), x_2(\omega), \cdots$ converges almost everywhere to $x(\omega)$, it also converges in probability to $x(\omega)$, but the converse need not hold. We sometimes use

$$x_n \overset{p}{\to} x \quad \text{or} \quad p \lim_{n \to \infty} x_n = x$$

to indicate that $x_n(\omega)$ converges in probability to $x(\omega)$.

Convergence in r*th* Mean. The sequence $x_1(\omega), x_2(\omega), \cdots$ is said to *converge in the rth mean*, for $r > 0$, if

$$\lim_{n \to \infty} E(|x_n - x|^r) = 0. \tag{1.49}$$

The most important special case is $r = 2$. Convergence in the r*th* mean for $r = 2$ is also called *quadratic-mean* convergence. If a sequence converges in the r*th* mean, it also converges in the r'*th* mean where $0 < r' \leq r$. Moreover, it also converges in probability. This last assertion can be easily demonstrated for $r = 2$ by use of the Tchebycheff inequality, which says for any random variable that

$$E(y^2) = \int_{-\infty}^{\infty} Y^2 dP_y(Y) \geq \varepsilon^2 \int_{|Y| > \varepsilon} dP_y(Y) = \varepsilon^2 \Pr(|y| > \varepsilon).$$

Thus, $\Pr(|x_n - x| > \varepsilon) \leq \varepsilon^{-2} E(|x_n - x|^2)$, which shows that if x_n converges in the quadratic mean to x, it also converges in probability to x.

Convergence in Distribution. The sequence $x_1(\omega), x_2(\omega), \cdots$ is said to *converge in distribution* to $x(\omega)$ if

$$\lim_{n \to \infty} P_{x_n}(X) = P_x(X) \tag{1.50}$$

for $-\infty < X < \infty$. Convergence in distribution is the weakest form of convergence as it is implied by convergence in probability, which, in turn, is implied by both almost everywhere and r*th* mean convergence. We will encounter convergence in distribution in two contexts. The first occurs in Ch. 5 where we study the response of linear systems to a point process excitation. We show that under certain conditions that the response tends to have a normal distribution as the rate at which input points occur tends to infinity. The second context arises in Ch. 6 where we study the effect of pooling the points of many point processes. We show that under certain conditions that the resulting pooled point process tends to have the Poisson distribution as the number of component processes tends to infinity.

There are several instances in the book where we require conditions for the validity of the relation

$$\lim_{n \to \infty} E(x_n) = E\left(\lim_{n \to \infty} x_n\right) = E(x), \qquad (1.51)$$

in which $x_1(\omega), x_2(\omega), \cdots$ is a sequence of random variables converging to a limit $x(\omega)$ in one of the above senses. These conditions validate the operation of interchanging the order of taking a limit and expectation. The simplest assertion to establish is that if $x_n(\omega)$ converges in the mean-square sense to $x(\omega)$, then (1.51) holds. This is an immediate consequence of the inequality

$$|E(x_n - x)| \leq E(|x_n - x|). \qquad (1.52)$$

As convergence in the rth mean for $r = 2$ implies convergence for $r = 1$, we see using (1.52) that $\lim_{n \to \infty} |E(x_n - x)| = \lim_{n \to \infty} |E(x_n) - E(x)| = 0$ provided x_n converges to x in the quadratic mean. The conditions given in the following theorem for the validity of (1.51) will be used often in the book.

Theorem 1.4.1 (*Generalized Bounded Convergence*). Let $\{g_n\}$ be a sequence of nonnegative random variables converging with probability one to a random variable g. Assume for each n that $E(g_n) < \infty$ and that $E(g) < \infty$. Let $\{f_n\}$ be a sequence of random variables converging with probability one to a random variable f. Assume for each n that $E(|f_n|) < \infty$ and that $|f_n| < g_n$ with probability one. If

$$E(g) = E\left(\lim_{n \to \infty} g_n\right) = \lim_{n \to \infty} E(g_n),$$

then

$$E(f) = E\left(\lim_{n \to \infty} f_n\right) = \lim_{n \to \infty} E(f_n) \qquad (1.53)$$

and

$$E(f \mid y) = E\left(\lim_{n \to \infty} f_n \mid y\right) = \lim_{n \to \infty} E(f_n \mid y), \tag{1.54}$$

where y is a random variable.

Proof. If $g_n = g$ for all n, then (1.53) follows from the bounded convergence theorem proved, for example, by H. Royden [17, p. 88], and (1.54) follows from a bounded convergence theorem for conditional expectations proved by J. Doob [4, p. 23]. If $g_n \neq g$ for all n, then (1.53) follows from a generalized bounded convergence theorem proved by H. Roydan [17, p. 89]. We slightly modify Doob's procedure to prove (1.54) when $g_n \neq g$ for all n. Thus, define $\hat{f}_n = \sup_{j \geq n} |f_j - f|$. Then, with probability one, $\hat{f}_1 \geq \hat{f}_2 \geq \ldots \geq 0$, $f \lim_{n \to \infty} \hat{f}_n = 0$, and $\hat{f}_n \leq \hat{g}_n$, where we define $\hat{g}_n = g + \sup_{j \geq n} g_j$. We note that $E(\hat{g}_n) < \infty$, $\hat{g}_n \leq \hat{g}_1$, and $\lim_{n \to \infty} \hat{g}_n = 2g$ with probability one. Consequently, by the bounded convergence theorem, there holds $E(2g) = E(\lim_{n \to \infty} \hat{g}_n) = \lim_{n \to \infty} E(\hat{g}_n)$. We conclude from the generalized bounded convergence theorem (1.53) that

$$0 = E\left(\lim_{n \to \infty} \hat{f}_n\right) = \lim_{n \to \infty} E(\hat{f}_n).$$

Now observe that

$$0 \leq |E(f \mid y) - E(f_n \mid y)| = |E(f - f_n \mid y)| \leq E(|f - f_n| \mid y) \leq E(\hat{f}_n \mid y).$$

Thus, to establish (1.54) it is sufficient to prove that $\lim_{n \to \infty} E(\hat{f}_n \mid y) = 0$ with probability one. As $\hat{f}_1 \geq \hat{f}_2 \geq \cdots \geq 0,$, we have that $E(\hat{f}_1 \mid y) \geq E(\hat{f}_2 \mid y) \geq \cdots \geq 0$ with probability one. Thus, with probability one, the sequence $E(\hat{f}_n \mid y)$ converges to a nonnegative random variable. Denote the limit by w; then, by definition $w = \lim_{n \to \infty} E(\hat{f}_n \mid y)$, and we have $E(w) \leq E(E(\hat{f}_n \mid y)) = E(\hat{f}_n)$. The right side tends to zero as n tends to infinity. Hence $E(w) = 0$, and as w is almost surely nonnegative, there holds $w = 0$ with probability one. This proves (1.54). $\square$

1.5 References

1.   J. Besag, "Spatial Interaction and the Statistical Analysis of Lattice Systems," *J. Royal Statistical Society*, Ser. B, Vol. 36, No. 2, pp.192-237, 1974.

2.   A. T. Bharucha-Reid, *Elements of the Theory of Markov Processes and Their Applications*, McGraw-Hill, New York, 1960.

3.   R. L. Dobrushin, "The Description of a Random Field by means of Conditional Probabilities and Conditions of Its Regularity," *Theory Prob. Appl.*, Vol. 13, pp. 197-224, 1968.

4.   J. L. Doob, *Stochastic Processes*, Wiley, New York, 1953.

5.   J. Evans, "Preliminary Analysis of ELF Noise," Tech. Note 1969-18, M.I.T. Lincoln Laboratory, Lexington, MA, March 1969.

6.   P. M. Fishman and D. L. Snyder, "Statistical Inference for Time Space Point Processes," *IEEE Transactions on Information Theory*, IT-22, 3, May 1976.

7.   I. I. Gikman and A. V. Skorokhod, *Introduction to the Theory of Random Processes*, Saunders, Philadelphia, PA, 1965.

8.   K. Huang, *Statistical Mechanics*, Wiley, New York, 1965.

9.   H. J. Kushner, "On the Differential Equations Satisfied by Conditional Probability Densities of Markov Processes with Applications," *J. SIAM Control*, Ser. A. 2, No. 1, pp. 106-119, 1964.

10.   R. S. Lipster and A. N. Shiryayev, *Statistics of Random Processes*, Springer-Verlag, NY, 1977.

11.   M. Loeve, *Probability Theory*, Van Nostrand, New York, 1963.

12.   J. E. Moyal, "The General Theory of Stochastic Population Processes," *Acta Math*, Vol. 108, pp. 1-31, 1962.

13.   J. Neveu, *Mathematical Foundations of the Calculus of Probability*, Holden-Day, San Francisco, CA, 1965.

14.   A. Papoulis, *Probability, Random Variables, and Stochastic Processes*, McGraw-Hill, New York, 1984.

15.   E. Parzen, *Modern Probability Theory and Its Applications*, Wiley, New York, 1960.

16.   E. Parzen, *Stochastic Processes*, Holden-Day, San Francisco, CA, 1962.

17.   H. L. Royden, *Real Analysis*, Macmillan, London, 1968

18.   D. Ruelle, "Thermodynamic Formalism," in: *Encyclopedia of Mathematics and Its Applications*, Vol. 5, Addison-Wesley, Reading, MA, 1978.

19.   R. L. Stratonovich, "On the Theory of Optimal Nonlinear Filtering of Random Functions," *Theory Probability Appl.*, Vol. 4, pp. 223-225, 1959.

20.   D. Vere-Jones, "Stochastic Models for Earthquake Occurrence," *J. Roy. Stat. Soc. B*, Vol. 32, No. 1, pp. 1-62, 1970.

21.   E. Wong and B. Hajek, *Stochastic Processes in Engineering Systems*, Springer-Verlag, New York, 1985.

CHAPTER TWO

POISSON PROCESSES

2.1 Introduction

The Poisson process is the simplest process associated with counting random numbers of points. We begin our study of these processes when the space where the points occur is a one-dimensional, semiinfinite, real line. While there is no mathematical reason to do so, we refer to this space as "time" because temporal phenomena seem to predominate in applications. The study of temporal Poisson-processes permits many of the properties of Poisson processes to be exhibited, but Poisson processes on multidimensional spaces are also important in applications. These are developed in Sec. 2.5.

Its simplicity notwithstanding, the Poisson counting process is important in at least three respects. First, it is found to be an accurate model in many and varied applications. One may well wonder why. A partial explanation is provided in this chapter in the form of some often well satisfied, relatively weak qualitative conditions under which a point process is a Poisson process. Further explanation is provided in Ch. 6 when we discuss the pooling of points from numerous subsidiary point processes. A second reason that the Poisson counting process is important is that it is a natural introductory topic in the investigation of more complicated counting processes. This will be evident when we get to Ch's. 6 and 7 where the restrictive conditions we impose here for the Poisson process are relaxed substantially. A third reason the Poisson counting process is important is that it is an essential building block for other processes having sample paths strikingly different from those of a counting process. These include discontinuous processes having nonidentical, nonunit jumps as well as the most general independent-increment processes.

A temporal Poisson-process for times $t \geq t_0$ is by definition a counting process $\{N(t): t \geq t_0\}$ with the following three properties:

i. $\Pr[N(t_0) = 0] = 1$;

ii. for $t_0 \leq s < t$, the increment $N(s,t) = N(t) - N(s)$ is Poisson distributed with parameter $\Lambda(t) - \Lambda(s)$,

$$\Pr[N(s,t) = n] = \frac{1}{n!}(\Lambda(t) - \Lambda(s))^n e^{-(\Lambda(t) - \Lambda(s))}, \qquad (2.1)$$

for $n = 0, 1, 2, \cdots$, where $\Lambda(t)$ is a finite-valued, nonnegative, nondecreasing function of t ;

iii. $\{N(t): t \geq t_0\}$ has independent increments.

Property (iii) is the distinguishing property. It means that for a Poisson counting process, the numbers of points in nonoverlapping intervals are statistically independent no matter how large or small the intervals and no matter how close or distant they may be; more precisely, if $[t_i, u_i)$, for $i = 1, 2, \cdots, k$, are k disjoint intervals on $[t_0, \infty)$, then

$$\Pr[N(t_1, u_1) = n_1, N(t_2, u_2) = n_2, \cdots, N(t_k, u_k) = n_k]$$

$$= \prod_{i=1}^{k} \Pr[N(t_i, u_i) = n_i] \tag{2.2}$$

for $k = 1, 2, \cdots$. Moreover, properties (i)-(iii) provide a complete statistical characterization for the Poisson counting process in the sense that the joint counting probability

$$\Pr[N(t_1) = n_1, N(t_2) = n_2, \cdots, N(t_k) = n_k] \tag{2.3}$$

can be determined for any collection of times $t_1 < t_2 < \cdots < t_k$ in $[t_0, \infty)$ and any collection of nonnegative integers $n_1, n_2, \cdots, n_k$, where $k = 1, 2, \cdots$. To see this, note by property (i) that $N(t_1) = N(t_0) + N(t_0, t_1)$. Hence,

$$\Pr[N(t_1) = n_1, N(t_2) = n_2, \cdots, N(t_k) = n_k]$$

$$= \Pr[N(t_0, t_1) = n_1 - n_0, N(t_1, t_2) = n_2 - n_1, \cdots, N(t_{k-1}, t_k) = n_k - n_{k-1}]$$

$$= \prod_{i=1}^{k} \Pr[N(t_{i-1}, t_i) = n_i - n_{i-1}], \tag{2.4}$$

where we define $n_0 = 0$, and the last equality follows from the independent increments property (iii). Thus, the joint probability (2.3) can be expressed in terms of the increment probabilities given in property (ii). It is evident that the independent-increments property (iii) plays a crucial role in characterizing the Poisson process. The effort to eliminate this property will lead us to the point processes of Ch. 6 that do not possess it.

The function Λ in property (ii) is termed the *parameter function* of the Poisson counting process. According to property (ii), the only analytic restrictions required in general on this function are that it be nonnegative and nondecreasing. In particular, there is no requirement that Λ be continuous; even if continuous, Λ need not be differentiable. It is of interest, therefore, to investigate the influence of additional restrictions on this function. If $\Lambda(t) - \Lambda(s)$ is finite, then points do not occur with certainty on the interval $[s,t)$. This follows from the equality $\Pr[N(s,t) \geq 1] = 1 - \exp[-(\Lambda(t) - \Lambda(s))]$ derived from (2.1); $\Lambda(t) - \Lambda(s) < \infty$ implies that $\Pr[N(s,t) \geq 1] < 1$. But, also, $\Lambda(t) - \Lambda(s)$ finite implies that an infinite number of points do not occur on $[s,t)$ because from (2.1), $\Pr[N(s,t) = n]$ then tends to zero as n tends to infinity. A further examination of (2.1) indicates that the points of a Poisson process can occur at predetermined times where the parameter function has a jump discontinuity, and the number of points occurring at such times has the Poisson distribution with parameter equal to the size of the jump. As $\Lambda(t)$ is a finite-valued, non-decreasing function of t, the collection of these predetermined times in $[s,t)$ is denumerable, and the sum of the jump sizes is finite. A. Khinchin [17] shows that $N(s,t)$ can be decomposed into two statistically independent components $N_d(s,t)$ and $N_c(s,t)$, which he terms singular and non-singular, respectively. $N_d(s,t)$ is the contribution to $N(s,t)$ of points occurring only at the predetermined times in $[s,t)$ where Λ is discontinuous, and $N_c(s,t) = N(s,t) - N_d(s,t)$ is the contribution to $N(s,t)$ of points occurring at the times in $[s,t)$ where Λ is continuous. If the parameter function is continuous, points do not occur at predetermined times, and there holds $\lim_{\delta \downarrow 0} \Pr[N(t,t+\delta) = 0] = 1$ for all $t \geq t_0$; in other words, $\{N(t); t \geq t_0\}$ is stochastically continuous if its parameter function is continuous. Finally, let us note for a Poisson process that the derivative of the parameter function, if it exists, is the instantaneous average rate that points occur. This follows because the expected number of points on the interval $[s,t)$ has the evaluation

$$E[N(s,t)] = \sum_{n=0}^{\infty} n \Pr[N(s,t) = n] = \Lambda(t) - \Lambda(s) \qquad (2.5)$$

for a Poisson counting process. Consequently, the existence of the limit

$$\lim_{\delta \downarrow 0} \frac{\Lambda(t+\delta) - \Lambda(t)}{\delta} \qquad (2.6)$$

implies that the point process has an instantaneous average rate at time t. Thus, if Λ is an absolutely continuous function of t, so that it can be expressed as

$$\Lambda(t) = \int_{t_0}^{t} \lambda(\sigma) d\sigma$$

for all $t \geq t_0$, where $\lambda(t)$ is a nonnegative function, then the limit in (2.6) exists and equals $\lambda(t)$. We term this function the *intensity function* of the process; at any time $t \geq t_0$, the intensity function is the instantaneous average rate that points occur.

Our goal in this chapter is to study the Poisson process in detail. We begin in Sec. 2.2 by giving some sufficient conditions for a counting process to be a Poisson process. The notions of *orderliness* and *evolution without aftereffects* are introduced here. In Sec. 2.3, we examine the statistics of the point locations and interpoint spacings for a Poisson process. Two of the more important quantities introduced are the *sample-function density* and the *likelihood functional*. The sample-function density can be interpreted roughly as the probability of obtaining a particular realization of the Poisson counting process on a specific observation interval. It therefore plays a central role in solving estimation problems for observed Poisson processes, as discussed in Sec. 2.4. The sample-function density and likelihood functional will be generalized in later chapters for other counting processes. A unifying aspect of the generalizations is that the form of these quantities is essentially always the same as that for the Poisson counting-process.

2.2 Conditions for Temporal Poisson-Processes

We now give some qualitative conditions for an arbitrary counting process to be a Poisson counting process. A statement of such conditions is important because alternative counting processes can be obtained by making obvious modifications of them. Furthermore, whether or not a Poisson counting process is a reasonable model in a given application can be judged by the degree to which the qualitative conditions are met.

Two attributes a point process may possess will be prominent in specifying these conditions. The first is *orderliness*, which is defined both unconditionally and conditionally as follows.

Orderliness. A counting process $\{N(t): t \geq t_0\}$ is called orderly at time $t \geq t_0$ if for any given ε, there exists a $\delta \equiv \delta(t, \varepsilon) > 0$ such that

$$\Pr[N(t, t + \delta') > 1] \leq \varepsilon \Pr[N(t, t + \delta') = 1] \qquad (2.7)$$

for all $\delta' \in (0, \delta)$. A point process is orderly on an interval of time if it is orderly for each time in the interval, and it is uniformly orderly on the interval if a single $\delta \equiv \delta(\varepsilon)$ can be chosen independently of t so (2.7) holds simultaneously for all times in the interval.

Were it not for the possibility that $\Pr[N(t, t + \delta) = 1]$ may be zero, this somewhat complicated definition of orderliness could be written more compactly as

$$\lim_{\delta \downarrow 0} \frac{\Pr[N(t,t+\delta) > 1]}{\Pr[N(t,t+\delta) = 1]} = 0.$$

Orderliness according to (2.7) is sometimes referred to as *Khinchin orderliness* in honor of A. Khinchin [17], who first recognized its consequence. Other varieties of orderliness have been proposed, not all of which are equivalent; see D. Daley [7] and D. Daley and D. Vere-Jones [8] for a summary of these. For an orderly point process, the probability of there being more than one point in an interval can be made an arbitrary small fraction of there being one point by selecting the interval to be sufficiently small. The qualitative interpretation is that points do not occur simultaneously, although the fact that points are distinct does not guarantee orderliness (see Problem 2.2.1). Orderliness is frequently a valid assumption for modeling physical phenomena, but it is not universally so. For instance, it is inappropriate for modeling job arrivals in a computer queue when jobs arrive in batches. Similarly, an optical field incident on a photoelectron converter can induce the nearly simultaneous emission of a random number of photoelectrons, and the assumption of orderliness for the photoelectron process may be inappropriate. As we have already seen in Section 2.1, points can occur simultaneously in a Poisson process where the parameter function has step discontinuities. Thus, in general, a Poisson counting process need not be orderly everywhere; it is orderly at those times t where the parameter function $\Lambda(t)$ is continuous.

There is also the notion of *conditional orderliness*, which is somewhat stronger than orderliness.

Conditional Orderliness. Denote by P an arbitrary event determined by the random variables $\{N(\sigma): t_0 \leq \sigma < t \}$. Thus, P can be any of the following events: the number of points on $[t_0, t)$ is n ; the number of points on $[t_0, t)$ is n and they are located at times $W_1, W_2, \cdots, W_n$; the first point to occur in $[t_0, t)$ is at W_1; etc. If t denotes the present, then P is any event in the past. A counting process $\{N(t); t \geq t_0 \}$ is conditionally orderly at time $t \geq t_0$ if for any P and any $\varepsilon > 0$, there exists a $\delta \equiv \delta(t,\varepsilon) > 0$ such that $\Pr[N(t,t+\delta') > 1 | P] \leq \varepsilon \Pr[N(t,t+\delta') = 1 | P]$ for all $\delta' \in (0,\delta)$. A point process will be called conditionally orderly if it is conditionally orderly for all $t \geq t_0$.

A point process that is conditionally orderly at time t is also unconditionally orderly at time t, but the converse need not hold. By virtue of its independent increments, a Poisson process with a continuous parameter function is conditionally orderly.

The second attribute a point process may possess is that of *evolution without aftereffects*, which is defined as follows.

Evolution Without Aftereffects. A point process on $[t_0, \infty)$ is said to evolve without aftereffects if for any $t \geq t_0$, the realization of points during $[t, \infty)$ does not depend in any way on the sequence of points during the interval $[t_0, t)$. This expresses the independence of the past and future of the point process. The notion is defined mathematically as follows. Let P be any event that can be associated with the random variables $\{N(\sigma): t_0 \leq \sigma < t\}$. Similarly, let F be any event that can be associated with the random variables $\{N(\sigma): \sigma \geq t\}$. If t denotes the present, then P and F are arbitrary events in the past and future, respectively. A point process evolves without aftereffects if the conditional probability of F given P equals the unconditional probability of F for all $t \geq t_0$.

A process with independent increments evolves without aftereffects, and the converse is also true. It is evident that a Poisson process evolves without aftereffects. The hypothesis that a point process evolves without aftereffects can fail in applications when the occurrence of a point causes either the generation or elimination of subsequent points. An example of a phenomenon in which the occurrence of a point leads to the elimination of subsequent points occurs with gamma-ray detectors; an incident gamma photon can cause the detector to become inoperative for some brief period of time, called a dead time, during which subsequent gamma photons are not detected. Processes with aftereffects are studied in Ch. 6.

We now state a theorem that gives qualitative conditions for an arbitrary counting process to be a Poisson counting process. Much of the remainder of the book is based on conditions and techniques that most closely resemble those used in this theorem.

Theorem 2.2.1 (*Conditions for a Poisson Process*). Let $\{N(t): t \geq t_0\}$ be the counting process associated with a point process on $[t_0, \infty)$. Suppose that:

a. the point process is conditionally orderly;

b. for all $t \geq t_0$ and for an arbitrary event P associated with the random variables $\{N(\sigma): t_0 \leq \sigma < t\}$, the limit as δ tends to zero of $\delta^{-1} \Pr[N(t, t+\delta) = 1 | P]$ exists; and the limit is a finite, integrable function of t alone. Denote the limiting function by $\lambda(t)$; thus,

$$\lambda(t) = \lim_{\delta \downarrow 0} \frac{1}{\delta} \Pr[N(t, t+\delta) = 1 | P], \qquad (2.8)$$

and the integral

$$\int_{s}^{t} \lambda(\sigma)d\sigma$$

exists and is finite for all finite intervals $[s,t]$, $t_0 \le s \le t$;

c. $\Pr[N(t_0) = 0] = 1$.

Then $\{N(t): t \ge t_0\}$ is a Poisson counting process with an absolutely continuous parameter function

$$\Lambda(t) = \int_{t_0}^{t} \lambda(\sigma)d\sigma.$$

Hypothesis (b) implies not only that the limit exists but, also, that $\{N(t): t \ge t_0\}$ evolves infinitesimally without aftereffects. In proving the theorem, we will demonstrate that the process also evolves (globally) without aftereffects.

Proof. We need to show both that the increment $N(s,t) = N(t) - N(s)$ is Poisson distributed with parameter $\Lambda(t)$ and that $\{N(t); t \ge t_0\}$ has independent increments. The objective of the first part of the proof is to derive a differential equation for $\Pr[N(s,t) = n]$ and then to demonstrate that the solution is the Poisson distribution. For $n \ge 1$, the event $\{N(s,t+\delta) = n\}$ can occur in $n+1$ mutually exclusive ways; namely,

$$
\begin{array}{llll}
1. & N(s,t) = n & \text{and} & N(t,t+\delta) = 0 \\
2. & N(s,t) = n-1 & \text{and} & N(t,t+\delta) = 1 \\
\cdot & & & \\
\cdot & & & \\
n+1. & N(s,t) = 0 & \text{and} & N(t,t+\delta) = n.
\end{array}
$$

Consequently,

$$\Pr[N(s,t+\delta) = n] = \sum_{k=0}^{n} \Pr[N(s,t) = n-k, N(t,t+\delta) = k]$$

$$= \Pr[N(t,t+\delta) = 0 | N(s,t) = n]\Pr[N(s,t) = n] \qquad (2.9)$$

$$+ \Pr[N(t,t+\delta) = 1 | N(s,t) = n-1]\Pr[N(s,t) = n-1]$$

$$+ \sum_{k=2}^{n} \Pr[N(t,t+\delta) = k | N(s,t) = n-k]\Pr[N(s,t) = n-k].$$

The last two terms are absent for $n = 0$, and the last term is absent for $n = 1$. Let $\varepsilon > 0$ be given, and select $\delta > 0$ so that both

$$\Pr[N(t, t+\delta) > 1 \,|\, N(s, t) = n - k]$$

$$\leq \quad \varepsilon \Pr[N(t, t+\delta) = 1 \,|\, N(s, t) = n - k] \qquad (2.10)$$

and

$$-\varepsilon \leq \frac{1}{\delta} \Pr[N(t, t+\delta) = 1 \,|\, N(s, t) = n] - \lambda(t) \leq \varepsilon. \qquad (2.11)$$

Such a δ exists by hypotheses (a) and (b). Inequalities (2.10) and (2.11) can be used to upper bound the last term in (2.9) according to

$$\sum_{k=2}^{n} \Pr[N(t, t+\delta) = k \,|\, N(s, t) = n - k] \Pr[N(s, t) = n - k]$$

$$\leq \sum_{k=2}^{n} \Pr[N(t, t+\delta) > 1 \,|\, N(s, t) = n - k] \Pr[N(s, t) = n - k]$$

$$\leq \varepsilon \sum_{k=2}^{n} \Pr[N(t, t+\delta) = 1 \,|\, N(s, t) = n - k] \Pr[N(s, t) = n - k]$$

$$\leq \varepsilon \Pr[N(t, t+\delta) = 1]$$

$$\leq \varepsilon \lambda(t) \delta + \varepsilon^2 \delta.$$

From the relation

$$\Pr[N(t, t+\delta) = 0 \,|\, N(s, t) = n] = 1 - \Pr[N(t, t+\delta) = 1 \,|\, N(s, t) = n]$$

$$-\Pr[N(t, t+\delta) > 1 \,|\, N(s, t) = n]$$

and some manipulations involving (2.10) and (2.11), we obtain

$$1 - (1 + \varepsilon)(\lambda(t)\delta + \varepsilon\delta)$$

$$\leq \Pr[N(t, t+\delta) = 0 \,|\, N(s, t) = n] \leq 1 - \lambda(t)\delta + \varepsilon\delta. \qquad (2.12)$$

Upon substituting these inequalities into (2.9) and rearranging, we have

$$-\varepsilon(2+\lambda(t)+\varepsilon) \le \frac{1}{\delta}\{\Pr[N(t,t+\delta)=n]-\Pr[N(s,t)=n]\}$$

$$+\lambda(t)\Pr[N(s,t)=n]-\lambda(t)\Pr[N(s,t)=n-1]$$

$$\le \varepsilon(2+\lambda(t)+\varepsilon).$$

By virtue of the arbitrariness of ε and the finiteness of $\lambda(t)$ assumed in b, we conclude that the derivative of the counting probability $\Pr[N(s,t)=n]$ with respect to t exists and satisfies

$$\frac{d\Pr[N(s,t)=n]}{dt}=-\lambda(t)\Pr[N(s,t)=n]+\lambda(t)\Pr[N(s,t)=n-1] \quad (2.13)$$

for $n \ge 1$ and $t_0 \le s \le t$. Following a similar series of steps, we also conclude that

$$\frac{d\Pr[N(s,t)=0]}{dt}=-\lambda(t)\Pr[N(s,t)=0] \quad (2.14)$$

for $t_0 \le s \le t$. It is evident from (2.10), (2.11), and (2.12) that the initial conditions for these equations are given by $\lim_{t \downarrow s}\Pr[N(s,t)=n]=\delta_{0,n}$ for $n = 0, 1, 2, \cdots$, where $\delta_{0,n}$ denotes the Kronecker delta function, which is one for $n = 0$ and zero otherwise. As outlined below, these differential-difference equations can be solved sequentially to demonstrate that $N(s,t)$ is Poisson distributed,

$$\Pr[N(s,t)=n]=\frac{1}{n!}\left(\int_s^t \lambda(\sigma)d\sigma\right)^n e^{-\int_s^t \lambda(\sigma)d\sigma} \quad (2.15)$$

for $n = 0, 1, 2, \cdots$.

We show next that $\{N(t); t \ge t_0\}$ has independent increments. The salient observation for this is that the argument used above to derive (2.15) is unchanged if $\Pr[N(s,t)=n]$ is replaced throughout by $\Pr[N(s,t)=n \mid P]$, where P is any event determined by the random variables $\{N(\sigma); t_0 \le \sigma < s\}$. In this way, we deduce that

$$\Pr[N(s,t)=n \mid P]=\frac{1}{n!}\left(\int_s^t \lambda(\sigma)d\sigma\right)^n e^{-\int_s^t \lambda(\sigma)d\sigma},$$

and, hence, that $\Pr[N(s,t)=n \mid P]=\Pr[N(s,t)=n]$. In other words, the number of points occurring in the interval $[s,t)$ is independent of any event that transpires in the previous interval $[t_0,s)$. That $\{N(t):t \geq t_0\}$ has independent increments is an immediate consequence of this, and Theorem 2.2.1 is thereby established. $\square$

The sequential solution of the differential-difference equations, (2.13) and (2.14), for $\Pr[N(s,t)=n]$ can be accomplished as follows. For fixed s, define

$$f_n(t) \overset{\Delta}{=} \Pr[N(s,t)=n]e^{\int_s^t \lambda(\sigma)d\sigma}.$$

Then, it is easily concluded that $f_n(t)$ satisfies the following differential-difference equations and boundary conditions:

$$\frac{df_0(t)}{dt}=0, \qquad f_0(s)=1$$

and

$$\frac{df_n(t)}{dt}=\lambda(t)f_{n-1}(t), \qquad f_n(s)=0 \text{ for } n \geq 1.$$

Thus, $f_n(0)=1$ for $t \geq s$, and

$$f_n(t)=\int_s^t \lambda(\sigma)f_{n-1}(\sigma)d\sigma$$

for $t \geq s$ and $n \geq 1$. The equality

$$f_n(t)=\frac{1}{n!}\left(\int_s^t \lambda(\sigma)d\sigma\right)^n, \qquad n=0,1,2,\ldots$$

can be shown straightforwardly by induction because the expression clearly holds when $n=0$, and

$$\int_s^t \lambda(\sigma)\left(\int_s^\sigma \lambda(\zeta)d\zeta\right)^n d\sigma=\int_s^t [\Lambda(\sigma)-\Lambda(s)]^n d\Lambda(\sigma),$$

where

$$\Lambda(\sigma)=\int_{t_0}^\sigma \lambda(\sigma')d\sigma'.$$

The function $\lambda(t)$ defined as the limit in hypothesis (b) emerges according to (2.5) as the instantaneous average rate that points occur at time t. We have termed this rate the *intensity function*. When the intensity $\lambda(t)$ is a constant independent of time, the corresponding Poisson counting process is said to be *homogeneous*. In this case,

$$\int_s^t \lambda(\sigma)d\sigma$$

is proportional to $t - s$, which implies that the counting statistics on the interval $[s,t)$ are the same as those on $[s+\tau, t+\tau)$ for all τ such that $t_0 \leq s + \tau$. It further follows that a homogeneous Poisson counting process has stationary, independent increments. Whenever $\lambda(t)$ is not a constant, the corresponding Poisson counting process is said to be *inhomogeneous*.

Let us note from (2.15) that

$$\Pr[N(t) - N(s) > 0] = 1 - e^{-\int_s^t \lambda(\sigma)d\sigma}$$

and that for $t_0 \leq t < s$,

$$\Pr[N(s) - N(t) > 0] = 1 - e^{-\int_t^s \lambda(\sigma)d\sigma}.$$

Consequently, for each $t \geq t_0$, $N(t)$ approaches $N(s)$ as s approaches t from either the right or the left in the sense that $\Pr[|N(t) - N(s)| > 0]$ approaches zero regardless of how s approaches t; for this reason, $\{N(t) : t \geq t_0\}$ is said to be stochastically continuous for each $t > t_0$. Similarly, for $t_0 \leq t$,

$$\Pr[N(t) - N(t_0) > 0] = 1 - e^{-\int_{t_0}^t \lambda(\sigma)d\sigma},$$

and we conclude that $\{N(t) : t \geq t_0\}$ is stochastically continuous from the right at t_0. It is of interest to note that while an inhomogeneous Poisson process is stochastically continuous at each point in an interval, the probability is in general less than one that it is continuous simultaneously throughout the interval. This obtains because

$$\Pr[N(\sigma) = N(s) \text{ for all } \sigma \in [s,t)] = e^{-\int_s^t \lambda(\sigma)d\sigma},$$

which is less than one whenever the intensity function is nonzero on all or part of the interval $[s,t)$.

The characteristic function,

$$M_{s,t}(jv) \overset{\Delta}{=} E[e^{jvN(s,t)}],$$

for the number of points in $[s,t)$ for a Poisson process can be determined by straightforward calculation using (2.1). Thus,

$$M_{s,t}(jv) = e^{-(\Lambda(t)-\Lambda(s))} \sum_{n=0}^{\infty} \frac{1}{n!} (\Lambda(t)-\Lambda(s))^n e^{jvn}$$

$$= e^{-(\Lambda(t)-\Lambda(s))} \sum_{n=0}^{\infty} \frac{1}{n!} \{(\Lambda(t)-\Lambda(s))e^{jv}\}^n$$

$$= \exp[(e^{jv}-1)(\Lambda(t)-\Lambda(s))]. \tag{2.16}$$

From this and the moment generating property of characteristic functions, we obtain

$$E[N(s,t)] = M_{s,t}^{(1)}(j0) = \Lambda(t)-\Lambda(s) \tag{2.17}$$

and

$$E[N^2(s,t)] = M_{s,t}^{(2)}(j0) = [\Lambda(t)-\Lambda(s)] + [\Lambda(t)-\Lambda(s)]^2, \tag{2.18}$$

where $M^{(k)}(j0)$ denotes the $k th$ derivative of $M(jv)$ with respect to jv with $v=0$. A frequently used property that is evident from these relations is that the variance of the number of points on the interval $[s,t)$

$$\mathrm{var}[N(s,t)] = E[N^2(s,t)] - E^2[N(s,t)],$$

equals the expected number of points $E[N(s,t)]$ for a Poisson process. This property is generalized in Problem 2.2.3.

The condition $\Pr[N(t_0)=0]$ in the definition of a Poisson process can be easily modified to model counters that are not reset to zero at the initial time t_0. If $N(t_0)$ is a nonnegative, integer-valued random variable having value k with probability $\Pr[N(t_0)=k]$, then

$$\Pr[N(t)=n] = \sum_{k=0}^{\infty} \Pr[N(t_0,t)=n-k]\Pr[N(t_0)=k] \tag{2.19}$$

because, by definition, $N(t) = N(t_0) + N(t_0, t)$ and $\{N(t): t \geq t_0\}$ has independent increments.

Summary. A theorem providing conditions for a point process to be a Poisson process has been established in this section. The notion of a point process evolving without aftereffects is the prominent hypothesis of the theorem because this property distinguishes the Poisson process from all other point processes. It expresses the complete absence of any influence by points before some time on points occurring after that time. There are additional ways to arrive at the Poisson process from simple starting assumptions. Three constructive techniques are developed in Problems 2.2.2, 2.3.11, and 2.3.12. Additional techniques are stated by F. Haight [11, Ch. 2].

2.3 Point-Location Statistics

Two aspects of a point process are of interest. The first may be termed *interval characteristics* and relates to the number of points occurring in arbitrary intervals of time; we have investigated the interval, or counting, statistics for the Poisson process in the preceding section. The second may be termed the *location characteristics* and relates to the point locations and interpoint spacings. The development of the location, or time, statistics for the Poisson process is the topic of this section.

Two interrelated time sequences are of interest and indicated in Fig. 2.1. The sequence $\{w_n\}$ is called the *waiting-time* or *occurrence-time* sequence. The former terminology seems preferred in the study of queues where points correspond to service times. We will mostly use the latter terminology. The sequence $\{t_n\}$ is called the *interarrival-time* sequence; t_n is the n*th* interarrival time and is the random time between the $(n\text{-}1)$st and n*th* occurrence times. We also refer to t_n as the *forward-occurrence time* from w_{n-1}.

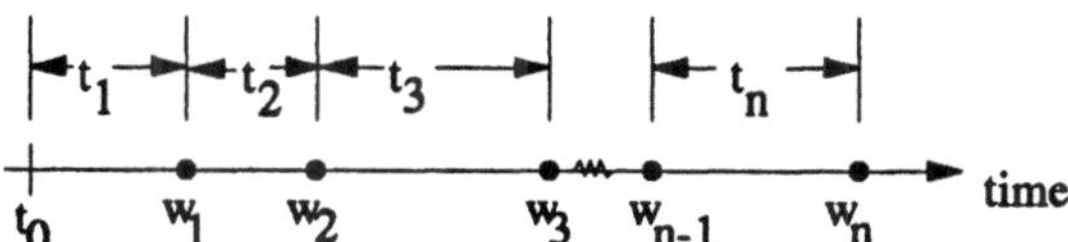

Figure 2.1. Waiting and interarrival time sequences.

Throughout this section, we assume that the conditions of Theorem 2.2.1 hold so that $\{N(t): t \geq t_0\}$ is a Poisson counting process with intensity function $\lambda(t)$. The parameter function for the process is

$$\Lambda(t) = \int_{t_0}^{t} \lambda(\sigma)d\sigma,$$

for $t \geq t_0$.

Our main objective is to derive an expression for the *sample-function density* of Poisson process. This important quantity is then applied in subsequent sections to problems of statistical inference for observed Poisson processes. In the course of deriving the sample-function density, we also obtain several other important statistical properties of the occurrence and interarrival times for a Poisson process.

We begin with the occurrence times. Denote by $p_w^{(n)}(W)$ the joint probability density for the first n occurrence times $w = (w_1, w_2, \cdots, w_n)$. For an inhomogeneous Poisson counting process with intensity $\lambda(t)$ for $t \geq t_0$, we have that

$$p_w^{(n)}(W) = \begin{cases} \left(\prod_{i=1}^{n} \lambda(W_i)\right)\exp\left(-\int_{t_0}^{W_n} \lambda(\sigma)d\sigma\right), & t_0 \leq W_1 \leq W_2 \leq \cdots \leq W_n \\ 0, & \text{otherwise}. \end{cases}$$

$$(2.20)$$

To establish this expression for the *joint-occurrence density*, consider the partitioning of time into the disjoint intervals shown in Fig. 2.2. We now introduce the counting statistics of the Poisson process by noting the identity of the event $\{w_i \in [W_i, W_i + \Delta W_i); i = 1, 2, \cdots, n\}$ with the event

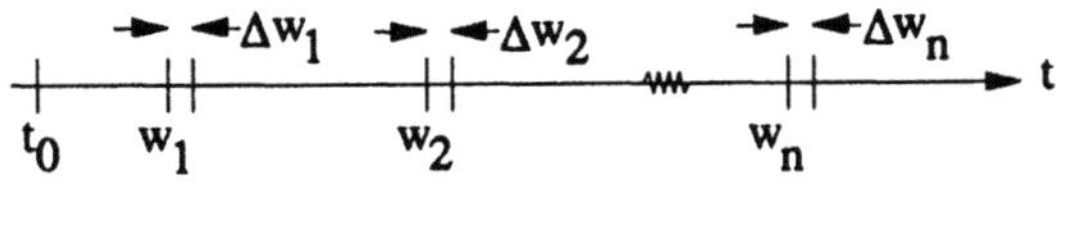

$\Delta w_1 \quad \Delta w_2 \quad \Delta w_n$

$t_0 \quad w_1 \quad w_2 \quad w_n \quad t$

Figure 2.2. Partitioning time into disjoint intervals.

$\{N(t_0, W_1) = 0, \quad N(W_1, W_1 + \Delta W_1) = 1, \quad N(W_1 + \Delta W_1, W_2) = 0, \quad \cdots, \quad N(W_n,$
$W_n + \Delta W_n) = 1\}$. By using the independence of the increments of N and the counting statistics, we then have

$$\Pr\{w_i \in [W_i, W_i + \Delta W_i); i = 1, 2, \cdots, n\}$$

$$= \left(\prod_{i=1}^{n} \int_{W_i}^{W_i + \Delta W_i} \lambda(\sigma) d\sigma \right) \exp\left(-\int_{t_0}^{W_n + \Delta W_n} \lambda(\sigma) d\sigma \right).$$

Consequently, the joint occurrence density,

$$p_w^{(n)}(W) \overset{\Delta}{=} \lim_{\max \Delta W_i \to 0} \frac{\Pr\{w_i \in [W_i, W_i + \Delta W_i); i = 1, 2, \cdots, n\}}{\prod_{i=1}^{n} \Delta W_i},$$

has the form given in (2.20).

The joint occurrence density for a homogeneous Poisson process with constant intensity λ is obtained from (2.20) as

$$p_w^{(n)}(W) = \begin{cases} \lambda^n e^{-\lambda(W_n - t_0)}, & t_0 \le W_1 \le W_2 \le \dots \le W_n \\ 0, & \text{otherwise}. \end{cases} \tag{2.21}$$

As an application of (2.20), let us demonstrate that the occurrence time sequence $w_1, w_2, \cdots, w_n$ for an inhomogeneous Poisson process is a Markov sequence with the transition density

$$p_{w_n | w_{n-1}}(W_n | W_{n-1}) = \lambda(W_n) e^{-[\Lambda(W_n) - \Lambda(W_{n-1})]}, \tag{2.22}$$

where

$$\Lambda(t) = \int_{t_0}^{t} \lambda(\sigma) d\sigma.$$

Using (2.20) and the definition of conditional probability densities, we have

$$p_{w_n | w_{n-1}, \cdots, w_1}(W_n | W_{n-1}, \cdots, W_1) = \frac{p_w^{(n)}(W)}{p_w^{(n-1)}(W)} = \lambda(W_n) e^{-\int_{W_{n-1}}^{W_n} \lambda(\sigma) d\sigma}.$$

Thus, the conditional density of w_n given that $w_i = W_i$ for $i = 1, 2, \cdots,$ $n - 1$, is not a function of W_i for $i = 1, 2, \cdots, n - 2$, and we deduce that

$$p_{w_n|w_{n-1}}(W_n \mid W_{n-1}) = \int \cdots \int p_{w_n|w_{n-1},\cdots,w_1}(W_n \mid W_{n-1},\cdots,W_1)$$

$$\times p_{w_{n-2},\cdots,w_1|w_{n-1}}(W_{n-2},\cdots,W_1 \mid W_{n-1})dW_{n-2}\cdots dW_1$$

$$= p_{w_n|w_{n-1},\cdots,w_1}(W_n \mid W_{n-1},\cdots,W_1), \qquad (2.23)$$

which, with the preceding equation, shows that the occurrence time sequence for an inhomogeneous Poisson counting process is a Markov sequence with the transition density given in (2.22).

We next examine a quantity termed the *forward-occurrence density*. This is the conditional probability density for the nth interarrival time, t_n, given the $n-1$ occurrence times $w_1, w_2, \cdots, w_{n-1}$, for $n = 2, 3, \cdots$. We also term the unconditional density for t_1 a forward-occurrence density. If points correspond to failures of some equipment, the forward-occurrence density is useful for determining the conditional probability that the equipment will be operative some period of time after the last failure given records of previous failure times. We will see that if such failures follow a Poisson process, it is unnecessary to record more than the most recent failure time for this determination.

The relation $w_n = w_{n-1} + t_n$ and the Markovian property (2.23) of the occurrence-time sequence imply

$$p_{t_n|w_{n-1},\cdots,w_1}(T \mid W_{n-1},\cdots,W_1) = p_{w_n|w_{n-1},\cdots,w_1}(W_{n-1}+T \mid W_{n-1},\cdots,W_1)$$

$$= p_{w_n|w_{n-1}}(W_{n-1}+T \mid W_{n-1}) \qquad (2.24)$$

for $n = 2, 3, \cdots$, and $T \geq 0$. We conclude from this and (2.22) that the forward-occurrence density for a Poisson process is given by

$$p_{t_n|w_{n-1},\cdots,w_1}(T \mid W_{n-1},\cdots,W_1) = \lambda(W_{n-1}+T)e^{-[\Lambda(W_{n-1}+T)-\Lambda(W_{n-1})]} \qquad (2.25)$$

for $n = 2, 3, \cdots$, and $T \geq 0$. This expression also holds for $n = 1$ by defining $w_0 = W_0 = t_0$. According to (2.24), the identical expression is the conditional density of t_n given only that $w_{n-1} = W_{n-1}$.

As an example, suppose that points correspond to times at which errors are made in decoding a message and that these errors occur as a homogeneous Poisson process with a rate of λ errors per second. Assume that errors are detected at time $W_1, W_2, \cdots, W_{n-1}$. As $\Lambda(t) = (t - t_0)\lambda$ in this instance, we have from (2.25) that the conditional probability density for the time to the nth error is given by

$$p_{t_n | w_{n-1}, \cdots, w_1}(T | W_{n-1}, \cdots, W_1) = \lambda e^{-\lambda T} \qquad (2.26)$$

for $n = 1, 2, \cdots$, and $T \geq 0$. Thus, t_n is exponentially distributed with parameter λ regardless of the number of previous errors and when they occurred. The mean time from the (n-1)*st* error to the *nth* error is $E(t_n) = \lambda^{-1}$. The conditional probability that no error is made for a period of T seconds following the (n-1)*st* error is given by

$$\Pr[t_n > T | w_1 = W_1, \cdots, w_{n-1} = W_{n-1}] = \int_T^\infty \lambda e^{-\lambda \sigma} d\sigma = e^{-\lambda T}. \qquad (2.27)$$

It is easy to see from (2.26) that the interarrival times for a homogeneous Poisson process with intensity λ are identically distributed with the common distribution being exponential with parameter λ. We will show, further, the important property that the interarrivals are independent. To establish this, first examine the probability that t_n exceeds T given that $t_i = T_i$ for $i = 1, 2, \cdots, n-1$. Thus,

$$\Pr[t_n > T | t_1 = T_1, t_2 = T_2, \cdots, t_{n-1} = T_{n-1}]$$

$$= \Pr[w_n > t_0 + T_1 + \cdots + T_{n-1} + T | w_1 = t_0 + T_1, \cdots, w_{n-1} = t_0 + T_1 + \cdots, + T_{n-1}]$$

$$= \Pr[w_n > t_0 + T_1 + \cdots + T_{n-1} + T | w_{n-1} = t_0 + T_1 + \cdots T_{n-1}],$$

where the last equality follows because the waiting time sequence is a Markov sequence. By differentiating with respect to T and using (2.22), we obtain

$$p_{t_n | t_{n-1}, \cdots, t_1}(T | T_{n-1}, \cdots, T_1) \qquad (2.28)$$

$$= \lambda(t_0 + T_1 + \cdots + T_{n-1} + T) e^{-[\Lambda(t_0 + T_1 + \cdots + T_{n-1} + T) - \Lambda(t_0 + T_1 + \cdots + T_{n-1})]}.$$

An immediate consequence of this is that interarrivals are independent for a homogeneous Poisson process. We summarize this result in the following theorem.

Theorem 2.3.1. For a homogeneous Poisson counting process with intensity k, the interarrival times $t_1, t_2, \cdots, t_n$ are independent and identically distributed with the common distribution being exponential with parameter λ.

Proof. Write the joint probability density for the first n interarrivals, $p_t^{(n)}(T)$, in the factored form

$$p_t^{(n)}(T)$$

$$= p_{t_n | t_{n-1}, \cdots, t_1}(T_n | T_{n-1}, \cdots, T_1) p_{t_{n-1} | t_{n-2}, \cdots, t_1}(T_{n-1} | T_{n-2}, \cdots, T_1) \cdots p_{t_1}(T_1)$$

and use (2.28) to obtain

$$p_t^{(n)}(T) = \lambda^n \exp\left(-\lambda \sum_{i=1}^{n} T_i\right), \qquad T_i \geq 0, \quad i = 1, \cdots, n.$$

The conclusion of Theorem 2.3.1 follows directly from this. $\square$

It is clear from (2.28) that interarrivals for an inhomogeneous Poisson process are not independent.

Occurrence times for a homogeneous Poisson process are gamma distributed. In particular, if the intensity is λ, then the nth occurrence time has the following probability density:

$$p_{w_n}(W) = \begin{cases} \dfrac{(\lambda W)^{n-1}}{(n-1)!} \lambda e^{-\lambda W}, & W \geq 0 \\ 0, & \text{otherwise.} \end{cases} \qquad (2.29)$$

That this is so is an immediate consequence of Theorem 2.3.1 and the fact that $w_n = t_1 + t_2 + \cdots + t_n$ because the density for w_n must then be the n-fold convolution of the exponential density with itself. It follows from this that the mean and variance of w_n are n/λ and n/λ^2, respectively. Somewhat more can be said about the distribution of w_n for n large because w_n is the sum of independent, identically distributed random variables with finite variance (if $\lambda \neq 0$). Specifically, the central limit theorem implies that for n large, the distribution function for w_n is approximately that of a normal random variable with mean n/λ and variance n/λ^2. See Problem 2.3.3 for an application of this asymptotic approximation.

2.3.1 Computer Simulation of Temporal Poisson-Processes

The converse of Theorem 2.3.1 is established in Problem 2.3.12 where it is shown that if $\{t_i\}$ is a sequence of independent, identically distributed, exponential random variables, and if the point process is constructed by assigning these successively as interarrival times, then the resulting point process is a homogeneous Poisson process with an intensity equal to the parameter of the exponential distribution. This construction is useful for simulating Poisson processes in Monte Carlo studies of situations too complicated for direct analytical investigation; this includes both homogeneous and inhomogeneous Poisson processes. The procedure is outlined in the following paragraphs.

Computer Simulation of a Homogeneous Poisson-Process. Algorithms for the generation of pseudorandom variables that are independent and uniformly distributed on [0,1] are given by B. Jansson [15], and A. Ralston and H. Wilf [26]. These are implemented in various ways on most computers. Suppose that $\{u_i\}$ is a sequence of such variables. Then a sequence of independent, exponentially distributed variables can be obtained from these by the transformation $t_i = -\lambda^{-1}\ln(u_i)$. It is easily verified that under this transformation, the probability density for t_i is $\lambda\exp(-\lambda T)$ for $T \geq 0$. A homogeneous Poisson process with intensity λ can therefore be simulated by: generating u_1, transforming to obtain t_1, then assigning t_1 as the time from t_0 to the first occurrence time; generating u_2, transforming to obtain t_2, then assigning t_2 as the time from the first to second occurrence time; and so forth. This method is generalized in Problem 2.3.15 to include inhomogeneous Poisson processes.

Computer Simulation of Counts of an Inhomogeneous Poisson-Process. Suppose that it is desired to generate a realization of a random variable $N(s,t)$ that is Poisson distributed with parameter

$$\int_s^t \lambda(\sigma)d\sigma.$$

Define $\{M(t): t \geq t_0\}$ to be a homogeneous Poisson process with unit intensity. Then, $M(\tau)$ is Poisson distributed with parameter τ. $M(\tau)$ and $N(s,t)$ have the same distribution when

$$\tau = \int_s^t \lambda(\sigma)d\sigma.$$

A sequence of independent, unit-parameter exponentially distributed variables can be used to obtain a realization of $\{M(t): t \geq t_0\}$, as described above. Then, simply keep generating interarrivals until an occurrence time outside the interval $[0,\tau)$ is obtained. The number of points in $[0,\tau)$ is a realization of $N(s,t)$. An alternative, more efficient method is given in Problem 2.3.4. See, also, Problem 2.3.15.

Computer Simulation of an Inhomogeneous Poisson-Process. A homogeneous Poisson process can be transformed into an inhomogeneous Poisson process, and vice versa, by a rescaling of the time coordinate. The rescaling required to achieve an inhomogeneous process with a specified intensity function is developed in Problem 2.3.5. In this way, an inhomogeneous Poisson process can be simulated by first simulating a homogeneous process as outlined above and then rescaling time.

Other methods for simulating Poisson processes in time and space are developed in Pbms. 2.2.2, 2.3.4, 2.3.11, and 2.3.12 and elsewhere in the text.

2.3.2 Sample-Function Density and Loglikelihood Function

The sample-function density and its logarithm, the loglikelihood function, are among the more important quantities we shall encounter. They will play a central role in Sec. 2.4 and in later chapters where we study problems of statistical estimation for observed Poisson-processes. Moreover, the sample-function density for several non-Poisson processes will be developed in later chapters, and its role in statistical estimation will be the same. A not easily anticipated result is that for a very broad class of non-Poisson counting-processes, the form of the sample-function density is essentially identical to that for Poisson processes. Consequently, the interpretation of the sample-function density developed here, as well as the techniques indicated for its use, will carry over unchanged in later chapters.

Let $\Omega = \{W_1, W_2, \cdots, W_n, N(t) = n\}$ be a particular realization of the random Poisson-process $\omega = \{N(\sigma): t_0 \leq \sigma < t\}$ on the interval $[t_0, t)$, as shown in Fig. 2.3. The *sample-function density*, denoted by $p(\Omega)$, is defined in terms of this realization according to:

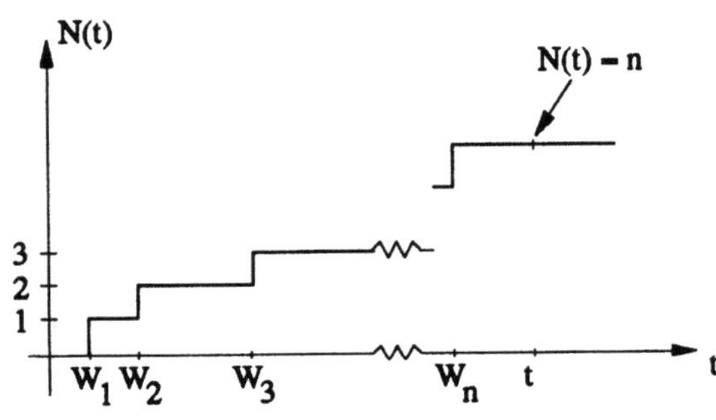

Figure 2.3 A particular sample function of *N*.

$$p(\Omega) \overset{\Delta}{=} \begin{cases} \Pr[N(t)=0], & N(t)=0 \\ p_w(W,N(t)=n), & N(t)=n \geq 1, \end{cases} \tag{2.30}$$

where

$$p_w(W,N(t)=n) \overset{\Delta}{=} \Pr[N(t)=n \mid w_1=W_1, w_2=W_2, \cdots, w_n=W_n] p_w^{(n)}(W).$$

Roughly speaking, the sample-function density determines the probability of obtaining this particular sample function of the point process on $[t_0, t)$ having $N(t)=n$ points occurring at times $w_1 = W_1, \cdots, w_n = W_n$. This interpretation follows from the approximation

$$p_w(W,N(t)=n)\left(\prod_{i=1}^{n} \Delta W_i\right)$$

$$\cong \Pr[N(t)=n, w_i \in [W_i, W_i+\Delta W_i); i=1,2,\cdots,n].$$

The sample-function density for a temporal Poisson-process is given in the following theorem.

Theorem 2.3.2. (*Sample-Function Density for a Temporal Poisson Process*) Let $\{N(t): t \geq t_0\}$ be an inhomogeneous Poisson counting process in time with intensity function $\lambda(t)$ for $t \geq t_0$. Then the sample-function density for N defined in (2.30) has the evaluation

$$p(\Omega) = \begin{cases} \exp\left(-\int_{t_0}^{t} \lambda(\sigma)d\sigma\right), & N(t)=0 \\ \left(\prod_{i=1}^{n} \lambda(W_i)\right)\exp\left(-\int_{t_0}^{t} \lambda(\sigma)d\sigma\right), & N(t)=n \geq 1. \end{cases} \tag{2.31}$$

Proof. Equation (2.31) follows immediately from the definition of the sample-function density in (2.30), the joint-occurrence density in (2.20), and the relation

$$\Pr[N(t)=n \mid w_1=W_1, w_2=W_2, \cdots, w_n=W_n]$$

$$= \Pr[N(t)-N(W_n^+)=0] = \exp\left(-\int_{W_n}^{t} \lambda(\sigma)d\sigma\right).$$

$\square$

The sample-function density is defined in (2.30) in terms of the particular sample function shown in Fig. 2.3. For this sample function, (2.31) can be alternatively written as

$$p(\Omega) = \exp\left[-\int_{t_0}^{t} \lambda(\sigma)\,d\sigma + \int_{t_0}^{t} \ln(\lambda(\sigma))N(d\sigma) \right], \qquad (2.32)$$

where the second integral can be interpreted as a Riemann-Stieltjes integral with the evaluation

$$\int_{t_0}^{t} \ln(\lambda(\sigma))N(d\sigma) = \begin{cases} 0, & N(t)=0, \\ \sum_{i=1}^{N(t)} \ln[\lambda(w_i)], & N(t) \ge 1. \end{cases}$$

We term integrals of this type *counting integrals*. They occur often in later chapters. In this form, it is evident that the sample-function density is a functional of the random Poisson-process $\{N(t); t \ge t_0\}$. If we consider an arbitrary, random sample-function $\omega = \{N(\sigma): t_0 \le \sigma < t\}$ on the interval $[t_0, t)$, it is seen that the sample-function density $p(\omega)$ may itself be regarded as a random variable.

The logarithm of the sample-function density, called the *loglikelihood function*, is given by

$$\mathcal{L} \overset{\Delta}{=} \ln[p(\omega)] = -\int_{t_0}^{t} \lambda(\sigma)\,d\sigma + \int_{t_0}^{t} \ln(\lambda(\sigma))N(d\sigma). \qquad (2.33)$$

This function of $\omega = \{N(\sigma): t_0 \le \sigma < t\}$ will be used in Sec. 2.4 when parameter estimation for observed Poisson processes is studied.

We now give several applications where the sample-function density proves to be useful.

2.3.3 Conditional Distribution of Occurrence Times

The first application that will be made of the sample-function density is to derive a useful additional property of the occurrence times. Under the condition $N(T)=n$, where $n \ge 1$, the n occurrence times $w_1, w_2, \cdots, w_n$ for an inhomogeneous Poisson-process with intensity $\lambda(t)$ have the same distribution as the order statistics of n independent, identically distributed random variables $u_1, u_2, \cdots, u_n$ with the common probability density being

$$p_u(U) = \begin{cases} \dfrac{\lambda(U)}{\displaystyle\int_{t_0}^{T}\lambda(\sigma)\,d\sigma}, & t_0 \leq U \leq T \\[4mm] 0, & \text{otherwise}. \end{cases} \tag{2.34}$$

The following is meant by *order statistics*. The u_i, being distributed throughout the interval $[t_0, T]$, are unordered. In terms of the u_i, define n random variables ordered in magnitude; for instance, the first and last order variables are $\min(u_1, u_2, \cdots, u_n)$ and $\max(u_1, u_2, \cdots, u_n)$, respectively. The assertion is that these new variables have the same distribution as the occurrence times given that $N(T) = n$.

To verify this assertion, let us note first that by using the expression for the sample-function density in (2.31) and the Poisson distribution for $N(T)$, the following is obtained

$$p_w(W \mid N(T)=n) \;=\; \frac{p_w(W, N(T)=n)}{\Pr[N(T)=n]} \;=\; n! \prod_{i=1}^{n} \left[\frac{\lambda(W_i)}{\displaystyle\int_{t_0}^{T}\lambda(\sigma)\,d\sigma} \right]$$

$$\tag{2.35}$$

for $t_0 \leq W_1 \leq \cdots \leq W_n \leq T$. Second, the joint probability density for the unordered variables $u_1, u_2, \cdots, u_n$ is

$$p_u(U) = \prod_{i=1}^{n} \left[\frac{\lambda(U_i)}{\displaystyle\int_{t_0}^{t}\lambda(\sigma)\,d\sigma} \right]$$

for $t_0 \leq U_i \leq T$ and $i = 1, 2, \cdots, n$. Let $o_1, o_2, \cdots, o_n$ denote the ordered random-variables associated with $u_1, u_2, \cdots, u_n$ as shown in Fig. 2.4. There are $n!$ permutations of the unordered variables that result in the same ordered variables, and these correspond to mutually exclusive events in the realization of the unordered variables. Since probabilities for mutually exclusive events add, we have the following joint density for the ordered variables:

$$p_o(O) = n! \prod_{i=1}^{n} \left[\frac{\lambda(O_i)}{\displaystyle\int_{t_0}^{t}\lambda(\sigma)\,d\sigma} \right]$$

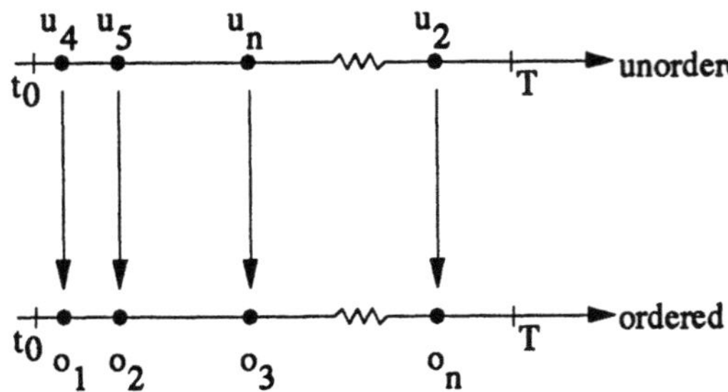

Figure 2.4. A realization of the unordered variables $u_1, u_2, \cdots, u_n$ and the corresponding ordered variables $o_1, o_2, \cdots, o_n$.

for $t_0 \le W_1 \le \cdots \le W_n \le T$. Comparison of this with (2.35) establishes the assertion.

Examination of (2.34) shows that under the condition that $N(T) = n$, the n occurrence times w_1, w_2, $\cdots$, w_n for a homogeneous Poisson counting process with intensity λ have the same distribution as the order statistics of n independent, uniformly distributed random variables on $[t_0, T]$ regardless of the particular value of λ. This fact provides one means to test whether or not a set of points observed experimentally is reasonably modeled as a homogeneous Poisson process. A simple test based on the central limit theorem is developed in Problem 2.3.3. E. Parzen [24, p. 141] describes some more powerful tests. This fact is also useful in the computer simulation of an homogeneous Poisson process; see Prob. 2.3.12.

As an example demonstrating the usefulness of these conditional properties of the occurrence times, let us determine the *characteristic functional* for an inhomogeneous Poisson counting process. A characteristic functional is a generalization of the idea of a characteristic function; it is important because it provides a complete statistical description of its associated random process. The characteristic functional associated with the random variables $\{N(\sigma): t_0 \le \sigma < T\}$ is defined by the expectation

$$\phi_N(jv) = E\left[\exp\left(j \int_{t_0}^{T} v(\sigma) N(d\sigma) \right) \right], \qquad (2.36)$$

where v is a real-valued function. The integral is a Riemann-Stieltjes integral having the evaluation

$$\int_{t_0}^{T} v(\sigma)N(d\sigma) = \begin{cases} 0, & N(T)=0 \\ \sum_{i=1}^{N(T)} v(w_i), & N(T) \geq 1. \end{cases}$$

The expectation in (2.36) is straightforward to evaluate using the properties of conditional expectation and (2.34). Thus,

$$\phi_N(jv) = \sum_{n=0}^{\infty} \Pr[N(T)=n]E\left[\exp\left(j\int_{t_0}^{T} v(\sigma)N(d\sigma)\right) \Big| N(T)=n\right]$$

$$= \Pr[N(T)=0] + \sum_{n=1}^{\infty} \Pr[N(T)=n]E\left[\exp\left(j\sum_{i=1}^{N(T)} v(w_i)\right) \Big| N(T)=n\right].$$

$$(2.37)$$

The following observations are useful for evaluating the conditional expectation in this expression. Suppose that $N(T)=n$ points occur in $[t_0, T)$. The occurrence times $w_1, w_2, \cdots, w_n$ represent one labeling of these n points; this labeling is ordered. Suppose that the points are relabeled by assigning the labels $u_1, u_2, \cdots, u_n$ in a completely random manner; this labeling is unordered. We note for this relabeling that the sum appearing in the conditional expectation remains unchanged; that is,

$$\sum_{i=1}^{n} v(w_i) = \sum_{i=1}^{n} v(u_i).$$

Consequently, the expectation in (2.37) can be evaluated by relabeling the occurrence times in this fashion and using the fact for a Poisson process that, as unordered variables, the point locations are independent and identically distributed with the common probability density being that in (2.34). Doing this, we obtain

$$E\left[\exp\left(j\sum_{i=1}^{N(T)} v(w_i)\right) \Big| N(T)=n\right] = E\left[\exp\left(j\sum_{i=1}^{N(T)} v(u_i)\right) \Big| N(T)=n\right]$$

$$= \left(\frac{\int_{t_0}^{T} \lambda(\sigma)e^{jv(\sigma)}d\sigma}{\int_{t_0}^{T} \lambda(\sigma)d\sigma}\right)^n.$$

$$(2.38)$$

Substitution of this expression into (2.37) and use of the Poisson distribution for $N(T)$ then shows that the characteristic functional for an inhomogeneous Poisson process on the interval $[t_0, t)$ is given by

$$\phi_N(jv) = \exp\left[\int_{t_0}^{T} \lambda(\sigma)(e^{jv(\sigma)} - 1)d\sigma\right]. \tag{2.39}$$

The characteristic function can be obtained by specialization of the characteristic functional. For illustration of this, take the arbitrary function v in (2.36) to have the form

$$v(\sigma) = \begin{cases} 0, & t_0 \leq \sigma < s \\ \alpha, & s \leq \sigma < t \\ 0, & t \leq \sigma < T. \end{cases}$$

Then,

$$\int_{t_0}^{T} v(\sigma)N(d\sigma) = \alpha N(s,t),$$

and the characteristic functional becomes the characteristic function for the increment $N(s,t) = N(t) - N(s)$. Equation (2.39) becomes

$$\phi_N(jv) = \exp\left[(e^{j\alpha} - 1)\int_{s}^{t} \lambda(\sigma)d\sigma\right].$$

Comparison of this expression with (2.16) shows for this choice of v that

$$\phi_N(jv) = M_{s,t}(j\alpha).$$

As a further illustration, suppose v has the piecewise constant form shown in Fig. 2.5, where the intervals $[t_i, u_i)$, $i = 1, 2, \cdots, k$, are mutually disjoint on $[t_0, T)$. For this choice of v,

$$\int_{t_0}^{T} v(\sigma)N(d\sigma) = \sum_{i=1}^{k} \alpha_i N(t_i, u_i),$$

and the characteristic function is the joint characteristic function for the k increments $N(t_i, u_i)$, $i = 1, 2, \cdots, k$. Equation (2.39) becomes

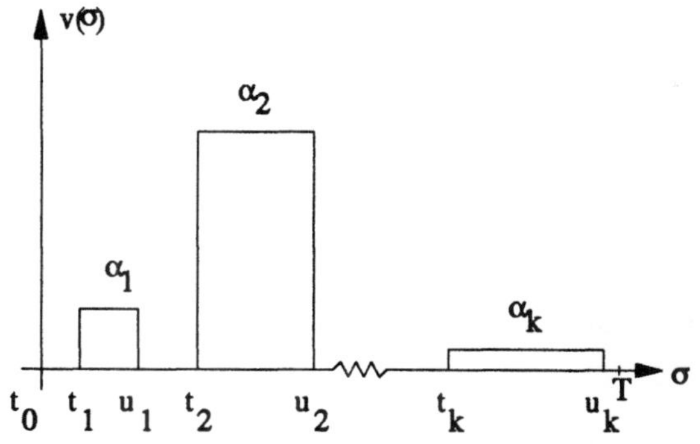

Figure 2.5. A particular choice of v for establishing increment statistics.

$$\phi_N(jv) = \prod_{i=1}^{k} \exp\left[\left(e^{j\alpha_i} - 1\right)\int_{t_i}^{u_i} \lambda(\sigma)d\sigma\right],$$

which demonstrates that N has independent, Poisson distributed increments, as we already knew. These two specializations of the characteristic functional of (2.36) show that it provides a complete statistical characterization of an inhomogeneous Poisson counting process since we have been able to deduce from it both the increment distribution of N and the fact that N has independent increments.

We have shown that under the condition $N(T) = n$, the n occurrence times $w_1, w_2, \cdots, w_n$ for an inhomogeneous Poisson process with intensity $\lambda(t)$ have the same distribution as the order statistics of n independent, identically distributed random variables with the common probability density given in (2.34). The converse is also true and sometimes useful in the simulation of Poisson processes. Specifically, a point process constructed on $[t_0, T)$ in the following manner is an inhomogeneous Poisson process with intensity $\lambda(t)$. First, select a nonnegative integer N according to the Poisson distribution

$$\Pr[N(T) = n] \;=\; \frac{1}{n!}\left(\int_{t_0}^{T}\lambda(\sigma)d\sigma\right)^n \exp\left(-\int_{t_0}^{T}\lambda(\sigma)d\sigma\right).$$

This can be accomplished using the procedure outlined above for simulating counts for an inhomogeneous Poisson process. Secondly, select N numbers $u_1, u_2, \cdots, u_n$ independently over the interval $[t_0, T)$, each with probability density

$$p_u(U) = \frac{\lambda(U)}{\displaystyle\int_{t_0}^{T} \lambda(\sigma)\,d\sigma}, \qquad t_0 \le U \le T.$$

The point process obtained by assigning these numbers as coordinates for points in $[t_0, T)$ is an inhomogeneous Poisson process with intensity $\lambda(t)$. This assertion can be proven by showing that the characteristic functional for the point process specified by the construction has the form in (2.39). That this is indeed the form is evident starting from (2.37) and paralleling the derivation of (2.39); see Problem 2.3.11.

2.3.4 Summary

In this section, the statistical properties of the occurrence and inter-arrival times for a Poisson counting process have been developed. The important results can be summarized as follows:

a.   the joint occurrence time density is given for an inhomogeneous process in (2.20) and for a homogeneous process in (2.21);

b.   the occurrence times form a Markov sequence;

c.   interarrival times for a homogeneous process are independent and identically distributed exponential random variables;

d.   occurrence times for a homogeneous process are gamma distributed;

e.   given the number of points in an interval is n, the n occurrence times have the same distribution as the order statistics for n independent and identically distributed random variables with the common distribution being that in (2.34)

One of the most important quantities developed in the section is the sample-function density, which completely characterizes the statistical properties of paths of the counting process. The sample-function density and its logarithm, the loglikelihood function, will be used in the following sections where we study problems of statistical inference for observed Poisson processes. Generalizations of the sample-function density for other counting processes will appear throughout the text. The results of this section are used in Problems 2.3.11 and 2.3.12 to develop constructive derivations of the Poisson process, which complement the derivation in Sec. 2.2.

2.4 Parameter Estimation for Temporal Poisson-Processes

The desire to know parameters that cannot be directly observed often occurs in practice. In the typical situation, data influenced by the parameters are collected and analyzed to estimate them. Two issues arise. The first is that of designing the analysis to be performed and the second that of predicting the accuracy of the resulting estimate. The purpose of this section is to address these issues when the collected data are reasonably modeled as a Poisson process. In later chapters, these issues for more complicated point-process models and their generalizations will be reencountered.

It is not our intention to give a complete theory of statistical parameter estimation here because it is extensive and readily available in numerous other texts, including those by L. Savage [30], M. Fisz [10], and H. Van Trees [36]. The monograph by D. Cox and P. Lewis [6] contains specialized estimation techniques for the Poisson process. Our attention will be restricted in this section to maximum-likelihood estimation for Poisson processes.

In general terms, the problem of statistical estimation for inhomogeneous Poisson counting processes can be stated as follows. Let $\{N(t): t \geq t_0\}$ be an inhomogeneous Poisson-process with intensity $\lambda(t:\mathbf{X})$, where λ is a known function of t and $\mathbf{X}$, and where $\mathbf{X}$ is a vector of unknown parameters having values in the space $\mathcal{X}$ of possible parameter values. Let $D(t_0, T)$ denote data available from observations of N on the interval $[t_0, T)$; for example, $D(t_0, T)$ can be simply the number of counts observed on the entire interval, the number of counts observed on several subintervals, or the entire counting path $\{N(t): t_0 \leq t < T\}$. Let $P[D(t_0, T):\mathbf{X}]$ denote the statistics of $D(t_0, T)$, where in this notation the statistical dependency on the parameters $\mathbf{X}$ of the intensity function is explicitly indicated. The problem is as follows: knowing $P[D(t_0, T):\mathbf{X}]$ for every $\mathbf{X} \in \mathcal{X}$, and having observed a particular set of data, guess the value of the parameters $\mathbf{X}$. We denote the function $\hat{\mathbf{X}} = \hat{\mathbf{X}}(D(t_0, T))$ of $D(t_0, T)$ as the estimate of $\mathbf{X}$ derived from the data $D(t_0, T)$.

Some specific situations where parameter estimation problems of this type arise are given in the following examples.

Example 2.4.1 *Radioactive Decay* ————————————————————————
As shown by R. Evans [9, p. 470], the emission of gamma or annihilation photons by a radioactive source can be modeled to a close approximation as an inhomogeneous Poisson process with an intensity of the form $X_1 \exp(-X_2 t)$ for $t \geq 0$, where $X_1 > 0$ depends on the quantity of the source material and $X_2 > 0$ is the reciprocal of the mean life of the source. The problem of estimating X_1 and X_2 from observed radiation for various sources is a problem of standing interest in nuclear physics, nuclear medicine, geochronology, and numerous other disciplines. ∎

Example 2.4.2 *Nuclear Medicine* ────────────────────────
The use of radioactive tracers in medical imaging applications has developed rapidly in recent years because it provides relatively noninvasive diagnostic procedures for clinical medicine and a basic tool in biochemical and physiological research. Radiotracers are employed in both static and dynamic studies. In static studies, the desire is to obtain an image of the spatial distribution of the tracer within the body. Dynamic studies are designed to obtain quantitative information about parameters relating to time-dependent phenomena, such as blood flow-rate, metabolic exchange rates, and lung ventilation [35, 22, 37]. Dynamic radiotracer data are typically collected with some type of scintillation detector and have the form of a series of discrete events each of which is associated with the emission of a light pulse in the crystal of the detector. To a first approximation, the series of light pulses can be modeled as an inhomogeneous Poisson process. As discussed by J. Jacques [14] and C. Sheppard [32], a form of intensity function that is frequently assumed is

$$\lambda(t:\mathbf{X}) = X_0 + \sum_{i=1}^{n} X_i e^{-X_{n+i}t} \qquad (2.40)$$

for $t \geq 0$, where $X_i > 0$ for $i = 0, 1, \cdots, 2n$. In practice, n is typically between one and about five. This intensity reflects physiological transport phenomena; for instance, the intensity decreases with time because the radioactive source is being removed from the field of view of the detector by blood flow. Thus, the $2n + 1$ parameters $X_0, \cdots, X_{2n}$ are of interest in both clinical and physiological studies. ∎

Example 2.4.3 *Optical Detection* ────────────────────────
The stream of photoelectrons produced when coherent light is focused onto a photosensitive surface has been shown by L. Mandel [19, 20] to be modeled to a good approximation by an inhomogeneous Poisson process. The model is also reviewed by S. Karp, E. O'Neill, and R. Gagliardi [16]. Suppose that the incident light can be characterized classically by the scalar electric field

$$s(t,\vec{r}) = \sqrt{2}\,\mathrm{Re}[S(t,\vec{r})e^{j2\pi\nu t}] \qquad (2.41)$$

at time t and position $\vec{r}$, where ν is the optical frequency, $S(t,\vec{r})$ is a complex space-time envelope that is slowly varying in comparison to ν, and $Re(\cdot)$ denotes the real-part operation. The light intensity at time t and position $\vec{r}$ is $|S(t,\vec{r})|^2$. The rate (or intensity) at which electrons emanate from the photodetector is related to the light intensity incident on its sensitive surface according to

$$\lambda(t) = \frac{\eta}{h\nu} \int_A |S(t,\vec{r})|^2 d^2r \;\; + \;\; \lambda_0, \tag{2.42}$$

where η is the quantum efficiency of the detector (i.e., the probability that a quantum of light energy is converted into an electron), h is Planck's constant, the integral is over the sensitive surface A of the detector, and λ_0 is the rate of generation of extraneous electrons in the detector and accounts for dark current. Parameter estimation problems arise in a number of ways with the use of photodetectors. For simplicity in describing these, we will assume that spatial variations of the field on A are negligible and write the parameterized rate function as

$$\lambda(t{:}X) = \alpha \, |\, S(t,\vec{r}) \,|^2 + \lambda_0,$$

where α is a constant. Some special cases of interest in optical communication and radar systems are:

1. *Pulse-Amplitude Modulation.* Suppose that the intensity of a pulsed light source is adjusted in accordance with a parameter X. The electron generation rate at the output of a photodetector used to observe the light is then $\lambda(t,X) = \alpha X \,|\, S(t)\,|^2 + \lambda_0$. The problem of estimating X from observations of the photodetector output has been studied by I. Bar-David [5].

2. *Pulse-Position Modulation, Optical Range-Finding.* Suppose that the timing of a pulse from a light source is adjusted in accordance with a parameter X. The electron generation rate at the output of a photodetector used to observe the light is then $\lambda(t,X) = \alpha \,|\, S(t-X)\,|^2 + \lambda_0$. The problem of estimating X from observations of the photodetector output has been studied by I. Bar-David [5] and E. Hoversten and D. Snyder [13]

3. *Pulse-Frequency Modulation, Optical Range-Rate Finding.* In an effort to measure the velocity of an object, the intensity of a light beam directed toward it is modulated sinusoidally. The light reflected has all its frequencies shifted because of the Doppler effect; the frequency of the modulation is shifted by an amount proportional to the modulation frequency and the range-rate of the object. The electron generation rate at the output of a photodetector used to observe the reflected light is then of the form

$$\lambda(t{:}X) = \alpha\{1 + m \cos[2\pi(f_m + X)t]\} + \lambda_0, \tag{2.43}$$

where α and m are constants ($\alpha > 0, |m| < 1$), f_m is the modulation frequency, and X is the Doppler shift. The problem of estimating X has been studied by C. Helstrom [12], who attributes it originally to N. McEvoy. ∎

Example 2.4.4 *Auditory Electrophysiology* ────────────
A procedure used in auditory electrophysiology is to insert a micro-electrode into an exposed auditory-nerve fiber and observe the electrical activity in the nerve in response to an acoustic stimulus applied to the outer ear. An example of the electrical signals obtained this way is shown in Fig. 1.1; the narrow pulses are called spike discharges. W. Siebert [33, 34] argues that for a restricted class of stimuli - sinusoidal tones of limited duration, amplitude, and frequency - the spike discharge sequence can be modeled to a first approximation as an inhomogeneous Poisson process with an intensity function of the form

$$\lambda(t:X) = X_1 \exp[X_2 \cos(2\pi f t + X_3)], \qquad (2.44)$$

where f is the frequency of the applied stimulus and the parameters X_1, X_2, and X_3 reflect the physiological mechanisms involved in converting the pressure stimulus into electrical nerve activity. W. Littlefield [18] has conducted extensive measurements to estimate X_1, X_2, and X_3 in order to make inferences about the physiological mechanisms involved. ∎

A form of data that are frequently collected for observed point processes is obtained by partitioning an observation interval $t_0, T)$ into a set of disjoint subintervals $[t_0, t_1)$, $[t_1, t_2)$, $\cdots$, $[t_{k-1}, t_k = T)$ and then collecting counts in each subinterval. In practice, the subintervals are often not of equal duration but, rather, are selected so that roughly the same number of counts is obtained in each. For instance, the interval durations may be increased systematically for the exponentially decaying intensities encountered in Ex's. 2.4.1 and 2.4.2 above. We refer to data of this type as *histogram data*. These histogram data are the subinterval counts n_1, n_2, $\cdots$, n_k, where $n_i = N(t_{i-1}, t_i)$ is the number of points observed in the interval $[t_{i-1}, t_i)$.

A *count-rate histogram*, formed from histogram data, is a type of estimate of the intensity function that is frequently used. It is constructed from the points

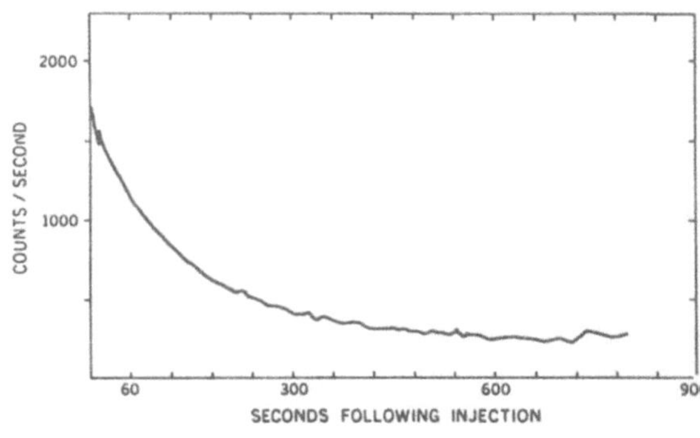

Figure 2.6. Count-rate histogram of radioactivity following injection of oxygen-15 labeled water.

$$\left(\frac{n_i}{t_i - t_{i-1}}, \frac{t_i + t_{i-1}}{2}; \quad i = 1, 2, ..., k \right). \tag{2.45}$$

An example of such a count-rate histogram is shown in Fig. 2.6. The data shown were obtained by monitoring radioactivity over the brain following the rapid injection of oxygen-15 labeled water into the internal carotid artery. For a discussion of this procedure, see Ter-Pogossian *et al.* [35]. The 780-second observation interval is divided into subintervals that increase systematically from 0.1 to 20 seconds. Counts were collected in the subintervals and are plotted as histogram points connected by straight lines. An advantage of this type of estimate of the intensity function is that it is nonparametric; no assumptions need to be made about the form of the intensity function. In what follows, however, we assume that the form of the intensity is known to within a set of parameters to be determined. The form of the intensity may be known in practice either by having a good understanding of the physical mechanism giving rise to the observed point process or by having constructed a sufficient number of count-rate histograms to postulate it.

2.4.1 Parametric Maximum-Likelihood Estimation

The method of maximum-likelihood will now be applied to problems in which the form of the intensity is known to within a finite set of parameters $\mathbf{X}$, which we term *parametric maximum-likelihood estimation.*[1] The maximum-likelihood estimate $\hat{\mathbf{X}}_{ML} = \hat{\mathbf{X}}_{ML}(n_1, n_2, \cdots, n_k)$ of $\mathbf{X}$ in terms of the observed subinterval counts $n_1, n_2, \cdots, n_k$ is by definition the value of $\mathbf{X}$ that maximizes the probability of having observed $n_1, n_2, \cdots, n_k$. This maximization is to be performed subject to any constraints that may exist on the possible parameter values, such as their being nonnegative or

---

[1] In Ch. 3, we will relax this restriction and discuss the estimation of a nonparametric intensity-function using maximum-likelihood techniques.

their being in some other specified set. The probability of observing n_1, n_2, $\cdots$, n_k for a given parameter vector $\mathbf{X}$ can be easily evaluated for an inhomogeneous Poisson process by using (2.1) and (2.2). Thus,

$$\Pr[N(t_{i-1},t_i)=n_i; i=1,\cdots,k:\mathbf{X}] \tag{2.46}$$

$$=\prod_{i=1}^{k}\left[\frac{1}{n_i!}\left(\int_{t_{i-1}}^{t_i}\lambda(\sigma:\mathbf{X})d\sigma\right)^{n_i}\exp\left(-\int_{t_{i-1}}^{t_i}\lambda(\sigma:\mathbf{X})d\sigma\right)\right].$$

The value of $\mathbf{X}$ that maximizes this probability also maximizes its logarithm because the logarithm is a monotonically increasing function of its argument. Consequently, the maximum-likelihood estimate of $\mathbf{X}$ in terms of the observed counts $n_1, n_2, \cdots, n_k$ maximizes the *loglikelihood function* $\mathcal{L}(\mathbf{X})$, defined by

$$\mathcal{L}(\mathbf{X})=-\int_{t_0}^{T}\lambda(\sigma:\mathbf{X})d\sigma+\sum_{i=1}^{k}n_i\ln\left(\int_{t_{i-1}}^{t_i}\lambda(\sigma:\mathbf{X})d\sigma\right), \tag{2.47}$$

where this maximization is subject to any constraints there may be on the possible values of $\mathbf{X}$.

Example 2.4.5 *Scaled Inhomogeneous Poisson Process* ————
Suppose that $\{N(t): t \geq t_0\}$ is an inhomogeneous Poisson process with a parameterized intensity function $\lambda(t:X)=X\mu(t)$ for $t \geq t_0$, where $\mu(t)$ is a known function of t, and $X > 0$ is an unknown scale factor. An observation of $N(\cdot)$ on the interval $[t_0, T)$ is made in order to determine X. Suppose that the observations are in the form of histogram data. Thus, the observations result in k subinterval counts $n_1, n_2, \cdots, n_k$ from which X can be estimated; here, n_i is the number of points observed to occur in the subinterval $[t_{i-1}, t_i)$. The loglikelihood function (2.47) becomes

$$\mathcal{L}(X)=-X\int_{t_0}^{T}\mu(\sigma)d\sigma+N_T\ln X+\sum_{i=1}^{k}n_i\ln\left(\int_{t_{i-1}}^{t_i}\mu(\sigma)d\sigma\right), \tag{2.48}$$

where $N(T)=n_1+n_2+\cdots+n_k$ is the total number of observed points. The maximum-likelihood estimate of X in terms of the histogram data maximizes this expression as a function of X, so

$$\hat{X}_{\text{ML}} = \frac{N(T)}{\displaystyle\int_{t_0}^{T} \mu(\sigma)\,d\sigma}. \tag{2.49}$$

Observe for this example that only the total number of points is needed to form the maximum-likelihood estimate and not the separate subinterval counts; in other words, $N(T)$ is a *sufficient statistic* for estimating X. From (2.49), it follows by taking $\mu(t) = 1$ that the maximum-likelihood estimate of the intensity of a homogeneous Poisson process is the observed time-average count rate $N(T)/(T - t_0)$. $\blacksquare$

In situations that are more general than that in Ex. 2.4.5 and, indeed, in almost all applications, it is not possible to obtain an analytic solution for the value of X maximizing the loglikelihood function (2.47). Thus, the maximum-likelihood estimate of X usually has to be obtained numerically or by some other indirect means. The numerical problem is to find the value of X that maximizes $\mathcal{L}(X)$, subject to any constraints, given n_1, n_2, $\cdots$, n_k . A number of algorithms are described in the literature for accomplishing this search [2, 23]. The efficiency and accuracy of several algorithms (steepest descent, conjugate gradient, variable metric, Newton-Raphson, gradient with respect to a non-Euclidean norm, and weighted least squares) are compared by J. Markham, D. Snyder, and J. Cox, Jr. [21] for exponentially decreasing intensities in the form of (2.40). Some alternative methods are discussed in Ch. 3.

Histogram data suppresses any useful information about the parameters that may be contained in the occurrence times of the observed point process. For this reason, our main focus in later chapters is on the use of the occurrence times rather than count-rate histograms derived from them. This implies a greater data gathering requirement in practice, but it makes possible the extraction of all available information about the parameters from the time series of detected events. Therefore, we also consider as observed data on the interval $[t_0, T)$ an entire counting path $\{N(\sigma) : t_0 \leq \sigma < T\}$. We refer to such data as *count-record* data. Count-record data are entirely equivalent to a record $\{N(T); w_1, w_2, \cdots, w_{N(T)}\}$ of the number of points and their occurrence times in $[t_0, T)$. It is clear that count-record data contain more information about the parameters than histogram data because histogram data can be obtained from count-record data but not vice versa.

Let $p(\omega : X)$ denote the sample-function density for the Poisson process $\{N(t) : t_0 \leq t < T\}$ when the intensity function $\lambda(t : X)$ is parameterized by X. Thus, from (2.32)

$$p(\omega{:}\mathbf{X}) = \exp\left[-\int_{t_0}^{T}\lambda(\sigma{:}\mathbf{X})\,d\sigma + \int_{t_0}^{T}\ln\lambda(\sigma{:}\mathbf{X})N(d\sigma)\right]. \qquad (2.50)$$

From the interpretation of the sample-function density developed in Sec. 2.3, it is clear that the value of $\mathbf{X}$ that maximizes (2.50) for a particular observed count record is the value of $\mathbf{X}$ that maximizes the probability of observing that record. This value of $\mathbf{X}$ is by definition the maximum-likelihood estimate $\hat{\mathbf{X}}_{\mathrm{ML}} = \hat{\mathbf{X}}_{\mathrm{ML}}(\{N(t){:}\,t_0 \le t < T\})$ of $\mathbf{X}$. Consequently, the maximum-likelihood estimate of $\mathbf{X}$ maximizes the sample-function density (2.50). Since the logarithm is a monotonically increasing function of its argument, the maximum-likelihood estimate also maximizes the logarithm of the sample-function density, which is the loglikelihood function of (2.33),

$$\mathcal{L}(\mathbf{X}) = -\int_{t_0}^{T}\lambda(\sigma{:}\mathbf{X})\,d\sigma + \int_{t_0}^{T}\ln\lambda(\sigma{:}\mathbf{X})N(d\sigma). \qquad (2.51)$$

This maximization is subject to any constraints that may exist on the possible values of $\mathbf{X}$. As with histogram data, the maximization will often need to be accomplished by a numerical search or some other indirect means.

Example 2.4.6 *Pulse-Position Modulation* ————————————

Suppose in an optical pulse-position modulation system or optical-ranging system that electrons emanate from the photodetector at a rate $\lambda(t{:}X) = \mu(t - X) + \lambda_0$, where $\mu(t)$ and λ_0 are the photoelectron and thermoelectron generation rates, respectively. The parameter to estimate is the delay X on the basis of electron emissions observed at the output of the photodetector during the interval $[t_0, T)$. This interval may be assumed to be sufficiently large to capture all the signal energy for all possible delays. The loglikelihood function (2.51) becomes

$$\mathcal{L}(X) = -\int_{t_0}^{T}[\mu(\sigma - X) + \lambda_0]\,d\sigma + \int_{t_0}^{T}\ln[\mu(\sigma - X) + \lambda_0]N(d\sigma). \qquad (2.52)$$

The first term on the right can be disregarded because it does not depend on X when the observation interval is sufficiently large, which we have already assumed. The search for the value of X that maximizes this loglikelihood can be mechanized by processing the photodetector output with a filter having an impulse response $h(t) = \ln[\mu(-t) + \lambda_0]$. According to (2.52), the instant at which the output of the filter achieves a maximum is the maximum-likelihood estimate of X. ∎

Example 2.4.7 *Auditory Electrophysiology* ──────────────

Suppose that spike discharges of the form shown in Fig. 1.1 and discussed in Ex. 2.4.4 are observed with a microelectrode in an auditory nerve fiber during the interval $[0, T)$. Assume that a sinusoidal acoustic stimulus is applied and that the discharge sequence can be modeled as an inhomogeneous Poisson process with the parameterized intensity function given in (2.44). It is convenient to assume that the observation interval contains an integral number of cycles of the stimulation. The goal is to estimate X_1, X_2, and X_3 from the observed spike-discharge sequence. After some manipulation, the loglikelihood function (2.51) becomes

$$\mathcal{L}(\mathbf{X}) = -X_1 I_0(X_2)T + N(T)\ln(X_1) + X_2 \xi \cos(X_3 + \phi), \qquad (2.53)$$

where $I_0(\cdot)$ is a modified Bessel function of the first kind of zero order, and where we define $\xi^2 = A_c^2 + A_s^2$ and $\phi = \arctan(A_s/A_c)$ in which A_c and A_s depend on the number $N(T)$ and occurrence times $\{w_i, i = 1, 2, \cdots, N(T)\}$ of the spikes according to

$$A_c = \int_0^T \cos(2\pi f\sigma)N(d\sigma) = \sum_{i=1}^{N(T)} \cos(2\pi fw_i), \qquad (2.54a)$$

and

$$A_s = \int_0^T \sin(2\pi f\sigma)N(d\sigma) = \sum_{i=1}^{N(T)} \sin(2\pi fw_i). \qquad (2.54b)$$

It is evident by inspection of (2.53) that

$$\hat{X}_{3,\mathrm{ML}} = -\phi = -\tan^{-1}\left(\frac{A_s}{A_c}\right). \qquad (2.55)$$

By setting the derivatives of $\mathcal{L}(\mathbf{X})$ with respect to X_1 and X_2 equal to zero, we find that

$$\hat{X}_{1,\mathrm{ML}} = \frac{N(T)}{T I_0(\hat{X}_{2,\mathrm{ML}})}, \qquad (2.56)$$

and that the maximum-likelihood estimate of X_2 is the unique value of X_2 satisfying the transcendental equation

$$\frac{I_1(X_2)}{I_0(X_2)} = \frac{\xi}{N(T)} = \frac{1}{N(T)}\sqrt{A_c^2 + A_s^2}, \qquad (2.57)$$

where $I_1(x) = \epsilon_0 (x)/dx$. ∎

2.4.2 Parameter-Estimation Accuracy

The maximum-likelihood estimate of some parameters $\mathbf{X}$ in terms of a particular record of observed data maximizes the probability of observing that data. Usually, the estimate will not be identically equal to the actual and unobservable value of $\mathbf{X}$; call this value $\mathbf{X}_a$. Thus, in any given experiment, there is a nonzero error $\tilde{\mathbf{X}} = \hat{\mathbf{X}}_{ML} - \mathbf{X}_a$. We now seek a characterization of this error so that judgements about the accuracy of the maximum-likelihood estimate can be made. This characterization will be on the ensemble of errors that result with repetitions of the experiment to collect data and generate the estimate; that is, with the error viewed as a random variable.

Determining the statistical properties of the error is one of the more demanding tasks in applications. The following approaches are used:

a. *evaluation of performance analytically.* The most complete characterization of the error is its probability distribution. Unfortunately, the error distribution is usually difficult, if not impossible, to determine analytically. One reason for this is that an explicit expression for the estimate as a function of the data is not usually available, but rather the estimate is specified as the limit point of some numerical algorithm that performs a search for maximizer of the loglikelihood function. Therefore, we are forced to seek less complete characterizations such as the moments of the error. The first and second moments are usually sought. The first moment

$$\mathbf{b}(\mathbf{X}) = E[\hat{\mathbf{X}}_{ML} - \mathbf{X}{:}\mathbf{X}] \tag{2.58}$$

is termed the *bias* of the maximum-likelihood estimate, and the second moment

$$\Sigma(\mathbf{X}) = E[(\hat{\mathbf{X}}_{ML} - \mathbf{X})(\hat{\mathbf{X}}_{ML} - \mathbf{X})'{:}\mathbf{X}] \tag{2.59}$$

the *mean square-error matrix.* The estimate is said to be *unbiased* if $\mathbf{b}(\mathbf{X}) = 0$ for all permissible values of $\mathbf{X}$. The mean square-error matrix is also termed the *error-covariance matrix* when the estimate is unbiased. The expectations in (2.58) and (2.59) are with respect to the distribution of the observed data (e.g., histogram or count-record data) used to generate the estimate for an arbitrary but fixed value of $\mathbf{X}$. Even the first two moments, (2.58) and (2.59), of the error are generally difficult to determine. Thus, rather than approach their evaluation directly, it is convenient first to derive a lower bound on the mean square-error matrix for an arbitrary estimate of $\mathbf{X}$ in terms of the available data. The mean square-error matrix for the maximum-likelihood estimate can then be compared to this bound.   Under favorable circumstances the maximum-likelihood estimate actually achieves the lower bound so that its mean square-error matrix is the smallest possible using any other conceivable estimate with the same data. Such a bound is given below in Theorem 2.4.1.

b.   *evaluation of performance by computer simulation*. Computer simulations are often used to predict the performance that would be experienced in estimating parameters from real data. The methods described in Sec. 2.3.1, and elsewhere in the text, for simulating Poisson processes are useful for this purpose. Performance is predicted by simulating the data, based on a mathematical model of the physical situation under study, and then producing the parameter estimates from the simulated data. Parameter estimates are compared to their true values used in the simulation. Repeated trials permit the bias, mean square-error matrix, and other performance measures to be estimated. Computation time increases as the mathematical model used in the simulation is made more complex to reflect the actual physical situation more accurately, so there is always a trade off between computation time and accuracy in the prediction of what the performance would be for real data.

c.   *evaluation of performance by controlled experiments*. Parameter estimation accuracy can be studied using real data collected in a controlled experiment.   A physical model is constructed that mimics the actual situation but with known parameter values, data are collected, and parameter estimates formed. Comparison of the estimates to the true model parameters permits the bias, mean square-error matrix, and other performance measures to be predicted.

Generally, some combination of the above three approaches is used to evaluate the statistical properties of the error in estimating parameters from data modeled by a Poisson process.

A lower bound on the mean square-error matrix for an arbitrary estimate $\mathbf{X}^*$ of $\mathbf{X}$ is given in the following theorem for both histogram and count-record data. For histogram data, the observations are of the counts $N(t_{i-1}, t_i)$ occurring in k subintervals $[t_{i-1}, t_i)$, $i = 1, 2, \cdots, k$ of the observation interval $[t_0, T)$, where $t_k = T$. For count-record data, the observations are of the random variables $\{N(\sigma): t_0 \leq \sigma < T\}$ or, equivalently, $\{N(T): w_i, i = 1, 2, \cdots, N(T)\}$. The lower bound matrix we give for these Poisson distributed observations is a special case of a more general bound called the Cramér-Rao bound. Consult C. Rao [27, p. 265] and H. Van Trees [36, p. 79] for a discussion of the bound in a general setting.

Theorem 2.4.1 (*Cramér-Rao Lower Bound*) Let $\{N(t): t \geq t_0\}$ be an inhomogeneous Poisson process with a parameterized intensity function $\{\lambda(t:X): t \geq t_0\}$. Let X^* denote an arbitrary estimate of X based on histogram or count-record observations of $N(\cdot)$ on the interval $[t_0, T)$. Then, the mean square-error matrix $\Sigma(X) = E[(X^* - X)(X^* - X)':X]$ satisfies[1]

$$\Sigma(X) \geq b(X)b'(X) + \left[I + \frac{\partial b(X)}{\partial X} \right] F^{-1}(X) \left[I + \frac{\partial b(X)}{\partial X} \right]', \qquad (2.60)$$

where $b(X) = E(X^* - X:X)$ is the bias of X^*, $\partial b(X)/X$ is the Jacobian matrix of the bias (the i,j-element is $\partial b_i(X)/\partial X_j$), I denotes the identity matrix, and $F(X)$ is the Fisher information matrix given for histogram data by

$$F(X) = \sum_{i=1}^{k} \left[\int_{t_{i-1}}^{t_i} \lambda(\sigma:X) d\sigma \right]^{-1} \left[\int_{t_{i-1}}^{t_i} \frac{\partial \lambda(\sigma:X)}{\partial X} d\sigma \right] \left[\int_{t_{i-1}}^{t_i} \frac{\partial \lambda(\sigma:X)}{\partial X} d\sigma \right]'$$

$$(2.61)$$

and for count-record data by

$$F(X) = \int_{t_0}^{T} \frac{1}{\lambda(\sigma:X)} \left[\frac{\partial \lambda(\sigma:X)}{\partial X} \right] \left[\frac{\partial \lambda(\sigma:X)}{\partial X} \right]' d\sigma. \qquad (2.62)$$

In (2.61) and (2.62), $\partial \lambda(\sigma:X)/\partial X$ denotes the gradient vector of $\lambda(\sigma:X)$ with respect to X (the ith element is $\partial b_j(X)/\partial X_i$). Equality holds in (2.60) if and only if the estimate X^* satisfies

$$X^* = X + b(X) + \left[I + \frac{\partial b(X)}{\partial X} \right] F^{-1}(X) \left[\frac{\partial L(X)}{\partial X} \right] \qquad (2.63)$$

for all X.

Proof. Define

---

[1]The matrix inequality is in the sense that the left side minus the right side in (2.60) is a nonnegative-definite matrix.

$$\mathbf{z}(\mathbf{X}) = \begin{bmatrix} \mathbf{X}^* - \mathbf{X} - \mathbf{b}(\mathbf{X}) \\ \dfrac{\partial \mathcal{L}(\mathbf{X})}{\partial \mathbf{X}} \end{bmatrix},$$

where $\mathcal{L}(\mathbf{X})$ is the loglikelihood function defined in (2.47) for histogram data and in (2.51) for count-record data. The bound (2.60) will follow from the expectation $E[\mathbf{z}(\mathbf{X})\mathbf{z}'(\mathbf{X}):\mathbf{X}]$. To evaluate this expectation, we need first to evaluate the expectations

$$E\left\{ [\mathbf{X}^* - \mathbf{X} - \mathbf{b}(\mathbf{X})]\left[\frac{\partial \mathcal{L}(\mathbf{X})}{\partial \mathbf{X}} \right]':\mathbf{X}\right\}$$

and

$$E\left\{ \left[\frac{\partial \mathcal{L}(\mathbf{X})}{\partial \mathbf{X}} \right]\left[\frac{\partial \mathcal{L}(\mathbf{X})}{\partial \mathbf{X}} \right]':\mathbf{X}\right\}$$

for both histogram and count-record data. For brevity, denote the count-record probability (2.46) by $P(n_1, n_2, \cdots, n_k:\mathbf{X})$. Then, for histogram data,

$$E\{[\mathbf{X}^* - \mathbf{X} - \mathbf{b}(\mathbf{X})]:\mathbf{X}\}$$

$$= \sum_{n_1} \sum_{n_2} \cdots \sum_{n_k} [\mathbf{X}^*(n_1, n_2, \ldots, n_k) - \mathbf{X} - \mathbf{b}(\mathbf{X})]P(n_1, n_2, \ldots, n_k:\mathbf{X})$$

$$= 0. \tag{2.64}$$

Therefore, by taking the gradient of (2.64) with respect to $\mathbf{X}$ and using $\partial P/\partial \mathbf{X} = P\partial L/\partial \mathbf{X}$, we have for histogram data that

$$E\left\{ [\mathbf{X}^* - \mathbf{X} - \mathbf{b}(\mathbf{X})]\left[\frac{\partial \mathcal{L}(\mathbf{X})}{\partial \mathbf{X}} \right]':\mathbf{X}\right\} = \mathbf{I} + \frac{\partial \mathbf{b}(\mathbf{X})}{\partial \mathbf{X}}. \tag{2.65}$$

The analog to (2.64) for count-record data is

$$E\{[\mathbf{X}^* - \mathbf{X} - \mathbf{b}(\mathbf{X})]:\mathbf{X}\}$$

$$= [\mathbf{X}^*(0) - \mathbf{X} - \mathbf{b}(\mathbf{X})]\Pr(N(T) = 0:\mathbf{X})$$

$$+ \sum_{n=1}^{\infty} \int \cdots \int [\mathbf{X}^*(n;W_1, \ldots, W_n) - \mathbf{X} - \mathbf{b}(\mathbf{X})]p_w(W, N(T) = n)\,dW_1 \cdots dW_n$$

$$= 0, \tag{2.66}$$

where the integrals are over the range $t_0 \le W_1 \le \cdots \le W_n \le T$, and where $p_w(W, N(T) = n)$ is defined in (2.30). The gradient of this expression with respect to $\mathbf{X}$ again yields (2.65) for count-record data. From (2.66), we have

$$\frac{\partial L(\mathbf{X})}{\partial \mathbf{X}} = \sum_{i=1}^{k} \int_{t_{i-1}}^{t_i} \frac{\partial \lambda(\sigma:\mathbf{X})}{\partial \mathbf{X}} d\sigma \left[-1 + \frac{N_i}{\int_{t_{i-1}}^{t_i} \lambda(\sigma:\mathbf{X}) d\sigma} \right],$$

where we define $N_i = N(t_{i-1}, t_i)$. By using

$$E(N_i N_j | \mathbf{X}) = \begin{cases} \int_{t_{i-1}}^{t_i} \lambda(\sigma:\mathbf{X}) d\sigma \int_{t_{j-1}}^{t_j} \lambda(\sigma:\mathbf{X}) d\sigma, & i \ne j \\ \int_{t_{i-1}}^{t_i} \lambda(\sigma:\mathbf{X}) d\sigma + \left[\int_{t_{i-1}}^{t_i} \lambda(\sigma:\mathbf{X}) d\sigma \right]^2, & i = j \end{cases}$$

we see after some manipulation that

$$E\left\{ \left[\frac{\partial L(\mathbf{X})}{\partial \mathbf{X}} \right] \left[\frac{\partial L(\mathbf{X})}{\partial \mathbf{X}} \right]' : \mathbf{X} \right\} = \mathbf{F}(\mathbf{X}), \qquad (2.67)$$

where $\mathbf{F}(\mathbf{X})$ is the Fisher information-matrix for histogram data, as given in (2.61). For count-record data, this same procedure yields (2.67) with $\mathbf{F}(\mathbf{X})$ as given in (2.62). By collecting these results together, we have for both histogram and count-record data that

$$\Lambda(\mathbf{X}) \stackrel{\Delta}{=} E[\mathbf{z}(\mathbf{X})\mathbf{z}'(\mathbf{X}):\mathbf{X}] = \begin{bmatrix} [\Sigma(\mathbf{X}) - \mathbf{b}(\mathbf{X})\mathbf{b}'(\mathbf{X})] & \left[\mathbf{I} + \dfrac{\partial \mathbf{b}(\mathbf{X})}{\partial \mathbf{X}} \right] \\ \left[\mathbf{I} + \dfrac{\partial \mathbf{b}(\mathbf{X})}{\partial \mathbf{X}} \right]' & \mathbf{F}(\mathbf{X}) \end{bmatrix}.$$

It is easy to see for an arbitrary choice of a vector $\mathbf{v}$ that $\mathbf{v}'\Lambda(\mathbf{X})\mathbf{v} = E[(\mathbf{v}'\mathbf{z}(\mathbf{X}))^2 : \mathbf{X}] \ge 0$. This inequality holds in particular for the choice

$$\mathbf{v} = \begin{bmatrix} \mathbf{v}_1 \\ -\mathbf{F}^{-1}(\mathbf{X}) \left[\mathbf{I} + \dfrac{\partial \mathbf{b}(\mathbf{X})}{\partial \mathbf{X}} \right]' v_1 \end{bmatrix}, \qquad (2.68)$$

where $\mathbf{v}_1$ is arbitrary. For this choice, we conclude that

$$\mathbf{v}_1\Sigma(\mathbf{X})\mathbf{v}_1 \geq \mathbf{v}_1'\left\{\mathbf{b}(\mathbf{X})\mathbf{b}'(\mathbf{X})+\left[\mathbf{I}+\frac{\partial\mathbf{b}(\mathbf{X})}{\partial\mathbf{X}}\right]\mathbf{F}^{-1}(\mathbf{X})\left[\mathbf{I}+\frac{\partial\mathbf{b}(\mathbf{X})}{\partial\mathbf{X}}\right]'\right\}\mathbf{v}_1.$$

This establishes (2.60) by virtue of the arbitrariness of $\mathbf{v}_1$. Furthermore, the arbitrariness of $\mathbf{v}_1$ also implies that

$$E[(\mathbf{v}'\mathbf{z}(\mathbf{X}))^2 \colon \mathbf{X}]$$

$$=E\left[\left(\mathbf{v}_1'\left\{\mathbf{X}^* - \mathbf{X} - \mathbf{b}(\mathbf{X}) - \left[\mathbf{I}+\frac{\partial\mathbf{b}(\mathbf{X})}{\partial\mathbf{X}}\right]\mathbf{F}^{-1}(\mathbf{X})\left[\frac{\partial\mathcal{L}(\mathbf{X})}{\partial\mathbf{X}}\right]\right\}\right)^2 \colon \mathbf{X}\right]$$

will be zero if and only if (2.63) holds for all permissible $\mathbf{X}$. This completes the proof of Theorem 2.4.1. $\square$

Some important conclusions can be drawn from the assertions of Theorem 2.4.1. These hold for both histogram and count-record data.

a.   Let $\mathbf{e}_i$ denote a unit vector with ith element equal to one and all other elements equal to zero. Then for arbitrary $\mathbf{X}^*$ there holds

$$E[(X_i^* - X_i)^2 \colon \mathbf{X}] = \mathbf{e}_i'\Sigma(\mathbf{X})\mathbf{e}_i \geq b_i^2(\mathbf{X})+\Delta_{ii},$$

where Δ_{ii} denotes the (i,i)-element of the matrix

$$\left[\mathbf{I}+\frac{\partial\mathbf{b}(\mathbf{X})}{\partial\mathbf{X}}\right]\mathbf{F}^{-1}(\mathbf{X})\left[\mathbf{I}+\frac{\partial\mathbf{b}(\mathbf{X})}{\partial\mathbf{X}}\right]'.$$

This shows that the mean square-error for any unbiased estimate of the parameter X_i is not less than a precomputable quantity that is defined independently of the method by which the estimate is selected.

b.   An estimate having a mean square-error matrix satisfying (2.60) with equality is said to be *efficient*. Suppose that there exists an efficient estimate $\mathbf{X}^*$. Then, in terms of this estimate, the maximum-likelihood estimate satisfies $\hat{\mathbf{X}}_{ML} + \mathbf{b}(\hat{\mathbf{X}}_{ML}) = \mathbf{X}^*$, where $\mathbf{b}(\hat{\mathbf{X}}_{ML})$ is the bias $\mathbf{b}(\mathbf{X})$ of $\mathbf{X}^*$ evaluated at the maximum-likelihood estimate. This follows immediately from (2.63) and the property that the gradient of the loglikelihood function $\mathcal{L}(\mathbf{X})$ is zero when $\mathbf{X}$ is the maximum-likelihood estimate. Thus, the estimate $\hat{\mathbf{X}}_{ML} + \mathbf{b}(\hat{\mathbf{X}}_{ML})$ is efficient. It follows that the maximum-likelihood estimate will be efficient only if the bias of $\mathbf{X}^*$ is zero for $\mathbf{X}$ equal to the maximum-likelihood estimate. The most important situation occurs when there exists an unbiased, efficient estimate because the maximum-likelihood estimate is also then unbiased and efficient.

c.   Suppose that $\mathbf{X}^*$ is an unbiased estimate of $\mathbf{X}$. Then from (2.60), the mean square-error matrix for $\mathbf{X}^*$ satisfies $\Sigma(\mathbf{X}) \geq \mathbf{F}^{-1}(\mathbf{X})$. Consider, now, that the experiment of collecting data to estimate $\mathbf{X}$ is repeated independently M times. Denote the estimate of $\mathbf{X}$ based on these M trials by $\mathbf{X}_M^*$. Then, the mean square-error matrix for $\mathbf{X}_M^*$ satisfies $\Sigma_M(\mathbf{X}) \geq (1/M)\mathbf{F}^{-1}(\mathbf{X})$; see Problem 2.4.4. The estimate $\mathbf{X}_M^*$ is said to be asymptotically efficient if $M\Sigma_M(\mathbf{X}) - \mathbf{F}^{-1}(\mathbf{X})$ tends to zero as M tends to infinity for all permissible $\mathbf{X}$. It is said to be asymptotically unbiased if $\mathbf{b}_M(\mathbf{X})$ tends to zero as M tends to infinity for all permissible $\mathbf{X}$. The maximum-likelihood estimate is both asymptotically efficient and unbiased under relatively weak conditions (see S. Wilks [38, p. 363]).

Example 2.4.8 *Scaled Inhomogeneous Poisson Process* ━━━━━━━
Consider again the model of Ex. 2.4.5 in which histogram data are collected in order to estimate the parameter X of the intensity $\lambda(t{:}X) = X\mu(t)$. It is easily seen from the expression for the maximum-likelihood estimate in (2.49) that $E(\hat{X}_{\text{ML}}{:}X) = X$, so the maximum-likelihood estimate is an unbiased estimate of X in terms of histogram data. From (2.48),

$$\frac{\partial L(X)}{\partial X} = -\int_{t_0}^{T} \mu(\sigma)d\sigma + \frac{N(T)}{X}, \tag{2.69}$$

and from (2.61),

$$F(X) = \frac{1}{X}\int_{t_0}^{T} \mu(\sigma)d\sigma. \tag{2.70}$$

Thus,

$$\hat{X}_{\text{ML}} = X + F^{-1}(X)\frac{\partial L(X)}{\partial X},$$

and the condition (2.63) is satisfied by the maximum-likelihood estimate. The maximum-likelihood estimate is both unbiased and efficient. As (2.60) is satisfied with equality, we have

$$E[(\hat{X}_{ML} - X)^2 : X] = F^{-1}(X) = \frac{X}{\displaystyle\int_{t_0}^{T} \mu(\sigma)\,d\sigma}. \tag{2.71}$$

This mean square-error in estimating X is the smallest that can be achieved with any conceivable estimate using histogram data. ∎

Example 2.4.9 *Pulse-Position Modulation* ────────────────

Assume as in Ex. 2.4.6 that $\lambda(t:X) = \mu(t-X) + \lambda_0$, where $\mu(t)$ is the photoelectron rate and λ_0 is the thermoelectron rate. According to Ex. 2.4.3, $\mu(t) = \alpha |S(t)|^2$, where $|S(t)|$ is the envelope of the optical field incident on the photodetector and $\alpha > 0$ is a constant assumed to be known. An estimate of X satisfying (2.63) cannot be found, so an efficient estimate does not exist. From (2.60) and (2.62), the mean square-error in estimating X using any unbiased estimate X^* satisfies the inequality

$$E[(X^* - X)^2 : X] \geq \left[\int_{t_0}^{T} \frac{\dot{\mu}^2(\sigma - X)}{\mu(\sigma - X) + \lambda_0}\, d\sigma \right]^{-1}$$

$$= \left[\int_{t_0}^{T} \frac{\dot{\mu}^2(t)}{\mu(t) + \lambda_0}\, dt \right]^{-1}, \tag{2.72}$$

where $\dot{\mu}(t) = d\mu(t)/dt$. In writing the last equality, it has been assumed that the observation interval $[t_0, T)$ is sufficiently large to capture all the signal energy for all possible values of X. Equation (2.72) summarizes all that can be said about the performance of X^* from the Cramér-Rao inequality when the efficiency of X^* cannot be established. The inequality provides only a lower bound and no assurance that it can be approached. The implication of remark (c) above is that with repeated illuminations of the reflector, the maximum-likelihood estimate will tend to become unbiased and have a mean square-error given by the right side of (2.72) divided by the number of illuminations. The lower bound in (2.72) can be given a qualitative interpretation in the limits of small and large thermoelectron rate λ_0. For this purpose, let $a(t)$ be a pulse in $[t_0, T)$ and $A(f)$ its Fourier transform. Define

$$B_a^2 = \frac{\displaystyle\int_{t_0}^{T} [da(t)/dt]^2\,dt}{\displaystyle\int_{t_0}^{T} a^2(t)\,dt} = \frac{\displaystyle\int_{-\infty}^{\infty} |fA(f)|^2\,df}{\displaystyle\int_{-\infty}^{\infty} |A(f)|^2\,df},$$

where the second equality follows from Parseval's theorem. Then, B_a can be interpreted qualitatively as the RMS bandwidth of $a(t)$. Also, let

$$E_a = \int_{t_0}^{T} a^2(t)\,dt$$

denote the energy of $a(t)$. In the limit $\lambda_0 \gg \mu(t)$, and for M repeated illuminations of the reflector, we have for the maximum-likelihood estimate that

$$E[(\hat{X}_{\mathrm{ML}} - X)^2 : X] \approx \frac{\lambda_0}{MB_\mu E_\mu}. \tag{2.73}$$

Thus, in the limit of high thermoelectron rate, the mean square-error for the maximum-likelihood estimate is given approximately by the noise-to-signal ratio λ_0 / E_μ reduced by a factor MB_μ dependent on the number of illuminations M and the RMS bandwidth B_μ of the square of the envelope $|S(t)|$ of the incident optical field. In the limit $\lambda_0 \ll \mu(t)$, and for M repeated illuminations of the reflector, we have

$$E[(\hat{X}_{\mathrm{ML}} - X)^2 : X] \approx \frac{1}{4MB_{|S|} E_{|S|}}. \tag{2.74}$$

Thus, in the limit of low thermoelectron rate, the mean square-error for the maximum-likelihood estimate is dependent on the RMS bandwidth and energy of the envelope $|S(t)|$. ∎

2.5 Multidimensional Poisson-Processes

In this section, we generalize the space X in which points occur from the semi-infinite real line $[t_0, \infty)$ to higher dimensions. Multidimensional spaces of interest include the two-dimensional plane $\mathcal{R}^2$, for which we term the point process a *quantum-limited image*, and the product space $[t_0, \infty) \times \mathcal{R}^2$, which is used to model time-dependent quantum-limited images. Quantum-limited images are discussed in Ch. 3. More generally, X can be an n-dimensional space $\mathcal{R}^n$, so the coordinate of each point is an n-dimensional vector $\mathbf{x}$. If A is a subset of X, then $N(A)$ is the number of points with coordinates in A.

A multidimensional Poisson-process $\{N(A) : A \subseteq X\}$ with intensity $\lambda(x)$ has the following properties:

a. for each $A \subseteq X$, $N(A)$ is Poisson distributed with parameter

$$\int_A \lambda(x)\,dx;$$

b. if $A_1, A_2, \cdots, A_k, \cdots$ are disjoint subsets of X, then $N(A_1)$, $N(A_2)$, $\cdots$, $N(A_k)$, $\cdots$ are statistically independent;

c. if $A_1, A_2, \cdots, A_k, \cdots$ are disjoint subsets of X, then

$$N\left(\bigcup_{k=1}^{\infty} A_k\right) = \sum_{k=1}^{\infty} N(A_k).$$

Property (b) is the analog of the independent-increments property of temporal Poisson processes.

Many of the statistical properties of temporal Poisson-processes can be readily extended to multidimensional Poisson-processes. The sample-function density and loglikelihood function of (2.32) and (2.33) are generalized as follows.

Let Ω denote a particular realization of the Poisson process on X with $N(X) = n$ points, as indicated in Fig. 2.7. For the realization shown, $N(X) = 12$ and $N(A) = 6$. The points can be given coordinate labels $\{X_1, X_2, \cdots, X_n\}$ in some arbitrary order to indicate their locations. Then Ω may be viewed as $\{X_1, X_2, \cdots, X_n, N(X) = n\}$ when the ordering of the elements of $\{X_1, X_2, \cdots, X_n\}$ is disregarded. Let $S(X_i) = \{x : \|x - X_i\| \le \rho_i\}$ be a sphere of radius ρ_i centered about the point labeled X_i. Then, for a given labeling of the points $\{X_1, X_2, \cdots, X_n\}$, the independent-increments property implies

$$\Pr[N(S(X_1)) = 1, N(S(X_2)) = 1, \cdots, N(S(X_n)) = 1, N(X) = n]$$

$$= \left[\prod_{i=1}^{n} \int_{S(X_i)} \lambda(x)\,dx \, \exp\left(-\int_{S(X_i)} \lambda(x)\,dx\right)\right] \exp\left(-\int_{x - \bigcup_{i=1}^{n} S(X_i)} \lambda(x)\,dx\right)$$

$$= \left[\prod_{i=1}^{n} \int_{S(X_i)} \lambda(x)\,dx\right] \exp\left(-\int_{x} \lambda(x)\,dx\right). \tag{2.75}$$

There are $n!$ different ways to label the points that yield this same probability of occurrence of the realization $\Omega = \{X_1, X_2, \cdots, X_n, N(X) = n\}$. Dividing both sides of (2.75) by the product of the volumes of the n spheres, letting the volumes tend to zero, and disregarding the ordering of the elements of $\{X_1, X_2, \cdots, X_n\}$ then yields the following expression for the sample-function density for the particular realization:

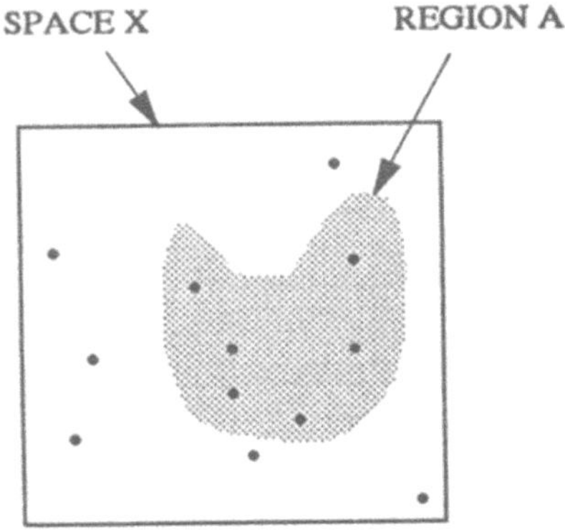

Figure 2.7 A realization of a multidimensional Poisson process.

$$p(\Omega) = \frac{1}{n!}\left[\prod_{i=1}^{n}\lambda(X_i)\right]e^{-\int_x \lambda(x)dx}, \tag{2.76}$$

where the value of the product on the right-hand side is invariant with any particular labeling. This may alternatively be rewritten for an arbitrary realization of the process as

$$p(\omega) = \frac{1}{N(X)!}\exp\left[-\int_x \lambda(x)dx + \int_x \ln[\lambda(x)]N(dx)\right], \tag{2.77}$$

where

$$\int_x \ln[\lambda(x)]N(dx) = \begin{cases} 0, & N(X)=0 \\ \sum_{i=1}^{N(X)} \ln[\lambda(x_i)], & N(X)\geq 1. \end{cases}$$

The quantity $p(\omega)$ is the sample-function density for a multidimensional Poisson-process. This is the generalization of (2.32) for multidimensional Poisson-processes; the factor $1/N(X)!$ results because points in the multi-dimensional space are unordered. Its logarithm,

$$L = -\int_x \lambda(x)dx + \int_x \ln[\lambda(x)]N(dx) - \ln[N(X)!], \tag{2.78}$$

is the corresponding loglikelihood function.

　　　The use of the sample-function density and loglikelihood function parallels that in Sec. 2.4 for temporal Poisson-processes. The following is a simple example illustrating its application for estimating image paramters from quantum-limited data.

Example 2.5.1 *Piecewise-Constant Intensity* ——————————————————
　　　Suppose that $\{N(A): A \in X\}$ is a Poisson process with an intensity that is piecewise constant over disjoint regions A_k, $k = 1, 2, \cdots, K$, in X:

$$\lambda(x) = \sum_{k=1}^{K} \lambda_k I_k(x),$$

where $I_k(x)$ is the indicator function for the kth region

$$I_k(x) = \begin{cases} 1, & x \in A_k \\ 0, & x \notin A_k \end{cases}.$$

Then, given a realization of the Poisson process, the loglikelihood function becomes

$$\mathcal{L}(\lambda) = -\sum_{k=1}^{K} \lambda_k ||A_k|| + \sum_{k=1}^{K} N(A_k) \ln \lambda_k - \ln[N(X)!],$$

where

$$||A_k|| = \int_X I_k(x)\,dx = \int_{A_k} dx,$$

and

$$N(X) = \sum_{k=1}^{K} N(A_k).$$

Maximization of the loglikelihood function with respect to λ yields the estimates

$$\hat{\lambda}_k = \frac{N(A_k)}{||A_k||},$$

for $k = 1, 2, \cdots, K$. Thus, the maximum-likelihood estimate of the constant intensity in a region is the number of points observed in the region divided by the size of the region. ∎

By paralleling the development of (2.34) [see Problem 2.5.2], it may be concluded for a multidimensional Poisson-process that the joint probability-density of the random locations of the points in X, given that $N(X) = n$, is given by

$$p_{x_1, \cdots, x_n | N(X)}(X_1, \cdots, X_n | n) = \prod_{i=1}^{n} \left[\frac{\lambda(X_i)}{\int_X \lambda(x)\, dx} \right]. \qquad (2.79)$$

Thus, given the number of points in X, their locations are independent, identically distributed random variables. Similarly, by paralleling the development of (2.39) [see Problem 2.5.3], it may be concluded that the characteristic functional for a multidimensional Poisson process is given by

$$\phi_N(jv) = \exp\left[\int_X \lambda(x)(e^{jv(x)} - 1)\, dx \right]. \qquad (2.80)$$

In Ch. 3, we will study various estimation problems involving multidimensional Poisson-processes, and the sample-function density and log-likelihood function will play an important role.

2.6 References

1.   M. Abramovitz and I. A. Stegun, *Handbook of Mathematical Functions*, National Bureau of Standards, U.S. Government Printing Office, Washington DC, 1964.

2.   M. Aoki, *Introduction to Optimization Techniques*, Macmillan, New York, 1971.

3.   L. Aroian, "Type-B Gram-Charlier Series," *Ann. Math. Stat.*, Vol. 8, pp. 183-192, 1937.

4.   E. W. Barankin, "Locally Best Unbiased Estimates," *Ann. Math. Stat.*, Vol. 20, p. 477, 1949.

5.   I. Bar-David, "Communication Under the Poisson Regime," *IEEE Trans. Information Theory*, Vol. IT-15, No. 1, pp. 31-37, Jan. 1969.

6.   D. R. Cox and P. A. W. Lewis, *The Statistical Analysis of Series of Events*, Methuen, London, 1966.

7.   D. J. Daley, "Various Concepts of Orderliness for Point Processes," in: *Stochastic Geometry and Analysis (Rollo Davidson Memorial Volume)*, Wiley, New York, 1974.

8.   D. J. Daley and D. Vere-Jones, "A Summary of the Theory of Point Processes," in: *Stochastic Point Processes: Statistical Analysis, Theory, and Applications* (P. A. W. Lewis, Ed.), Wiley, New York, 1972

9. R. D. Evans, *The Atomic Nucleus*, McGraw-Hill, New York, 1963.

10. M. Fisz, *Probability Theory and Mathematical Statistics*, Wiley, New York, 1967.

11. F. Haight, *Handbook of the Poisson Distribution*, Wiley, New York, 1967.

12. C. Helstrom, "Estimation of Modulation Frequency of a Light Beam," Appendix E in: *Optical Space communication*, Proc. of a Workshop held at Williams College (R. S. Kennedy and S. Karp, Ed's.), Williamstown, Mass. 1968.

13. E. Hoversten and D. L. Snyder, "On the Performance of Pulse-Position Modulation in Direct-Detection Optical Communication Systems: Mean-Square Error and Threshold," Proc. of the 1972 International Information Theory Symposium, Pacific Grove, Calif., Jan. 1972.

14. J. A. Jacques, "Tracer Kinetics" in: *Principles of Nuclear Medicine* (H. N. Wagner, Jr., Ed.), Saunders, Philadelphia, 1968.

15. B. Jansson, *Random Number Generators*, Victor Pettersons, Stockholm, 1966.

16. S. Karp, E. L. O'Neill, and R. M. Gagliardi, "Communication Theory for the Free Space Optical Channel," *Proc. IEEE*, Vol. 58, No. 10, pp. 1611-1626, Oct. 1970.

17. A. Y. Khinchin, "On Poisson Sequences of Chance Events," *Theory of Probability and Its Applications*, Vol. 1, No. 3, pp. 291-297, 1956.

18. W. M. Littlefield, "Investigation of the Linear Range of the Peripheral Auditory System," D.Sc. Thesis, Sever Institute of Technology, Washington University, St. Louis, Mo., Dec. 1973.

19. L. Mandel, "Fluctuations of Light Beams," in *Progress in Optics*, Vol. II, (E. Wolf, Ed.), Wiley, New York, 1963.

20. L. Mandel, "Fluctuations of Photon Beams and Their Correlations," *Proc. Phys. Soc.* (London), Vol. 72, No. 1, pp. 1037-1048, 1958.

21. J. Markham, D. L. Snyder, and J. R. Cox, Jr., "A Numerical Implementation of the Maximum-Likelihood Method of Parameter Estimation for Tracer-Kinetic Data," *J. Mathematical Biosciences*, Vol. 28, pp. 275-300, 1976.

22. M. A. Mintun, M. E. Raichle, W. R. W. Martin, and P. Herscovitch, "Brain Oxygen Utilization Measured with ^{15}O Radiotracers and Positron Emission Tomography," *J. Nuclear Medicine*, Vol. 25, No. 2, pp. 177-187, 1984.

23. J. Ortega and W. Rheinholdt, *Iterative Solution of Nonlinear Equations in Several Variables*, Academic Press, New York, 1970.

24. E. Parzen, *Stochastic Processes*, Holden-Day, San Francisco, 1962.

25. R. Peirls, "Statistical Errors in Counting Experiments," *Proc. Roy. Stat. Soc.* (London), Vol. A149, pp. 467-486, 1935.

26. A. Ralston and H. Wilf, *Mathematical Methods for Digital Computers*, Vol. 2, p. 249, Wiley, New York, 1967.

27. C. R. Rao, *Linear Statistical Inference and Its Applications*, Wiley, New York, 1965.

28. J. Riordan, "Moment Recurrence Relations for Binomial, Poisson, and Hypergeometric Frequency Distributions," *Ann. Math. Stat.*, Vol. 8, pp. 103-111, 1937.

29. E. Rutherford and H. Geiger, "The Probability Variations in the Distributions of a-Particles," *Phil. Mag. S6*, Vol. 20, p. 698, 1910.

30. J. L. Savage, *The Foundations of Statistics*, Wiley, New York, 1954.

31. H. E. Schaffer, "Algorithms 369: Generation of Random Numbers Satisfying the Poisson Distribution," *Communications of the Assoc. for Computing Machinery*, Vol. 13, No. 1, January 1970.

32. C. W. Sheppard, *Basic Principles of the Tracer Method*, Wiley, New York, 1954.

33. W. M. Siebert, "Stimulus Transformation in the Peripheral Auditory System, " in: *Recognizing Patterns* (P. A. Kolers and M. Eden, Ed's.), M.I.T. Press, Cambridge, Mass., 1968.

34. W. M. Siebert, "Frequency Discrimination in the Auditory System: Place or Periodicity Mechanisms?," *Proc. IEEE*, Vol. 58, No. 5, pp. 723-730, May 1970.

35. M. M. Ter-Pogossian, J. O. Eichling, D. O. Davis, and M. J. Welch, "The Measure *In Vivo* of Regional Cerebral Oxygen Utilization by means of Oxyhemoglobin Labelled with Radioactive Oxygen-15," *J. Clin. Invest.*, Vol. 49, pp. 381-391, 1970.

36. H. L. Van Trees, *Detection, Estimation, and Modulation Theory: Part I*, Wiley, New York, 1968.

37. J. B. West, C. T. Dollery, and P. Hugh-Jones, "The Use of Radioactive Carbon Dioxide to Measure Regional Blood Flow in the Lungs of Patients with Pulmonary Disease," *J. Clin. Invest.*, Vol. 40, pp. 1-12, Jan. 1961.

38. S. S. Wilks, *Mathematical Statistics*, Wiley, New York, 1962.

2.7 Problems

2.2.1. The fact that points are distinct and that only a finite number occur in any finite interval is insufficient to guarantee that a point process is orderly. To verify this, consider a point process on the interval $-1 \le t \le 1$ constructed as follows. The process has two distinct points, the first of which is selected uniformly at random on the interval. Call the location of this first point t_1. The second point is placed at $t = \frac{1}{2}t_1$ for $t_1 \ne 0$ and at $t = \frac{1}{2}$ for $t_1 = 0$.

a. Show that the point process is orderly at $t = \frac{1}{2}$.

b. Show that it is not orderly at $t = 0$.

2.2.2. (*Constructive Derivation of the Poisson Process*). Suppose a process consisting of unit-area pulses of duration Δ is constructed in the following fashion. At time $t = 0$, a decision is made as to whether or not a pulse is to occur in the interval $[0,\Delta)$. An affirmative decision is made if heads occurs on the toss of a biased coin for which the probability of getting heads is p and of getting tails is $q = 1 - p$. At time $t = \Delta$, a decision is made as to whether or not a pulse is to occur in $[\Delta,2\Delta)$. The decision is made by tossing the coin independently again. This procedure is repeated at the beginning of each interval. The decision for the interval $[k\Delta,(k+1)\Delta)$ depends only on the coin toss at time $t = k\Delta$ and not any previous tosses. Thus, pulse decisions are made without aftereffects.

a. Let $s = k\Delta$ and $t = (k+K)\Delta$, and denote by the number of pulses in $[s,t)$. Assume that k and K are nonnegative integers. Evaluate the mean and variance of $N(s,t)$. Determine the probability that $N(s,t) = n$ for n = 0, 1, 2, $\cdots$, K.

b. Now examine what happens as Δ tends to zero. Let k and K increase in such a way that $s = k\Delta$ and $t = (k+K)\Delta$ remain fixed as Δ tends to zero. Also, let $p = \lambda\Delta$ so that the expected number of pulses in $[s,t)$ remains fixed. Show that $\Pr[N(s,t)=n]$ tends to the Poisson distribution with parameter $\lambda(t-s)$ as Δ tends to zero. This is how in 1837 S. Poisson discovered this distribution now bearing his name. See F. Haight [11, p. 113] for an historical account of this discovery.

c. Define $N(t)=N(0)+N(0,t)$, where $\Pr[N(0)=0]=1$. Explain why $\{N(t):t \geq 0\}$ is a homogeneous Poisson process under the limiting conditions of part (b).

d. Discuss how these results can be used in a Monte Carlo simulation of an inhomogeneous Poisson-process.

2.2.3. The moments $m_k = E(x^k)$, for $k = 1, 2, \cdots$ are one set of parameters that characterize a random variable x. There are other sets that also arise in practice, one of which is the set of *cumulants* or *semiinvariants*, γ_k, defined by the Taylor-series expansion of the logarithm of the characteristic function for x,

$$\Gamma_x(jv) \overset{\Delta}{=} \ln E(e^{jvx}) = \sum_{k=1}^{\infty} \frac{(jv)^k}{k!} \gamma_k.$$

the function $\Gamma_x(jv)$ is termed the *semi-invariant generating function* for x.

a. Derive the following relationships between the moments and cumulants:

$$m_1 = \gamma_1,$$
$$m_2 = \gamma_2 + \gamma_1^2,$$
$$m_3 = \gamma_3 + 3\gamma_1\gamma_2 + \gamma_1^3,$$
$$m_4 = \gamma_4 + 3\gamma_2^2 + 4\gamma_1\gamma_3 + 6\gamma_1^2\gamma_2 + \gamma_1^4$$

Conversely,
$$\gamma_1 = m_1 = \text{mean}(x),$$
$$\gamma_2 = m_2 - m_1^2 = \text{variance}(x),$$
$$\gamma_3 = m_3 - 3m_1 m_2 + 2m_1^3,$$
$$\gamma_4 = m_4 - 3m_2^2 - 4m_1 m_3 + 12m_1^2 m_2 - 6m_1^4.$$

b. Show that all of the cumulants except γ_1 are invariant under translation; that is, show that all the cumulants for $y = x + a$ are identical to those of x except for the first.

c. Determine all of the cumulants when $x = N(s,t) = N(t) - N(s)$ is an increment of an inhomogeneous Poisson counting process with intensity $\lambda(t)$.

d. Let $y = x_1 + \cdots + x_n$ be the sum of n statistically independent random variables that are not necessarily identically distributed. Demonstrate that

$$\gamma_k(y) = \sum_{i=1}^{n} \gamma_k(x_i), \qquad k = 1, 2, \ldots .$$

That is, the kth cumulant of y is the sum of the kth cumulants of x.

e. Determine all of the cumulants of y when the elements in the sum of part (d) are identically distributed normal random variables having mean m and variance σ^2.

2.2.4. Let $\{N(t): t \geq t_0\}$ be an inhomogeneous Poisson process with intensity function $\{\lambda(t): t \geq t_0\}$.

a. Show that the first and second moments for the increment $N(s,t) = N(t) - N(s)$ are given by

$$E[N(s,t)] = \int_s^t \lambda(\sigma)d\sigma$$

and

$$E[N^2(s,t)] = \int_s^t \lambda(\sigma)d\sigma + \left(\int_s^t \lambda(\sigma)d\sigma\right)^2$$

b. Show that higher-order moments of $N(s,t)$ can be determined using the recursion formula

$$E[N^{m+1}(s,t)] = \mu E[N^m(s,t)] + \mu\frac{dE[N^m(s,t)]}{d\mu},$$

where

$$\mu = \int_s^t \lambda(\sigma)d\sigma.$$

This recursion formula is attributed by F. Haight [11, p. 6] to L. Aroian [3] and J. Riordan [28].

2.2.5. Let $N(\cdot)$ be a Poisson-distributed random variable with mean Λ. Define the function $f(N)$ according to

$$f(N) = \begin{cases} \dfrac{1}{N}, & N \geq 1 \\ 0, & N = 0 \end{cases}.$$

a. Evaluate the expectation $E[f(N)]$ of $f(N)$. Hint: express your answer in terms of the exponential-integral function $Ei(\cdot)$, which is tabulated by M. Abromowitz and I. A. Stegun [1].

b.   Under what circumstances, if any, is Λ^{-1} a good approximation to $E[f(N)]$?

2.2.6. (*Birth Processes*). Let $\{N(t): t \geq t_0\}$ be a counting process described by the following assumptions:

   i. $\{N(t): t \geq t_0\}$ is conditionally orderly;

   ii.   for all $t \geq t_0$, the limit of $\delta^{-1}\Pr[N(t,t+\delta) = 1 \mid N(t) = n]$ exists as δ tends to zero, and the limit is a finite, integrable function of t and a function of n. Denote the limit by $\lambda_n(t)$; thus,

$$\lambda_n(t) = \lim_{\delta \downarrow 0} \frac{1}{\delta} \Pr[N(t, t+\delta) = 1 | N(t) = n];$$

iii.　$\Pr[N(0) = 1] = 1$.

The difference between this process and a Poisson counting process of Theorem 2.2.1 is contained in assumption (ii). Here, we allow the intensity to depend on the number of events, so the process does not evolve without aftereffects. This process is called a *pure birth-process*. If N is the number of members of a population (of bacteria, particles in a chain reaction, etc.), then we allow the rate of population growth to vary with the size of the population. Note that initially there is one member in the population (we could just as easily have two), and there are no deaths (a birth-death process allows for this).

a.　Let $P_n(t) = \Pr[N(t) = n]$. Determine a differential equation with appropriate boundary conditions that $P_n(t)$ satisfies. Verify your equation using the known results when $\lambda_n(t) = \lambda(t)$ is not a function of n.

b.　The simplest birth process is the homogeneous, linear birth-process in which the intensity is not a function of time and is proportional to the population size; that is, $\lambda_n(t) = \lambda n$, where $\lambda > 0$. Determine $P_1(t)$ for the homogeneous, linear birth-process.

c.　Demonstrate that $P_n(t)$ for the homogeneous, linear birth-process satisfies the following recursion equation

$$P_n(t) = (n-1)\lambda \int_0^t P_{n-1}(\xi) e^{-\lambda n(t-\xi)} d\xi$$

for $n = 2, 3, \cdots$.

d.　Let

$$\tilde{P}_n(s) = \int_0^\infty P_n(t) e^{-st} dt$$

be the Laplace transform of $P_n(t)$ for the homogeneous, linear birth-process. Demonstrate that

$$\tilde{P}_n(s) = \sum_{k=0}^{n-1} (-1)^k \binom{n-1}{k} \frac{1}{s + (k+1)\lambda},$$

where

$$\binom{n-1}{k} = \frac{(n-1)!}{k!(n-1-k)!}$$

is the binomial coefficient.

e.   By using either the result of (c) and induction or by inverting the result of (d), show that

$$P_n(t) = e^{-\lambda t}(1 - e^{-\lambda t})^{n-1} \qquad \text{for} \quad n = 1, 2, \ldots$$

This is called the *Yule-Furry distribution*.

f.   The mean and variance of a homogeneous, linear birth-process N can be found by using $P_n(t)$ given in (e). However, in most general cases, $P_n(t)$ is not easily determined. It is then convenient to use the following alternative procedure. Multiply the differential equation for $P_n(t)$ in (a) by n (or n^2) and sum on n to obtain a differential equation for $E[N(t)]$ (or $E[N^2(t)]$). Solve the differential equation for the case of a homogeneous, linear birth-process in order to evaluate the mean and variance of $N(t)$.

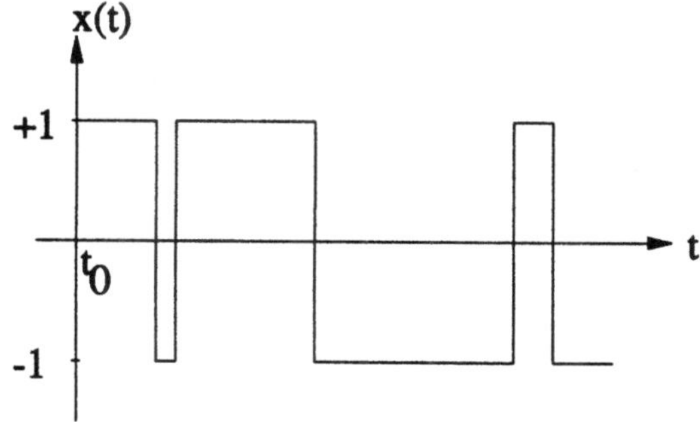

Figure P.2.2.6. Typical sample function of a random telegraph-wave.

2.2.7. (*Random-Telegraph Wave*). Let $\{N(t): t \geq t_0\}$ be an inhomogeneous Poisson counting process with intensity function $\{\lambda(t): t \geq t_0\}$. Let $\{x(t): t \geq t_0\}$ be the process defined by

$$x(t) = x_0(-1)^{N(t)},$$

where the initial state x_0 is either $+1$ or -1 with equal probability and independently of $\{N(t): t \geq t_0\}$. A typical sample function for x is shown in Fig. P.2.2.6. This process is called a *random telegraph-wave*. Let

$$M(ju,jv) = E[e^{j(ux(s)+vx(t))}]$$

be the joint characteristic function for $x(s)$ and $x(t)$, where $t_0 \leq s \leq t$.

a.   Show that the correlation function for x is given by

$$R(s,t) \overset{\Delta}{=} E[x(t)x(s)] = \exp\left(-2\int_s^t \lambda(\sigma)d\sigma\right),$$

where $t_0 \leq s \leq t$. HINT: establish and use the relation:

$$\sum_{n \text{ even}} \frac{1}{n!}\left(\int_s^t \lambda(\sigma)d\sigma\right)^n = \frac{1}{2}\left[\exp\left(+\int_s^t \lambda(\sigma)d\sigma\right)+\exp\left(-\int_s^t \lambda(\sigma)d\sigma\right)\right].$$

b.   Show that

$$M(ju,jv) = \cos(u)\cos(v) - R(s,t)\sin(u)\sin(v).$$

2.2.8.   Suppose that the points of an inhomogeneous Poisson process with intensity $\lambda(t)$ are deleted independently with probability p. Show that the resulting counting process is Poisson, and specify its intensity. The counting process for the deleted point process is a possible model for a faulty counter in which points are not counted with probability p. Give an argument demonstrating that a Poisson point process subjected to systematic deletions (for example, every other point deleted) is no longer Poisson.

2.3.1.   Let $\{N(t): t \geq t_0\}$ be a homogeneous Poisson process with intensity λ, and define $x(t)$ and $g(t)$ by:

$$x(t) = \begin{cases} 1, & N(t)=1 \\ 0, & N(t) \neq 1. \end{cases}$$

and

$$g(t) = \int_0^t x(\tau)d\tau.$$

Notice that if $t > w_2$, then $g(t) = t_2$, where w_2 and t_2 are the second arrival and interarrival times, respectively.

  a. Determine the probability that $x(t) = 0$.
  b. Evaluate the expected value of $g(t)$.
  c. What happens as $t \to \infty$?

2.3.2. Suppose that jobs submitted for processing at a computer facility are classified as long runs or short runs. Suppose, further, that the arrival of long and short runs are independent and that each can be modeled as a homogeneous Poisson process with arrival rates λ_L and λ_S, respectively. Show that the probability of exactly m short runs arriving in a time interval between the arrival of two long runs is

$$\frac{\lambda_L}{\lambda_L + \lambda_S}\left(\frac{\lambda_S}{\lambda_L + \lambda_S}\right)^m$$

for $m = 0, 1, 2, \cdots$.

2.3.3. (*Tests for a Poisson Process*). Consider an interval $[0, T)$ in which n events are observed to occur at the sequence of instants $w_1, \cdots, w_n$. This problem outlines two simple tests for deciding whether or not the events obey a homogeneous Poisson distribution with unknown mean rate λ.

  a. *Waiting Time Test*: Using a central limit theorem argument, justify the assertion that *if* the events were governed by a Poisson law, then the probability is approximately 0.95 that

$$S_n = \sum_{j=1}^{n} w_j$$

lies in the range $\frac{1}{2}[n + 11.96\sqrt{n/3}]$. Discuss how this observation can be used as a possible test of whether or not the events obey a Poisson law.

  b. *Interarrival Time Test*: Let $t_1 = w_1, t_2 = w_2 - w_1, \cdots, t_n = w_n - w_{n-1}$ be the interarrival times for the observed sequence of events. Let N be the number of interarrival times of duration exceeding U. Demonstrate that *if* the events were governed by a Poisson law with parameter λ, then the mean and variance of the random variable N satisfy

$$E(N) = ne^{-\lambda U}$$

and

$$E(N) \pm \sigma_N = E(N)\left(1 \pm \sqrt{\frac{1}{n}(e^{\lambda U} - 1)}\right).$$

Discuss how a sketch of N versus U for the observed events and of the two equations derived from the Poisson assumption can be used to estimate λ and as a possible test of whether or not the events obey a Poisson law.

2.3.4. Let $u_i, i = 1, 2, \cdots$, be a sequence of independent, identically distributed random-variables, each uniformly distributed over $(0,1]$. Let z be the random variable generated by H. Schaffer's procedure [31]:

STEP 1.   set: $i = 0, T = 1, z = -1$
STEP 2. $i \leftarrow i + 1, z \leftarrow z + 1$
STEP 3.   generate u_i
STEP 4. $T \leftarrow T u_i$
STEP 5.   if $T < e^{-\lambda}$ go to STEP 2, else STOP.

Prove that z is Poisson distributed with parameter λ.

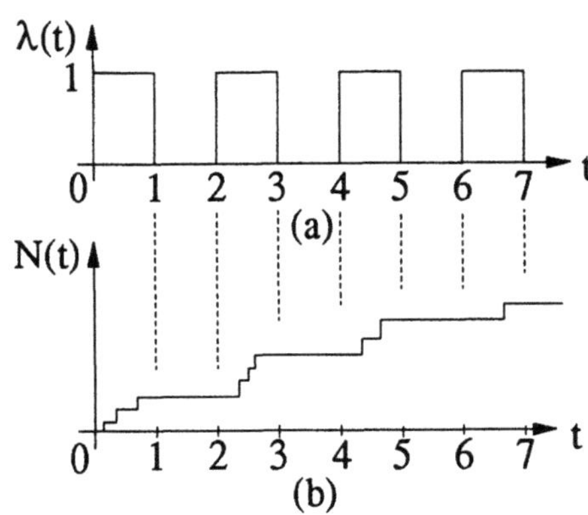

Figure P.2.3.5. Periodic intensity function for an inhomogeneous Poisson process.

2.3.5. By a suitable time scaling, an inhomogeneous Poisson process can be converted into a homogeneous Poisson process, and vice versa. Let $\{N(t): t \geq t_0\}$ be an inhomogeneous Poisson process with intensity $\lambda(t)$ such that the mean-value function

$$\Lambda(t) = \int_0^t \lambda(\sigma) d\sigma$$

is nondecreasing. Let $\Lambda^{-1}(u)$ be the inverse function of $\Lambda(t)$; that is, for $u > 0$, $\Lambda^{-1}(u)$ is the smallest value of t for which $\Lambda(t) \geq u$.

a. Show that the counting process defined by $M(u) = N(\Lambda^{-1}(u))$ is a homogeneous Poisson process for $u \geq 0$ with unit intensity.

b. Suppose that $\lambda(t)$ is the periodic function shown in Fig. P.2.3.5a. Sketch $\Lambda^{-1}(u)$. Also sketch the realization of $M(u)$ corresponding to the realization of $N(t)$ shown in Fig. P.2.3.5b.

c. Conversely, suppose that $M(u)$ for $u \geq 0$ is a homogeneous Poisson process with unit intensity. Show that $N(t) = M(\Lambda(t))$ for $t \geq 0$ is an inhomogeneous Poisson process with intensity $\lambda(t)$.

2.3.6. Let $\{N(t): t \geq 0\}$ be a homogeneous Poisson process. Show that the conditional probability density for the kth occurrence time w_k given that $N(T) = n \geq k$ is given by

$$p_{w_k}(\tau | N(T) = n) \;=\; \frac{n!}{(k-1)!(n-k)!}\left(\frac{\tau}{T}\right)^{k-1}\left[1-\left(\frac{\tau}{T}\right)\right]^{n-k}\frac{1}{T}$$

for $0 \leq \tau < T$. Thus, conditionally, w_k is a beta-distributed random variable given $N(T) = n$.

2.3.7. Let $\{N(t): t \geq t_0\}$ be an inhomogeneous Poisson process with intensity function $\lambda(t)$. Define the random variable y by

$$y \stackrel{\Delta}{=} \int_{t_0}^{T} v(\sigma)N(d\sigma) = \begin{cases} 0, & N(T) = 0 \\ \sum_{i=1}^{N(T)} v(w_i), & N(T) \geq 1. \end{cases}$$

Show that the mean and variance of y are given by

$$\int_{t_0}^{T} v(\sigma)\lambda(\sigma)d\sigma \quad \text{and} \quad \int_{t_0}^{T} v^2(\sigma)\lambda(\sigma)d\sigma,$$

respectively. Hint: use the conditional properties of the occurrence times given that $N(T) = n$.

2.3.8. The superposition of independent Poisson processes is a Poisson process. Let $\{N_i(t): t \geq t_0, i = 1, 2, \cdots, M\}$ be M independent Poisson counting processes with intensity functions $\lambda_i(t)$. Show that

$$z(t) = \sum_{i=1}^{M} N_i(t)$$

is a Poisson counting process and determine its intensity function.

2.3.9. The difference of two Poisson processes obviously is not a Poisson process because negative integers can occur. Let $\{N_1(t); t \geq t_0\}$ and $\{N_2(t); t \geq t_0\}$ be independent, homogeneous Poisson counting processes with constant intensities λ_1 and λ_2, respectively. Let $\{z(t): t \geq t_0\}$ be their difference $z(t) = N_1(t) - N_2(t)$.

    a. Show that z has stationary, independent increments.
    b. Evaluate $\Pr[z(t) - z(s) = k]$ for $0 \leq s \leq t$ and for $k = 0, \pm 1, \pm 2, \cdots$.
    c. Evaluate the limit $\lim_{t \to \infty} \Pr(|z(t)| < c)$ for $c > 0$.

2.3.10. (*Khinchin Decomposition*). Let $\{N(t): t \geq t_0\}$ be a Poisson counting process with parameter $\Lambda(t)$. Here, the function Λ may not possess a derivative, so N may not have an intensity function.

    a. Show that the characteristic functional for N on the interval $[t_0, T)$ is given by

$$\phi_N(jv) \overset{\Delta}{=} E\left[\exp\left(j \int_{t_0}^{T} v(\sigma) N(d\sigma) \right) \right] = \exp\left[\int_{t_0}^{T} (e^{jv(\sigma)} - 1) d\Lambda(\sigma) \right].$$

    b. The function $\Lambda(t)$ can be expressed as the sum $\Lambda(t) = \Lambda^{(c)}(t) + \Lambda^{(d)}(t)$, where $\Lambda^{(c)}(t)$ is a continuous function of t, and $\Lambda^{(d)}(t)$ is a piecewise-constant function having only jump discontinuities. Use this decomposition and the characteristic functional of part (a) to conclude that N can be decomposed into the sum of two independent Poisson processes as $N(t) = N^{(c)}(t) + N^{(d)}(t)$, where the parameter functions for $N^{(c)}(t)$ and $N^{(d)}(t)$ are $\Lambda^{(c)}(t)$ and $\Lambda^{(d)}(t)$, respectively.

2.3.11. (*Constructive Derivation of the Poisson Process*). Suppose that a point process on $[t_0, T)$ is constructed in the following manner. First, select a nonnegative integer N according to the Poisson distribution with parameter

$$\int_{t_0}^{T} \lambda(\sigma) d\sigma.$$

Then select N numbers $u_1, \cdots, u_N$ independently over the interval $[t_0, T)$, each with probability density

$$p_{u_i}(U) = \frac{\lambda(U)}{\displaystyle\int_{t_0}^{T} \lambda(\sigma) d\sigma}.$$

Assign these numbers as coordinates for N points in $[t_0, T)$. Show that the resulting point process is an inhomogeneous Poisson process on $[t_0, T)$ with intensity function $\lambda(t)$.

2.3.12. (*Constructive Derivation of the Poisson Process*). The converse of Theorem 2.3.1 is established in this problem. Suppose that $\{t_i\}$ is a sequence of independent, identically distributed random variables with the common distribution being exponential with parameter λ. Construct a point process by assigning these variables successively as interarrival times starting from $t = 0$. The conclusion to be drawn from this problem is that the point process so constructed is a homogeneous Poisson process with intensity λ.

a. Let $w_k = t_1 + \cdots + t_k$ be the k*th* occurrence time of the constructed point process for $k = 1, 2, \cdots$. Show that the joint probability density for w_n and w_{n+1} is given by

$$p_{w_i, w_{i+1}}(X, Y) = \frac{1}{(n-1)!} \lambda^{n+1} X^{n-1} e^{-\lambda Y},$$

for $0 \leq X \leq Y$.

b. By using the result of part (a) and the identity of the events $\{N(t) = n\}$ and $\{w_n < t, w_{n+1} \geq t\}$, show that $N(t)$ is Poisson distributed with parameter λt.

c. Show that

$$p_{w_1, w_2, \ldots, w_n}(W_1, W_2, \ldots, W_n | N(t) = n) = \frac{n!}{t^n},$$

for $0 \le W_1 \le \cdots \le W_n \le t$. Conclude, therefore, that the occurrence times for the constructed point process have the same distribution given $N(t) = n$ as the order statistics on n independent random variables uniformly distributed on $[0, t]$.

d. Combine the results of Problem 2.3.11 and parts (b) and (c) to conclude that the constructed point process is a homogeneous Poisson process with intensity λ.

2.3.13. Let

$$\tilde{N}(t) = N(t) - \int_{t_0}^{t} \lambda(\sigma) d\sigma,$$

where N is an inhomogeneous Poisson process with intensity function $\lambda(t)$. This process is a centered (i.e., zero mean) version of N.

a. Show that the characteristic functional is given by

$$\phi_{\tilde{N}}(jv) = \exp\left[\int_{t_0}^{T} \lambda(\sigma)(e^{jv(\sigma)} - 1 - jv(\sigma)) d\sigma\right].$$

b. Let $v(\sigma) = \varepsilon_1 v_1(\sigma) + \varepsilon_2 v_2(\sigma)$, where ε_1 and ε_2 are constants. Demonstrate for this choice of v that

$$\frac{1}{j^2} \frac{\partial^2 \phi_{\tilde{N}}(jv)}{\partial \varepsilon_1 \partial \varepsilon_2}\bigg|_{\varepsilon_1 = \varepsilon_2 = 0} = E\left\{\int_{t_0}^{T} v_1(\sigma)\tilde{N}(d\sigma) \int_{t_0}^{T} v_2(\sigma)\tilde{N}(d\sigma)\right\}$$

$$= \int_{t_0}^{T} \lambda(\sigma)v_1(\sigma)v_2(\sigma) d\sigma,$$

where the first equality can be deduced from the definition of the characteristic functional and the second from the evaluation of the characteristic functional in part (a).

c. Now for $i = 1$ and 2, let $v_i(\sigma) = 1$ for $\sigma \in [t_0, t)$ and $v_i(\sigma) = 0$ otherwise. Use the result of part (b) to conclude that

$$E[\tilde{N}(t_1)\tilde{N}(t_2)] = \text{cov}(N(t_1), N(t_2))$$

$$= \min\left(\int_{t_0}^{t_1} \lambda(\sigma) d\sigma, \int_{t_0}^{t_2} \lambda(\sigma) d\sigma\right).$$

d. By taking

$$v(\sigma) = \sum_{i=1}^{k} \varepsilon_i v_i(\sigma),$$

use this procedure to show that

$$E[\tilde{N}(t_1)\tilde{N}(t_2)\tilde{N}(t_3)] = \min\left(\int_{t_0}^{t_1}\lambda(\sigma)d\sigma, \int_{t_0}^{t_2}\lambda(\sigma)d\sigma, \int_{t_0}^{t_3}\lambda(\sigma)d\sigma\right)$$

and

$$E[\tilde{N}(t_1)\tilde{N}(t_2)\tilde{N}(t_3)\tilde{N}(t_4)]$$

$$= \min\left(\int_{t_0}^{t_1}\lambda(\sigma)d\sigma, \int_{t_0}^{t_2}\lambda(\sigma)d\sigma, \int_{t_0}^{t_3}\lambda(\sigma)d\sigma, \int_{t_0}^{t_4}\lambda(\sigma)d\sigma\right).$$

Thus, conclude that the covariance function for a homogeneous Poisson counting process is identical to that of a Wiener process, but that this is not so for higher-order covariances.

2.3.14. Let $p(\omega)$ be the sample-function density for an inhomogeneous Poisson-process on $[t_0, T)$ with intensity $\lambda(t)$. Demonstrate that the k*th* moment $[p^k(\omega)]$ of the sample-function density is given by

$$E[p^k(\omega)] = \exp\left[-(k+1)\int_{t_0}^{T}\lambda(t)dt + \int_{t_0}^{T}\lambda^{k+1}(t)dt\right].$$

2.3.15. Let u be a random variable that is uniformly distributed on $[0, 1]$, and define w as the solution to

$$u = \exp\left[-\int_{W_{n-1}}^{w}\lambda(\sigma)d\sigma\right],$$

where $\lambda(\cdot)$ is the intensity function of an inhomogeneous Poisson process; w can be determined numerically, for example with Newton's method, when given u, W_{n-1}, and $\lambda(\cdot)$.

a. Determine $p_{w|w_{n-1}}(W \mid W_{n-1})$, the probability density of w given $w_{n-1} = W_{n-1}$.

b.   Discuss how the result in (a) can be used to simulate an inhomogeneous Poisson-process. Hint: consider (2.22).

2.4.1. Reproduced in the table below are data obtained in the historic experiment of E. Rutherford and H. Geiger [29] on the emission of α particles from a radioactive source.

k	0	1	2	3	4	5	6	7	8	9	10	11	total
n_k	57	203	383	525	532	408	273	139	49	27	10	6	2612

In the table, k is the number of α-particles they observed in a unit time interval of duration $T = 1/8$ minute, and n_k is the number of such intervals in which k α-particles were observed.

a.   Assume that α-particles are emitted as a homogeneous Poisson process with an intensity λ emission per minute. Determine the maximum-likelihood estimate of λ given the data observed by Rutherford and Geiger.

b. As a qualitative test of the assumption of Poisson distributed emissions, add two rows to the table. In one row, evaluate and enter the quantity

$$\frac{1}{k!}(\hat{\lambda}_{\mathrm{ML}}T)^k e^{-\hat{\lambda}_{\mathrm{ML}}T}$$

for $k = 0, 1, \cdots$. In the second row, enter the relative frequency of k α-particles, for $k = 0, 1, \cdots$,

$$\frac{n_k}{\sum\limits_{k=0}^{11} n_k}.$$

c.   Write a computer program to simulate Rutherford and Geiger's experiment. Use it to create a table of (k, n_k) values, and repeat parts (a) and (b). Discuss your results.

2.4.2. Consider the scaled inhomogeneous Poisson process of Example 2.4.5, but assume now that count-record data $\{N(\sigma): t_0 \leq \sigma < T\}$ are observed rather than histogram data.

a. Determine the maximum-likelihood estimate for X.

b. Demonstrate that the maximum-likelihood estimate is efficient, and evaluate its mean square-error.

2.4.3. Consider the model of Example 2.4.7, which is used in auditory electrophysiology.

a. Evaluate the mean and variance for the random variables A_c and A_s defined in (2.54). Hint: use the conditional properties of the occurrence times given $N(T)$ that are developed in Section 2.3.

b. Suppose that $\mathbf{X}^*$ is an unbiased estimate of $\mathbf{X}$. Show that $E[(\mathbf{X}^* - \mathbf{X})(\mathbf{X}^* - \mathbf{X})':\mathbf{X}] \geq \mathbf{F} - 1(\mathbf{X})$, where

$$
\mathbf{F}(\mathbf{X}) = \begin{bmatrix} X_1^{-1}I_0(X_2) & I_1(X_2) & 0 \\[2mm] I_1(X_2) & \frac{1}{2}X_1(I_0(X_2)+I_2(X_2)) & 0 \\[2mm] 0 & 0 & \frac{1}{2}X_1X_2^2(I_0(X_2)-I_2(X_2)) \end{bmatrix},
$$

where $I_n(X_2)$ is a modified Bessel function of the first kind of order n.

2.4.4. Suppose that $\mathbf{X}^*$ is an unbiased estimate of $\mathbf{X}$ in terms of either histogram or count-record data. The mean square-error matrix then satisfies $\Sigma(\mathbf{X}) \geq \mathbf{F}^{-1}(\mathbf{X})$. Suppose, now, that the experiment of collecting data is repeated independently M times. Show that the mean square-error matrix for an unbiased estimate $\mathbf{X}_M^*$ of $\mathbf{X}$ in terms of the pooled data from the M experiments is lower bounded by $(1/M)\mathbf{F}^{-1}(\mathbf{X})$.

2.4.5. Suppose that $\mathbf{X}^*$ is an unbiased estimate of $\mathbf{X}$ in terms of either histogram or count-record data. The mean square-error matrix then satisfies $\Sigma(\mathbf{X}) \geq \mathbf{F}^{-1}(\mathbf{X})$. Let $\sigma_{ii}(\mathbf{X})$ denote the ith diagonal element of $\Sigma(\mathbf{X})$; this is the mean square-error in estimating the ith element of $\mathbf{X}$. Show that $\sigma_{ii}(\mathbf{X}) \geq [\mathbf{F}^{-1}(\mathbf{X})]_{ii} \geq f_{ii}^{-1}(\mathbf{X})$, where $f_{ii}(\mathbf{X})$ is the ith diagonal element of $\mathbf{F}(\mathbf{X})$. This second inequality is often not very tight, but it can be useful when the inverse of $\mathbf{F}(\mathbf{X})$ is difficult to determine.

2.4.6. Let $\{N(t): t > -\infty\}$ be an inhomogeneous Poisson process with intensity $\lambda(t,X) = \mu(t - X)$, where $\mu(t)$ is the Gaussian pulse

$$\mu(t) = \frac{\alpha}{\sqrt{2\pi\tau^2}} \exp\left[-\frac{t^2}{2\tau^2}\right], \qquad -\infty < t < \infty.$$

Suppose that the number N and occurrence times $\{w_1, \cdots, w_N\}$ for all points in $(-\infty, \infty)$ are observed.

a. Show that the maximum-likelihood estimate of X is given by

$$\hat{X}_{ML} = \begin{cases} 0, & N = 0 \\ \dfrac{1}{N}\sum_{i=1}^{N} w_i, & N \geq 1. \end{cases}$$

Thus, conclude that the maximum-likelihood estimate of the pulse delay X is the arithmetic average of the observed data.

b. Determine the mean and variance of the maximum-likelihood estimate.

c. Is the maximum-likelihood estimate unbiased? Is it efficient?

d. Evaluate $F(X)$, the Fisher information matrix.

e. Use the results of Problem 2.3.11 in writing a computer program to simulate this inhomogeneous Poisson-process for $X = 1$ various values of α and τ. Form the maximum-likelihood estimate of X. By averaging results of repeated trials, estimate the bias and mean square-error of the estimate of X. Discuss your results.

2.4.7. At time $t = 0$, a quantity of a radioactive isotope is placed in the field of view of an ideal detector, and the number N and occurrence times $\{w_1, \cdots, w_N\}$ of detected photons are measured for all $t \geq 0$. Assume that detected photons form an inhomogeneous Poisson process on $[0, \infty)$ with a parameterized intensity $\lambda(t, X) = X_1 \exp(-t/X_2)$, where X_1 depends on the quantity of the isotope and the detector sensitivity, and X_2 is a physical constant characteristic of the isotope.

a. Determine expressions for the maximum-likelihood estimates of X_1 and X_2.

b. Determine lower bounds on the mean square-errors in estimating X_1 and X_2 using unbiased estimates.

R. Peirls [25] was evidently first to develop the expression of part (a). His interesting paper also contains isotope on the effect of a finite observation interval $[0, T)$, on the selection of T in repeated experiments, and on certain nonideal effects found in real detectors.

2.4.8. (*Barankin Bound for Histogram Data*). The Cramér-Rao inequality of Theorem 2.4.1 provides a lower bound on the mean square-error. An alternative inequality developed by E. Barankin [4] provides the *greatest* lower bound on the mean square-error. In this problem, we develop the Barankin bound in estimating a single parameter using an unbiased estimate and histogram data. The method can be extended to count-record data and multiple parameters.

Let $\{N(t): t \geq t_0\}$ be an inhomogeneous Poisson process with intensity $\lambda(t, X)$. Observations taken to estimate X are in the form of counts $N(t_{i-1}, t_i)$ occurring in k subintervals $[t_{i-1}, t_i)$ of an observation interval $[t_0, T)$, where $t_k = T$. Denote the counting probability in (2.46) by $P(n_1, \cdots, n_k \mid X)$. Let $X^* = X^*(n_1, \cdots, n_k)$ be an unbiased estimate of X in terms of the histogram data. Then

$$\sum_{n_1, n_2, \cdots, n_k} X^* P(n_1, n_2, \cdots, n_k \mid X) = X$$

for all permissible values of X. Now, choose M permissible values $X_1, \cdots, X_M$ for X. Also, choose M numbers $\alpha_1, \cdots, \alpha_M$. It follows that

$$\sum_{n_1, n_2, \cdots, n_k} X^* \left[\sum_{m=1}^{M} \alpha_m P(n_1, n_2, \cdots, n_k \mid X_m) \right] = \sum_{m=1}^{M} \alpha_m X_m.$$

a. Let X_a denote the actual value of X. By subtracting

$$\sum_{m=1}^{M} \alpha_m X_a$$

from both sides of the last expression, conclude that

$$\sum_{n_1, n_2, \cdots, n_k} (X^* - X_a) \left[\frac{\sum_{m=1}^{M} \alpha_m P(n_1, n_2, \cdots, n_k \mid X_m)}{P^{1/2}(n_1, n_2, \cdots, n_k \mid X_a)} \right] P^{1/2}(n_1, n_2, \cdots, n_k \mid X_a)$$

$$= \sum_{m=1}^{M} \alpha_m (X_m - X_a).$$

b. Using the result of part (a) and the Schwarz inequality, show that

$$E[(X^* - X_a)^2 \mid X_a] \geq \frac{\left(\displaystyle\sum_{m=1}^{M} \alpha_m (X_m - X_a) \right)^2}{\displaystyle\sum_{l=1}^{M} \sum_{m=1}^{M} \alpha_l G(X_l, X_m \mid X_a) \alpha_m},$$

where

$$G(X_l, X_m \mid X_a) \overset{\Delta}{=} \sum_{n_1, n_2, \cdots, n_k} \frac{P(n_1, n_2, \cdots, n_k \mid X_l) P(n_1, n_2, \cdots, n_k \mid X_m)}{P(n_1, n_2, \cdots, n_k \mid X_a)}.$$

This inequality holds for any choice of $X_1, \cdots, X_M$ and $\alpha_1, \cdots, \alpha_M$. Consequently, the tightest bound is obtained by selecting these $2M$ parameters to maximize the right side. The result is the Barankin bound, which has been shown by Barankin to be the greatest lower bound.

c. Show that

$$\ln G(X_l, X_m \mid X_a) = - \int_{t_0}^{T} [\lambda(\sigma, X_l) + \lambda(\sigma, X_m) - \lambda(\sigma, X_a)] d\sigma$$

$$+ \sum_{i=1}^{k} \left[\int_{t_{i-1}}^{t_i} \lambda(\sigma, X_l) d\sigma \right] \left[\int_{t_{i-1}}^{t_i} \lambda(\sigma, X_m) d\sigma \right] \left[\int_{t_{i-1}}^{t_i} \lambda(\sigma, X_a) d\sigma \right]^{-1}$$

d. The Cramér-Rao bound for an unbiased estimate can be obtained as a special case of the Barankin bound by selecting $M = 2, X_1 = X_a$, $X_2 = X_a + \delta$, $\alpha_1 = -\delta^{-1}$, and $\alpha_2 = \delta^{-1}$. Obtain the Cramér-Rao bound in this way by taking the limit as δ approaches zero.

2.5.1. A homogeneous Poisson counting process in a three-dimensional Euclidean space is such that the number of points in a region A is Poisson distributed with parameter $\lambda v(A)$, where $v(A)$ is the volume of A, and $\lambda > 0$,

$$\Pr[N(A) = n] = \frac{1}{n!} [\lambda v(A)]^n e^{-\lambda v(A)}.$$

Furthermore, the numbers of points in disjoint volumes are independent. The notion of interarrival times in a one-dimensional space is here replaced by the notion of *nearest-neighbor distance*. Determine the distribution function and the mean of the nearest-neighbor distance.

2.5.2. Let $\{N(x): x \in X\}$ be a multidimensional Poisson-process with intensity function $\lambda(x)$.

a. Derive the expression in (2.79) for the joint probability-density of the locations of the points given that $N(X) = n$.

b. Discuss how this result can be used for a Monte Carlo simulation of a multidimensional Poisson-process.

2.5.3. Let $\{N(x): x \in X\}$ be a multidimensional Poisson-process with intensity function $\lambda(x)$ on the space X.

a.   Derive the expression in (2.80) for the characteristic functional.

b.   By choosing $v(\sigma)$ appropriately, determine the joint characteristic function for $N(A_k)$, where $A_k \subseteq X$ for $k = 1, 2, \cdots, K$.

2.5.4. Suppose that X is a square region in the plane. This region is subdivided into M^2 square pixels to form an $M \times M$ image array. Let $\{N(x): x \in X\}$ be a two-dimensional Poisson-process with intensity function $\lambda(x)$ that is constant over each pixel.

a. Determine the loglikelihood function. Show that it depends only on the number of points in each pixel and not their locations within pixels.

b. Determine the maximum-likelihood estimate of the intensity function in terms of an observed realization of the two-dimensional Poisson-process.

c. Is this estimate biased? Is it efficient?

2.5.5. Let X be the square region $\{0 \leq x_1 \leq 1, 0 \leq x_2 \leq 1\}$ of the plane $\mathcal{R}^2$. Let A be an arbitrary subset of X. The area of A can be estimated as follows.   Let $\{N(x): x \in X\}$ be a two-dimensional, homogeneous Poisson-process with constant intensity-function λ.

a. Determine the loglikelihood function, and express it as a sum of two terms, one of which depends only on those points that fall in A and the other only on those points outside A.

b. Use the result of part (a) to obtain an expression for the maximum-likelihood estimate of the area of A.

c. Is the estimate of part (b) biased? Is it efficient?

d. How large should λ be so that the standard deviation of the estimate is within p % of the true area?

e. Write a computer program to simulate $\{N(x): x \in X\}$. Your program should have as an output the maximum-likelihood estimate of the area of A.

f. Suppose that the region A is that portion of X that lies under the exponential

$$x_2 = e^{-x_1}.$$

Use your computer program to estimate the area under this exponential to within 0.1% of its true value. Discuss your results.

2.5.6. Let $\{N(A): A \subseteq X\}$ be a multidimensional Poisson-process with intensity $\lambda(x)$. Suppose that points of this process are deleted independently with a spatially dependent probability $d(x)$ and that the remaining points are superimposed with those of another, independent Poisson process on X with intensity $\lambda_0(x)$. Show that the resulting point process is a Poisson process with intensity $[1 - d(x)]\lambda(x) + \lambda_0(x)$.

CHAPTER THREE

TRANSLATED POISSON-PROCESSES

3.1 Introduction

The point processes we consider in this chapter are useful as models for measured data acquired about an underlying, unobservable point-process when the measurements are imperfect and in the form of a point process. Such a measurement is illustrated in Fig. 3.1. Points of the underlying process, called the *input point-process*, occur on a space X.

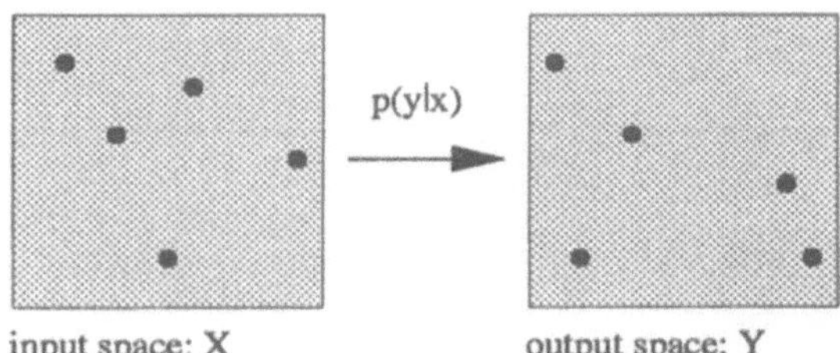

Figure 3.1 Translation of points on space X to points on space Y, with transition density $p(y \mid x)$.

Measurements dependent on the points of this process result in points on the space Y and form the *output point-process*. We think of the points on the input space X as being translated by the measurements to the output space Y. In some instances, X and Y may be the same space, and input points are moved about within the space to form output points. More generally, the two spaces need not be the same or even of the same dimension, as will be seen in the examples that follow.

An input point at location x is randomly translated to location y in the output space according to a conditional probability-density $p(y \mid x)$, called the *transition density*, as indicated in Fig. 3.1. This density may be usefully regarded as a *point-spread function* in the sense that $p(y \mid x)$ would be the blurred image formed by repeating a large number of experiments in which input points occur only at location x and output points are superimposed to form a composite image. In general, a collection of input points would be translated to become a collection of output points according to some joint transition density. However, in what follows, we shall only consider translations in which each input point is translated independently of others, in which case the joint density is the product of the transition densities for each translated point. Translations without this independence assumption are also of interest and are studied in Sec. 7.5.

Let $N(X)$ and $M(\mathcal{Y})$ be the numbers of points occurring in the input and output spaces, respectively. These numbers are equal for the translations in Fig. 3.1. However, they are unequal for many applications where a translated point-process model is useful. Some input points are missed in the measurements and do not appear as translated output points, and some extraneous or noise points that are not related to input points appear in the measurements. The model for a translated point-process developed in Sec. 3.2 includes independent deletions and insertions.

The following are four examples from quantum-limited imaging in which translated point-processes provide a useful model.

Example 3.1.1 *Electron Microscopic-Autoradiography* ─────────────

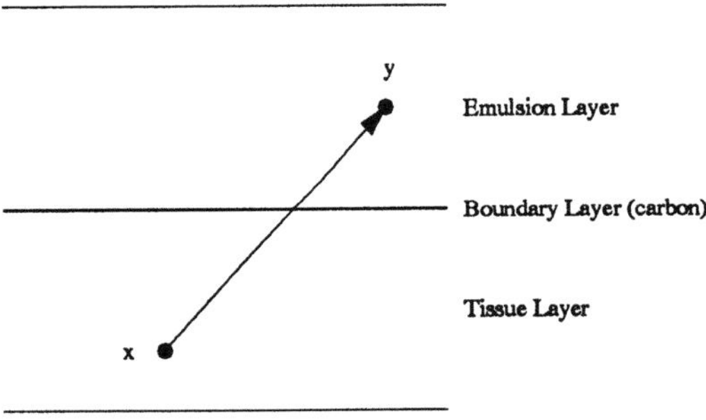

Figure 3.2   A section through an autoradiographic preparation showing a beta particle emitted from the radioactive source at x and traveling to y where it interacts with a silver halide.

Electron microscopic autoradiography is a technique for the quantitation of radioactive-tracer concentrations in subcellular structures [17, 18]. Ultrathin sections of tissue containing the tracer are coated with a thin layer of a photographic emulsion. Photons created in radioactive decays interact with the emulsion to yield latent silver grains that are subsequently developed photographically. An electron micrograph made following development yields an image of the grains superimposed on the various subcellular organelle structures containing radioactivity; see Fig. 1.5. As illustrated in Fig. 3.2, a silver grain may form at a site that is distant from the site of the radioactive decay that produced it. The decaying atom is at location x, and the grain is formed at location y. A radioactive decay in one organelle may result in a grain overlying some other organelle because the distances between x and y may be larger than the sizes of some subcellular structures. Thus, in contrast to the simple situation in Ex. 2.5.1, the radioactivity in a subcellular organelle cannot be estimated accurately by merely counting the grains on that organelle and dividing by the size of the organelle. The point-spread function $p(y \mid x)$ is of the form

$$p(y \mid x) = \frac{D}{2\pi(|y - x|^2 + D^2)^{3/2}}, \tag{3.1}$$

where D is the distance from the tissue midplane to the emulsion midplane [17, 18]. The radioactive emissions are well modeled as a multidimensional Poisson process having an intensity that is a constant within each subcellular organelle type but varying from one type to another. ∎

Example 3.1.2 *Positron-Emission Tomography* —————————————

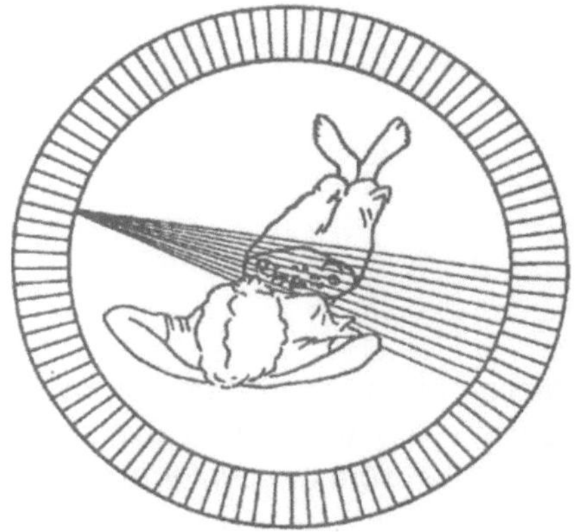

Figure 3.3 A positron-emission tomograph consisting of a ring of scintillation detectors in a plane transverse to the body. Each detector can sense annihilation photon-pairs with several opposing detectors. (From [26, ©1981 IEEE])

A positron-emission tomograph is an instrument used for the quantification of radioactive-tracer concentrations within the body [26]. A quantity of a compound labeled with a positron-emitting radionuclide is introduced into the body. The particular compound is selected because it is preferentially absorbed into the organ, tissue, or biochemical process of interest. Positrons are produced at the sites of radioactive decay within the body. Shortly after a positron is created, it annihilates with an electron, and two high energy photons are produced that propagate at the speed of light c in nearly opposite directions along a line. Annihilation photons are measured with the tomograph, which consists of scintillation detectors arranged in one or more planar rings that surround the body, as shown in Fig. 3.3. The plane of a ring defines a thin section through the body and the internal organs of interest. Two opposing detectors sensing photons

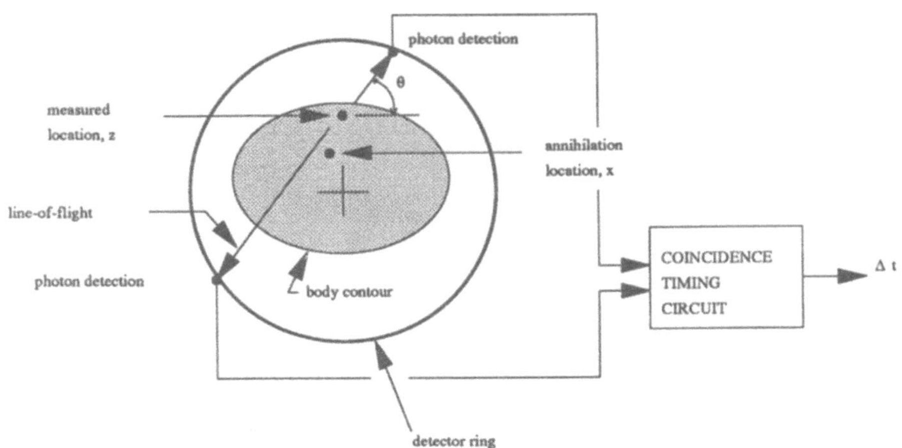

Figure 3.4 Detection of an annihilation that occurred at x. The measurement is $y = (z, \theta)$, where z is along the line-of-flight at the position $\Delta t/2c$ from the center of the flight line.

within a small time interval define a pencil-like cylindrical volume containing the line-of-flight along which an annihilation occurred. Each detector can be in coincidence with several opposing detectors permitting many flight lines occurring at random orientations to be measured. These measurements are subsequently processed to form an image of the concentration of the tracer, such as the image of oxygen-15 labeled water in Fig. 1.4. In addition to the line-of-flight of annihilation photons, the differential propagation-time is also measured using coincidence-timing circuitry, as shown in Fig. 3.4. If the line-of-flight and differential time-of-flight were known precisely, then the location of the annihilation would also be known precisely. However, the finite size of the detectors used and the finite resolving time of the coincidence-timing circuitry limits the precision of the measurement. An annihilation occurring at location x within the emission-space X results in a measurement at location $y = (z, \theta)$ in the measurement space $\mathcal{Y}$, where X is a two-dimensional planar region defined by the intersection of the plane of the detector ring with the body, and $\mathcal{Y}$ is a three-dimensional space in which a point defines a line-of-flight and a differential time-of-flight. From the physics describing radioactive decay, annihilations occur as a Poisson process with an intensity $\lambda(x)$ that is proportional to the concentration of the radiotracer [5]. The transition density $p(y \mid x)$ relating a measurement location to an annihilation location models the precision of the tomograph in locating annihilations. In practice, this density is well approximated by a two-dimensional Gaussian density of the form:

$$p(y \mid x) = p(z, \theta \mid x)$$

$$= p(z \mid \theta, x)p(\theta \mid x), \tag{3.2}$$

where

$$p(\theta \mid x) = \frac{1}{\pi}, \qquad 0 \le \theta \le \pi,$$

and

$$p(z \mid \theta, x) = \frac{1}{2\pi\sqrt{\det[R(\theta)]}} e^{-\frac{1}{2}(z-x)'R^{-1}(\theta)(z-x)}, \tag{3.3}$$

with the covariance matrix $R(\theta)$ given by

$$R(\theta) = T(\theta)\begin{bmatrix} \sigma_e^2 & 0 \\ 0 & \sigma_b^2 \end{bmatrix}T'(\theta),$$

where the prime denotes the matrix transpose operation, σ_e^2 and σ_b^2 are the error variances along and transverse to the line-of-flight, respectively, and $T(\theta)$ is the rotation matrix

$$T(\theta) = \begin{bmatrix} \cos\theta & \sin\theta \\ -\sin\theta & \cos\theta \end{bmatrix}.$$

To give some indication of practical values, the variance along the line of flight corresponds to a full-width-at-half-maximum (FWHM) of about 7.5 cm, and the variance in the transverse direction to a FWHM of about 1.15 cm in the Super PETT-I instrument described by Ter-Pogossian, *et al.* [31][1]. Random deletions and insertions can occur. Annihilation events can be deleted in the measurements if one or both of the photons fail to be detected [27], which can be due either to Compton scattering of photons in the tissues of the body through which they propagate or to the propagation of the photons outside the plane of the detec-

---

[1] For a Gaussian density with a variance of σ^2, FWHM $= 2\sigma\sqrt{2\ln2} \approx 2.355\sigma$.

tors; see Problem 3.1.1. Extraneous events in the measurements can occur when a pair of photons originating in two separate annihilations, located in or out of the plane of the detector array, are sensed within a small time interval and are counted as photons from a single annihilation [11]. These effects are large in practice and must be recognized to produce acceptable estimates of activity distributions; the deletion of 30% to 50% of input points, and upwards of 30% of output points being insertions are not uncommon.

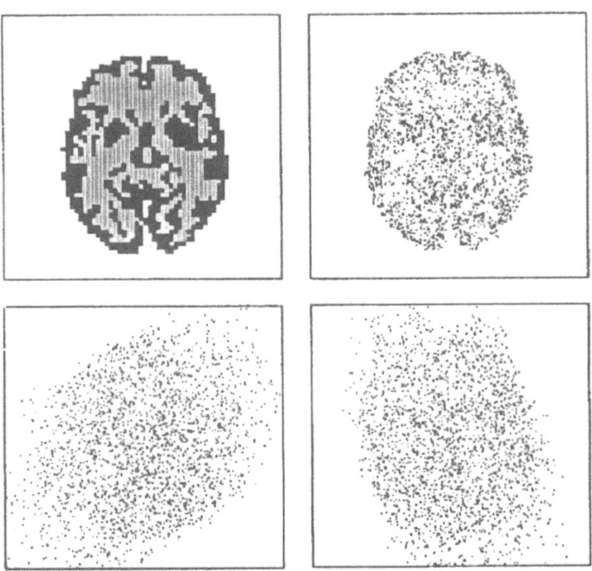

Figure 3.5 Computer simulation of data collected in positron-emission tomography.

Shown in Fig. 3.5 is a computer simulation corresponding to the distribution of radioactivity in a human brain. The upper left panel is a digitized Hoffman brain-phantom. A Hoffman brain-phantom is a plastic model with a known geometry having voids that can be filled with a radioactive liquid for performance studies on emission tomographs.[1] The black region of the phantom corresponds to gray matter and, for the simulation, contains 100% relative activity; the lighter region corresponds to white matter and contains 25% relative activity; and the white region external to the brain contains no radioactivity. Points in the upper right panel of the figure are at the locations of annihilations; these points were produced by simulating a spatial Poisson process with the Hoffman phantom as its intensity. Points in the lower two panels were obtained by translating annihilation points randomly and independently by random

---

[1] The Hoffman brain-phantom is made by the Data Spectrum Company.

variables with density (3.3) for the parameters of Super PETT-I given above. The lower left (resp. right) panel corresponds to measured data having lines-of-flight oriented at 45° (resp. 110°); in the instrument, lines-of-flight may be quantized to one of ninety-six angles, in which case the data form ninety-six images similar to the lower panels, with each containing the translations of about 1/96 of the total number of annihilations. The problem of estimating the distribution of radioactivity in the upper left panel from data in the form of the lower panels motivates the estimation methods we describe in this chapter and the next. An example of an actual image produced using these methods is in Fig. 1.4. ∎

Example 3.1.3 *Single-Photon-Emission Tomography* ————————

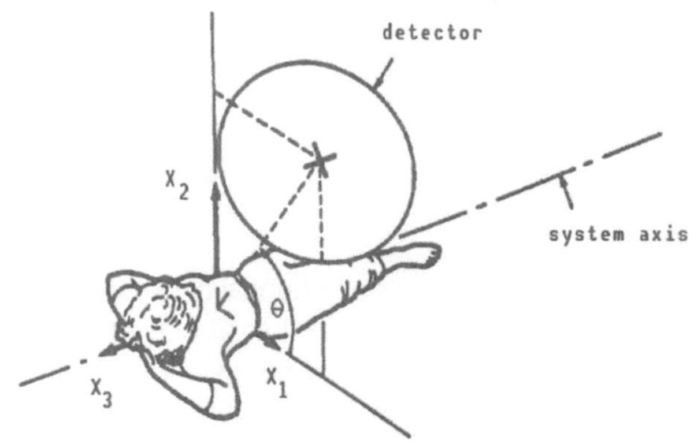

Figure 3.6 A single-photon tomograph consisting of a large scintillation detector, or gamma-ray camera, that rotates around the body. (From [19, ©1985 IEEE])

Another form of instrumentation for imaging radioactive-tracer concentrations in the body is shown in Fig. 3.6 [15, 19, 1]. This is used for tracers that produce a single photon at the site of a radioactive decay, so the instrument is called a *single-photon emission tomograph*. Photons are sensed with a large, circular scintillation-detector, which is rotated about the body to collect photons from many viewing positions. The space X in Fig. 3.1 is the three-dimensional volume scanned by the detector; a point at x in this space is at the site of a radioactive decay. The space Y is defined by the detector and its angle of rotation; a point $y = (u, \theta)$ indicates the location u where a photon strikes the detector when the detector is at angle θ. A number of factors influence the transition density $p(y \mid x)$ [19, 1]. One is the response of the detector to a point source of radioactivity;

this point response varies with the distance of the source from the surface of the detector, broadening as the distance increases. Another factor is the effect of attenuation of photons as they propagate through various tissues and bone on their flights from the decay sites towards the detector. These two effects are so pronounced that they must be taken into account for accurate images of radioactivity distributions to be produced. ∎

Example 3.1.4 *Low Light-Level Astronomy* ────────────────

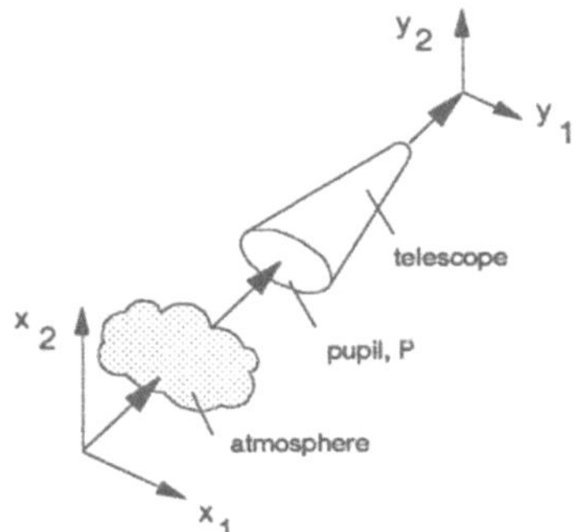

Figure 3.7 Imaging a faint astronomical object.

Short-exposure photographs of a faint astronomical-object are used to form an image of the object in the presence of atmospheric turbulence [10, 21, 3]. As many as 3×10^4 to 10^5 photographs may be used, with each taken during a period that is short compared to the time scale of variations in the propagation path of the light through the atmosphere. The geometry is shown in Fig. 3.7. Light propagates from the distant object through the atmosphere and is then collected with a telescope and focused onto photographic film or an electronic photodetector. The space X in Fig. 3.1 may be viewed as the object space with points in this space being light quanta emitted by the object. The space $\mathcal{Y}$ is the detector space with points in this space corresponding to detected light quanta in the form of silver grains in a film or photoconversion sites in a photodetector. If the atmosphere is homogeneous or quiescent, then the transition density $p(y\,|\,x)$ of Fig. 3.1 is the normalized point-spread function of the telescope, which would be an Airy pattern for a diffraction-limited circular pupil. The effects of turbulence begin to be exhibited as temperature variations along the propagation path increase. The object's light falling on the detector dances about in a random walk as these temperature inhomogenieties change with time; this is examined further in Ch. 7. For stronger turbulence, the transition density in Fig. 3.1 becomes quite complicated as the atmosphere exhibits randomly time-varying amplitude and phase distortions. ∎

In each of the above examples, the physics governing the generation of input points indicates that a Poisson process model is an appropriate description of the input point-process. Points are translated independently from the input to the output space, with a transition density that is determined by the characteristics of the instrumentation used to observe input points. In the next section, we shall investigate the statistical properties of the output point-process under these conditions.

3.2 Statistics of Translated Poisson-Processes

The following theorem establishes the statistical properties of the output point-process when the input point-process is a Poisson process and points are translated independently from the input space to the output space.

Theorem 3.2.1 (*A translated Poisson process is a Poisson process.*) Let $\{N(A): A \subseteq X\}$ be a Poisson process with an integrable intensity function $\{\lambda(x): x \in X\}$. Points of this input point-process are translated to the output space $\mathcal{Y}$ to form the output point-process $\{M(B): B \subseteq \mathcal{Y}\}$, where each point is independently translated according to the transition density $p(y \mid x)$. Then:

a. If there are no insertions and deletions, $\{M(B): B \subseteq \mathcal{Y}\}$ is a Poisson process with intensity

$$\mu(y) = \int_X p(y \mid x)\lambda(x)\,dx. \tag{3.4}$$

b. If, in addition, input points are deleted independently with probability $d(y \mid x)$ as they are being translated to the output space, and if the points of an independent Poisson-process with intensity $\mu_0(y)$ are superimposed in the output space with those points translated from the input space, then $\{M(B): B \subseteq \mathcal{Y}\}$ is a Poisson process with intensity

$$\mu(y) = \int_X p(y \mid x)s(y \mid x)\lambda(x)\,dx + \mu_0(y), \tag{3.5}$$

where $s(y \mid x) = 1 - d(y \mid x)$ is the probability that a point translating from input location x to output location y survives.

Proof. We prove part (a) of this theorem by demonstrating that the characteristic functional of the output point-process is that of a Poisson process with the intensity given in (3.4). The characteristic functional of the output point-process is given by:

$$E\left[\exp\left(j\int_{\mathcal{Y}} v(y)M(dy)\right)\right] = E\left[\exp\left(j\sum_{i=0}^{M(\mathcal{Y})} v(y_i)\right)\right]$$

$$= E\left\{E\left[\exp\left(j\sum_{i=0}^{N(X)} v(y_i)\right)\Big| N(X)\right]\right\}, \qquad (3.6)$$

where $N(X)$ and $M(\mathcal{Y})$ are the numbers of points on the input and output spaces, respectively, which are equal without deletions and insertions, and $\{y_1, y_2, \cdots, y_{N(X)}\}$ are the locations of the output points. Given the number $N(X)$ of input points, their locations $\{x_1, x_2, \cdots, x_{N(X)}\}$ are, from (2.79), independent, identically distributed random variables with probability density

$$\frac{\lambda(x)}{\int_x \lambda(x)\,dx}.$$

Also, each input point is translated independently to become an output point, with the transition density $p(y\,|\,x)$. Consequently, given the number $N(X)$, the locations of the output points are independent random variables with density

$$\frac{\int_x p(y\,|\,x)\lambda(x)\,dx}{\int_x \lambda(x)\,dx}.$$

The inner expectation in (3.6) then becomes:

$$E\left[\exp\left(j\sum_{i=0}^{N(X)} v(y_i)\right)\Big| N(X)\right] = \{E[e^{jv(y)}|\,N(X)]\}^{N(X)}$$

$$= \left\{\frac{\int_{\mathcal{Y}}\int_x e^{jv(y)} p(y\,|\,x)\lambda(x)\,dx\,dy}{\int_x \lambda(x)\,dx}\right\}^{N(X)}.$$

Substituting this expression into (3.6), we find that the characteristic functional of the output point-process is given by

$$E\left[\exp\left(j\int_{\mathcal{Y}}v(y)M(dy)\right)\right]$$

$$=\sum_{n=0}^{\infty}\left[\frac{\int_{\mathcal{Y}}\int_{x}e^{jv(y)}p(y\mid x)\lambda(x)\,dx\,dy}{\int_{x}\lambda(x)\,dx}\right]^{n}\frac{\left(\int_{x}\lambda(x)\,dx\right)^{n}}{n!}e^{-\int_{x}\lambda(x)dx}$$

$$=\exp\left[\int_{\mathcal{Y}}(e^{jv(y)}-1)\mu(y)\,dy\right],\tag{3.7}$$

where $\mu(y)$ is given in (3.4). From (2.80), this is the characteristic functional of a Poisson process on $\mathcal{Y}$ with intensity (3.4), which establishes part (a) of the theorem. Independently deleting the points of a Poisson process with intensity $\lambda(x)$ and deletion probability $d(y\mid x)$ results in another Poisson process with intensity $[1-d(y\mid x)]\lambda(x)$, and superimposing the points of an independent Poisson-process with intensity $\mu(y)$ with those of a Poisson process with intensity $\mu_0(y)$ results in a Poisson process with intensity $\mu(y)+\mu_0(y)$; see Problem 2.5.6. It follows that the output point-process of a translated Poisson-process subject to independent deletions and insertions is a Poisson process with intensity (3.5), which establishes part (b) of the theorem. ❑

Theorem 3.2.1 indicates the solution to the *forward problem* of determining the statistics of points in the output space given the statistics of points in the input space. The important fact is that points on the output space form a Poisson process if those on the input do and if points are translated independently from the input to the output space. The ramifications of this are clear. The counting and point-location statistics on the output space are those of the Poisson processes of Ch. 2 with the intensity function determined by the intensity function on the input space, the transition density-function from the input to the output space, the deletion-probability function, and the insertion rate according to (3.5). The sample-function density and the likelihood function for determining maximum-likelihood estimates of parameters influencing any of these functions are those of the Poisson process on the output space. For example, without insertions and deletions, we have from (2.78) that the log-likelihood functional is

$$\mathcal{L}=-\int_{\mathcal{Y}}\mu(y)\,dy+\int_{\mathcal{Y}}\ln[\mu(y)]M(dy)-\ln[M(\mathcal{Y})!].$$

Thus, without insertions and deletions, we have from (3.4) that

$$\mathcal{L} = -\int_X \lambda(x)\,dx$$

$$+ \int_{\mathcal{Y}} \ln\left[\int_X p(y \mid x)\lambda(x)\,dx\right] M(dy) - \ln[M(\mathcal{Y})!], \qquad (3.8)$$

and with insertions and deletions, we have from (3.5) that

$$\mathcal{L} = -\int_X \beta(x)\lambda(x)\,dx + \int_{\mathcal{Y}} \mu_0(y)\,dy \qquad (3.9)$$

$$+ \int_{\mathcal{Y}} \ln\left[\int_X p(y \mid x)s(y \mid x)\lambda(x)\,dx + \mu_0(y)\right] M(dy) - \ln[M(\mathcal{Y})!],$$

where

$$\beta(x) = \int_{\mathcal{Y}} p(y \mid x)s(y \mid x)\,dy$$

is the average survival probability for points leaving location x in the input space X and translating towards anywhere in the output space $\mathcal{Y}$.

Of equal or even more importance to determining the statistics of points on the output space are *inverse problems* in which inferences about the points in the input space are to be made on the basis of points observed in the output space. Inverse problems are examined in the following section.

3.3 Estimation for Translated Poisson-Processes

Estimation for translated Poisson-processes is important in applications where points in the output space represent measured data, and inferences must be made about quantities associated with the input space. Determining the conditional statistics of points on the input space given the measured data is necessary in addressing these problems. The two theorems in the following subsection provide expressions for the conditional expectations and, as a result of (1.24), the minimum mean square-error estimates of the number of points in some subset of the input space given histogram and count-record data, respectively, derived from points in the output space. These theorems are then used in the subsequent subsection for addressing other important estimation problems, such as estimating parameters influencing the input intensity or, indeed, estimating the input intensity itself, given output histogram and count-record data.

3.3.1 Count Estimation

Theorem 3.3.2 (*conditional mean for histogram data*). Let $\{N(A);$ $A \subseteq X)\}$ be a Poisson process with an integrable intensity function

$\{\lambda(x): x \in X\}$. Points of this process are translated to the output space $\mathcal{Y}$ to form the output point-process $\{M(B): B \subseteq \mathcal{Y}\}$, where each point is independently translated according to the transition density $p(y \mid x)$. Let $\{\mathcal{Y}_1, \mathcal{Y}_2, \cdots, \mathcal{Y}_k, \cdots\}$ be disjoint sets in $\mathcal{Y}$ such that $\mathcal{Y} = \cup_{k=1}^{\infty} \mathcal{Y}_k$. Define $M(\mathcal{Y}_k)$ to be the number of points in $\mathcal{Y}_k$, and let $M = \{M(\mathcal{Y}_1), M(\mathcal{Y}_2), \cdots, M(\mathcal{Y}_k), \cdots\}$ denote histogram data derived from points in the output space. Then:

a. if there are no insertions and deletions, the conditional expectation $E[N(A) \mid M]$ of the number of points in a subset A of the input space X given the histogram data M is

$$E[N(A) \mid M] = \sum_{k=1}^{\infty} \left[\frac{\int_{\mathcal{Y}_k} \int_A p(y \mid x)\lambda(x)\,dx\,dy}{\int_{\mathcal{Y}_k} \int_X p(y \mid x)\lambda(x)\,dx\,dy} \right] M(\mathcal{Y}_k); \qquad (3.10)$$

b. if, in addition, input points are deleted independently with probability $d(y \mid x)$ as they are being translated to the output space, and if the points of an independent Poisson-process with intensity $\mu_0(y)$ are superimposed in the output space with those points that survived the translation from the input space, then

$$E[N(A) \mid M]$$

$$= \int_{\mathcal{Y}} \int_A d(y \mid x)\lambda(x)\,dx\,dy \qquad (3.11)$$

$$+ \sum_{k=1}^{\infty} \left[\frac{\int_{\mathcal{Y}_k} \int_A p(y \mid x)s(y \mid x)\lambda(x)\,dx\,dy}{\int_{\mathcal{Y}_k} \int_X p(y \mid x)s(y \mid x)\lambda(x)\,dx\,dy + \int_{\mathcal{Y}_k} \mu_0(y)\,dy} \right] M(\mathcal{Y}_k),$$

where $s(y \mid x) = 1 - d(y \mid x)$ is the probability that a point translating from the input location x to the output location y survives.

Proof. The proof of part (a) follows from that of part (b) by selecting both $d(y \mid x)$ and $\mu_0(y)$ to be zero. For part (b), let $M(A; \mathcal{Y}_k)$ be the number of points in $\mathcal{Y}_k$ that are translated from A. Since randomly deleting points of a Poisson process yields another Poisson process, points in the output space that are translated from the subset A of the input space form a Poisson process on $\mathcal{Y}$ with intensity

$$\int_A p(y \mid x)s(y \mid x)\lambda(x)dx.$$

Also, let $N_d(A)$ be the number of points from A that are deleted; these points form a Poisson process on X with intensity

$$\int_A d(y \mid x)\lambda(x)dx.$$

Then,

$$N(A) = N_d(A) + \sum_{k=1}^{\infty} M(A;\mathcal{Y}_k),$$

from which it follows that

$$E[N(A) \mid M(\mathcal{Y}_1), M(\mathcal{Y}_2), \cdots, M(\mathcal{Y}_k), \cdots]$$

$$= E[N_d(A) \mid M(\mathcal{Y}_1), M(\mathcal{Y}_2), \cdots, M(\mathcal{Y}_k), \cdots] \tag{3.12}$$

$$+ \sum_{k=1}^{\infty} E[M(A;\mathcal{Y}_k) \mid M(\mathcal{Y}_1), M(\mathcal{Y}_2), \cdots, M(\mathcal{Y}_k), \cdots].$$

Since deleted points are independent of those appearing in the output space, it follows that

$$E[N_d(A) \mid M(\mathcal{Y}_1), M(\mathcal{Y}_2), \cdots, M(\mathcal{Y}_k), \cdots]$$

$$= E[N_d(A)] = \int_{\mathcal{Y}} \int_A d(y \mid x)\lambda(x)dxdy. \tag{3.13}$$

For use in the last term on the right side of (3.12), we have

$$E[M(A;\mathcal{Y}_k) \mid M(\mathcal{Y}_1), M(\mathcal{Y}_2), \cdots, M(\mathcal{Y}_k), \cdots]$$

$$= E[M(A;\mathcal{Y}_k) \mid M(\mathcal{Y})_k]$$

$$= \frac{\mu(A;\mathcal{Y}_k)}{\mu(A;\mathcal{Y}_k) + \mu(A^c;\mathcal{Y}_k) + \int_{\mathcal{Y}_k} \mu_0(y)dy} M(\mathcal{Y}_k), \tag{3.14}$$

where

$$\mu(A;\mathcal{Y}_k) = E[M(A;\mathcal{Y}_k)] = \int_{\mathcal{Y}_k} \int_A p(y \mid x)s(y \mid x)\lambda(x)dxdy$$

is the average number of points in $\mathcal{Y}_k$ that were translated from A, and

$$\mu(A^c;\mathcal{Y}_k) = E[M(A^c;\mathcal{Y}_k)] = \int_{\mathcal{Y}_k} \int_{A^c} p(y \mid x)s(y \mid x)\lambda(x)\,dx\,dy$$

is the average number of points in $\mathcal{Y}_k$ that were translated from the complement $A^c = X - A$ of A. Substituting these expressions into (3.12) yields

$$E[N(A) \mid M(\mathcal{Y}_1), M(\mathcal{Y}_2), \cdots, M(\mathcal{Y}_k), \cdots]$$

$$= \int_{\mathcal{Y}} \int_A d(y \mid x)\lambda(x)\,dx\,dy \tag{3.15}$$

$$+ \sum_{k=1}^{\infty} \left[\frac{\mu(A;\mathcal{Y}_k)}{\mu(A;\mathcal{Y}_k) + \mu(A^c;\mathcal{Y}_k) + \int_{\mathcal{Y}_k} \mu_0(y)\,dy} \right] M(\mathcal{Y}_k),$$

which is (3.11). □

In the following theorem we give expressions analogous to (3.10) and (3.11) for count-record data.

Theorem 3.3.3 (*conditional mean for count-record data*). Let $\{N(A): A \subseteq X\}$ be a Poisson process with an integrable intensity function $\{\lambda(x): x \in X\}$. Points of this process are translated to the output space $\mathcal{Y}$ to form the output point-process $\{M(B): B \subseteq \mathcal{Y}\}$, where each point is independently translated according to the transition density $p(y \mid x)$. Denote the count-record data by $M = \{y_1, y_2, \cdots, y_m, M(\mathcal{Y}) = m\}$, where y_i is the location of an observed point in the output space, and a total of m points are observed. Then:

a.   if there are no insertions and deletions, the conditional expectation $E[N(A) \mid M]$ of the number of points in a subset A of the input space X given the count-record data M is

$$E[N(A) \mid M] = \int_{\mathcal{Y}} \left[\frac{\int_A p(y \mid x)\lambda(x)\,dx}{\int_X p(y \mid x)\lambda(x)\,dx} \right] M(dy); \tag{3.16}$$

b. if, in addition, input points are deleted independently with probability $d(y \mid x)$ as they are being translated to the output space, and if the points of an independent Poisson-process with intensity $\mu_0(y)$ are superimposed in the output space with those points that survived the translation from the input space, then

$$E[N(A) \mid M] = \int_{\mathcal{Y}} \int_A d(y \mid x)\lambda(x)\,dx\,dy$$

$$+ \int_{\mathcal{Y}} \left[\frac{\int_A p(y \mid x)s(y \mid x)\lambda(x)\,dx}{\int_X p(y \mid x)s(y \mid x)\lambda(x)\,dx + \mu_0(y)} \right] M(dy),$$

$$(3.17)$$

where $s(y \mid x) = 1 - d(y \mid x)$ is the probability that a point translating from the input location x to the output location y survives.

Proof. For the proof of this theorem, we note that the histogram-data of Theorem 3.3.2 become equivalent to count-record data as the size of each of the subsets of $\mathcal{Y}$ in $\{\mathcal{Y}_1, \mathcal{Y}_2, \cdots, \mathcal{Y}_k, \cdots\}$ tends towards zero. Thus, if $v_k = \|\mathcal{Y}_k\|$ is the volume of the kth subset, then (3.16) and (3.17) follow from (3.10) and (3.11), respectively, in the limit as $\max_k(v_k)$ tends to zero. $\square$

3.3.2 Parameter Estimation

The estimation of parameters on the input space given data determined by points on the output space is in general much more difficult than estimating the same parameters given data on the input space. Any combination of random translations, insertions and deletions that affect measurements of a point process can complicate parameter estimation. This is illustrated by the following two examples.

Example 3.3.1 *Estimating an Intensity Function* ————————

Suppose that the input space X is partitioned into K disjoint subsets $X_1, X_2, \cdots, X_K$ such that $X = \cup_{k=1}^K X_k$ and

$$\lambda(x) = \sum_{k=1}^K \lambda_k I_k(x), \qquad (3.18)$$

where

$$I_k(x) = \begin{cases} 1, & x \in X_k \\ 0, & x \notin X_k. \end{cases}$$

Thus, the intensity on the input space is piecewise constant, as occurs in Ex. 3.1.1 for electron microscopic-autoradiography for which X_k is the region occupied by the kth type of subcellular structure, and λ_k is proportional to the concentration of radioactivity in that structure. We wish to estimate the intensity function $\{\lambda(x){:}x \in X\}$ or, equivalently, the parameters $\lambda = [\lambda_1 \quad \lambda_2 \quad \cdot \quad \cdot \quad \cdot \quad \lambda_K]'$, where $\lambda_k \geq 0$ for each k. It is readily verified that if points on the input space can be measured directly without translations, insertions and deletions, then the maximum-likelihood estimate of the intensity is given by

$$\hat{\lambda}(x) = \sum_{k=1}^{K} \hat{\lambda}_k I_k(x), \tag{3.19}$$

where

$$\hat{\lambda}_k = \frac{N(X_k)}{\| X_k \|},$$

in which $N(X_k)$ is the number of points observed in X_k, and $\| X_k \|$ is the size of the kth subregion. Thus, estimating the piecewise-constant intensity-function is straightforward in terms of points on the input space. Suppose, however, that there are random translations, as occurs in Ex. 3.1.1, and that count-record data on the output space are measured. From (3.8), the log-likelihood function is given by

$$\mathcal{L}(\lambda) = -\sum_{k=1}^{K} \lambda_k \| X_k \| + \int_{\mathcal{Y}} \ln\left[\sum_{k=1}^{K} \lambda_k f_k(y) \right] M(dy) - \ln[M(\mathcal{Y})!], \tag{3.20}$$

where

$$f_k(y) = \int_{X_k} p(y \mid x)dx.$$

Forming $\partial \mathcal{L}(\lambda)/\partial \lambda_k = 0$ for each k yields the following necessary condition that the maximum-likelihood estimate $\hat{\lambda}$ must satisfy

$$\int_{\mathcal{Y}} \left[\frac{f_k(y)}{\| X_k \| \sum_{k=1}^{K} \hat{\lambda}_k f_k(y)} \right] M(dy) = 1. \tag{3.21}$$

There is in general no closed-form solution of this nonlinear equation for $\hat{\lambda}$, so some numerical procedure for maximizing $\mathcal{L}(\lambda)$ must be used to determine the desired intensity-estimate. If we substitute (3.19) into (3.21) and note that $f_k(y)\| X_k\|^{-1}$ approaches $p(y\,|\,x)$ as $\max_k\| X_k\|$ tends to zero, then (3.21) becomes

$$\int_{\mathcal{Y}}\left[\frac{p(y\,|\,x)}{\int_X p(y\,|\,x)\hat{\lambda}(x)\,dx}\right]M(dy)=1. \tag{3.22}$$

Evidently, this is a nonlinear integral-equation that must be solved to determine a maximum-likelihood estimate of an intensity function that is not piecewise constant, as occurs in Ex. 3.1.2 for positron-emission tomography. Again, there is no closed-form solution, so some numerical method is needed. ∎

Example 3.3.2 *Pulse Position Modulation, Optical Range Finding* ———
        Consider the model in Problem 2.4.5. The input space is $X = \{x: x > -\infty\}$. Points on X occur as a Poisson process with intensity

$$\lambda(x:\delta)=\frac{\Lambda}{\sqrt{2\pi\tau^2}}\exp\left[-\frac{(x-\delta)^2}{2\tau^2}\right]. \tag{3.23}$$

If the number of points and their locations $\{X_1, X_2, \cdots, X_n, N(X)=n\}$ on the input space are available as data, then the maximum-likelihood estimate of δ is straightforwardly seen to be the centroid of these data,

$$\delta_{ML}=\frac{1}{n}\sum_{i=1}^{n}X_i. \tag{3.24}$$

Suppose, however, that points on the input space cannot be measured directly but, rather, only after random translations to an output space $\mathcal{Y}$, with a transition density

$$p_{y|x}(Y\,|\,X)=\frac{1}{\sqrt{2\pi\sigma^2}}\exp\left[-\frac{(Y-X)^2}{2\sigma^2}\right], \tag{3.25}$$

and after extraneous points are inserted with an intensity $\mu_0(y)$. The output space corresponds to the surface of a detector having some finite extent, and the extraneous points are due to detector noise. Suppose that the number of points and their locations $\{Y_1, Y_2, \cdots, Y_n, M(\mathcal{Y}) = m\}$ on the output space are measured. From (3.9), the loglikelihood in terms of these data is

$$\mathcal{L}(\delta) = -\Lambda + \int_{\mathcal{Y}} \mu_0(y)\,dy \tag{3.26}$$

$$+ \int_{\mathcal{Y}} \ln\left[\frac{\Lambda}{\sqrt{2\pi(\tau^2 + \sigma^2)}} \exp\left[-\frac{(y-\delta)^2}{2(\tau^2 + \sigma^2)} \right] + \mu_0(y) \right] M(dy) - \ln[m!].$$

Taking the derivative of this loglikelihood with respect to δ yields a transcendental equation with no closed-form solution in general. If, however, there are no insertions, $\mu_0(y) = 0$, then it is readily verified that the maximum-likelihood estimate of δ is at the centroid of the points on the output space. ∎

Example 3.3.3 *Photon Differencing, Phase Retrieval* ———————————
    Suppose that propagation through the atmosphere causes the light of a distant object to move about in a random walk when viewed through a telescope, as described in Ex. 3.1.4. Let the point-spread function of the telescope be $p(y \mid x)$, and assume that the atmosphere may be regarded as frozen while a short exposure or snapshot of the object is taken. Let the total exposure time for collecting J snapshots be

$$T = \sum_{j=1}^{J} (\tau_j - \tau_{j-1}),$$

where $[\tau_{j-1}, \tau_j)$ is the jth exposure interval, $\tau_0 = t_0$, and $\tau_J = T + t_0$. The photodetection process can be modeled as a translated Poisson-process in time and space, for which the intensity $\{\mu(t, y): t \geq t_0, y \in \mathcal{Y}\}$ of photodetections as a functional of the object's intensity is given by $\mu(t, y) = \mu(y - m_j)$ for t in the jth exposure interval $[\tau_{j-1}, \tau_j)$, where

$$\mu(y) = \int_x p(y \mid x)\lambda(x)\,dx$$

is the photoelectron intensity in the absence of atmospheric effects, and m_j is an unknown displacement due to the state of the atmosphere during the jth exposure. Photoevents within an exposure interval evolve as a homogeneous Poisson process in time, with a rate of $r = \int_{\mathcal{Y}} \mu(y)\,dy$ conversions per second, and a translated Poisson-process in space, with intensity $(\tau_j - \tau_{j-1})\mu(y - m_j)$ during the jth exposure. One way that data can be collected is to preselect exposure intervals, in which case the number of photoconversions in the jth exposure is a Poisson-distributed random variable with mean $(\tau_j - \tau_{j-1})r$. An alternative data collection strategy is to preselect the number of photoconversions in each exposure, in which case the duration of an exposure is a random variable having the gamma distribution in (2.29) with parameter r.

Photon differencing is a method used to combat the unknown effects of the atmosphere [2, 10, 21]. For this, the location of one photoconversion in an exposure is selected as a reference from which all other photoconversions in that exposure are measured. Let $\{y_{1,j}, y_{2,j}, \cdots, y_{N_p,j}, N_j\}$ denote the locations and number of photoconversions in the jth exposure, ordered by the times $\{t_{1,j}, t_{2,j}, \cdots, t_{N_p,j}\}$ at which they occur, and form the difference data

$$z_{i,j} = y_{i,j} - y_{N_p,j}, \qquad i = 1,2,\cdots,N_j-1; \quad j = 1,2,\cdots,J. \qquad (3.27)$$

The joint probability-density of the photon differences for the jth exposure, given the number of photoconversions in the exposure, follows from the conditional independence of the locations and (2.79). For preselected exposure intervals, T. Schulz [23] observes that

$$p_{z_1,\cdots,z_{n-1}|N_j}[Z_1,\cdots,Z_{n-1} \mid n] = r^{-n}R_\mu^{(n)}(Z_1,\cdots,Z_{n-1}), \qquad (3.28)$$

where

$$R_\mu^{(n)}(Z_1,\cdots,Z_{n-1}) = \int_{\mathcal{Y}} \mu(y)\mu(y+Z_1)\cdots\mu(y+Z_{n-1})\,dy$$

is the nth-order autocorrelation function of $\mu(\cdot)$. Independence of increments then yields the loglikelihood functional for preselected intervals as

$$\mathcal{L} = \ln\left\{ \prod_{j=1}^{J} r^{-n_j}R_\mu^{(n_j)}\big(Z_{1,j},\cdots,Z_{n_j-1,j}\big)\Pr[N_j = n_j]\right\}, \qquad (3.29)$$

which has the following evaluation when terms that do not depend on the object's intensity are suppressed

$$\mathcal{L} = -T\int_{\mathcal{Y}} \mu(y)\,dy + \sum_{j=1}^{J} \ln R_\mu^{(n_j)}\big(Z_{1,j},\cdots,Z_{n_j-1,j}\big).$$

A similar expression is obtained when the number of conversions in an exposure is preselected. Let n be the number of conversions per exposure interval, and assume that the time per exposure and the $n-1$ photon differences in it are measured. The loglikelihood becomes

$$\mathcal{L} = \ln\left\{ \prod_{j=1}^{J} r^{-n} R_\mu^{(n)}(Z_{1,j}, \cdots, Z_{n-1,j}) p_{w_n}(T_j - T_{j-1}) \right\},$$

which becomes

$$\mathcal{L} = -T \int_{\mathcal{Y}} \mu(y)\,dy + \sum_{j=1}^{J} \ln R_\mu^{(n)}(Z_{1,j}, \cdots, Z_{n-1,j}) \tag{3.30}$$

when terms that are not a function of the object's intensity are dropped. This can be written more compactly by letting z_j denote an $n-1$ dimensional vector of the differences observed in the jth exposure, regarding the sequence of such vectors $z_1, z_2, \cdots, z_J$ for the series of J exposures as a multidimensional point process, and letting $M(A)$ denote the number of points of this process in the subset A of the set Z of all possible $n-1$ dimensional vectors of differences. Then,

$$\mathcal{L} = -T \int_{\mathcal{Y}} \mu(y)\,dy + \int_Z \ln[R_\mu^{(n)}(z)] M(d z). \tag{3.31}$$

Suppose, as an example, that we wish to estimate the function $\mu(\cdot)$ based upon J snapshots having two photoconversions each. By paralleling the development in Ex. 3.3.1, we see that the following nonlinear integral equation must be solved for the function $\hat{\mu}(\cdot)$ that maximizes $\mathcal{L} \equiv \mathcal{L}(\mu)$

$$\frac{1}{T} \int_Z \frac{\hat{\mu}(y+z) + \hat{\mu}(y-z)}{R_\mu^{(2)}(z)} M(dz) = 1. \tag{3.32}$$

As noted by T. Schulz [23] and T. Schulz and D. Snyder [24], the problem of solving (3.32) for $\hat{\mu}(\cdot)$ can be related to the important problem of determining a nonnegative function from the magnitude of its Fourier transform, which is called the *phase-retrieval problem* [6, 7]. If many exposures are collected, so that the total exposure time T is large, then $M(dz)/T$ can be approximated by its asymptotic expected-value $(1/2r)R_\mu^{(2)}(z)\,dz$, so that $\hat{\mu}(\cdot)$ must then satisfy

$$\frac{1}{2r}\int_z \frac{\hat{\mu}(y+z)+\hat{\mu}(y-z)}{R_\mu^{(2)}(z)}R_\mu^{(2)}(z)\,dz = 1. \tag{3.33}$$

Clearly, any function $\hat{\mu}(\cdot)$ having a second-order autocorrelation $R_\mu^{(2)}(z)$ that is equal to $R_\mu^{(2)}(z)$ satisfies this equation. Thus, solving this equation for $\hat{\mu}(y)$ is equivalent to finding a function $\mu(y)$ given its second-order autocorrelation $R_\mu^{(2)}(z)$. Since the Fourier transform of the second-order autocorrelation of a function is the squared magnitude of the Fourier transform of the function, this is also equivlalent to solving the phase-retrieval problem. Similar results hold for finding a function from its nth-order autocorrelation if n photoconversions per exposure are collected; the most important example is $n = 3$, corresponding to recovering a function from its triple correlation or bispectrum [21, 24].

There is no closed-form solution to (3.32), so some numerical method is needed. This is pursued further in Ex. 3.3.6. ∎

Ex's. 3.3.1 to 3.3.3 are representative of the fact that using data derived from the output space to estimate parameters influencing the input space generally leads to intractable equations that can only be solved numerically. A numerical method that has proven to be effective in a variety of applications is the *expectation-maximization algorithm* of A. Dempster, N. Laird, and D. Rubin [4]. L. Shepp and Y. Vardi [25] were first to suggest this method for emission-tomography applications. Besides the intractability of the equations satisfied by parameter estimates, there is also a fundamental issue that must be addressed. If the number of parameters to be estimated becomes large, as in nonparametric density-estimation problems, estimates can become unstable, exhibiting a very rough and undesirable behavior. This effect is not evident in the argument we used for obtaining (3.22), but it is present and will be demonstrated in Sec. 3.2.3. Regularization of parameter estimates through the use of constraints is necessary to alleviate these instabilities and is the subject of Sec. 3.4. It will there be seen that the expectation-maximization algorithm also accommodates various regularization approaches.

The Expectation-Maximization Algorithm

The expectation-maximization (EM) algorithm is based on the concepts of an incomplete-data space $\mathcal{Y}$, a complete-data space $\mathcal{Z}$, and a mapping $h:\mathcal{Z}\to\mathcal{Y}$ between them.

incomplete-data space. The incomplete-data space is the space $\mathcal{Y}$ in which measured data takes its values. This is the output space of Fig. 3.1 when data are in the form of a measured point process subject to random translations, insertions, and deletions.

complete-data space. The complete-data space Z is a hypothetical space that is contrived to accomplish two goals: 1, make the expectation and maximization steps of the EM algorithm analytically tractable; and 2, make the resulting computations required for the EM algorithm feasible for numerically producing estimates. The complete-data space for a given problem is not unique, and the EM algorithm that results can be more or less complicated depending on the choice made. Often, an appropriate choice is suggested by an understanding of the mechanisms of distortion and noise governing data measured on the incomplete-data space.

mapping from Z to $\mathcal{Y}$. The complete-data space is *larger* than the incomplete-data space in the sense that complete data must determine incomplete data. There must be a known function $h(\cdot)$ mapping complete data into incomplete data. This is generally a mapping with many points of Z yielding the same point in $\mathcal{Y}$. This mapping function and the incomplete data y place a constraint, determined by $h(z) = y$, on the values that the complete data z may have.

As an example, let $y = x + n$ describe some measured data, where $x \in \mathcal{R}^1$ is a "signal" and $n \in \mathcal{R}^1$ is an additive "noise." Then, $\mathcal{Y} = \mathcal{R}^1$. If we select the complete data as $z = (x, n)$, then $Z = \mathcal{R}^2$, and the mapping function is $h(z) = h(x, n) = x + n$. If a particular measurement yields the incomplete data $y = Y$, then the complete data must lie in the subset $Z(Y)$ of the complete-data space defined by the line $\{z: z \in Z, h(z) = x + n = Y\}$. Alternatively, selecting the complete data as $z = (x, y) \in \mathcal{R}^2$ requires the mapping $h(x, y) = y$, and the measurement $y = Y$ restricts z to lie on the line $\{z: z \in Z, h(x, y) = y = Y\}$.

Denote the collection of parameters to be estimated by θ, and assume that this vector lies in a set $\Theta \subseteq \mathcal{R}^n$ of possible parameter values. An estimate of θ is termed *admissible* if it is in Θ. The parameters θ may be nonrandom, or they may be random. We will use maximum-likelihood (ML) estimation for nonrandom parameters, and for random parameters, we assume that there is a known prior probability density $p(\theta)$, and we use maximum *a posteriori* probability (MAP) estimation.

Two loglikelihood functions are important for estimating the parameters θ from measured data $y \in \mathcal{Y}$ when the EM algorithm is used. The *incomplete-data loglikelihood* for nonrandom parameters $\mathcal{L}_{id}(\theta) =$

$\ln[p_y(Y:\theta)]$, where $p_y(Y:\theta)$ denotes the density[1] of the incomplete data as a function of the parameters, is just the loglikelihood we have used heretofore in parameter estimation. The maximum-likelihood estimate $\hat{\theta} \equiv \hat{\theta}(y)$ of θ is an admissible vector-valued function of the incomplete data $y \in \mathcal{Y}$ that maximizes the incomplete-data loglikelihood $\hat{\theta} = \text{argmax}_{\theta \in \Theta}[\mathcal{L}_{id}(\theta)]$. If the parameters are random, then the quantity $\ln[p(\theta)]$ should be added to $\ln[p_y(Y:\theta)]$ in forming the incomplete-data loglikelihood.

The second loglikelihood that is important is new. It is called the *complete-data loglikelihood* and is defined by $\mathcal{L}_{cd}(\theta) = \ln[p_z(Z:\theta)]$, where $p(Z:\theta)$ denotes the density of the complete data as a function of the parameters. The quantity $\ln[p_\theta(\theta)]$ should be added to the right-hand side if the parameters are random. It is clear that, in general, $\hat{\theta}$ is not a maximizer of the complete-data loglikelihood $\mathcal{L}_{cd}(\theta)$. Nevertheless, $\mathcal{L}_{cd}(\theta)$ is important in the EM algorithm for determining $\hat{\theta}$ numerically.

The EM algorithm is iterative. Starting from an initial, admissible estimate $\hat{\theta}^{(0)}$ of the parameters, a sequence of admissible estimates $\hat{\theta}^{(1)}$, $\hat{\theta}^{(2)}$, ..., $\hat{\theta}^{(k)}$, ... is produced for which the corresponding sequence of incomplete-data loglikelihoods is nondecreasing $\mathcal{L}_{id}(\hat{\theta}^{(0)}) \leq \mathcal{L}_{id}(\hat{\theta}^{(1)}) \leq \mathcal{L}_{id}(\hat{\theta}^{(2)}) \leq \cdots \leq \mathcal{L}_{id}(\hat{\theta}^{(k)}) \leq \cdots$. Two steps are required at each stage of the iteration to reach the next stage, an expectation (E) step and a maximization (M) step.

E-step. Determine the conditional expectation of the complete-data loglikelihood,

$$Q(\theta \mid \hat{\theta}^{(k)}) = E[\mathcal{L}_{cd}(\theta) \mid y, \hat{\theta}^{(k)}], \qquad (3.34)$$

given the incomplete data and assuming that $\theta = \hat{\theta}^{(k)}$. The conditional expectation is evaluated according to

$$E[\mathcal{L}_{cd}(\theta) \mid y = Y, \hat{\theta}^{(k)}] = \int_Z p_{z|y}(Z \mid Y:\hat{\theta}^{(k)})\ln[p_z(Z:\theta)]dZ, \qquad (3.35)$$

where

---

[1] "Density" is used here in a general sense. This may be either the probability density or the sample-function density, whichever is appropriate in describing the data.

$$p_{z|y}(Z \mid Y : \theta) = \begin{cases} \dfrac{p_z(Z:\theta)}{\displaystyle\int_{Z(Y)} p_z(Z:\theta)\,dZ}, & Z \in Z(Y) \\[6pt] 0, & Z \notin Z(Y), \end{cases} \tag{3.36}$$

and $Z(Y) = \{z : z \in Z, h(z) = Y\}$. The denominator on the right side of this equation is the density of the incomplete data,

$$\int_{Z(Y)} p_z(Z:\theta)\,dZ = p_y(Y:\theta). \tag{3.37}$$

The quantity $\ln[p(\theta)]$ must be added to the right side of (3.35) if the parameters are random.

M-step. Determine the stage $k+1$ parameter estimate as an admissible maximizer of $Q(\theta \mid \hat{\theta}^{(k)})$,

$$\hat{\theta}^{(k+1)} = \underset{\theta \in \Theta}{\operatorname{argmax}}\,[Q(\theta \mid \hat{\theta}^{(k)})]. \tag{3.38}$$

It is established in the following theorem that repeated application of these E and M steps produces a nondecreasing sequence of incomplete-data loglikelihoods.

Theorem 3.3.4 (*EM algorithm*) Let the sequence $\{\hat{\theta}^{(k)} : k = 1, 2, \cdots\}$, be defined according to the E and M steps, (3.34) and (3.38). Then, the corresponding sequence of incomplete-data loglikelihoods $\{L_{id}(\hat{\theta}^{(k)}) : k = 1, 2, \cdots\}$ is nondecreasing.

Proof. For the proof of this theorem, note from (3.36) and (3.37) that for $Z \in Z(Y)$ there holds

$$\ln p_y(Y:\theta) = \ln p_z(Z:\theta) - \ln p_{z|y}(Z \mid Y : \theta). \tag{3.39}$$

The left side of this equation is the incomplete-data loglikelihood, and the first term on the right is the complete-data loglikelihood. Multiplying both sides by $p_{z|y}(Z \mid Y : \hat{\theta}^{(k)})$, and integrating with respect to Z yields

$$L_{id}(\theta) = Q(\theta \mid \hat{\theta}^{(k)}) - \int_{Z(Y)} p_{z|y}(Z \mid Y : \hat{\theta}^{(k)}) \ln p_{z|y}(Z \mid Y : \theta)\,dZ. \tag{3.40}$$

Evaluating (3.40) for both $\theta = \hat{\theta}^{(k+1)}$ and $\theta = \hat{\theta}^{(k)}$, and then subtracting, results in

$$L_{id}(\hat{\theta}^{(k+1)}) - L_{id}(\hat{\theta}^{(k)}) = Q[\hat{\theta}^{(k+1)} \mid \hat{\theta}^{(k)}] - Q[\hat{\theta}^{(k)} \mid \hat{\theta}^{(k)}] \tag{3.41}$$

$$- \int_{Z(Y)} p_{z|y}(Z \mid Y : \hat{\theta}^{(k)}) \ln \left[\frac{p_{z|y}(Z \mid Y : \hat{\theta}^{(k+1)})}{p_{z|y}(Z \mid Y : \hat{\theta}^{(k)})} \right] dZ.$$

Application of the inequality $\ln(x) \leq x - 1$, for which equality holds if and only if $x = 1$, then yields

$$L_{id}(\hat{\theta}^{(k+1)}) - L_{id}(\hat{\theta}^{(k)}) \geq Q[\hat{\theta}^{(k+1)} \mid \hat{\theta}^{(k)}] - Q[\hat{\theta}^{(k)} \mid \hat{\theta}^{(k)}]$$

$$- \int_{Z(Y)} p_{z|y}(Z \mid Y : \hat{\theta}^{(k)}) \left[\frac{p_{z|y}(Z \mid Y : \hat{\theta}^{(k+1)})}{p_{z|y}(Z \mid Y : \hat{\theta}^{(k)})} - 1 \right] dZ$$

$$= Q[\hat{\theta}^{(k+1)} \mid \hat{\theta}^{(k)}] - Q[\hat{\theta}^{(k)} \mid \hat{\theta}^{(k)}], \tag{3.42}$$

with equality if and only if $p_{z|y}(Z \mid Y : \hat{\theta}^{(k+1)}) = p_{z|y}(Z \mid Y : \hat{\theta}^{(k)})$, which holds if $\hat{\theta}^{(k+1)} = \hat{\theta}^{(k)}$. The M-step (3.38) implies that the right side of (3.42) is nonnegative in general and equal to zero if $\hat{\theta}^{(k+1)} = \hat{\theta}^{(k)}$. Hence,

$$L_{id}(\hat{\theta}^{(k+1)}) \geq L_{id}(\hat{\theta}^{(k)}), \tag{3.43}$$

which establishes Theorem 3.3.4. $\square$

Establishing the convergence properties of the sequence of parameter estimates for the EM algorithm is difficult. This issue is addressed by C. Wu [34] in general and by Y. Vardi, L. Shepp, and L. Kaufman [33] and K. Lange and R. Carson [14] in the particular instance of estimating the intensity of a Poisson process. There are a number of issues, some of which we summarize following C. Wu [34]. If the incomplete-data log-likelihood is bounded at each stage, $L_{id}(\hat{\theta}^{(k)}) < \infty$, then (3.43) implies that $L_{id}(\hat{\theta}^{(k)})$ converges monotonically towards some limiting value L^*; however, there is no guarantee that L^* is the global maximum of $L(\theta)$ over the admissible set Θ. In general, convergence can be towards a global maximum, a local maximum, or a stationary point depending on the initial estimate $\hat{\theta}^{(0)}$ that is selected.[1] The convergence of $\{L_{id}(\hat{\theta}^{(k)})\}$ to L^* does not necessarily imply the convergence of $\{\hat{\theta}^{(k)}\}$ to an admissible limit θ^*. According to C. Wu [34, Corollary 1], if $L_{id}(\theta)$ is unimodal for $\theta \in \Theta$ and

---

[1] A stationary point of $L_{id}(\theta)$ is a θ for which the gradient $\partial L_{id}(\theta)/\partial \theta$ is zero.

θ^* denotes the only stationary point, and if $Q(\theta \mid \theta')$ is continuously differentiable in θ and continuous in θ', then the EM sequence $\{\hat{\theta}^{(k)}\}$ converges to the unique maximizer θ^* for any $\hat{\theta}^{(0)}$ in the interior of Θ. Stronger statements about convergence can be made for particular applications. We mention, especially, the results of Y. Vardi, L. Shepp, and L. Kaufman [33] and K. Lange and R. Carson [14] establishing the convergence to a unique maximum of the iteration sequence $\{\hat{\theta}^{(k)}\}$ for estimating Poisson-process intensities.

Parameter Estimation via the EM Algorithm

We now give two examples in using the EM algorithm for estimating a finite number of parameters for translated Poisson-processes.

Example 3.3.4 *Estimating a Signal in Additive Noise.* ─────────

Let n_s and n_n be independent, Poisson-distributed random variables corresponding to signal and noise counts, respectively, and let their sum $n = n_s + n_n$ be the measured, or incomplete, data. Assume that λ_s, where $\lambda_s \geq 0$, is the unknown mean parameter of the signal counts and that λ_n is the known mean parameter of the noise. The goal is to estimate λ_s. The incomplete-data loglikelihood is

$$\mathcal{L}_{id}(\lambda_s) = -(\lambda_s + \lambda_n) + n \ln(\lambda_s + \lambda_n) - \ln(n!). \tag{3.44}$$

The maximum-likelihood estimate of λ_s is the admissible maximizer of $\mathcal{L}_{id}(\lambda_s)$; $\hat{\lambda}_s = \text{argmax}_{\lambda_s \geq 0} \mathcal{L}_{id}(\lambda_s)$. Thus,

$$\hat{\lambda}_s = \max(0, n - \lambda_n). \tag{3.45}$$

The analytical determination of the maximum-likelihood estimate of the desired parameter is straightforward in this example. As we have noted, however, this is not typical in applications, so numerical methods such as the EM algorithm are usually required. It is instructive to use the EM algorithm in this example because the resulting estimates can be compared to the analytical result (3.45). Let (n_s, n_n) be the complete data, so that the mapping from the complete to the incomplete data is $h(n_s, n_n) = n_s + n_n$. Since the components of the complete data are independent, the complete-data loglikelihood is

$$\mathcal{L}_{cd}(\lambda_s) = -(\lambda_s + \lambda_n) + n_s \ln(\lambda_s) + n_n \ln(\lambda_n), \tag{3.46}$$

where only terms depending on λ_s and λ_n have been retained. For the E-step (3.34), we have

$$Q(\lambda_s \mid \hat{\lambda}_s^{(k)}) = -\lambda_s + \hat{n}_s^{(k)} \ln(\lambda_s) - \lambda_n + \hat{n}_n^{(k)} \ln(\lambda_n), \qquad (3.47)$$

where

$$\hat{n}_s^{(k)} = E[n_s \mid n, \hat{\lambda}_s^{(k)}] = \frac{\hat{\lambda}_s^{(k)}}{\hat{\lambda}_s^{(k)} + \lambda_n} n, \qquad (3.48)$$

and $\hat{n}_n^{(k)} = E[n_n \mid n, \hat{\lambda}_s^{(k)}] = n - \hat{n}_s^{(k)}$ are conditional mean and, therefore, minimum mean square-error estimates of the signal and noise counts given the total counts and the stage-k estimate of λ_s. The last two terms on the right in (3.47) are not a function of λ_s and so do not affect the maximization of $Q(\lambda_s \mid \hat{\lambda}_s^{(k)})$ in the M-step (3.38). Equivalently, the M-step is therefore

$$\hat{\lambda}_s^{(k+1)} = \underset{\lambda_s \geq 0}{\operatorname{argmax}} \, [-\lambda_s + \hat{n}_s^{(k)} \ln(\lambda_s)]. \qquad (3.49)$$

Hence,

$$\hat{\lambda}_s^{(k+1)} = \hat{n}_s^{(k)} = \frac{\hat{\lambda}_s^{(k)}}{\hat{\lambda}_s^{(k)} + \lambda_n} n. \qquad (3.50)$$

Stable points $\hat{\lambda}_s^*$ of the sequence (3.50) satisfy the quadratic equation $(\hat{\lambda}_s^*)^2 + \hat{\lambda}_s^* \lambda_n = n \hat{\lambda}_s^*$. The admissible solutions are $\hat{\lambda}_s^* = 0$ and, provided that $n \geq \lambda_n$, $\hat{\lambda}_s^* = n - \lambda_n$. Checking the values of the incomplete-data loglikelihood at these two candidate solutions shows that the maximizing stable point is the maximum-likelihood estimate (3.45). Shown in Fig. 3.8 is a computer simulation demonstrating the behavior of $\hat{\lambda}_s^{(k)}$ and $L_{id}(\hat{\lambda}_s^{(k)})$ as a function of the iteration number k and initial value $\hat{\lambda}_s^{(0)}$. For this simulation, $\lambda_s = 10$, $\lambda_n = 5$, and the method of Problem 2.3.4 was used to simulate n, with the result $n = 22$. The result after 10 iterations of the EM algorithm is shown for two initial conditions. The sequence of incomplete-data loglikelihoods is nondecreasing for both initial conditions, as it must be for all initial conditions. The sequence of estimates for both initial conditions converge towards the maximum-likelihood estimate, which from (3.45) equals 17. Note from (3.50), however, that if the initial condition had been selected as $\hat{\lambda}_s^{(0)} = 0$, then $\hat{\lambda}_s^{(k)} = 0$ for all k, which is far from the maximum-likelihood estimate. This shows that care must be exercised in using the EM algorithm when there are multiple stable points because the

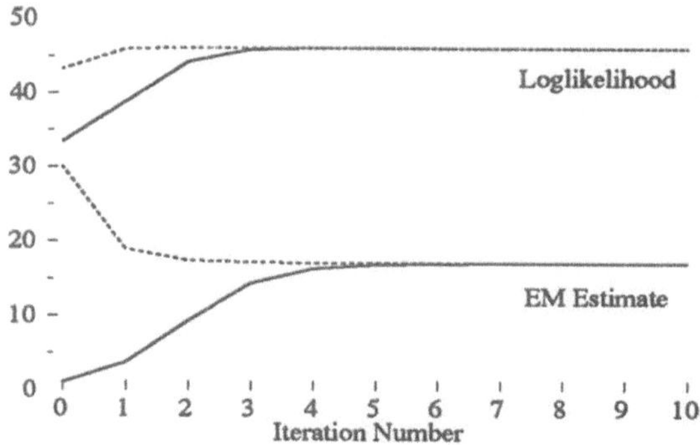

Figure 3.8 Sequence of EM Estimates and Corresponding Incomplete-Data Loglikelihoods (solid line: $\hat{\lambda}_s^{(0)} = 1$, dashed line: $\hat{\lambda}_s^{(0)} = 30$).

convergence point of the EM sequence can depend on the initial condition that is chosen. It is readily verified for this example that $\partial^2 \mathcal{L}_{id}(\lambda_s)/\partial\lambda_s^2 = -n/(\lambda_s + \lambda_n)^2 < 0$ and $Q(\lambda_s \mid \lambda_s')$ is continuously differentiable in λ_s and continuous in λ_s'. Consequently, according to the results of C. Wu [34] and K. Lange and R. Carson [14], $\{\hat{\lambda}_s^{(k)}\}$ converges to the maximum-likelihood estimate (3.45) for any choice of $\hat{\lambda}_s^{(0)}$ that is greater than zero. ∎

Example 3.3.5 *Estimating an Intensity via the EM Algorithm* ————————
Consider the model of Ex. 3.3.1 again. We now regard the quantity $\mathcal{L}(\lambda)$ in (3.20) as the incomplete-data loglikelihood. Let the complete data be the concatenation of the point locations and their number on the input space, $\{x_1, x_2, \cdots, x_{N(X)}, N(X)\}$, with the point locations and their number on the output space, $\{y_1, y_2, \cdots, y_{N(Y)}, N(Y)\}$. In applying the EM algorithm, the only terms of the complete-data loglikelihood that are significant are those dependent on the parameters to be estimated because others do not affect the M-step. If $\tilde{\mathcal{L}}_{cd}(\lambda)$ denotes the complete-data loglikelihood with all terms that are not a function of λ suppressed, we have

$$\tilde{\mathcal{L}}_{cd}(\lambda) = -\sum_{j=1}^{K} \lambda_j \| x_j \| + \sum_{j=1}^{K} N(X_j)\ln(\lambda_j), \tag{3.51}$$

which is simply the loglikelihood on the input space. The E-step yields

$$\tilde{Q}(\lambda \mid \hat{\lambda}^{(k)}) = -\sum_{j=1}^{K} \lambda_j \| X_j \| + \sum_{j=1}^{K} E[N(X_j) \mid M, \hat{\lambda}^{(k)}] \ln(\lambda_j), \qquad (3.52)$$

where M denotes the measured incomplete-data. From (3.16), we then have

$$\tilde{Q}(\lambda \mid \hat{\lambda}^{(k)}) = -\sum_{j=1}^{K} \lambda_j \| X_j \| + \sum_{j=1}^{K} \left\{ \int_{\mathcal{Y}} \left[\frac{\hat{\lambda}_j^{(k)} f_j(y)}{\sum_{i=1}^{K} \hat{\lambda}_i^{(k)} f_i(y)} \right] M(dy) \right\} \ln(\lambda_j).$$

$$(3.53)$$

The function $\tilde{Q}(\lambda \mid \hat{\lambda}^{(k)})$ must now be maximized over the set of admissible λ for the M-step. Using $\partial \tilde{Q}/\partial \lambda_j = 0$ for $\lambda = \hat{\lambda}^{(k+1)}$ yields

$$\hat{\lambda}_j^{(k+1)} = \hat{\lambda}_j^{(k)} \int_{\mathcal{Y}} \left[\frac{f_j(y)}{\| X_j \| \sum_{i=1}^{K} \hat{\lambda}_i^{(k)} f_i(y)} \right] M(dy). \qquad (3.54)$$

Eq. (3.54) defines a sequence of estimates of λ for which the corresponding sequence of incomplete-data loglikelihoods is nondecreasing. Let H be the Hessian matrix of the incomplete-data loglikelihood, with elements $h_{ij} = \partial^2 \mathcal{L}_{id}(\lambda)/\partial \lambda_i \partial \lambda_j$. It is readily verified that

$$v'Hv = -\int_{\mathcal{Y}} \left[\frac{(v'f(y))^2}{(\lambda'f(y))^2} \right] M(dy) \leq 0, \qquad (3.55)$$

where $v \in \mathcal{R}^K$ and $f(y)$ is a vector with $f_j(y)$ as its jth element. Consequently, H is nonpositive definite. If $M(\mathcal{Y}) \neq 0$ and $f(y)$ is such that the inequality in (3.48) is strict, so the Hessian is negative definite, then $\mathcal{L}_{id}(\lambda)$ is unimodal. Furthermore, the function $\tilde{Q}(\lambda \mid \lambda')$ in (3.53) is continuously differentiable in the elements of λ and continuous in the elements of λ'. Consequently, according to the results of C. Wu [34] and K. Lange and R. Carson [14], the sequence $\{\hat{\lambda}_s^{(k)}\}$ defined by (3.54) converges to the maximum-likelihood estimate of λ for any choice of $\hat{\lambda}_s^{(0)}$ having positive elements. ∎

Example 3.3.6 *Photon Differencing, Phase Retrieval* ————————
Consider Ex. 3.3.3 again. T. Schulz [23] and T. Schulz and D. Snyder [24] demonstrate that the limit points of the sequence of functions produced by the following EM iteration satisfy the necessary conditions for maximizing (3.31) for $n = 2$ photoconversions per exposure

$$\hat{\mu}^{(k+1)}(y) = \hat{\mu}^{(k)}(y)\frac{1}{T}\int_z \frac{\hat{\mu}^{(k)}(y+z)+\hat{\mu}^{(k)}(y-z)}{R^{(2)}_{\hat{\mu}^{(k)}}(z)}M(dz). \tag{3.56}$$

As the total exposure time becomes large, a sequence of functions converging towards the solution of (3.33), and hence towards the solution to the phase-retrieval problem, is produced,

$$\hat{\mu}^{(k+1)}(y) = \hat{\mu}^{(k)}(y)\frac{1}{2r}\int_z \frac{\hat{\mu}^{(k)}(y+z)+\hat{\mu}^{(k)}(y-z)}{R^{(2)}_{\hat{\mu}^{(k)}}(z)}R^{(2)}_{\mu}(z)\,dz. \tag{3.57}$$

∎

3.4 Constrained Estimation

Estimating an intensity function is important in many applications, including emission tomography and low light-level imaging. Often the domain of the function is partitioned into quantization elements, such as pixels or voxels, in which case the number of parameters to be estimated equals the number of elements. This number can be very large, and it grows as the sizes of the quantization elements are selected to be small in order to achieve high resolution, eventually reaching infinity as the quantization of the domain is refined towards the continuous domain. The estimation of an intensity function, or, equivalently, a very large or infinite number of parameters, can result in very pronounced instabilities when the methods of the previous section are used. Estimates produced with the EM algorithm can be quite encouraging at first, having better resolution, signal-to-noise ratio, and contrast compared to other methods. However, as iterations proceed towards increasing likelihood, the disturbing instabilities emerge. Sharp transitions in the underlying intensity that is being estimated are greatly accentuated and emerge in the successive iterates with a substantial, quite objectionable overshoot. Also, estimates appear to become more "noisy" as the successive iterates are produced, with high peaks and low valleys, seemingly randomly distributed throughout the estimate, masking structures of interest. These *edge* and *noise artifacts* are fundamental when estimates of a large number of parameters are sought from noisy, ill conditioned data.

Let points on the input space X of Fig. 3.1 occur as a Poisson process with an intensity function $\{\lambda(x): x \in X\}$. Points of this process are observed after random translations to the output space $\mathcal{Y}$. Points on the output space occur as a Poisson process with an intensity function $\{\mu(y): y \in \mathcal{Y}\}$ given by

$$\mu(y) = \int_X p(y \mid x)\lambda(x)\,dx, \tag{3.58}$$

where $p(y \mid x)$ is the transition density governing the random translation of a point from X to $\mathcal{Y}$.

There are two problems that complicate the estimation of $\{\lambda(x): x \in X\}$ from points observed on $\mathcal{Y}$. The first is present even if so much data are available that the function $\{\mu(y): y \in \mathcal{Y}\}$ can be regarded as known perfectly. Then, the integral equation (3.58) must be solved for $\{\lambda(x): x \in X\}$ for a given kernel $p(y \mid x)$. This is termed a *deterministic inverse problem* in general and a *deconvolution problem* if the kernel as a function of x and y depends only on the difference $y - x$. The second problem is that the point-process data available are random, so estimating $\{\lambda(x): x \in X\}$ from measurements on the output space $\mathcal{Y}$ is a *stochastic inverse problem*.

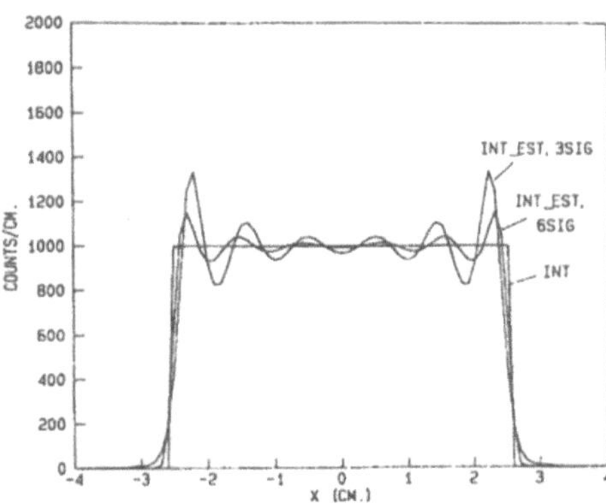

Figure 3.9 Estimate of a rectangular intensity function from mean-value data, demonstrating the edge artifact when the EM kernel is truncated to zero at 3 (INT_EST.3SIG) and 6 (INT_EST.6SIG) standard deviations. (From [28, ©1987 IEEE])

Deterministic inverse-problems are notoriously unstable. Small variations in $\mu(\cdot)$, which may occur due to estimation errors inevitably present with a finite amount of data, can result in large variations in the solution $\lambda(\cdot)$ to (3.58). Computational methods for solving (3.58) can be exquisitely sensitive to the numerical precision and implementation used. One manifestation of this instability is a Gibbs' overshoot that occurs at sharp transitions in the true $\{\lambda(x): x \in X\}$. Fig. 3.9 shows a demonstration of this effect. The ideal data $\{\mu(y): y \in Y\}$ were simulated according to (3.58) for a rectangular intensity $\{\lambda(x): x \in X\}$ given by

$$\lambda(x) = \begin{cases} 5100, & |x| \leq 2.55 \\ 0, & \text{otherwise}' \end{cases} \tag{3.59}$$

and $p(y \mid x)$ was selected as a normal density with mean x and a full width at half maximum of 1.0. Here, $\mu(\cdot)$ can be expressed in terms of error functions and, therefore, can be determined numerically with high precision. Then, the following EM iterations were performed to solve (3.58)

$$\bar{\lambda}^{(k+1)}(x) = \bar{\lambda}^{(k)}(x) \int_Y \left[\frac{p(y \mid x)}{\int_x p(y \mid x')\bar{\lambda}^{(k)}(x')dx'} \right] \mu(y)dy. \tag{3.60}$$

This equation was obtained from (3.54) by letting $\max_k \| X_k \|$ tend to zero, replacing $M(dy)$ by its mean value $\mu(y)dy$, and denoting the resulting sequence of estimates by $\{\bar{\lambda}^{(k)}\}$, where the overbar indicates that the estimates are based on mean-value data. A total of 5000 iterations of (3.60) were performed starting from an initial estimate set to the constant value of 1. The graph labeled INT_EST.3SIG is the estimate of the intensity when the EM kernel $p(y \mid x)$ in (3.60) is truncated to zero beyond 3 standard deviations for the purpose of the numerical implementation; this introduces a mismatch between the "true" normal density $p(y \mid x)$ producing the data and the $p(y \mid x)$ used as the EM kernel, but the mismatch appears to be slight because of the rapid decrease towards zero of the normal density, and it is certainly slight compared to the mismatches experienced in practice for real instrumentation. The *edge artifact* can be seen quite clearly as an overshoot, slightly in excess of 30 percent at the discontinuities of $\lambda(x)$. For the mean-value data used in this simulation, the observed magnitude of the edge artifact is somewhat dependent on the number of iterations employed as well as on parameter details, including the number of standard deviations beyond which the values for the kernel of the EM algorithm are set to zero. The latter effect is demonstrated by the graph in Fig. 3.9 labeled INT_EST.6SIG, which is the estimate when the EM kernel is truncated to 6 standard deviations; the overshoot has been reduced to about 18 percent. The 18 percent overshoot is sustained, approximately, for truncation points well beyond 6 standard deviations. This sensitivity to the seemingly innocuous choice of truncation point is a manifestation of the instability of the deterministic inverse problem (3.58). No method for solving (3.58) can reproduce high frequency components present in an intensity function with sharp discontinuities if the data available have suppressed high frequency information too much due to the nature of $p(y \mid x)$; here, the Gaussian convolution-kernel suppresses high frequencies exponentially quadratically fast, and they cannot be recovered, resulting in the Gibbs' phenomenon of overshoot. This effect is particularly objectionable in intensity estimation because these functions are nonnegative; the usual 9 percent Gibbs' undershoot and overshoot becomes an 18 percent overshoot due to the nonnegativity constraint.

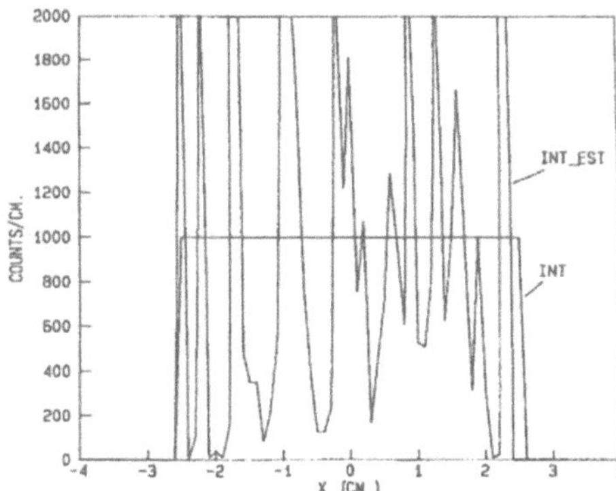

Figure 3.10 Estimate of a rectangular intensity function from random data, demonstrating the noise artifact. (From [28, ©1987 IEEE])

The random nature of the point processes on the input and output spaces is a source of the *noise artifact*. This objectionable effect is demonstrated in Fig. 3.10 where the results are displayed of a computer simulation in which the rectangular intensity function of Fig. 3.9 was estimated from random data rather from mean-value data. The estimate is very rough, with rapid variations between small and large values. The basic problem may be seen by imagining that points on the input space can be measured directly without translations. The loglikelihood functional is then

$$\mathcal{L}(\lambda) = -\int_X \lambda(x)\,dx + \int_X \ln[\lambda(x)]N(dx). \tag{3.61}$$

Define $\lambda^*(x)$ according to

$$\lambda^*(x) = \sum_{n=1}^{N(x)} p_\Delta(x - x_n), \tag{3.62}$$

where $N(X)$ and $x_1, x_2, \cdots$ are the number of points on X and their locations, and $p_\Delta(x)$ is defined by

$$p_\Delta(x) = \begin{cases} \dfrac{1}{\Delta}, & |x| \leq \dfrac{\Delta}{2} \\[2ex] 0, & |x| > \dfrac{\Delta}{2} \end{cases}.$$

Consider the loglikelihood (3.61) evaluated at $\lambda^*(x)$,

$$\mathcal{L}(\lambda^*) = -N(\mathcal{X})[1 + \ln(\Delta)]. \tag{3.63}$$

As Δ tends towards zero, $\lambda^*(x)$ tends towards a superposition of $N(\mathcal{X})$ impulse functions, with an impulse located at each of the points x_n in $\mathcal{X}$, and $\mathcal{L}(\lambda^*)$ tends towards infinity. Translation errors are present in the simulation performed for Fig. 3.9, so these ideal impulses will not be seen, but the EM iterations obtained from (3.54) by letting $\max_k\|\mathcal{X}_k\|$ tend to zero,

$$\hat{\lambda}^{(k+1)}(x) = \hat{\lambda}^{(k)}(x) \int_{\mathcal{Y}} \left[\frac{p(y\,|\,x)}{\int_X p(y\,|\,x')\hat{\lambda}^{(k)}(x')dx'} \right] M(dy), \tag{3.64}$$

evidently solves the stochastic inverse problem well enough that the rough behavior of a collection of impulses is approximated. The loglikelihood functional tending towards infinity as Δ tends to zero demonstrates that there is no well defined maximizer over the set of nonnegative functions.

The noise and edge artifacts seen separately in Fig's. 3.9 and 3.10 are both present in practice, but one or the other may dominate depending on the number of data points available and other parameters of the problem. These effects are present when an attempt is made to estimate a large number of parameters in a stochastic inverse problem. They are especially pronounced in the infinite-dimensional problem we have examined where an intensity-function is to be estimated. The artifacts are not caused by the use of the EM algorithm but would be experienced with any method used to maximize the loglikelihood functional because no well behaved maximizer exists *subject only to a nonnegativity constraint*. The last phrase is important because it suggests that the instabilities can be controlled through the introduction of constraints into the estimation problem, a process termed *regularization*.

3.4.1 Regularization of Estimates

Four methods of regularization are described in the following sections: Grenander's method of sieves, the method of penalties, the method of Markov random fields, and the method of resolution kernels. The first three can be very effective in reducing noise artifacts, and a combination of these and the use of resolution kernels can reduce both noise and edge artifacts.

Grenander's Method of Sieves

The method of sieves introduced by U. Grenander [9] can be used to suppress the noise artifact. The idea is to constrain the estimate of $\lambda(\cdot)$ to be in a smooth subset, called a *sieve*, of the set of nonnegative functions. The sieve is selected in a manner that depends on the number of measurement points; the more measurement points there are, the larger the sieve. By allowing the size of the sieve to grow in an appropriate manner with the number of measurement points, it is hoped that the constrained estimate is consistent in the sense that it converges to the true intensity as the number of measured points tends to infinity. Establishing consistency is a difficult task in general.

A sieve that is a subset of the set of nonnegative functions can be defined according to

$$S = \left\{ \lambda : \lambda(x) = \int_Z s(x \mid z)\xi(z)\,dz \right\}, \tag{3.65}$$

where $\{\xi(z) : z \in Z\}$ is an integrable intensity-function, and the kernel of the sieve, $\{s(x \mid z) : x \in X, z \in Z\}$, is a nonnegative function that is normalized as a conditional probability density, $\int_X s(x \mid \cdot)\,dx = 1$. As an example when $X = Z = \mathcal{R}^1$, the kernel can be selected to be a Gaussian density function

$$s(x \mid z) = \frac{1}{\sqrt{2\pi\sigma_s^2}} \exp\left[-\frac{(x-z)^2}{2\sigma_s^2} \right].$$

The sieve then consists of all the nonnegative functions that can be produced by convolutions of all the integrable intensity-functions with a Gaussian density of zero mean and variance σ_s^2. This variance is made smaller as the number of measurement points increases, so the sieve includes all integrable intensities in the limit as σ_s^2 tends to zero.

A sieve-constrained maximum-likelihood estimate of an intensity function $\lambda(\cdot)$ is obtained by maximizing the loglikelihood functional (3.8) over the sieve set S defined in (3.65),

$$\hat{\lambda}_{ML}(x) = \underset{\lambda \in S}{\mathrm{argmax}}\, L.$$

We see from (3.8) and (3.65) that for intensities $\lambda(\cdot)$ that are sieve members, the loglikelihood can be rewritten as a functional of the intensities $\xi(\cdot)$ that generate the sieve,

$$L = -\int_Z \xi(x)\,dx + \int_Y \ln\left[\int_Z k(y \mid z)\xi(z)\,dz \right] M(dy) - \ln[M(\mathcal{Y})!], \tag{3.66}$$

where $k(\cdot \mid \cdot)$ is the concatenation of the kernel of the sieve and the transition density from the input to the output space,

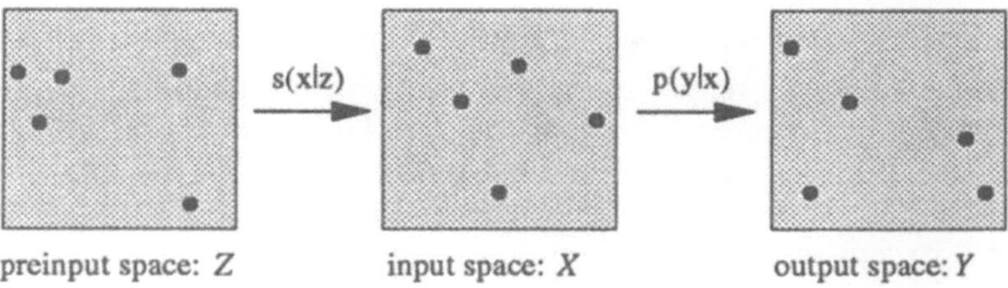

Figure 3.11   Random translations between the preinput, input, and measurement spaces, with transition densities s(x|z) and p(y|x).

$$k(y \mid z) = \int_X p(y \mid x)s(x \mid z)\,dx. \tag{3.67}$$

Consequently, the constrained estimate is that member of the sieve defined by

$$\hat{\lambda}_{\mathrm{ML}}(x) = \int_Z s(x \mid z)\hat{\xi}(z)\,dz, \tag{3.68}$$

where $\hat{\xi}(\cdot)$ is an unconstrained maximizer of the loglikelihood functional (3.66),

$$\hat{\xi}(z) = \operatorname*{argmax}_{\xi} L. \tag{3.69}$$

No direct solution for this maximization problem is known. It can be addressed numerically using the expectation-maximization algorithm. For this purpose, we interpret the loglikelihood (3.66) in terms of translations of the points of a fictitious Poisson process we define on a *preinput* space Z, as shown in Fig. 3.11. Let the intensity function for this process be $\{\xi(z): z \in Z\}$. A point at location z of this hypothetical Poisson-process is randomly translated to the input space X with the kernel of the sieve serving as the transition density. This point is further translated randomly, with the transition density $p(y \mid x)$, to the output space Y. The composite kernel $k(y \mid z)$ in (3.67) may then be regarded as a transition density governing translations of points from Z to Y. Then, from Theorem 3.2.1, the point process on Y is a Poisson process with intensity $\int_Z k(y \mid z)\xi(z)dz$. Since $\int_Z k(y \mid z)\xi(z)dz = \int_X p(y \mid x)\lambda(x)dx$, the Poisson process produced in this way on the output space Y is equal, with probability one, to the original Poisson process forming the measurements. The problem in (3.69) of forming an unconstrained maximizer of the incomplete-data loglikelihood is then seen to be identical to that leading to the iterations in (3.64) except that the transition density is $k(y \mid z)$ rather than $p(y \mid x)$. Hence, the sequence defined by

$$\hat{\xi}^{(k+1)}(z) = \hat{\xi}^{(k)}(z) \int_Y \left[\frac{k(y \mid z)}{\int_Z k(y \mid z')\hat{\xi}^{(k)}(z')dz'} \right] M(dy) \qquad (3.70)$$

produces a corresponding sequence of loglikelihoods that is nondecreasing, $\mathcal{L}(\hat{\xi}^{(k+1)}) \geq \mathcal{L}(\hat{\xi}^{(k)})$. The constrained estimate of $\lambda(\cdot)$ is obtained by performing the iterations (3.70) until a stable point $\hat{\xi}^{(\infty)}(\cdot)$ is reached and then determining the corresponding member of the sieve according to

$$\hat{\lambda}_{\mathrm{ML}}(x) = \int_Z s(x \mid z)\hat{\xi}^{(\infty)}(z)dz. \qquad (3.71)$$

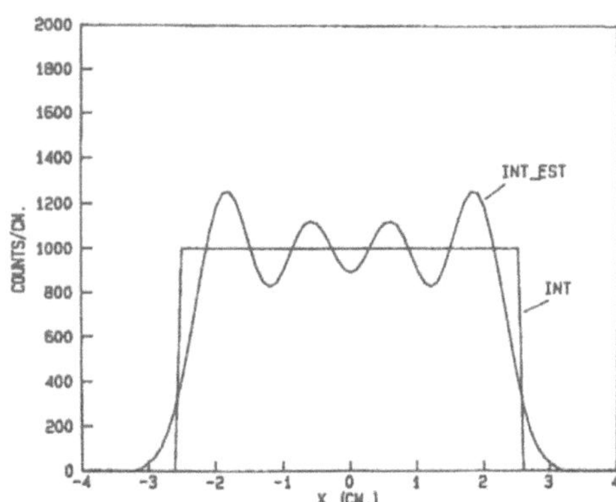

Figure 3.12 A sieve-constrained estimate showing reduced noise artifact compared to the unconstrained estimate in Fig. 3.10. (From [28, ©1987 IEEE])

The effectiveness of a sieve constraint in reducing the noise artifact is demonstrated in Fig. 3.12. The graph labeled INT_EST was produced by selecting the sieve kernel as a Gaussian density with a mean of zero and a full width at half maximum (FWHM) of 1.0 and then performing 5000 iterations of (3.70) followed by the evaluation of (3.71) on the $5000th$ iterate. The same data were used as for the unconstrained maximum-likelihood estimate shown in Fig. 3.10. For this, the transition density $p(y \mid x)$ governing the translation of points from the input space to the output space was Gaussian with a mean of x and a FWHM of 1.0.

Consequently, from (3.67), the kernel $k(y\,|\,z)$ used in (3.70) was a Gaussian density with a mean of z and a FWHM of $\sqrt{2}$. The rough behavior seen in the unconstrained estimate of Fig. 3.10 is greatly reduced in the sieve-constrained estimate of Fig. 3.12.

The Method of Penalties

As has been suggested by Good and Gaskins [8] another way to remedy the apparent difficulty is to perform the maximization of the likelihood with a roughness penalty $\mathcal{E}(\lambda)$, where we may view this penalty functional as describing our prior knowledge about the smoothness of the intensity λ. Specifically, the approach is to maximize the functional $G(\lambda)$ defined by

$$G(\lambda) = L(\lambda) - \mathcal{E}(\lambda),\tag{3.72}$$

where $L(\lambda)$ is the log-likelihood functional given by (3.61), and $\mathcal{E}(\lambda)$ is the penalty that describes our prior knowledge of the intensity λ. As I. Good and R. Gaskins [8] have noted, since λ is a nonnegative function, G must be maximized subject to the constraint that $\lambda \geq 0$. This is equivalent to finding the real function γ, where $\gamma(x) = \sqrt{\lambda(x)}$, that maximizes the functional $G(\gamma^2)$ given by

$$G(\gamma^2) = L(\gamma^2) - \mathcal{E}(\gamma^2).\tag{3.73}$$

Finding the function γ that maximizes $G(\gamma^2)$ is more straightforward than finding the λ that maximizes $G(\lambda)$ because we have removed the difficult constraint that λ is nonnegative as this is automatically satisfied by $\lambda = \gamma^2$.

All that is left is to choose a penalty that expresses any prior knowledge on the smoothness of λ or expresses some property that the estimate of λ should have. As suggested by Good and Gaskins [Goo], the roughness of a distribution can be measured by determining the difficulty of discriminating it from a shifted version of itself and is quantified in terms of the amount of information associated with the process of discrimination. Following S. Kullback [13], the *information divergence* between a distribution $\lambda(x)$ and its shifted version $\lambda(x + \varepsilon)$ is given by

$$\int_x [\lambda(x) - \lambda(x + \varepsilon)] \ln\left\{ \frac{\lambda(x)}{\lambda(x + \varepsilon)} \right\} dx.\tag{3.74}$$

Using a Taylor series expansion about $\varepsilon = 0$ and keeping only the most significant term, the roughness measure becomes proportional to

$$\int_x \frac{|\lambda'(x)|^2}{\lambda(x)}\,dx = \int_x |\gamma'(x)|^2 dx, \tag{3.75}$$

with $\gamma(x) = \sqrt{\lambda(x)}$. We call (3.75) *Good's roughness penalty*.

Another view of the roughness prior as a measure of discriminability may be formulated by examining the detection limit of the position of a waveform of known shape with unknown position. Assuming that the position of the pulse is uniformly distributed over the observation interval, the Cramér-Rao mean-square error bound on the position of the waveform when given Poisson data is given by the inverse of the Fisher information, which is precisely Good's roughness measure (see Ex. 2.4.9 with $\lambda_0 = 0$). Viewing the positioning uncertainty as a measure of resolution, the introduction of Good's penalty becomes a direct trade off between smoothness and resolution in the reconstruction.

The principle property of this particular choice of penalty that has motivated us in [29] to select it for intensity estimation in emission tomography is that previously there did not appear to be any natural way to introduce bandwidth limitations into the maximum-likelihood estimation procedure. This is in sharp contrast to the bandwidth limitations that are readily introduced with the filtered back-projection and confidence weighted algorithms of conventional and time-of-flight tomography, respectively. In these approaches, the bandwidth of the reconstruction filter is selected so as to be consistent with the sampling plan used during data collection. That Good's roughness (3.75) is such a bandwidth constraint may be seen by examining the set $S(B)$ of functions γ defined by

$$S(B) = \left\{ \gamma : \int_x \left| \frac{d\gamma(x)}{dx} \right|^2 dx \leq (2\pi B)^2 \int_x \gamma^2(x)dx \right\}, \tag{3.76}$$

where B parameterizes the set and determines the second central moment of the energy-spectrum of γ. Then, maximizing likelihood over the family $S(B)$ is identical to adding Good's roughness penalty according to (3.73) and yields the constrained maximum-likelihood problem of finding the function $\hat{\gamma}$ that maximizes the penalized log-likelihood functional given by

$$G(\gamma^2) = -\int_x \gamma^2(x)dx + \int_x \ln[\gamma^2(x)]N(dx) \tag{3.77}$$

$$- \alpha \left[\int_x \left| \frac{d\gamma(x)}{dx} \right|^2 dx - (2\pi B)^2 \int_x \gamma^2(x)dx \right].$$

The desired estimate is $\hat{\lambda} = \hat{\gamma}^2$, where $\hat{\gamma}$ maximizes $G(\gamma^2)$. This becomes our constrained, or penalized maximum-likelihood estimation problem. Notice that $\mathcal{S}(B)$ can be regarded as a sieve in the method of sieves, with the parameter B controlling the size of the sieve.

By introducing the bandwidth constraint of (3.76) as a penalty function, we now show that the function $\hat{\gamma}$ that maximizes $G(\gamma^2)$ is an *exponential spline* of the data.

Theorem 3.4.1 (*exponential splines for Good's roughness penalty*). The function $\hat{\gamma}$ given by

$$\hat{\gamma} = \operatorname*{argmax}_{\gamma}\left[-\int_X \gamma^2(x)\,dx + \int_X \ln[\gamma^2(x)]N(dx) \right.$$
$$\left. -\alpha\left\{ \int_X \left| \frac{d\gamma(x)}{dx} \right|^2 dx - (2\pi B)^2 \int_X \gamma^2(x)\,dx \right\} \right] \tag{3.78}$$

satisfies the nonlinear equation

$$\hat{\gamma}(x) = \sum_{i=1}^{N} \frac{h(x - x_i)}{\hat{\gamma}(x_i)}, \tag{3.79}$$

with N the total number of points observed, x_i the location of the ith point, and $h(\cdot)$ is the exponential function

$$h(x) = \frac{1}{2\sqrt{\alpha\beta}} e^{-\sqrt{\beta/\alpha}\,|x|}, \tag{3.80}$$

with $\beta = 1 - \alpha(2\pi B)^2$.

Before proving this theorem, we note similar results with somewhat different constraints were first derived for estimating probability densities by R. Tapia and J. Thompson [30]. As has also been shown in [30] and [9], if the first derivative constraint of (3.76) is chosen to be less than some constant, and not a function of the energy in γ, then a polynomial spline for h occurs as in (3.80). The proof follows [29].

Proof. Using the generalized Kuhn-Tucker theorem [16], this is equivalent to finding the γ that maximizes the penalized log-likelihood functional of (3.73) where the penalty $\mathcal{E}(\gamma^2)$ is given by

$$\mathcal{E}(\gamma^2) = \alpha \left\{ \int_x \left| \frac{d\gamma(x)}{dx} \right|^2 dx - (2\pi B)^2 \int_x \gamma^2(x)\, dx \right\}, \qquad (3.81)$$

where the Lagrange multiplier α is nonnegative, and $\mathcal{E}(\hat{\gamma}^2) = 0$. Using the calculus of variations, we conclude that $\hat{\gamma}(x)$ is the solution to the nonlinear differential equation

$$\alpha \hat{\gamma}(x) \frac{d^2 \hat{\gamma}(x)}{dx^2} - \beta \hat{\gamma}^2(x) = -\sum_{i=1}^{N} \delta(x - x_i). \qquad (3.82)$$

Substituting

$$\hat{\gamma}(x) = \sum_{i=1}^{N} \frac{h(x - x_i)}{\hat{\gamma}(x_i)}$$

directly into the differential equation (3.82), where h is given in (3.80) shows that the exponential spline of (3.79) maximizes the log-likelihood of (3.73) subject to the bandwidth constraint of (3.76). $\square$

One important property of $\hat{\gamma}$ is that

$$\int_x \hat{\gamma}^2(x)\, dx = \int_x \hat{\lambda}(x)\, dx = N, \qquad (3.83)$$

showing that the integral of the intensity estimate $\hat{\lambda}$ is normalized to the total number of measured points. To see this, define $< a(x), b(x) >$ according to

$$< a(x), b(x) > = \alpha \int_x \frac{da(x)}{dx} \frac{db(x)}{dx}\, dx + \beta \int_x a(x) b(x)\, dx.$$

Using integration-by-parts and (3.82), we conclude that $< h(x - x_j), \hat{\gamma}(x) > = \hat{\gamma}(x_j)$. Furthermore, this equality and (3.79) yields $< \hat{\gamma}(x), \hat{\gamma}(x) > = N$. But

$$< \hat{\gamma}(x), \hat{\gamma}(x) > = \mathcal{E}(\hat{\gamma}^2) + \int_x \hat{\gamma}^2(x)\, dx,$$

which implies (3.83) because $\mathcal{E}(\gamma^2) = 0$ for $\gamma = \hat{\gamma}$. Since $\alpha = 0$ results in the unconstrained solution of (3.61), we conclude that α must be greater than zero or the bandwidth constraint of (3.76) will not be satisfied. To insure that the kernel h is an exponential and does not oscillate, we require $\alpha(2\pi B)^2 < 1$.

The Method of Markov Random Fields

For extending penalty methods into multiple dimensions, it is natural to associate penalties $\mathcal{E}(\cdot)$ with Markov random fields. This we do by changing our view of the penalized-likelihood approach only slightly to one in which we regard the intensity parameters λ as a random process with Markov random field priors. Then, the maximum-likelihood estimation problem is transformed into a maximum *a posteriori* estimation one. That is, defining the prior on λ as

$$p(\lambda) = \frac{1}{Z} e^{-\mathcal{E}(\lambda)},$$

where the normalizer is given by

$$Z = \int e^{-\mathcal{E}(\lambda)} d\lambda,$$

then the maximum *a posteriori* estimation problem becomes: maximize with respect to λ the posterior density $p(\lambda \mid \text{data})$ given by

$$p(\lambda \mid \text{data}) \propto e^{\mathcal{L}(\lambda) - \mathcal{E}(\lambda)}.$$

This is identical to the penalized likelihood functional $G(\lambda)$ of (3.72). The major advantage of this view is simply that in multiple dimensions the convenience of equations such as the exponential splines may be lost, while the power of conditional neighborhood relations of Markov fields is gained.

The crucial issue in applying this approach becomes choosing the random field prior as there are many available. Following B. Roysam, J. Shrauner, and M. Miller [22], we choose the prior induced by Good's roughness in two dimensions as follows. First, define $\lambda(x) = \gamma^2(x)$, where $x = [x_1 \quad x_2]'$ is the two dimensional coordinate in $\mathcal{R}^2$. Let $\varepsilon_\theta = [\varepsilon_1 \quad \varepsilon_2]'$ be a unit vector oriented at angle θ to the gradient vector $\nabla\gamma = [\partial\gamma/\partial x_1 \quad \partial\gamma/\partial x_2]'$, and define Good's roughness in the direction of ε_θ according to

$$g(\theta) = \int_{\mathcal{R}^2} <\varepsilon_\theta, \nabla\gamma>^2 dx = \cos^2\theta \int_{\mathcal{R}^2} |\nabla\gamma|^2 dx, \qquad (3.84)$$

which is a directionally dependent extension of (3.75) in two dimensions. For problems in $\mathcal{R}^2$ when there are no natural or preferred coordinate directions, the estimate of λ should not depend on the orientation of the coordinate system. This is equivalent to requiring that the processing be rotationally invariant. Applying this to $g(\theta)$ amounts to averaging over all θ in $[0, 2\pi)$, yielding a measure of roughness $\bar{g}$ that is independent of direction:

$$\bar{g} = \frac{1}{2\pi} \int_0^{2\pi} g(\theta)\, d\theta = \frac{1}{2} \int_{\mathcal{R}^2} |\nabla \gamma|^2 dx. \tag{3.85}$$

This is the directionally independent extension of (3.75) that we shall use. Following (3.73), we then adopt the following roughness penalty

$$\mathcal{E}(\gamma^2) = \alpha_1 \int_{\mathcal{R}^2} |\nabla \gamma|^2 dx + \alpha_2 \int_{\mathcal{R}^2} \gamma^2(x)\, dx \tag{3.86}$$

$$= \alpha_1 \int_{-\infty}^{\infty} \int_{-\infty}^{\infty} \left\{ \left| \frac{\partial \gamma}{\partial x_1} \right|^2 + \left| \frac{\partial \gamma}{\partial x_2} \right|^2 \right\} dx_1 dx_2 + \alpha_2 \int_{-\infty}^{\infty} \int_{-\infty}^{\infty} \gamma^2(x_1, x_2)\, dx_1 dx_2.$$

The prior density of the Markov random field defined on the discrete lattice (i, j) is induced by approximating derivatives in (3.86) via first differences, yielding the density $p(\gamma) = Z^{-1} \exp[-\mathcal{E}(\gamma^2)]$ with

$$\mathcal{E}(\gamma^2) = \alpha_1 \sum_{i,j} \left\{ |\gamma(i+1, j) - \gamma(i, j)|^2 + |\gamma(i, j+1) - \gamma(i, j)|^2 \right\}$$

$$+ \alpha_2 \sum_{i,j} \gamma^2(i, j). \tag{3.87}$$

To understand the kind of smoothing in two dimensions that is induced by Good's prior, we need only examine the nearest neighbor relations determined by the conditional probabilities of the random field on the lattice. Define γ as the collection of $\gamma(i, j)$ for all i and j, and denote by $\gamma \backslash \gamma(i, j)$ the set γ with the element $\gamma(i, j)$ removed. Then, the conditional probabilities of the field become

$$p(\gamma(i, j) \mid \gamma \backslash \gamma(i, j)) = \frac{p(\gamma)}{p(\gamma \backslash \gamma(i, j))}$$

$$= \frac{1}{Z} \exp[-\mathcal{E}_{N(i, j)}(\gamma^2(i, j))],$$

where

$$\mathcal{E}_{N(i,j)}(\gamma^2(i,j)) = \alpha_1 |\gamma(i+1,j) - \gamma(i,j)|^2 + |\gamma(i-1,j) - \gamma(i,j)|^2$$

$$+ |\gamma(i,j+1) - \gamma(i,j)|^2 + |\gamma(i,j-1) - \gamma(i,j)|^2 + \alpha_2 \gamma^2(i,j),$$

implying pixel (i,j) has neighborhood structure $N(i,j) = \{(i,j+1)$ $(i,j-1)\ (i+1,j)\ (i-1,j)\}$. Note that $\tilde{Z}$ normalizes the conditional density over $\gamma(i,j)$ and is different from Z. Therefore, Good's roughness prior can be expected to induce nearest neighbor smoothing.

Incorporating Penalties and Random-Field Priors into the Expectation-Maximization Algorithm

To demonstrate these effects on the translated Poisson-process model of Thm. 3.2.1 and Ex. 3.3.1, the log-posterior from (3.8) and (3.20) with added random field prior becomes

$$\int_{\mathcal{Y}} \ln\left[\sum_{j=1}^{K} \gamma^2(j) f_j(y)\right] M(dy)$$

$$- \ln[M(\mathcal{Y})!] - \alpha_1 \sum_{j=1}^{K} |\gamma(j+1) - \gamma(j)|^2 - \alpha_2 \sum_{j=1}^{K} \gamma^2(j), \quad (3.88)$$

with

$$f_j(y) = \int_{X_j} p(y \mid x)\, dx,$$

and X_j is pixel j in the lattice, with all pixels of equal size. The two-dimensional version is almost identical with the rotationally invariant version of Good's prior substituted for the last term above. As shown by A. Dempster, N. Laird, and D. Rubin [4], the expectation-maximization algorithm can be combined with a prior for deriving the maximum *a posteriori* estimator by simply adding the prior to the maximization at each stage of the algorithm. The complete-data log-posterior with prior added is

$$\sum_{j=1}^{K} N(X_j) \ln[\gamma^2(j)] - \alpha_1 \sum_{j=1}^{K} |\gamma(j+1) - \gamma(j)|^2 - \alpha_2 \sum_{j=1}^{K} \gamma^2(j). \quad (3.89)$$

Then, the $(k+1)st$ iterate $\lambda^{k+1} = (\gamma^{k+1})^2$ becomes the maximizer of the following expectation:

$$\gamma^{k+1}(j) = \underset{\gamma}{\mathrm{argmax}} \left\{ \sum_{j=1}^{K} E[N(X_j)|M,\lambda^k] \ln[\gamma^2(j)] \right. \tag{3.90}$$

$$\left. - \alpha_1 \sum_{j=1}^{K} |\gamma(j+1) - \gamma(j)|^2 - \alpha_2 \sum_{j=1}^{K} \gamma^2(j) \right\},$$

where M denotes the measured, or incomplete, data, and the conditional expectation is given by (3.16). The two-dimensional version is obtained by simply substituting the two-dimensional roughness into the model and applying the identical analysis.

Since the conditional mean of (3.90) is simply the solution derived at every iteration by the expectation-maximization algorithm, the addition of Good's prior corresponds to smoothing the unconstrained expectation-maximization algorithm solution at each iteration. As suggested by B. Roysam, J. Shrauner, and M. Miller [22], the equations are solved by performing several steps of a Jacobi like gradient iteration at each stage of the expectation-maximization algorithm to solve the nonlinear difference equation. The importance of the gradient implementation is only that it requires purely local updating determined by the nearest neighbors in the image. The gradient at any location i,j requires only the values of the nearest neighbors on the grid, and as a result the addition of the roughness prior adds an insignificant amount of computation while it provides an excellent method of regularizing the unconstrained solution.

Shown in Fig's. 3.13 and 3.14, from M. Miller [20], is a variance study of a pie phantom having six slices, demonstrating the use of Good's roughness in the context of emission tomography with the translation model. Fig. 3.13 shows the radioactivity distribution used for the pie phantom studied. The left column of Fig. 3.14 shows the $3000th$ iteration of the unconstrained maximum-likelihood algorithm applied to data consisting of 16 view angles and having 460K (top row) and 1,200K (bottom row) total counts. Notice the extremely sharp peaks and valleys in the reconstruction. The middle panels show the result of post-filtering with Good's roughness the final unconstrained result on the left, and on the right is shown the maximum *a posteriori* probability solution. The latter solution was generated by adding Good's penalty to each iteration of the expectation- maximization algorithm according to (3.90). The parameters were chosen so that the middle and right panels have identical roughness. Superimposed inside of each of the pie wedges are the sample variances obtained with 25 independent computer simulations. These results show that for the same resolution, the maximum *a posteriori* probability solution shows variability that is lower by a factor of two than the post-filter approach.

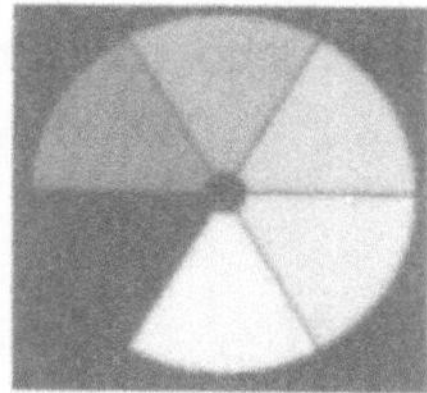

Figure 3.13 Pie phantom.

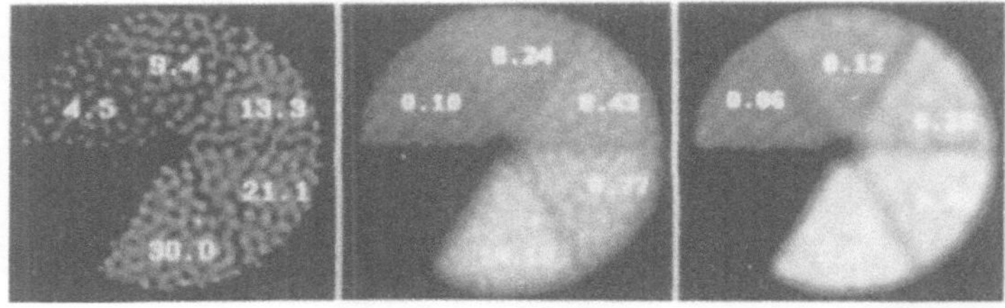

Figure 3.14 Variance in estimating pie phantom intensities, showing sample variances in each of the pie wedges, multiplied by a factor of 10^{-3}. They are in clockwise order starting from ten o'clock. *Left*: 4.5K, 9.4K, 13.3K, 21.2K, 30.0K. *Middle*: 0.1K, 0.24K, 0.43K, 0.77K, 1.41K. *Right*: 0.06K, 0.12K, 0.23K, 0.40K, 0.67K.

The Method of Resolution Kernels

Any intensity function $\lambda(\cdot)$ that is a solution to the linear inverse-problem defined by (3.58) is a stable point of the iteration sequence defined by (3.60). Hence, (3.60) may be regarded as a method for numerically solving (3.58) while enforcing the constraint that the solution must be nonnegative [28]. However, the inverse problem (3.58) is ill-posed for most kernels $p(y \mid x)$, and solutions produced with (3.60), or any other method, are unstable and exhibit the Gibbs' phenomenon and sensitivity to implementation choices, as illustrated in Fig. 3.9. A. Tikhonov [32] and L. Joyce and W. Root [12] address ill-posed linear inverse-problems of this type.

A common cause for the ill-posedness of (3.58) is that the kernels $p(y \mid x)$ encountered in practice suppress high frequencies that are present in $\lambda(\cdot)$ so much that they cannot be recovered practically from the data $\mu(\cdot)$. One approach for alleviating this difficulty is to seek a solution to (3.58) not for $\lambda(\cdot)$ but, rather, for some smoother function having a reduced high frequency content. Let $\{d(w): w \in \mathcal{W}\}$ denote such a *desired function* defined by

$$d(w) = \int_X r(w \mid x)\lambda(x)dx, \qquad (3.91)$$

where $\{r(w \mid x): w \in \mathcal{W}, x \in X\}$ is a given "resolution" kernel, which we assume to be normalized as a conditional probability density, $\int_{\mathcal{W}} r(w \mid \cdot)dw = 1$. Then, $d(\cdot)$ is a version of the intensity $\lambda(\cdot)$ seen with a resolution or point-spread function $r(\cdot \mid \cdot)$.

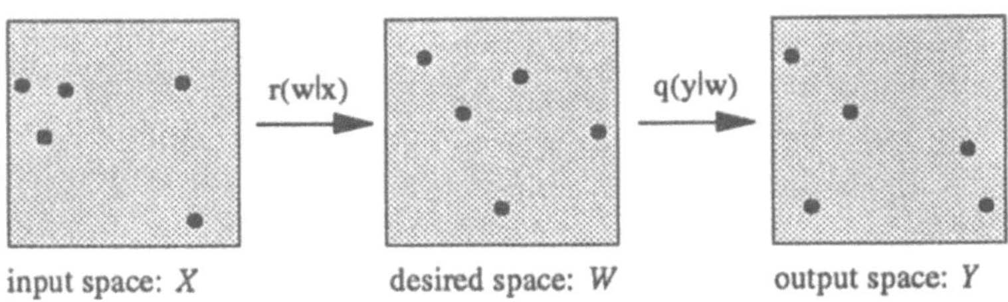

Figure 3.15 Random translations from the input space
X to the desired space W to the output space Y
with transition densities r(w|x) and q(y|w).

The problem of estimating $d(\cdot)$ given a resolution kernel $r(\cdot \mid \cdot)$ and given some measurements $M(\cdot)$ of points on the output space may be addressed by making use of the random translation property of Poisson processes, Thm. 3.2.1. For this purpose, we introduce a hypothetical Poisson process in a "desired" space $\mathcal{W}$ lying between the input space X and the output space $\mathcal{Y}$, as depicted in Fig. 3.15. Translations from X to $\mathcal{W}$ occur independently with density $r(w \mid x)$ and from $\mathcal{W}$ to $\mathcal{Y}$ independently with a density $q(y \mid w)$. The resolution kernel $r(\cdot \mid \cdot)$ is prespecified by knowing the degree and form of blurring that can be tolerated in the approximation of $\lambda(\cdot)$ by $d(\cdot)$. The density $q(\cdot\,|\,\cdot)$ is determined by the requirement that the intensity of the Poisson process on the output space $\mathcal{Y}$ in Fig. 3.15,

$$\int_w q(y\,|\,w)d(w)dw = \int_x\left[\int_w q(y\,|\,x)r(w\,|\,x)dw\right]\lambda(x)dx, \qquad y\in\mathcal{Y}$$

must be equal to the intensity $\{\mu(y)\colon y\in\mathcal{Y}\}$ of the original Poisson process on the output space so that the two Poisson processes are equal with probability one. Hence, we require that $p(\cdot\,|\,\cdot)$ and $r(\cdot\,|\,\cdot)$ be such that there exists a solution $q(\cdot\,|\,\cdot)$ to

$$p(y\,|\,x)=\int_w q(y\,|\,w)r(w\,|\,x)dw. \tag{3.92}$$

Then, by combining (3.58) and (3.92), we see that $d(\cdot)$ is the solution to

$$\mu(y)=\int_w q(y\,|\,w)d(w)dw. \tag{3.93}$$

This equation will generally be better posed than (3.58) for two reasons: $q(\cdot\,|\,\cdot)$ will have greater resolution than $p(\cdot\,|\,\cdot)$, and $d(\cdot)$ will have less high frequency content than $\lambda(\cdot)$. For example, if $p(y\,|\,x)$ is a Gaussian density $\mathcal{N}(x,K_p)$ on $\mathcal{R}^n$, and if $r(w\,|\,x)$ is selected as a Gaussian density $\mathcal{N}(x,K_r)$ on $\mathcal{R}^n$, then a solution $q(y\,|\,w)$ exists and is a Gaussian density $\mathcal{N}(w,K_p-K_r)$ provided that the matrix K_p-K_r is nonnegative definite.

The incomplete-data loglikelihood can be expressed, using (3.92) and (3.93), as a functional of $d(\cdot)$ according to

$$\mathcal{L}_{id}=-\int_{\mathcal{Y}}\int_x p(y\,|\,x)\lambda(x)dxdy + \int_{\mathcal{Y}}\ln\left[\int_x p(y\,|\,x)\lambda(x)dx\right]M(dy)$$

$$=-\int_{\mathcal{Y}}\int_w q(y\,|\,w)d(w)dwdy + \int_{\mathcal{Y}}\ln\left[\int_w q(y\,|\,w)d(w)dw\right]M(dy). \tag{3.94}$$

Thus, the problem of estimating $d(\cdot)$ in terms of measurements $M(\cdot)$ on the output space is entirely equivalent to that of estimating $\lambda(\cdot)$, but with the transition density being $q(\cdot\,|\,\cdot)$ instead of $p(\cdot\,|\,\cdot)$. It then follows from (3.64) that the sequence of functions $d^{(k)}(\cdot)$ defined by

$$\hat{d}^{(k+1)}(w) = \hat{d}^{(k)}(w) \int_{\mathcal{Y}} \left[\frac{q(y \mid w)}{\int_{w} q(y \mid w')\hat{d}^{(k)}(w')dw'} \right] M(dy), \qquad (3.95)$$

produces a corresponding nondecreasing sequence of loglikelihoods $\mathcal{L}_{id}(\hat{d}^{(k+1)}) \geq \mathcal{L}_{id}(\hat{d}^{(k)})$. Equation (3.95) is an expectation-maximization algorithm for producing an unconstrained maximum-likelihood estimate of $d(\cdot)$ from the measurements $M(\cdot)$.

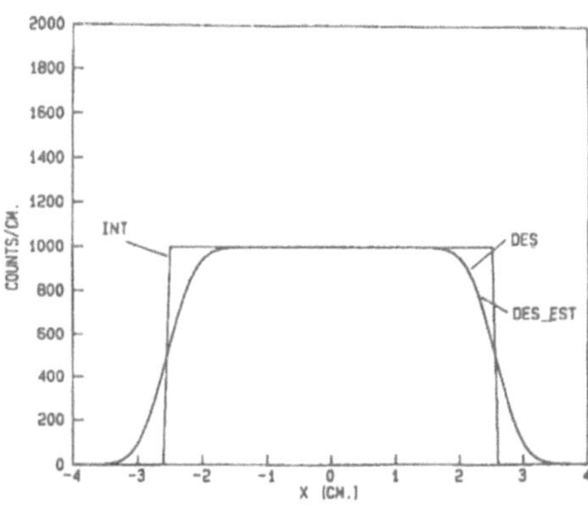

Figure 3.16 Estimate from mean-value data of a rectangular intensity function seen with a finite resolution of 0.8 FWHM, showing the edge artifact of Fig. 3.9 suppressed. (From [28 ©1987 IEEE])

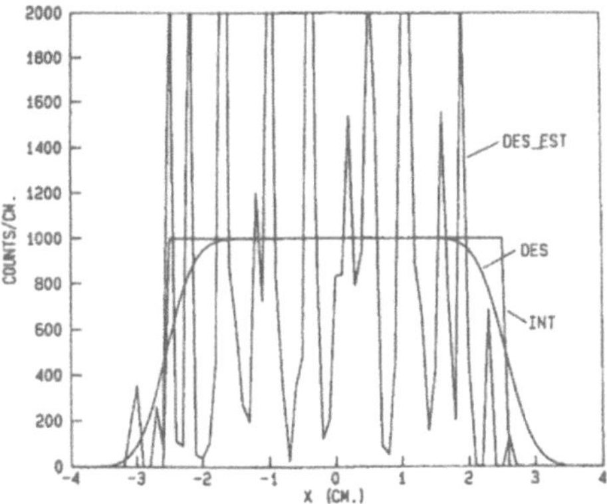

Figure 3.17 Estimate from random data of a rectangular intensity function seen with a finite resolution of 0.8 FWHM, showing the presence of the noise artifact. (From [28 ©1987 IEEE])

The effectiveness of a resolution constraint in reducing the edge artifact is demonstrated in Fig. 3.16. Shown is the same rectangular intensity $\lambda(x)$ as used in producing the results in Fig. 3.9. Also shown as the graph labeled DES is the desired function $d(w)$ obtained from (3.91) by viewing $\lambda(x)$ with a resolution kernel $r(w \mid x)$ selected as a Gaussian density mean of x and a FWHM of 0.8. For this choice, $q(y \mid w)$ is a Gaussian density with a mean of w and a FWHM of 0.6. The graph labeled DES_EST was produced by performing 5000 iterations (3.95) for mean value data (i.e., with $M(du)$ replaced by its mean $\mu(u)\,du$), which are the same data used for Fig. 3.9. The edge artifact seen in Fig. 3.9 is suppressed with the resolution constraint.

Shown in Fig. 3.17 as the graph labeled DES_EST is the result of using (3.95) to estimate $d(\cdot)$, with the same resolution constraint used for Fig. 3.16, but with the random data that was used to produce the intensity estimate in Fig. 3.10. The noise artifact is evident. The results of the previous section suggest that a means for overcoming the rough behavior of this estimate while still suppressing the edge artifact is to use a sieve constraint or a Markov random-field prior while estimating $d(\cdot)$.

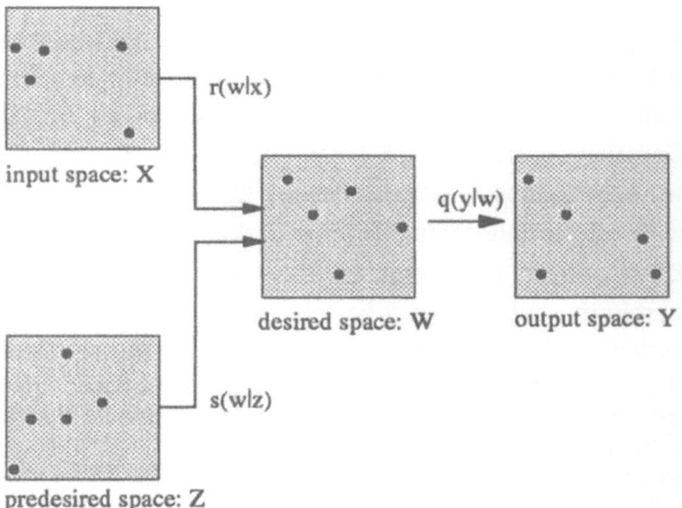

Figure 3.18 Random translations from the input space X to the desired space W with transition density $r(w\,|\,x)$, random translations from the predesired space Z to the desired space Z with transition density $s(w\,|\,z)$, and random translations from the desired space W to the output space Y with transition density $q(y\,|\,w)$.

The required processing for the sieve method becomes evident by examining Fig. 3.18 and using the random translation property of Poisson processes. The desired estimate is obtained by first estimating the preinput intensity $\xi(\cdot)$ on the preinput space Z. Then, the estimate of $d(\cdot)$ is that member of the sieve obtained by operating on the estimate of $\xi(\cdot)$ with the kernel of the sieve. The desired estimate can be produced numerically using the expectation-maximization algorithm according to the following equations:

$$\xi^{(k+1)}(z) = \xi^{(k)}(z) \int_{\mathcal{Y}} \left[\frac{k(y \mid z)}{\int_{z} k(y \mid z')\xi^{(k)}(z')dz'} \right] M(dy), \qquad (3.96)$$

and

$$d^{(k)}(w) = \int_{z} s(w \mid z)\xi^{(k)}(z)dz, \qquad (3.97)$$

where the kernel $k(\cdot \mid \cdot)$ in (3.96) is defined by

$$k(y \mid z) = \int_{w} q(y \mid w)s(w \mid z)dw, \qquad (3.98)$$

in which $q(\cdot \mid \cdot)$ is defined in (3.92), and $s(\cdot \mid \cdot)$ is the kernel of the sieve selected for suppressing the noise artifact. For example, if $p(y \mid x)$ is a Gaussian density $\mathcal{N}(x, K_p)$ on $\mathcal{R}^n$, and if $r(w \mid x)$ and $s(w \mid z)$ are selected as a Gaussian densities $\mathcal{N}(x, K_r)$ and $\mathcal{N}(z, K_s)$ on $\mathcal{R}^n$, respectively, then $q(y \mid w)$ exists and is a Gaussian density $\mathcal{N}(w, K_p - K_r)$ provided that the matrix $K_p - K_r$ is nonnegative definite, and $k(y \mid z)$ is a Gaussian density $\mathcal{N}(z, K_p + K_s - K_r)$. In general, the sieve and resolution kernels can be rather arbitrary, but they must be selected so that $k(\cdot \mid \cdot)$ is nonnegative. The particular choice of equal sieve and resolution kernels results in $k(\cdot \mid \cdot) = p(\cdot \mid \cdot)$, in which case the estimate of $d(\cdot)$ is produced by the unconstrained iterations (3.64) followed by an application of the sieve kernel according to (3.97).

Shown in Fig. 3.19 as the graph labeled DES_EST is the result of using (3.96) and (3.97) to estimate $d(\cdot)$ from the same random data that produced the rough estimate in Fig. 3.17. Gaussian resolution and sieve kernels were used with 0.8 and 1.0 FWHM, respectively. The use of both a resolution and sieve constraint results in an improved estimate in which both the noise and edge artifacts have been suppressed.

3.4.2 Producing Constrained Estimates on Parallel Architectures

Since the introduction by L. Shepp and Y. Vardi [25] of the expectation-maximization algorithm for the computational generation of maximum-likelihood images in emission tomography, there have been a number of investigators who have applied this method to a variety of quantum limited imaging problems. While this iterative approach for producing maximum-likelihood estimates is promising, it is computationally demanding and can result in reconstruction times that are not acceptable for routine applications on conventional computer architectures. We have addressed the computation issues for emission

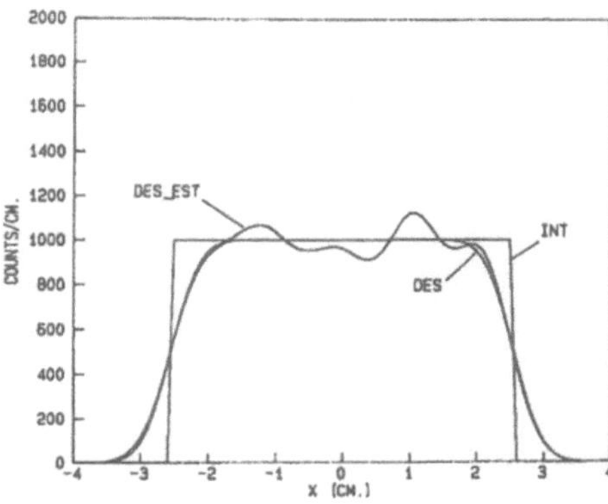

Figure 3.19 Estimate from random data of a rectangular intensity function seen with a finite resolution of 0.8 FWHM and a Gaussian sieve kernel of 1.0 FWHM, showing the noise and edge artifacts suppressed. (From [28 ©1987 IEEE])

tomography via implementations on the class of massively parallel, single-instruction, multiple-data (SIMD) architectures. With the advent of such massively parallel processors implemented as mesh-connected arrays of bit-serial processor elements on single integrated circuits, it is possible by proper restructuring of the computations to implement imaging algorithms with computation times that are several orders of magnitude lower than that obtained with conventional processors. By reformulating the superposition integrals required for projection and back-projection of the EM iterations as solutions of partial differential equations (PDEs), the speed-up required for applications can be attained. The key for achieving this resides in the fact that the PDE solution exploits the most important requirement of mesh-connected processors, which is local-data passage. Computation times on the order of 1 second per iteration have been demonstrated for emission tomography using the Model 610 64×64 Discrete Array Processor made by the Active Memory Technology Co. [20, 22, 35].

3.5 Conclusions

In this chapter, we have developed a model for measurements of a point process that contain translation, insertion, and deletion errors. This model is an important tool in addressing estimation problems in a wide variety of applications involving point-process data. Various constraints are introduced to control instabilities seen in estimates produced with unconstrained maximum-likelihood estimation.

3.6 References

1. E. S. Chornoboy, C. J. Chen, M. I. Miller, T. R. Miller, and D. L. Snyder, "An Evaluation of Maximum-Likelihood Reconstruction for SPECT," *IEEE Transactions on Medical Imaging*, Vol. NS-38, 1989.

2. J. C. Dainty and A. H. Greenaway, "Estimation of Spatial Power Spectra in Speckle Interferometry," *J. Optical Society of America*, Vol. 69, pp. 786-790, May 1979.

3. L. C. de Freitas and J. C. Dainty, "Object Reconstruction from Photon-Limited Centroided Data of Randomly Translating Images," *Optics Letters*, Vol. 13, pp. 264-266, April 1988.

4. A. P. Dempster, N. M. Laird, and D. B. Rubin, "Maximum Likelihood from Incomplete Data via the EM Algorithm," *J. R. Statistical Society B*, Vol. 39, pp. 1-37, 1977.

5. R. D. Evans, *The Atomic Nucleus*, McGraw-Hill, New York, 1955.

6. J. R. Fienup, "Reconstruction of an Object from the Modulus of Its Fourier Transform," *Optics Letters*, pp. 27-29, 1978.

7. J. R. Fienup, "Phase Retrieval Algorithms: A Comparison," *Applied Optics*, Vol. 21, pp. 2758-2769, August 1982.

8. I. J. Good and R. A. Gaskins, "Nonparametric Roughness Penalties for Probability Densities," *Biometrica*, Vol. 58, pp. 255-277, 1971.

9. U. Grenander, *Abstract Inference*, Wiley, New York, 1981.

10. N. Hetterich and G. Weigelt, "Speckle Interferometry Observations of Pluto's Moon Cheron," *Astronomy and Astrophysics*, Vol. 125, pp. 246-248, 1983.

11. T. J. Holmes, D. C. Ficke, and D. L. Snyder, "Modeling of Accidental Coincidences in Both Conventional and Time-of-Flight Positron-Emission Tomography," *IEEE Transactions on Nuclear Science*, Vol. NS-31, pp. 627-631, February 1984.

12. L. S. Joyce and W. L. Root, "Precision Bounds in Superresolution Processing," *Journal of the Optical Society of America*, Vol. 1, pp. 149-168, February 1984.

13. S. Kullback, *Information Theory and Statistics*, Wiley, New York, 1968.

14. K. Lange and R. Carson, "EM Reconstruction Algorithms for Emission and Transmission Tomography," *Journal of Computer Assisted Tomography*, Vol. 8, pp. 306-316, April 1984.

15. S. A. Larsson, *Gamma Emission Tomography*, Acta Radiologica, Supplementum 363, ISSN 0365-5954, P. O. Box 7449, S-103 91, Stockholm, Sweden, 1980.

16. D. Luenberger, *Optimization by Vector Space Techniques*, Wiley, New York, 1969.

17. M. I. Miller, K. B. Larson, J. E. Saffitz, D. L. Snyder, and L. J. Thomas, Jr., "Maximum-Likelihood Estimation Applied to Electron-Microscopic Autoradiography," *J. Electron Microscope Technique*, Vol. 2, pp. 611-636, 1985.

18. M. I. Miller, B. Roysam, J. E. Saffitz, K. B. Larson, D. Fuhrmann, and L. J. Thomas, Jr., "A New Method for the Analysis of electron Microscopic Autoradiographs," *BioTechniques*, Vol. 5, No. 4, pp. 322-328, 1987.

19. M. I. Miller, D. L. Snyder, and T. R. Miller, "Maximum-Likelihood Reconstruction for Single-Photon Emission Computed Tomography," *IEEE Transactions on Nuclear Science*, Vol. NS-32, pp. 769-778, February 1985.

20. M. I. Miller and B. Roysam, "Bayesian Image Reconstruction for Emission Tomography - Incorporating Good's Roughness Prior on Massively Parallel Processors," Proc. National Academy of Sciences, April 1991.

21. M. J. Northcott, G. R. Ayers, and J. C. Dainty, "Algorithms for Image Reconstruction from Photon-Limited Data Using Triple Correlation," *J. Optical Soc. of Amer. A*, Vol. 5, pp. 986-992, July 1988.

22. B. Roysam, J. A. Shrauner, and M. I. Miller, "Bayesian Imaging Using Good's Roughness Measure - Implementation on a Massively Parallel Processor," IEEE ICASSP-88, Vol. M4.21, pp. 932-935, March 1988.

23. T. J. Schulz, "Image Recovery for Randomly Moving Objects," D.Sc. Thesis, Department of Electrical Engineering, Washington University, St. Louis, MO, May 1990.

24. T. J. Schulz and D. L. Snyder, "Imaging a Randomly Moving Object from Quantum Limited Data: Applications to Image Recovery from Second and Third Order Autocorrelation," *J. Optical Society of America A*, Vol. 8, pp. 801-807, May 1991.

25. L. A. Shepp and Y. Vardi, "Maximum Likelihood Reconstruction for Emission Tomography," *IEEE Trans. on Medical Imaging*, Vol. MI-1, pp. 113-121, October 1982.

26. D. L. Snyder, L. J. Thomas, Jr., and M. M. Ter-Pogossian, "A Mathematical Model for Positron-Emission Tomography Systems Having Time-of-Flight Measurements," *IEEE Transactions on Nuclear Science*, Vol. NS-28, No. 3, pp. 3575-3583, 1981.

27. D. L. Snyder, "Utilizing Side Information in Emission Tomography," *IEEE Transactions on Nuclear Science*, Vol. NS-31, No. 1, pp. 533-537, February 1984.

28. D. L. Snyder, M. I. Miller, L. J. Thomas, Jr., and D. G. Politte, "Noise and Edge Artifacts in Maximum-Likelihood Reconstructions for Emission Tomography," *IEEE Transactions on Medical Imaging*, Vol. MI-6, pp. 228-238, September 1987.

29. D. L. Snyder and M. I. Miller, "The Use of Sieves to Stabilize Images Produced with the EM Algorithm," *IEEE Transactions on Nuclear Science*, Vol. NS-32, pp. 3864-3872, October 1985.

30. R. A. Tapia and J. R. Thompson, *Nonparametric Probability Density Estimation*, Johns Hopkins University Press, Baltimore, MD, 1978.

31. M. M. Ter-Pogossian, D. C. Ficke, M. Yamamoto, and J. T. Hood, Sr., "Super PETT I: A Positron Emission Tomograph Utilizing Photon Time-of-Flight Information," *IEEE Transactions on Medical Imaging*, Vol. MI-1, pp. 179-187, 1982.

32. A. Tikhonov, *Solutions to Ill-Posed Problems*, Winston and Sons, N.Y., 1977.

33. Y. Vardi, L. A. Shepp, and L. Kaufman, "A Statistical Model for Positron Emission Tomography," *J. American Statistical Association*, Vol. 80, pp. 8-35, March 1985.

34. C. F. J. Wu, "On the Convergence Properties of the EM Algorithm," *The Annals of Statistics*, Vol. 11, pp. 95-103, 1983.

35.   A. W. McCarthy, R. C. Barrett, and M. I. Miller, "Systolic Implementation of the EM Algorithm for Emission Tomography on a Mesh Connected Processor," Proc. of the 22*nd* Annual Conference on Information Sciences, pp. 373-374, Princeton Univ., Princeton NJ, 1988.

3.7 Problems

3.1.1. Suppose in Fig. 3.4 that the two photons created in the annihilation of a positron emanate from location x and propagate in opposite directions along a line towards detectors located at d_1 and d_2. Denote the line segments from x to each detector by $\ell(x, d_i)$ for $i = 1, 2$. A photon propagating from x will encounter various tissues, bone, and air as it flies towards a detector. As it does so, there is a possibility that it will undergo Compton scatter, with its energy so attenuated as a result that it is not sensed by the detector. If $\{\alpha(z): z \in \mathcal{R}^2\}$ denotes the two-dimensional attenuation density of the medium, as determined by the constituents of the medium, and if η denotes the efficiency of a detector (i.e., fraction of photons striking a detector that are sensed), then the survival probability that a photon flying along $\ell(x, d_i)$ is sensed is

$$s(d_i \mid x) = \eta \exp\left[-\int_{\ell(x, d_i)} \alpha(z)\, dz \right].$$

Assume that photons are attenuated independently and that both must be sensed for an annihilation event to be recorded. Determine the probability that one or both photons created in an annihilation are not sensed. Show that the probability that an annihilation event is recorded is not a function of the position of the annihilation along the flight line.

3.2.1. Let $\{N(A): A \subseteq X\}$ be a Poisson process on a multidimensional space X with intensity $\lambda(x)$. Points of this process are translated randomly and independently to form a point process $\{M(B): B \subseteq \mathcal{Y}\}$ on the space $\mathcal{Y}$. Assume that the two spaces are identical, $\mathcal{Y} = X$, and that points are translated according to a transition density $p(y \mid x)$. A third point process $\{K(C): C \subseteq Z\}$ is formed by superimposing the points in X and $\mathcal{Y}$, where $Z = \mathcal{Y} = X$. There is a point in Z wherever there is a point in either X or $\mathcal{Y}$.

a.   By noting that $K(C) = N(C) + M(C)$, determine $E[K(C)]$, the expected number of points in C, in terms of $\lambda(x)$ and $p(y \mid x)$.

b. Determine the characteristic functional for the process $\{K(C): C \subseteq Z\}$,

$$E\left\{\exp\left[j\int_Z v(z)K(dz)\right]\right\}.$$

c. Determine $E[K(C)]$ using the characteristic functional.

3.3.1. Let the points of a Poisson process with intensity $\lambda(x)$ on an input space X be translated randomly and independently to an output space Y with transition density $p(y \mid x)$. Suppose that $M(Y)=m$ output points are observed at locations $y_1, y_2, \cdots, y_m$. Show that

$$\Pr[N(A)=n \mid y_1, y_2, \ldots, y_m, M(Y)=m]$$

$$= \sum_{\substack{i_1=0 \\ i_1+i_2+\ldots+i_m=n}}^{1} \sum_{i_2=0}^{1} \cdots \sum_{i_m=0}^{1} \prod_{k=1}^{m} (p_k)^{i_k}(1-p_k)^{1-i_k},$$

where $A \subset X$, n is an integer between or equal to 0 and m, and

$$p_k = \frac{\displaystyle\int_A p(y_k \mid x)\lambda(x)\,dx}{\displaystyle\int_X p(y_k \mid x)\lambda(x)\,dx}.$$

3.3.2 Let $\{N(A): A \subset \mathcal{R}^1\}$ be a Poisson process with intensity given by

$$\lambda(x) = \begin{cases} \dfrac{1}{2}\Lambda, & -1 \le x \le 1 \\[2mm] 0, & \text{otherwise,} \end{cases}$$

where Λ is a constant. A second point process $\{M(B): B \subset \mathcal{R}^1\}$ is created by randomly and independently translating each point of N; a point of N at x is translated to become a point of M at $y = x + \varepsilon$, where ε is a Gaussian random variable with zero mean and unit variance.

a. Give an expression for the probability $\Pr(y \in B)$ that a point is translated into the set $B \subset \mathcal{R}^1$.

b. Write a computer program to simulate points of N and M. As a test of your program, compare $\Pr(y \in B)$ and $M(B)/M(\mathcal{R}^1)$ for several choices of B and Λ. Discuss.

c. Let $n = N([-0.5, 0.5])$ be the number of points of N in the interval $[-0.5, 0.5]$. Give an expression for the minimum mean square-error estimate $\hat{n}$ of n in terms of the locations y_1, y_2, $\cdots$ of measured points of M.

d. Use your computer program to form values of n and $\hat{n}$ for several values of Λ. Repeat this for K independent trials, and let $(n_k, \hat{n}_k)$ be the values of these quantities on the kth trial. Estimate the bias and mean square-error of $\hat{n}$ using

$$\hat{\text{bias}} = \frac{1}{K} \sum_{k=1}^{K} (n_k - \hat{n}_k)$$

and

$$\hat{\text{mse}} = \frac{1}{K} \sum_{k=1}^{K} (n_k - \hat{n}_k)^2.$$

Make a graph of the estimated bias and mean square-error as a function of Λ. Discuss.

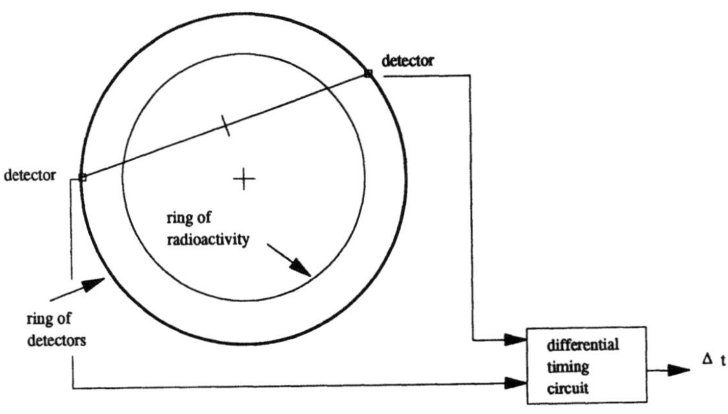

Figure P.3.3.3. Detector ring for emission tomography.

3.3.3. Shown in Fig. P.3.3.3 are two of the detectors in an array of detectors having a ring geometry surrounding a positron-emitting radionuclide. These detectors are precisely located at distances $+L$ and $-L$ from the center of the line connecting them. A radioactive decay produces a positron, which annihilates with an electron thereby producing two high energy photons that propagate at the speed of light, c, in opposite directions along the line connecting the detectors. If the annihilation occurs at position x along the line, and if the timing circuit measures the differential propagation-time Δt of the two annihilation photons perfectly, then x is given by $c\Delta t/2$. However, the timing circuit is imperfect and introduces an error such that the quantity measured is not the true position of the annihilation but rather $y = x + \tau + n$, where τ is a nonrandom offset due to bias misadjustment in the timing circuitry, and n is a zero-mean Gaussian random variable with variance σ^2 due to random noise in the circuitry. In practice, a different but fixed offset can be expected for every pair of opposing detectors around the ring, with each detector being in opposition with several detectors. A method used for calibrating the timing circuits of all the detector pair so as to know and compensate for offset errors is to place a ring of radioactivity within and concentric with the detector array. For the detector pair shown in the figure, such a ring results in a Poisson process modeling the radioactivity distribution and having an intensity of the form

$$\lambda(x) = \frac{\Lambda}{2}[\delta(x + D) + \delta(x - D)],$$

where the ring intersects the line connecting the detector pair at distances $+D$ and $-D$ from the center of the line, and Λ is the expected total-number of annihilations sensed by the detector pair.

a. Assume that annihilations occur as a Poisson process on the input space $X = (-\infty, \infty)$ and that measured locations are modeled as a translated Poisson process on an output space $\mathcal{Y} = (-\infty, \infty)$. The number m and locations $(y_1, y_2, \cdots, y_m)$ of output points are to be used to estimate the offset τ. On the basis of the above description, show that the maximum-likelihood estimate of the offset τ satisfies the following transcendental equation:

$$\hat{\tau} = \frac{1}{m}\sum_{i=1}^{m} y_i - \frac{D}{m}\sum_{i=1}^{m} \tanh\left[\frac{D}{\sigma^2}(y_i - \hat{\tau})\right].$$

This equation is not easily solved for the maximum-likelihood estimate of d except in the special case of Problem 2.4.6 when $D = 0$.

b. Adopt $\{y_1, y_2, \cdots, y_m, d_1, d_2, \cdots, d_m, m\}$ as complete data for use in an expectation-maximization algorithm for numerically producing the solution to the above transcendental equation, where $d_k = +D$ when the annihilation occurs from location $+D$ in the ring, and $d_k = -D$ when the it occurs at $-D$. Show that the complete-data loglikelihood function is:

$$L_{cd}(\tau) = -\frac{m}{2}\ln(2\pi\sigma^2) - \frac{1}{2\sigma^2}\sum_{i=1}^{m}(y_i - d_i - \tau)^2.$$

c. By performing an E-step to determine

$$Q(\tau \mid \tau^{(k)}) = E[L_{cd}(\tau) \mid m, y_1, y_2, \ldots, y_m, \hat{\tau}^{(k)}]$$

and then an M-step by maximizing Q with respect to τ, show that the following is an EM sequence for producing the maximum-likelihood estimate of τ

$$\hat{\tau}^{(k+1)} = \frac{1}{m}\sum_{i=1}^{m}y_i - \frac{D}{m}\sum_{i=1}^{m}\tanh\left[\frac{D}{\sigma^2}(y_i - \hat{\tau}^{(k)})\right].$$

3.3.4. Generalize (3.21) and (3.22) to include insertions and deletions as well as translations. Assume that insertions occur as an independent Poisson process with intensity $\mu_0(y)$ on the output space and that deletions occur with probability $d(y \mid x)$.

3.3.5 Let $r_i = s_i + n_i$, $i = 1, 2, \cdots, N$ be a sequence of N independent, noisy measurements of a signal s that is a Gaussian random-variable with a mean of zero and a variance of v_s. The noise samples are mutually independent and independent of the signal samples, and they are identically distributed Gaussian random variables, each with a mean of zero and a variance of v_n. Assume that v_n is known and that v_s is to be estimated from the incomplete data $\mathbf{r} = [r_1 \ r_2 \ \cdot \ \cdot \ \cdot \ r_N]'$. Let $\hat{v}_s$ denote the estimate of v_s.

a. Show that the maximum-likelihood estimate of v_s is given by

$$\hat{v}_s = \max\left(0, \frac{1}{N}\sum_{i=1}^{N}r_i^2 - v_n\right).$$

b. Select $(s_1, s_2, \cdots, s_N, n_1, n_2, \cdots, n_N)$ as the complete data for an EM algorithm for estimating v_s. Perform the E-step to determine

$$Q(v_s \mid v_s^{(k)}) \overset{\Delta}{=} E[\mathcal{L} \mid \mathbf{r}, v_s^{(k)}].$$

c. Perform the M-step to determine

$$v_s^{(k+1)} \overset{\Delta}{=} \underset{v_s \geq 0}{\operatorname{argmax}} \; Q(v_s \mid v_s^{(k)}).$$

d. Show that stable points of the EM iterations equal the estimate of part (a).

e. Perform a computer simulation to study the behavior of the EM sequence versus iteration number. Graph and discuss your results.

3.3.6 Let

$$\lambda(x) = \sum_{k=1}^{5} \lambda_k I_k(x)$$

be the intensity function of a Poisson process on $\mathcal{R}^1$, where

$$I_k(x) = \begin{cases} 1, & k-1 \leq x < k \\ 0, & \text{otherwise} \end{cases}$$

and where $\lambda_k = 10k$. Let x_i denote the location of the ith point of this process.

a. Write a computer program to simulate the points of this process. Complete the second and third rows of the following table, in which

$$N_k = \sum_{i=1}^{\infty} I_k(x_i)$$

is the number of points realized in $k - 1 \leq x < k$, and $\hat{\lambda}_k$ is the maximum-likelihood estimate of λ_k given the point locations $x_1, x_2, \cdots$.

k	1	2	3	4	5
λ_k	10	20	30	40	50
N_k	.	.	.	.	.
$\hat{\lambda}_k$	.	.	.	.	.
$\hat{\lambda}_k$	.	.	.	.	.

b. Modify your computer program to produce a new set of points on $\mathcal{R}^1$ by translating each point x_i by an independent Gaussian random variable n_i with zero mean and a full width at half maximum of 1 (or standard deviation of $1/\sqrt{8\ln 2}$). Denote the locations of the translated points by $y_1, y_2, \cdots$, where $y_i = x_i + n_i$. Use (3.54) to complete the last row of the above table, where $\hat{\lambda}_k$ is now the maximum-likelihood estimate of λ_k given the point locations $y_1, y_2, \cdots$. Compare the two estimates and discuss your results.

CHAPTER FOUR

COMPOUND POISSON-PROCESSES

4.1 Introduction

One motivation for the model we develop in this chapter is provided by the atmospheric-noise data shown in Fig. 1.3. It is evident that a point process model can account for the occurrence times of the pulses. However, these times alone do not reflect all of the significant features. The amplitudes of the pulses exhibit wide variation and have a strong influence on a radio receiver operating at low frequencies. Even a first-approximation model for low-frequency atmospheric noise should, therefore, include the amplitude as well as occurrence time of each pulse. It is this procedure of endowing each temporal point with an ancillary variable, an amplitude in this instance, which characterizes the models of this chapter.

A *marked point-process* is a point process with an auxiliary variable, called a *mark*, associated with each point. These processes were studied in detail by D. König and K. Matthes [4] and K. Matthes [6, 7]. A mark is used to identify random quantities associated with the point it accompanies.

The notation to be used is indicated in Fig. 4.1.

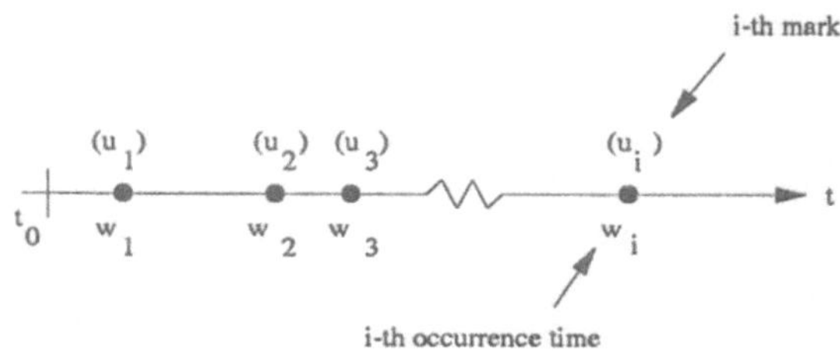

Figure 4.1 A marked point-process.

Each occurrence time w_i in a temporal point-process has associated with it a mark u_i having values in a specified space $\mathcal{U}$ of allowed values. In our development, the mark space will be one of the following:

a.   A denumerable collection $\{U_1, U_2, \cdots, U_n, \cdots\}$, where each U_n is a finite-dimensional vector of real numbers. A typical example is the collection $\{1, 2, \cdots, n, \cdots\}$ of positive integers. The mark n on the ith point can indicate that n events occur simultaneously at the instant w_i. Certain point processes that fail to satisfy the conditions of orderliness can be included in this way as marked point-processes with integer-valued marks.

b.   A region of a finite-dimensional Euclidean space. An example is a semiinfinite real line representing the energies that may be released in a lightning discharge. Another example is a region in two-dimensions or in three-dimensions representing possible positions of some point events of a time-space point-process, such as the spatial location of a decay in a radioactive substance.

More abstract mark spaces can be included, but the full generality possible will not be developed here (see [4]).

There is a useful analog of a counting process for marked point-processes. We term this a *mark-accumulator process*. It is defined as follows. Let $\{N(t): t \geq t_0\}$ be a counting process indicating the total number of points in the interval $[t_0, t)$ regardless of their marks. The mark-accumulator process $\{x(t): t \geq t_0\}$ is defined by

$$x(t) = \sum_{i=0}^{N(t)} u_i, \qquad (4.1)$$

where we identify $u_0 = 0$. Thus, $x(t)$ is the vector sum of the marks on all points occurring in $[t_0, t)$. The paths of $x(\cdot)$ are piecewise constant functions of t with jumps occurring at the random occurrence-times of $N(\cdot)$; the jump at the ith occurrence time w_i is the mark u_i. An example of a marked point-process, its associated counting process $N(\cdot)$, and accumulator process $x(\cdot)$ is shown in Fig. 4.2; the mark space is the real line. It is evident that a marked point-process and its accumulator process are equivalent whenever 0 is not a permissible mark.

In the general theory, the marks $\{u_i\}$ need not form an independent sequence of random variables. Nor is it required that the marks be independent of the counting process $N(\cdot)$ or the occurrence-time sequence $\{w_i\}$. Thus, general marked point-processes encompass a very rich class of interesting and useful models. Our objective in this chapter is more limited. We shall study a particular marked point-process with the following properties:

a. $\{N(t): t \geq t_0\}$ is an inhomogeneous Poisson process with intensity function $\{\lambda(t): t \geq t_0\}$;

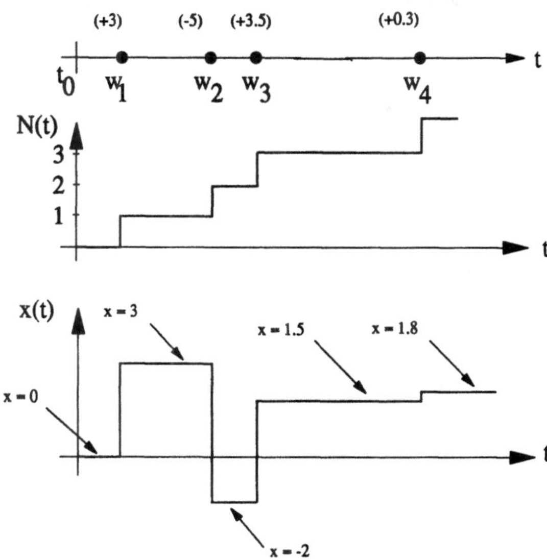

Figure 4.2   A marked point-process and its mark-accumulator process.

b. $\{u_i\}$ is a sequence of mutually independent, identically distributed random variables that are also independent of $N(\cdot)$.

A marked point-process satisfying (a) and (b) is termed a *compound Poisson-process*. We use this same terminology for the associated mark-accumulator process $\{x(t): t \geq t_0\}$. The independence assumptions in (b) are used repeatedly throughout the chapter. They imply that $x(\cdot)$ has independent increments.

Our discussion of compound Poisson-processes is divided into several topics. Statistics for a compound Poisson-process are discussed in Sec. 4.2. The characteristic functional, which completely characterizes the process statistically, is derived, and some of its applications are given. The first- and second-order statistics are derived. In Sec. 4.3, we give some representations for a compound Poisson-process in terms of sums of independent Poisson processes. These representations are useful not only for the insight they provide into the structure of the compound Poisson-process but also in many applications where this process is used as a model. The sample-function density for a compound Poisson-process is also derived. Its interpretation and use are entirely analogous to those of Ch. 2 for the sample-function density of a Poisson process. Problems of statistical inference for observed compound Poisson-processes are examined as an application of the representation and the sample-function density. The relationship of the compound Poisson-process to multi-dimensional Poisson-processes is developed.

4.2 Statistics of Compound Poisson-Processes

Let $\{x(t): t \geq t_0\}$ be a compound Poisson-process. Thus, $x(t)$ is given by (4.1), where $\{N(t): t \geq t_0\}$ and $\{u_i\}$ satisfy the above properties (a) and (b). In this section, we determine the statistics of $x(\cdot)$. These include the characteristic functional of the process and its first- and second-order statistics.

4.2.1 Characteristic Functional for a Compound Poisson-Process

The characteristic functional for a Poisson process is defined in (2.36). We now generalize this to include a sequence of independent random variables assigned as marks to the points of the process. The characteristic functional associated with the random variables $\{x(\tau): t_0 \leq \tau < T\}$ is defined by the following expectation

$$\phi_x(jv) = E\left[\exp\left(j \int_{t_0}^{T} v'(t)dx(t) \right) \right], \tag{4.2}$$

where the prime denotes the transpose operation, and where $\{v(t): t_0 \leq t < T\}$ is an arbitrary vector-valued function. The integral is to be interpreted as a Riemann-Stieltjes integral having the evaluation

$$\int_{t_0}^{T} v'(t)dx(t) = \begin{cases} 0, & N(T) = 0 \\ \sum_{i=1}^{N(T)} v'(w_i)u_i, & N(T) \geq 1. \end{cases}$$

The expectation in (4.2) is easily evaluated using: the properties of conditional expectation, the property of a Poisson process that the occurrence times have the same distribution given $N(T)$ as the order statistics of $N(T)$ independent, identically distributed random variables with the common distribution given in (2.34), and the independence of the marks $\{u_i\}$. Thus,

$$\phi_x(jv) = \sum_{n=0}^{\infty} \Pr(N(T) = n)E\left[\exp\left(j \int_{t_0}^{T} v'(\sigma)dx(\sigma) \right) | N(T) = n \right] \tag{4.3}$$

$$= \Pr(N(T) = 0) + \sum_{n=1}^{\infty} \Pr(N(T) = n)E\left[\exp\left(j \sum_{i=1}^{N(T)} v'(w_i)u_i \right) | N(T) = n \right].$$

By conditioning on the occurrence times and using both the independence and identically distributed properties of the marks, we obtain

$$E\left[\exp\left(j\sum_{i=1}^{N(T)} v'(w_i)u_i\right)\Big|N(T)=n\right]$$

$$=E\left\{E\left[\exp\left(j\sum_{i=1}^{N(T)} v'(w_i)u_i\right)\Big|N(T)=n;w_1,w_2,\ldots,w_{N(T)}\right]\Big|N(T)=n\right\}$$

$$=E\left\{\prod_{i=1}^{N(T)} M_u(jv(w_i))\Big|N(T)=n\right\}, \tag{4.4}$$

where $M_u(j\alpha)$ is the characteristic function for the marks.

$$M_u(j\alpha)\stackrel{\Delta}{=}E(e^{j\alpha'u}), \tag{4.5}$$

We now parallel the argument leading from (2.37) to (2.38) to obtain

$$E\left[\exp\left(j\sum_{i=1}^{N(T)} v'(w_i)u_i\right)\Big|N(T)=n\right]=\left[\frac{\int_{t_0}^{T}\lambda(t)M_u(jv(t))\,dt}{\int_{t_0}^{T}\lambda(t)\,dt}\right]^n. \tag{4.6}$$

Upon substituting this expression into (4.3) and using the Poisson distribution for $N(T)$, we conclude that the characteristic functional for a compound Poisson-process is given by

$$\phi_x(jv)=\exp\left[\int_{t_0}^{T}\lambda(t)(M_u(jv(t))-1)\,dt\right]. \tag{4.7}$$

We now illustrate the use of this characteristic functional with several examples. These follow by making particular selections of the function $v(\cdot)$ and of the space in which marks take values.

Example 4.2.1 *Characteristic Function for $x(t)$* ——————————
The characteristic function of a compound Poisson process at a particular instant can be obtained by specialization of the characteristic functional. For illustration of this, take the function $v(\cdot)$ to be

$$v(\sigma)=\begin{cases}0, & t_0\le\sigma<s\\ \alpha, & s\le\sigma<t\\ 0, & t\le\sigma<T.\end{cases}$$

Then

$$\int_{t_0}^{T} v'(\sigma)dx(\sigma) = \alpha'(x(t) - x(s)),$$

and the characteristic functional of (4.2) becomes the characteristic function for the increment $x(t)$-$x(s)$. Equation (4.7) becomes

$$\phi_x(jv) = M_{x(t)-x(s)}(j\alpha) = \exp\left[\int_s^t \lambda(\sigma)d\sigma(M_u(j\alpha) - 1)\right]. \tag{4.8}$$

The characteristic function for $x(t)$ is obtained upon setting $s = t_0$. Of course, the moments of $x(t) - x(s)$ can be determined by differentiating (4.8) with respect to α and then setting α equal to zero. This is pursued in Sec. 4.2.2 below. It is generally difficult analytically to Fourier invert (4.8) to determine the probability distribution for the increment. ∎

Example 4.2.2 *$x(t)$ has independent increments* ────────────

A compound Poisson-process has independent increments. This is evident from its construction and can be established mathematically by use of the characteristic functional (4.7). For this purpose, let $[s_1, t_1)$, $[s_2, t_2)$, $\cdots$, $[s_k, t_k)$ be a specified collection of k disjoint time intervals, and set

$$v(\sigma) = \begin{cases} \alpha_k, & s_k \leq \sigma < t_k \\ 0, & \text{otherwise,} \end{cases}$$

where $\alpha_1, \alpha_2, \cdots, \alpha_k$ are given constants. This choice for $v(\cdot)$ is a vector analog of that shown in Fig. 2.3. Then (4.7) becomes

$$\phi_x(jv) = \prod_{i=1}^{k} \exp\left[(M_u(j\alpha_i) - 1)\int_{s_i}^{t_i} \lambda(\sigma)d\sigma\right], \tag{4.9}$$

which demonstrates that the compound Poisson process $x(\cdot)$ has independent increments. ∎

Example 4.2.3 *Poisson Process with Nonunit Jumps* ────────────

A Poisson counting process has unit jumps and can be viewed as a special case of a compound Poisson process when the mark space consists of a single value of unity. In this case, $M_u(j\alpha) = \exp(j\alpha)$, and (4.7) reduces to the characteristic functional of a Poisson counting process (2.39). A Poisson process having all jumps of nonunit size U occurs when the mark space consists of the single value U. For this process, we have

$$\phi_x(jv) = \exp[\int_{s_i}^{t_i} \lambda(\sigma)(e^{jv''(\sigma)U} - 1)d\sigma. \tag{4.10}$$

An alternative way to construct this process is to identify it with $UN(t, U)$, where $\{N(t, U): t \geq t_0\}$ is a Poisson counting process that counts jumps of size U in $x(\cdot)$. This alternative construction leads to the representations in Sec. 4.3 for compound Poisson-processes and to space-time Poisson processes. ∎

Example 4.2.4 *Generalized Poisson Counting-Processes* ─────────
Suppose that the mark space is denumerable, $u \in \{U_1, U_2, \cdots, U_k, \cdots\}$ and that the probability a given mark has value U_k is p_k for $k = 1, 2, \cdots$. Then

$$M_u(j\alpha) = \sum_{k=1}^{\infty} p_k e^{j\alpha'U_k}, \tag{4.11}$$

and (4.7) becomes

$$\phi_x(jv) = \exp\left[\int_{t_0}^{T} \lambda(\sigma)\left(\sum_{k=1}^{\infty} p_k e^{jv'(\sigma)U_k} - 1\right)d\sigma\right]. \tag{4.12}$$

Let us examine two special cases of this. For the first, take the mark space to be $\{0,1\}$. The resulting compound Poisson-process can model a defective counter in which points are missed and not counted with probability p. The mark 0 labels these deleted points. The mark accumulator process counts the number of points with unit mark. We see from (4.12) that

$$\phi_x(jv) = \exp\left[\int_{t_0}^{T} (1-p)\lambda(\sigma)(e^{jv(\sigma)} - 1)d\sigma\right]. \tag{4.13}$$

We conclude from this and (2.39) the important fact that the compound Poisson-process resulting from random deletions of points of a Poisson process with intensity function $\lambda(\cdot)$ is a Poisson process with intensity $(1 - p)\lambda(\cdot)$. For the second special case, take the mark space to be $\{1, 2, \cdots, k, \cdots\}$ so that marks are positive integers. The corresponding compound Poisson-process is called a *generalized* Poisson-process and is used as a model when points can occur simultaneously. Here, k points can occur at the same instant with probability p_k. In this way, the orderliness restriction of Ch. 2 is relaxed. We have

$$M_u(j\alpha) = \sum_{k=1}^{\infty} p_k e^{j\alpha k}, \tag{4.14}$$

and the characteristic functional for the mark accumulator process, which counts the total number of points including those which occur simultaneously, is given by

$$\phi_x(jv) = \exp\left[\int_{t_0}^{T} \lambda(\sigma)\left(\sum_{k=1}^{\infty} p_k e^{jv(\sigma)k} - 1 \right) d\sigma \right]. \tag{4.15}$$

The characteristic function for $x(t)$, the number of points in $[t_0, t)$, is

$$M_{x(t)}(j\alpha) = \exp\left[\int_{t_0}^{T} \lambda(\sigma) d\sigma \left(\sum_{k=1}^{\infty} p_k e^{j\alpha k} - 1 \right) \right]. \tag{4.16}$$

The characteristic function of (4.16) is usually difficult to invert to obtain an explicit expression for the counting distribution of $x(t)$. Of course, this inversion is easy when $p_k = \delta_{1k}$, where $\delta_{1k} = 0$ for $k \neq 0$ and $\delta_{11} = 1$, because $x(\cdot)$ is then a Poisson counting-process. Another example where this inversion is possible occurs when $p_k = (-k \ln p) - (1-p)^k$ for $0 < p < 1$ and $k = 1, 2, \cdots$. It is then seen that

$$M_{x(t)}(j\alpha) = \left(\frac{p}{1 - (1-p)e^{j\alpha}} \right)^{\beta(t)}, \tag{4.17}$$

where

$$\beta(t) = \frac{1}{\ln p} \int_{t_0}^{t} \lambda(\sigma) d\sigma;$$

see Problem 4.2.3. This is the characteristic function for the negative binomial distribution. Consequently,

$$\Pr[x(t) = n] = \frac{\Gamma(\beta(t) + n)}{n! \Gamma(\beta(t))} p^{\beta(t)} (1-p)^n \tag{4.18}$$

for $n = 0, 1, 2, \cdots$, where $\Gamma(\cdot)$ is the gamma function. ∎

4.2.2   First- and Second-Order Statistics for a Compound Poisson-Process

The first and second moments of a compound Poisson-process $\{x(t): t \geq t_0\}$ are readily determined using the moment generating properties of the characteristic function (4.8); namely

$$E[x(t)] = \frac{\partial M_{x(t)}(j\alpha)}{j\partial\alpha}\bigg|_{\alpha=0},$$

and

$$E[x(t)x'(t)] = \frac{\partial^2 M_{x(t)}(j\alpha)}{j^2\partial\alpha^2}\bigg|_{\alpha=0},$$

where $\partial M/\partial\alpha$ denotes the gradient vector of M with respect to α (the ith element is $\partial M/\partial\alpha_i$), $\partial^2 M/\partial\alpha^2$ denotes the Hessian matrix of M with respect to α (the i,jth element is $\partial^2 M/\partial\alpha_i\partial\alpha_j$), and the gradient and Hessian are evaluated at $\alpha = 0$. By direct calculation using these relations and (4.8), we conclude that

$$E[x(t)] = E(u)\int_{t_0}^{t}\lambda(\sigma)d\sigma, \tag{4.19}$$

and

$$\Sigma(t) \overset{\Delta}{=} E[x(t)x'(t)] - E[x(t)]E[x'(t)] = E(uu')\int_{t_0}^{t}\lambda(\sigma)d\sigma. \tag{4.20}$$

We term $\Sigma(t)$ the *instantaneous covariance-matrix* of $x(t)$. Thus, the expected value of $x(t)$ is the product of the mean value of the marks and the mean number of points in $[t_0, t)$. Similarly, the instantaneous covariance-matrix of $x(t)$ is the product of the mean square value of the marks and the variance of the number of points in $[t_0, t)$.

The covariance matrix

$$K_x(s,t) \overset{\Delta}{=} E[x(s)x'(t)] - E[x(s)]E[x'(t)]. \tag{4.21}$$

can be calculated using the independent increments property of the compound Poisson-process. Thus, for $t_0 \leq s < t$, we have

$$K_x(s,t) = E[x(s)(x(t)-x(s))'] + E[x(s)x'(s)] - E[x(s)]E[x'(t)].$$

$$(4.22)$$

As $K_x(s,t) = K_x(t,s)$, we obtain for $t_0 \leq t < s$ that $K_x(s,t) = \Sigma(t)$. Consequently, by combining these results and using the property that $\Lambda(t) = \int_{t_0}^{t} \lambda(\sigma)d\sigma$ is a nondecreasing function of t, we have

$$K_x(s,t) = \Sigma(\min(s,t)) = E(uu')\min[\Lambda(t),\Lambda(s)].\qquad(4.23)$$

The first- and second-moment statistics for inhomogeneous and homogeneous Poisson-processes follow from those of a compound Poisson-process by specializing the mark space. These are summarized in Table 4.1.

Table 4.1. First- and Second-Order Statistics for Poisson Processes

Type of Poisson Process	Compound	Inhomogeneous	Homogeneous
Mean	$E(u)\Lambda(t)$	$\Lambda(t)$	$\lambda(t-t_0)$
Instantaneous Covariance, $\Sigma(t)$	$E(uu')\Lambda(t)$	$\Lambda(t)$	$\lambda(t-t_0)$
Covariance, $K_x(s,t)$	$E(uu')$ $\times\min[\Lambda(s),\Lambda(t)]$	$\min[\Lambda(s),\Lambda(t)]$	$\lambda\min(s-t_0, t-t_0)$

Summary. The statistics of a compound Poisson-process have been developed in this section. The important points are:

a. The characteristic functional completely characterizes a random process statistically. The evaluation of this quantity is given in (4.2) for a compound Poisson-process.

b. The characteristic function for $x(t)$ is given in (4.8). The mean and instantaneous covariance-matrix for $x(t)$ follow directly from this and are given in (4.19) and (4.20), respectively.

c. The probability distribution function for $x(t)$ is generally difficult to determine.

d.   The characteristic functional for a generalized Poisson process is given in (4.15). This process relaxes the orderliness restriction of Chapter 2 in that points can occur simultaneously.

4.3 Representation of Compound Poisson-Processes

A compound Poisson-process can be represented as a superposition of independent Poisson counting-processes. This representation not only provides insight into the structure of the process, but also it proves useful in applications. The representation will also be important in Ch. 5 where we study the influence of compound Poisson-processes on dynamical systems.

We will first develop the representation when the mark space is denumerable. Thus, let $\{x(t): t \ge t_0\}$ be a compound Poisson-process with a mark space $\mathcal{U} = \{U_1, U_2, \cdots, U_k, \cdots\}$, where the kth mark U_k occurs with probability p_k. Suppose that $t_0 = t_{0n} < t_{1n} < t_{2n} < \cdots < t_{nn} = t$ is a sequence of partitions of $[t_0, t)$ such that

$$\lim_{n \to \infty} \max_{1 \le j \le n} (t_{jn} - t_{j-1,n}) = 0.$$

Define the processes $\{x(t, U_k): t \ge t_0\}$, for $k = 1, 2, \cdots$ by

$$x(t, U_k) = \lim_{n \to \infty} \sum_{j=1}^{n} (x(t_{jn}) - x(t_{j-1,n})) I_{\{U_k\}}[x(t_{jn}) - x(t_{j-1,n})], \qquad (4.24)$$

where $I_{\{U\}}[x]$ is a mark indicator function defined by

$$I_{\{U\}}[x] = \begin{cases} 1, & x = U \\ 0, & x \ne U. \end{cases}$$

Thus, $x(t, U_k)$ is the sum of the marks that occur during $[t_0, t)$ and have value U_k. It is evident that for $t \ge t_0$ we almost surely have

$$x(t) = \sum_{k=1}^{\infty} x(t, U_k) = \sum_{k=1}^{\infty} U_k N(t, U_k), \qquad (4.25)$$

where $N(t, U_k)$ is the number of points that occur during $[t_0, t)$ and have mark U_k. This leads to the decomposition and reconstruction of $x(\cdot)$ shown in Fig. 4.3. The important features of this figure are that $\{x(t): t \ge t_0\}$ can be decomposed into the counting processes $\{N(t, U_1): t \ge t_0\}$, $\{N(t, U_2): t \ge t_0\}$, $\cdots$, $\{N(t, U_k): t \ge t_0\}$, $\cdots$ and that

$\{x(t): t \geq t_0\}$ can be reconstructed in terms of these counting processes according to (4.25). The question is: *How are these counting processes characterized statistically?* The answer is in the following theorem.

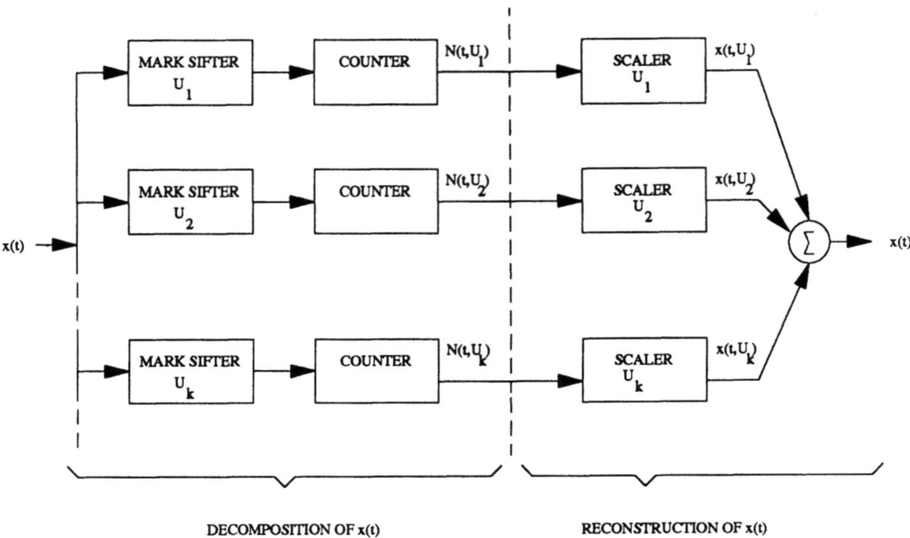

Figure 4.3. Decomposition and reconstruction of a compound Poisson-process.

Theorem 4.3.1 (*Representation for a Compound Poisson-Process Having a Denumerable Mark Space*) Let $\{x(t): t \geq t_0\}$ be a compound Poisson-process with a denumerable mark space $\{U_1, U_2, \cdots, U_k, \cdots\}$. Points occur as a Poisson process with integrable intensity $\{\lambda(t): t \geq t_0\}$, and the probability that mark U_k occurs on a particular point is p_k independently of marks on other points. Let $\{N(t, U_k): t \geq t_0\}$ count points having mark U_k. Then the counting processes $\{N(t, U_k): t \geq t_0\}$, for $k = 1, 2, \cdots$, are mutually independent Poisson counting-processes for which the kth process has intensity $\{p_k \lambda(t): t \geq t_0\}$. In terms of these, we have the representation for $x(\cdot)$ as

$$x(t) = \sum_{k=1}^{\infty} U_k N(t, U_k), \tag{4.26}$$

for $t \geq t_0$.

Proof. We parallel a constructive argument due to L. Breiman [1, p. 311] to demonstrate that the processes defined in (4.24) are mutually independent Poisson processes of the type in Ex. 4.2.3 such that $\{x(t,U_k): t \geq t_0\}$ has jumps of size U_k and intensity $\{p_k\lambda(t): t \geq t_0\}$. The representation in (4.26) is an immediate consequence of this. Thus, construct processes $\{x^*(t,U_k): t \geq t_0\}$, for $k = 1,2,\cdots$, which are mutually independent and with characteristic functional on $[t_0, t)$ given by

$$\exp\left[\int_{t_0}^{T} p_k\lambda(\sigma)\left(e^{jv'(o)U_k} - 1\right)d\sigma\right].$$

Hence, from (4.10), $\{x^*(t,U_k): t \geq t_0\}$ is a Poisson process with jumps of size U_k and intensity $p_k\lambda(t)$. It follows that $\{x(t,U_k): t \geq t_0\}$ equals $\{x^*(t,U_k): t \geq t_0\}$ for each k *separately* because they have the same characteristic functional. Our objective is to show that this equality holds jointly for all k *simultaneously*. For this purpose, define

$$x^*(t) = \sum_{k=1}^{\infty} x^*(t,U_k).$$

From this construction and (4.12), it is evident that $\{x^*(t,U_k): t \geq t_0\}$ has the same characteristic functional on $[t_0, T)$ as $\{x(t,U_k): t \geq t_0\}$ for every $T \geq t_0$. Therefore, $\{x^*(t,U_k): t \geq t_0\}$ and $\{x(t,U_k): t \geq t_0\}$ have the same finite distributions. But, by the construction

$$x^*(t,U_k) = \lim_{n \to \infty} \sum_{j=1}^{\infty} (x^*(t_{jn}) - x^*(t_{j-1,n}))I_{\{U_k\}}(x^*(t_{jn}) - x^*(t_{j-1,n})).$$

Hence, $\{x^*(t,U_k): t \geq t_0\}$ is the same function on $\{x^*(t): t \geq t_0\}$ as $\{x(t,U_k): t \geq t_0\}$ is on $\{x(t): t \geq t_0\}$. It follows that the processes $\{x(t): t \geq t_0\}$, $\{x(t,U_k): t \geq t_0\}$, $k = 1,2,\cdots$, have the same joint distributions as the processes $\{x^*(t): t \geq t_0\}$, $\{x^*(t,U_k): t \geq t_0\}$, $k = 1,2,\cdots$. $\square$

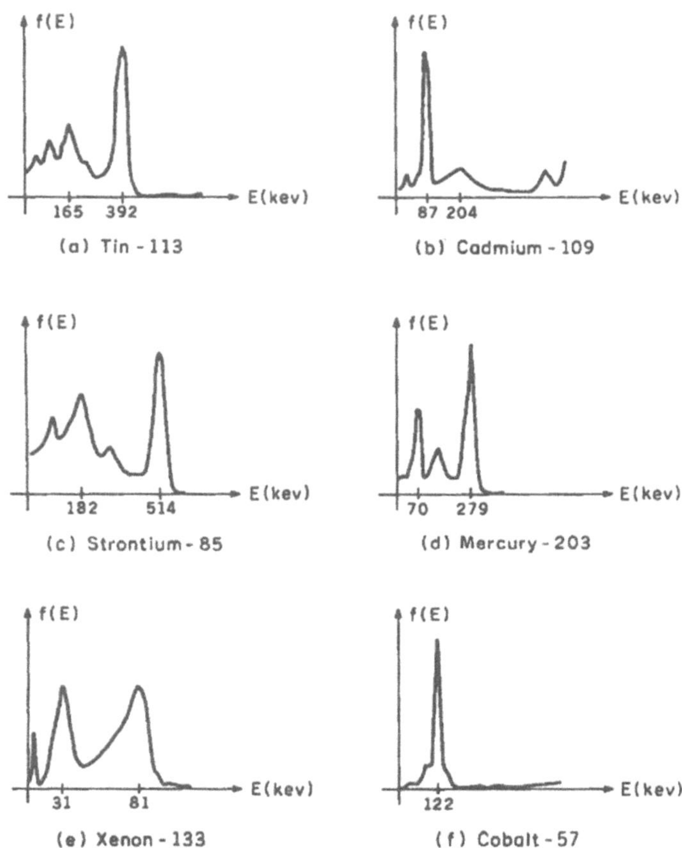

Figure 4.4. Energy spectra for several radioisotopes observed with an Anger scintillation camera.

Example 4.3.1 *Multichannel Analysis* ━━━━━━━━━━━━━━━

Suppose that a quantity of radioactive material is observed using a scintillation detector. Light flashes in the scintillating crystal of such a detector have intensities that depend on the energies of the incident gamma photons emitted from the source. The spectrum of energies for these light pulses in a given detector is characteristic of the radioactive element. Examples of energy spectra for several isotopes are shown in Fig. 4.4; these were obtained using an *Anger scintillation camera* by N. Mullani and H. Huang [8].

Denote the energy spectrum of an isotope by $f(E)$, and partition energy into disjoint energy bins $[E_k, E_k + \Delta E_k)$, $k = 1, 2, \cdots$. We then model the sequence of light flashes in idealized form as a marked point-process

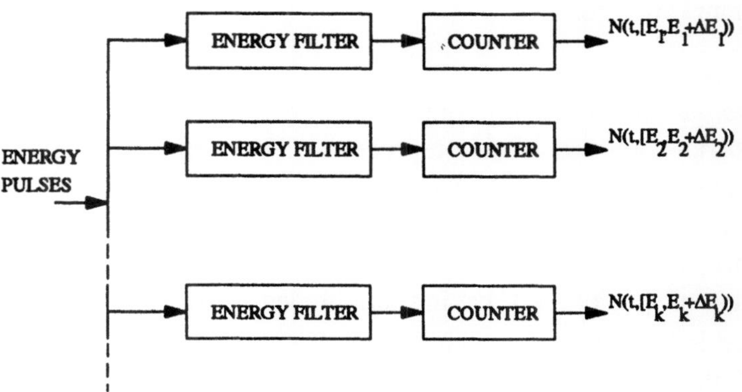

Figure 4.5. Multichannel analyzer.

in which marks are mutually independent and integer valued with integer k indicating a photon having energy in the k*th* energy bin. The probability that a photon arrives having energy in the k*th* bin is

$$p_k = \frac{\displaystyle\int_{E_k}^{E_k+\Delta E_k} f(E)\,dE}{\displaystyle\int_0^\infty f(E)\,dE}.$$

Suppose that light flashes occur at instants corresponding to a Poisson process with intensity $\{\lambda(t): t \ge t_0\}$. Here, $\lambda(t)$ will be an exponentially decreasing function of time. The marked point-process defined in this way is a compound Poisson-process, and Theorem 4.3.1 implies that the counting processes $\{N(t,[E_k,E_k+\Delta E_k)): t \ge t_0\}$, $k=1,2,\cdots$, shown in Fig. 4.5 are independent Poisson counting processes with corresponding intensities $p_k\lambda(\cdot)$. In practice, the decomposition shown in Fig. 4.5 is implemented using a *multichannel analyzer*; see D. Ross and C. Harris [9, p. 160]. The device works by first generating narrow pulses having amplitudes proportional to energy. The energy filter is then easily realized with an analog-to-digital converter, and for each k, the counts $N(t,[E_k,E_k+\Delta E_k))$ are accumulated in a register for display or they are assigned a memory location in a digital computer for numerical analysis.

In this way, $2^{13} = 8192$ parallel channels are easily achieved. The important consequence of Theorem 4.3.1 is that the collection of Poisson counting-processes is, to within the quantization implied by the finite energy bins, equivalent to the marked point-process at the input to the multichannel analyzer. ∎

We now remove the restriction that the mark space be denumerable. Our objective is to obtain a continuous analog of (4.26) as a representation of a general compound Poisson-process. Thus, let $\{x(t): t \geq t_0\}$ be a compound Poisson-process with mark space $\mathcal{U}$. For our purposes, $\mathcal{U}$ may be taken as a possibly unbounded region of a finite-dimensional Euclidean space; then, each mark $u \in \mathcal{U}$ is a vector of real numbers.

Jumps of $\{x(t): t \geq t_0\}$ having amplitudes in disjoint regions of $\mathcal{U}$ contribute independent compound Poisson components to $x(\cdot)$. This is seen as follows. Partition $\mathcal{U}$ into disjoint regions $B_1, B_2, \cdots, B_k, \cdots$ such that $\mathcal{U} = \cup_{k=1}^{\infty} B_k$. Also, let $t_0 = t_{0n} < t_{1n} < t_{2n} < \cdots < t_{nn} = t$ be a sequence of partitions of $[t_0, t)$ such that

$$\lim_{n \to \infty} \max_{1 \leq j \leq n} (t_{jn} - t_{j-1,n}) = 0.$$

Now define the processes $\{x(t, B_k): t \geq t_0\}$, for $k = 1, 2, \cdots$, according to

$$x(t, B_k) = \lim_{n \to \infty} \sum_{j=1}^{n} (x(t_{jn}) - x(t_{j-1,n})) I_{\{B_k\}}[x(t_{jn}) - x(t_{j-1,n})], \tag{4.27}$$

where $I_{\{B_k\}}[x]$ is a mark indicator function defined by

$$I_{\{B_k\}}(x) = \begin{cases} 1, & x \in B_k \\ 0, & x \notin B_k. \end{cases}$$

These definitions generalize (4.24). The process $\{x(t, B_k): t \geq t_0\}$ is a mark-accumulator process for those points of the marked point-process having a mark in B_k. It is evident that for $t \geq t_0$, we almost surely have

$$x(t) = \sum_{k=1}^{\infty} x(t, B_k) \tag{4.28}$$

for $t \geq t_0$. By an argument entirely analogous to that used for the proof of Theorem 4.3.1, it follows that the processes $\{x(t, B_k): t \geq t_0\}$, $k = 1, 2, \cdots$, are mutually independent, compound Poisson-processes with corresponding intensities $\{p_k\lambda(t): t \geq t_0\}$, where p_k is the probability that a mark takes on a value in B_k.

We now obtain a continuous analog of (4.26) for a compound Poisson-process with a nondenumerable mark space. We will assume that the mark space is a finite-dimensional Euclidean space. I. Gikhman and A. Skorokhod [2, p. 270] develop a similar representation when the mark space is a more general, measurable space.

Let $B_1, B_2, \cdots, B_k, \cdots$ be disjoint regions whose union equals the mark space $\mathcal{U}$, as before. Assume that the diameters of these regions do not exceed δ, where the diameter of B_k is the maximum distance between any two points in B_k. Denote by $N(t, B_k)$ the number of points in $[t_0, t)$ having marks in B_k. By the above argument, the processes $\{N(t, B_k): t \geq t_0\}$, for $k = 1, 2, \cdots$, are mutually independent Poisson counting-processes with corresponding intensities $\{p_k\lambda(t): t \geq t_0\}$. Let U_k denote an arbitrary point in B_k, and denote by $\| U_k \|$ the largest distance from U_k to any other point in B_k. Then $\| U_k \| < \delta$ and (4.28) imply that

$$|x(t) - \sum_{k=1}^{\infty} U_k N(t, B_k)| \leq \sum_{k=1}^{\infty} \| U_k \| N(t, B_k) \leq \delta N(t),$$

where $N(t)$ is the total number of points occurring in $[t_0, t)$ regardless of their mark. We note for the last term on the right side that

$$E[\delta N(t)] = \delta \int_{t_0}^{t} \lambda(\sigma) d\sigma \quad \text{and} \quad \text{var}[\delta N(t)] = \delta^2 \int_{t_0}^{t} \lambda(\sigma) d\sigma.$$

We assume that $\lambda(\cdot)$ is integrable. Consequently, for a partition $B_1, B_2, \cdots, B_k, \cdots$ of $\mathcal{U}$ such that $\delta \to 0$, we have that

$$\lim_{\delta \to 0} E\left(|x(t) - \sum_{k=1}^{\infty} U_k N(t, B_k)|^2 \right) = 0.$$

Thus, the sum $\sum_{k=1}^{\infty} U_k N(t, B_k)$ converges to a limit in the mean-square sense as δ tends to zero. We denote this limit by $\int_{\mathcal{U}} U N(t, dU)$ and conclude that a compound Poisson-process with a nondenumerable mark space can be represented by

$$x(t) = \int_{\mathcal{U}} U N(t, dU), \tag{4.29}$$

which is the continuous analog of (4.26).

The interpretation of the term $N(t,dU)$ appearing in (4.29) is that $\{N(t,dU): t \geq t_0\}$ is a Poisson process that counts the number of points in $[t_0,t)$ having marks in the elemental volume $[U,U+dU)$. The corresponding intensity function is $[P_u(U+dU)-P_u(U)]\lambda(t)$, where $P_u(U)$ is the probability-distribution function for the mark random variable. The processes $\{N(t,dU_j): t \geq t_0\}$, $j=1,2,\cdots$, are mutually independent for disjoint elemental volumes. Finally,

$$N(t,B_k) = \int_{B_k} N(t,dU) \tag{4.30}$$

is a Poisson counting-process associated with counting points having marks in B_k. The intensity for $\{N(t,B_k): t \geq t_0\}$ is

$$\lambda(t)\int_{B_k} P_u(dU) \stackrel{\Delta}{=} \lambda(t)p_k.$$

An important additional interpretation for the representation in (4.29) can be given in terms of the multidimensional Poisson-processes of Ch. 2. Here, we view marked points as occurring in a multidimensional space $[t_0,\infty)\times \mathcal{U}$; each point has a time coordinate in $[t_0,\infty)$ and a spatial coordinate in $\mathcal{U}$. Let T be an interval of time in $[t_0,\infty)$ and B a subset of $\mathcal{U}$, and denote by $M(T,B)$ the number of points in $T\times B$; this is the number of points with time coordinates in T and marks in B. It follows from the above discussion that $\{M(T,B): T \in [t_0,\infty), B \in \mathcal{U}\}$ is a Poisson process on the multidimensional space $[t_0,\infty)\times \mathcal{U}$. That is, the numbers of points in disjoint regions of $[t_0,\infty)\times \mathcal{U}$ are independent, and the number of points in a region $T\times B$ is Poisson distributed with parameter

$$\int_T \int_B \lambda(t)\,dt\,P_u(dU).$$

This process M is a particular multidimensional, time-space Poisson-process in which the general intensity $\lambda(t,u)$ factors into the product of a function only of t and a function only of u. If the mark random-variable is continuous so that a probability density $p_u(U)$ exists, the intensity for this time-space Poisson process is $\lambda(t)p_u(U)$. Furthermore,

$$N(t,dU) = \int_{t_0}^{t} M(d\sigma, dU),$$

so that (4.29) can also be written as

$$x(t) = \int_{t_0}^{t} \int_{\mathcal{U}} UM(d\sigma, dU).$$

Thus, a compound Poisson-process can be represented in terms of a time-space Poisson process, as summarized in the following theorem.

Theorem 4.3.2 (*Representation for a Compound Poisson-Process Having a Nondenumerable Mark Space*). Let $\{x(t): t \geq t_0\}$ be a compound Poisson process with a nondenumerable mark space $\mathcal{U}$. Points occur as a Poisson process with integrable intensity $\{\lambda(t): t \geq t_0\}$, and the probability distribution function for the marks is $P_u(U)$. Then, $x(t)$ can be represented as

$$x(t) = \int_{t_0}^{t} \int_{\mathcal{U}} UM(d\sigma, dU), \qquad (4.31)$$

where $\{M(T,B): t \in [t_0, \infty), B \in \mathcal{U}\}$ is a time-space Poisson-process with a factorable intensity such that

$$E[M(T,B)] = \int_{T} \int_{B} \lambda(t)\, dt\, P_u(dU).$$

If the marks are continuous random variables with probability density $p_u(U)$, then the intensity for M is $\lambda(t)p_u(U)$.

4.4 Estimation for Compound Poisson-Processes

Problems of statistical inference for compound Poisson-processes arise frequently in practice when these processes are used as models for observed data. We will now examine such problems briefly to indicate the usefulness of the representations of Theorems 4.3.1 and 4.3.2. Only maximum-likelihood parameter estimation is considered. The basic framework for solving these problems was developed in Sec. 2.4 of Ch. 2. Recall that the sample-function density played a central role in parameter estimation for observed Poisson-processes and that the likelihood function is determined by it. The same is true here for the sample-function density of a compound Poisson-process, so we begin by developing expressions for this important quantity.

Consider, first, a marked point-process $\{x(t): t \geq t_0\}$ with a denumerable mark-space $\mathcal{U} = \{U_1, U_2, \cdots\}$. Let $\{N(t): t \geq t_0\}$ count points of this process regardless of their mark. The quantity $p[\{x(\sigma): t_0 \leq \sigma < t\}]$ defined below is by definition the sample-function density for this process on the interval $[t_0, t)$

$$p[\{x(\sigma): t_0 \leq \sigma < t\}]$$

$$= \begin{cases} \Pr[N(t) = 0], & N(t) = 0 \\ p_w(W, u_1 = \xi_1, \ldots, u_n = \xi_n, N(t) = n), & N(t) = n \geq 1, \end{cases} \quad (4.32)$$

where

$$p_w(W, u_1 = \xi_1, \ldots, u_n = \xi_n, N(t) = n) = \Pr(N(t) = n \mid w = W, u_1 = \xi_1, \ldots, u_n = \xi_n)$$

$$\times \Pr(u_1 = \xi_1, \ldots, u_n = \xi_n \mid w = W) p_w^{(n)}(W),$$

and where $\xi_i \in \mathcal{U} = \{U_1, U_2, \cdots\}$ for each i, $1 \leq i \leq n$. The first factor in the product on the right-hand side is the conditional probability that $N(t) = n$ given the n occurrence times and marks in $[t_0, t)$; the second factor is the conditional probability of the n marks given the n occurrence times. As with (2.30), the sample-function density (4.32) may be interpreted roughly as the probability of obtaining a realization of the marked point-process on $[t_0, t)$ with $N(t) = n$ points located at instants $w_1 = W_1$, $\cdots, w_n = W_n$ and corresponding marks $u_1 = \xi_1, \cdots, u_n = \xi_n$.

The sample-function density for a compound Poisson-process with a denumerable mark-space is readily determined from the general definition in (4.32) because of the independence properties in the construction of this process. Let $P(U_k)$ denote the probability that mark U_k occurs on a point. Then,

$$p_w(W, u_1 = \xi_1, \ldots, u_n = \xi_n, N(t) = n) = \Pr(N(t) = n \mid w = W) p_w^{(n)}(W) \prod_{i=1}^{n} P(\xi_i).$$

$$(4.33)$$

Furthermore, from (2.30), (2.31), and the definition of a compound Poisson-process, we have that

$$p[\{x(\sigma): t_0 \le \sigma < t\}]$$

$$= \begin{cases} \exp\left[-\int_{t_0}^{t} \lambda(\sigma)d\sigma \right], & N(t)=0 \\[2ex] \left[\prod_{i=1}^{n} \lambda(W_i)P(\xi_i) \right] \exp\left[-\int_{t_0}^{t} \lambda(\sigma)d\sigma \right], & N(t)=n \ge 1. \end{cases} \qquad (4.34)$$

We find it convenient to incorporate into (4.34) a particular sample function of the marked point-process on $[t_0, t)$; namely, the sample function for which $N(t)=n$, and, for $n \ge 1$, $w_1 = W_1, \cdots, w_n = W_n$, $u_1 = \xi_1, \cdots$, $u_n = \xi_n$. Let $N(t, U_i)$ be the number of marks among $\xi_1, \cdots, \xi_n$ equal to U_i, for $i = 1, 2, \cdots$, where $\sum_{i=1}^{\infty} N(t, U_i) = n$. Then, for this sample function, (4.34) can be written alternatively as

$$p[x(\sigma): t_0 \le \sigma < t\}] \qquad (4.35)$$

$$= \exp\left[-\int_{t_0}^{t} \lambda(\sigma)d\sigma + \int_{t_0}^{t} \ln\lambda(\sigma)N(d\sigma) + \sum_{i=1}^{\infty} \ln[P(U_i)]N(t, U_i) \right],$$

where the counting integral is defined in (2.32). Thus, the sample-function density for a compound Poisson process splits into two factors, $\exp\left[-\int_{t_0}^{t} \lambda(\sigma)d\sigma + \int_{t_0}^{t} \ln[\lambda(\sigma)]N(d\sigma) \right]$ and $\exp\{\sum_{i=1}^{\infty} \ln[P(U_i)]N(t, U_i)\}$. The first is contributed by the number of points and their occurrence times. This factor is just the sample-function density of (2.32) for a Poisson process. The second term is contributed by the marks; it is the probability of the marks $\xi_1, \xi_2, \cdots, \xi_n$ occurring in the realization.

There is another way to write (4.35) that provides a useful interpretation of the sample-function density for a compound Poisson-process. By using $N(t) = \sum_{i=1}^{\infty} N(t, U_i)$ and $\sum_{i=1}^{\infty} P(U_i) = 1$, we rewrite (4.35) in the form

$$p[x(\sigma): t_0 \le \sigma < t\}] = \prod_{i=1}^{\infty} p[\{N(\sigma, U_i): t_0 \le \sigma < t\}], \qquad (4.36)$$

where

$$p[\{N(\sigma, U_i): t_0 \le \sigma < t\}]$$

$$= \exp\left[-\int_{t_0}^{t} P(U_i)\lambda(\sigma)d\sigma + \int_{t_0}^{t} \ln[P(U_i)\lambda(\sigma)]N(d\sigma, U_i) \right].$$

Thus, the sample-function density for a compound Poisson-process factors into the product of the sample-function densities for each of the independent components in the representation for $\{x(t): t \geq t_0\}$ in Theorem 4.3.1.

The use of the sample-function density for estimating unknown parameters that influence the intensity of a compound Poisson-process parallels its use in Ch. 2 for a Poisson process. Let $\{x(t): t \geq t_0\}$ be a compound Poisson-process with a denumerable mark-space $\mathcal{U} = \{U_1, U_2, \cdots\}$. Suppose that points occur with intensity $\{\lambda(t:\theta): t \geq t_0\}$ and that the probability mark U_k occurs is $P(U_k:\theta)$, where θ is a collection of unknown parameters. Observations of $x(\cdot)$ are made on the interval $[t_0, T)$ in order to estimate these parameters. Suppose that the data collected are the number, occurrence times, and marks for all points with occurrence times in $[t_0, T)$. From Sec. 2.4 of Ch. 2, the maximum-likelihood estimate of θ in terms of observations $\{N(T) = n, w_1 = W_1, \cdots, w_n = W_n, u_1 = \xi_1, \cdots, u_n = \xi_n\}$ is the value of θ that maximizes the probability of having observed these data. Just as in Sec. 2.4, the maximum-likelihood estimate of θ maximizes the sample-function density or, equivalently, its logarithm evaluated at the given data. Thus, from (4.35), the estimate maximizes

$$L(\theta) = \ln p[\{x(\sigma): t_0 \leq \sigma < T\}:\theta] \tag{4.37}$$

$$= -\int_{t_0}^{T} \lambda(\sigma, \theta)d\sigma + \int_{t_0}^{t} \ln\lambda(\sigma, \theta)N(d\sigma) + \sum_{i=1}^{\infty} \ln[P(U_i:\theta)]N(T, U_i),$$

as a function of θ, where the right-hand side is evaluated at the observed realization of $\{x(\sigma): t_0 \leq \sigma < T\}$. Except in especially simple situations, this maximization will need to be performed numerically.

We see from (4.37) that the marks can be disregarded and not measured if the mark distribution $P(U_k:\theta)$ is not a function of θ. Similarly, the occurrence times can be disregarded if $\lambda(t:\theta)$ is not a function of θ.

An alternative expression for $L(\theta)$ that follows from (4.36) is

$$L(\theta) = \sum_{i=1}^{\infty} L_i(\theta), \tag{4.38}$$

where

$$L_i(\theta) \overset{\Delta}{=} -\int_{t_0}^{T} P(U_i:\theta)\lambda(\sigma, \theta)d\sigma + \int_{t_0}^{T} \ln[P(U_i:\theta)\lambda(\sigma, \theta)]N(d\sigma, U_i).$$

Thus, the loglikelihood function $\mathcal{L}(\theta)$ is the sum of the loglikelihood functions $\mathcal{L}_i(\theta)$ for each independent component in the representation of Theorem 4.3.1. This decomposition leads to the implementation for generating $\mathcal{L}(\theta)$ shown schematically in Fig. 4.6. It is worth noting that this decomposition will be especially useful for reducing the complexity of determining the maximum-likelihood estimate of the vector θ if each component loglikelihood depends on one or a few of the parameters independently of all the other components because the estimate of θ can then be determined by finding the estimates of elements of this vector individually or a few at a time rather than all of them, jointly, together.

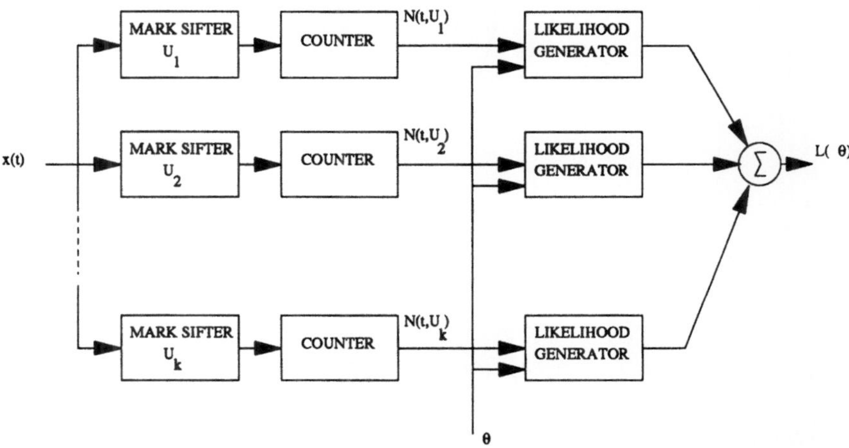

Figure 4.6. Decomposition of the likelihood function for a compound Poisson-process with a denumerable mark space.

As discussed in Ex. 4.3.1, the initial sifting of points according to their mark is often accomplished in practice using a multichannel analyzer. An alternative way in which this sifting can occur is illustrated in the next example.

Example 4.4.1 *Optical Position-Sensing* ————————————

An optical communication system with a requirement for active tracking in order to ensure that the receiver looks directly at the transmitter needs a sensor to detect the position of the light beam arriving at the receiver. The electrical output of the sensor is used to generate feedback signals for controlling the orientation of telescopes and mirrors to maintain receiver-transmitter alignment even though the receiver and transmitter may be in relative motion. A simple device for sensing position consists of a photodetector having a photoemissive surface that is divided into four independent quadrants as illustrated in Fig. 4.7.

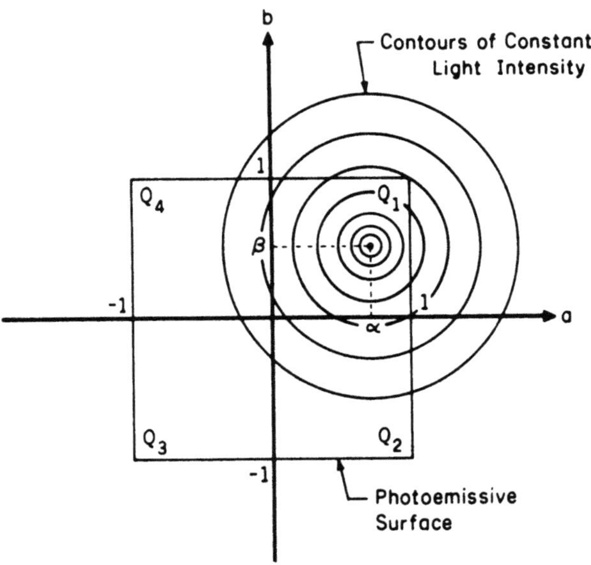

Figure 4.7. Four-quadrant optical position-sensor.

Photoelectron emissions are monitored in each quadrant during an observation interval $[0, T)$ and used to sense the position of an incident beam of light. We assume that the intensity of light falling on the device has a circularly symmetric Gaussian distribution given by

$$I(a,b:\alpha,\beta) = I_0 \exp\left\{-\frac{1}{2\rho^2}[(a-\alpha)^2 + (b-\beta)^2]\right\}. \tag{4.39}$$

The parameters α and β are the beam-position parameters, which are unknown and desired. The parameters I_0 and ρ characterize the beam's peak-intensity and width; these are assumed to be known. We assume that photoemissions in a quadrant form a homogeneous Poisson-process with an intensity that is proportional to the integral of $I(a,b{:}\alpha,\beta)$ over that quadrant (see (2.42)). We incorporate the proportionality constant into the peak intensity I_0. The four quadrants of the detector are assumed to operate independently. Photoemissions for the entire sensor can, therefore, be viewed as a compound Poisson-process with a denumerable mark-space $\mathcal{U} = \{Q_1, Q_2, Q_3, Q_4\}$, where the mark Q_i identifies a point as originating in quadrant i. For use in (4.38), we have

$$P(Q_i{:}\alpha,\beta)\lambda(t{:}\alpha,\beta) = \int\int_{A_i} I(a,b{:}\alpha,\beta)\,da\,db, \qquad (4.40)$$

where $A_1 = \{a,b \mid 0 \le a \le 1, 0 \le b \le 1\}$, $A_2 = \{a,b \mid 0 \le a \le 1, -1 \le b \le 0\}$, $A_3 = \{a,b \mid -1 \le a \le 0, -1 \le b \le 0\}$, and $A_4 = \{a,b \mid -1 \le a \le 0, 0 \le b \le 1\}$. Substitution of these expressions into (4.38) results in

$$I(\alpha,\beta) = -T \int_{-1}^{1}\int_{-1}^{1} I(a,b{:}\alpha,\beta)\,da\,db$$

$$+ \sum_{i=1}^{4} N(T,Q_i)\ln\left[\int\int_{A_i} I(a,b{:}\alpha,\beta)\,da\,db\right]. \qquad (4.41)$$

The numbers of photoevents in each quadrant are sufficient statistics for estimating α and β; occurrence times need not be measured, which results because of the assumed homogeneity of the processes. The values of α and β that maximize (4.41) give the maximum-likelihood estimate of the beam position and might be used with a feedback controller to recenter the beam over the position sensor. ∎

Let us now remove the restriction that marks be denumerable. Thus, suppose that $\{x(t){:}\, t \ge t_0\}$ is a marked point-process with mark space $\mathcal{U}$, where $\mathcal{U}$ is a region of a finite-dimensional Euclidean space. We assume that marks are continuous random-variables with probability density $p_u(U)$. Let $\{N(t){:}\, t \ge t_0\}$ count points of this process regardless of their mark. The sample function density for $x(\cdot)$ on the interval $[t_0, t)$ is defined by

$$p[\{x(\sigma){:}\, t_0 \le \sigma < t\}]$$

$$= \begin{cases} \Pr[N(t) = 0], & N(t) = 0 \\ p_{w,u_1,\ldots,u_n}(W, U_1, \ldots, U_n, N(t) = n), & N(t) = n \ge 1, \end{cases} \qquad (4.42)$$

where

$$p_{w, u_1, \ldots, u_n}(W, U_1, \ldots, U_n, N(t) = n) = \Pr[N(t) = n \mid w = W, u_1 = U_1, \ldots, u_n = U_n]$$

$$\times p_{u_1, \ldots, u_n}(U_1, \ldots, U_n \mid w = W) p_w^{(n)}(W),$$

where the first factor in the product on the right-hand side is the conditional probability that $N(t) = n$ given the n occurrence times and marks on $[t_0, t)$; the second factor is the conditional joint probability-density of the first n marks given the n occurrence times; and the last factor is the joint occurrence-density for the first n occurrence times. It is evident for a compound Poisson-process that (4.42) becomes

$$p[\{x(\sigma): t_0 \le \sigma < t\}]$$

$$= \begin{cases} \exp\!\left[-\int_{t_0}^{t} \lambda(\sigma) d\sigma\right], & N(t) = 0 \\[2ex] \left[\prod_{i=1}^{n} \lambda(W_i) p_u(U_i)\right] \exp\!\left[-\int_{t_0}^{t} \lambda(\sigma) d\sigma\right], & N(t) = n \ge 1. \end{cases} \tag{4.43}$$

This is the analog of (4.34) for a nondenumerable mark space. In place of (4.35), we have

$$p[\{x(\sigma): t_0 \le \sigma < t\}] \tag{4.44}$$

$$= \exp\!\left[-\int_{t_0}^{t} \lambda(\sigma) d\sigma + \int_{t_0}^{t} \ln[\lambda(\sigma)] N(d\sigma) + \int_{u} \ln[p_u(U)] N(t, dU)\right],$$

where $N(t, dU)$ denotes the number of points in $[t_0, t)$ having marks in $[U, U + dU)$; see (4.30). Finally, the analog of (4.36) for a nondenumerable mark-space is

$$p[\{x(\sigma): t_0 \le \sigma < t\}] \tag{4.45}$$

$$= \exp\!\left[-\int_{t_0}^{t} \lambda(\sigma) d\sigma + \int_{t_0}^{t} \int_{u} \ln[p_u(U) \lambda(\sigma)] M(d\sigma, dU)\right],$$

where M is the multidimensional, time-space Poisson-process of the representation in Theorem 4.3.2. Equations (4.44) and (4.45) have interpretations analogous to those for (4.35) and (4.36), respectively. Furthermore, these equations are used in the same way as (4.35) and (4.36) for solving problems of statistical inference for observed compound Poisson-processes having a nondenumerable mark-space.

Example 4.4.2 *Optical Position Sensing, continued* ————————

Consider once again the model of Ex. 4.4.1, but suppose now that the photoemissive surface of the detector is not partitioned into four quadrants. Rather, the number, time of occurrence, and position of each photoevent are measured to determine the position of the light beam. The mark space corresponds to positions in the photoemissive surface: $\mathcal{U} = \{a,b: -1 \le a \le 1, -1 \le b \le 1\}$. Photoevents form a compound Poisson-process in which marks indicate the position of the photoevent in the detector. For use in (3.45), we have

$$\lambda(t:\alpha,\beta) = \int_{-1}^{1} \int_{-1}^{1} I(a,b:\alpha,\beta)\,da\,db,$$

and

$$p_u(a,b:\alpha,\beta) = \frac{I(a,b:\alpha,\beta)}{\displaystyle\int_{-1}^{1}\int_{-1}^{1} I(U_1,U_2:\alpha,\beta)\,dU_1\,dU_2},$$

where $I(a,b:\alpha,\beta)$ is defined in (4.39). From (4.45), the loglikelihood function is

$$\mathcal{L}(\alpha,\beta) = \ln p[\{x(\sigma): t_0 \le \sigma < t\}] \tag{4.46}$$

$$= -T \int_{-1}^{1} \int_{-1}^{1} I(a,b:\alpha,\beta)\,da\,db + \sum_{i=1}^{N(T)} \ln I(a_i,b_i:\alpha,\beta),$$

where $N(T)$ is the number of photoevents observed in $[0,T)$, and (a_i,b_i), for $i = 1,2,\cdots,N(T)$, are their positions. The maximum-likelihood estimate of the beam position is given by the values of α and β that maximize $\mathcal{L}(\alpha,\beta)$. The search for these values will generally require numerical implementation, but there is one situation where an analytic solution is possible. This obtains when the beam size and range of possible positions are small in comparison to the size of the photoemissive surface. Then, the first term in (4.46) will be independent of α and β to a good approximation and can be disregarded. It is then straightforward to verify that the maximum-likelihood estimate of the beam position is given by

$$(\hat{\alpha}_{ML},\hat{\beta}_{ML}) = \left(\frac{1}{N(T)}\sum_{i=1}^{N(T)} a_i, \frac{1}{N(T)}\sum_{i=1}^{N(T)} b_i \right). \tag{4.47}$$

Thus, the estimated position is at the arithmetic average of the observed photoevents (or the center-of-gravity of the photoevents if each is imagined to carry unit mass). ∎

Summary. In this section, we have developed a representation for compound Poisson-processes. Theorems 4.3.1 and 4.3.2 give the representations for denumerable and nondenumerable mark-spaces, respectively. These representations provide insight into the structure of data processing needed to solve parameter estimation problems for observed compound Poisson-processes. For example, they lead naturally to the use of multi-channel analysis, which is commonly employed in practice when the marks correspond to energies or pulse amplitudes. Expressions were derived in (4.35), (4.36), and (4.45) for the sample-function density of a compound Poisson-process. This quantity plays a central role in solving problems of statistical inference.

4.5 Statistical Inference for Mixed Poisson-Processes

A Poisson process having an intensity function $\lambda(\cdot)$ that can be expressed as a sum,

$$\lambda(t) = \sum_{k=1}^{K} \lambda_k(t), \tag{4.48}$$

is called a *mixed Poisson-process*. Such a Poisson process can be thought of as arising from a mixing or pooling of the points from independent Poisson-processes, with the kth component process having intensity $\lambda_k(\cdot)$. The points contributed to the mixture by the component processes are indistinguishable from one another as there is no mark indicating their origin. The probability p_k that a point occurring at time t in the mixed process was contributed by the kth component process is

$$p_k = \frac{\lambda_k(t)}{\displaystyle\sum_{i=1}^{K} \lambda_i(t)} = \frac{\lambda_k(t)}{\lambda(t)}, \tag{4.49}$$

as may be seen from the approximation

$$p_k = \Pr(A_k \mid A) = \Pr(A \mid A_k) \times \frac{\Pr(A_k)}{\Pr(A)} \approx 1 \times \frac{\lambda_k(t)\Delta t}{\lambda(t)\Delta t},$$

where A is the event that there is a point of the mixture in $[t, t + \Delta t)$ and A_k is the event that there is a point of component k in $[t, t + \Delta t)$.

Example 4.5.1 *Multiexponential Models* ——————————————
Multiexponential models are used in nuclear medicine to describe the time-dependent behavior of a radioactive tracer in physiochemical studies [3], such as in the measurement of oxygen utilization in the human brain [5]. Data collected form a mixed Poisson-process with an intensity in the form of (4.48) with the intensity of the kth modal compartment being

$$\lambda_k(t) = a_k e^{-b_k(t-t_0)}, \tag{4.50}$$

for $t \geq t_0$, where t_0 is the instant the radiotracer is introduced. The parameters $x = \{a_k, b_k : k = 1, \cdots, K\}$ are the important quantities; a_k is influenced by the initial amount of radiotracer and the exchange of radiotracer between the kth and the other modal compartments, and b_k determines the time constant of the kth compartment. The problem is to estimate all the parameters x in terms of data $\{N(t): t_0 \leq t < T\}$. The log-likelihood function is

$$L(x) = -\sum_{k=1}^{K} \int_{t_0}^{T} \lambda_k(t)\,dt + \int_{t_0}^{T} \ln\left[\sum_{k=1}^{K} \lambda_k(t)\right] N(dt). \tag{4.51}$$

Equating the gradient $\partial L(x)/\partial x$ to zero yields a set of $2K$ coupled, nonlinear equations for the maximum-likelihood estimate of x. These equations cannot be solved analytically, and their numerical solution requires a difficult search in a $2K$-dimensional parameter space. A useful observation to make is that if the points also had an auxiliary mark indicating the modal compartment from which it arose, then the parameter estimation problem is simplified to that of estimating the parameters of each mode separately and independently of the others. The representation in Fig. 4.6 indicates that the maximum-likelihood estimate of (a_k, b_k) maximizes the kth mode loglikelihood function

$$L_k(a_k, b_k) = -\int_{t_0}^{T} \lambda_k(t)\,dt + \int_{t_0}^{T} \ln[\lambda_k(t)] N(dt, k), \tag{4.52}$$

where $N(t, k)$ counts the number of points from the kth mode alone. While the gradient of L_k also yields nonlinear, coupled equations for the estimate of (x_k, x_{k+1}), only two parameters are involved, so a numerical search for the estimate is simplified compared to that for (4.51) where there are $2K$ parameters. This observation suggests that the expectation-maximization algorithm could be helpful as an approach for determining the estimate of x numerically, as developed further in Ex. 4.5.3. ∎

The expectation-maximization algorithm introduced in Ch. 3 can be useful for producing estimates of parameters that influence the intensity of a mixed Poisson-process. We imagine that a mark is affixed to each point, where the mark on a point indicates the component process from which it came. The complete data may then be taken to be the marked Poisson-process, and the incomplete data are formed from the marked Poisson-process by simply ignoring the marks, which just yields the original mixed Poisson-process. Let the mark space be $\mathcal{U} = \{1, 2, \cdots, K\}$. Then, from (4.38), the complete-data loglikelihood is

$$\mathcal{L}_{\text{CD}}(x) = \sum_{k=1}^{K} \left\{ -\int_{t_0}^{T} \lambda_k(\sigma : x)\, d\sigma + \int_{t_0}^{t} \ln \lambda_k(\sigma : x) N(d\sigma, k) \right\}$$

$$= -\int_{t_0}^{T} \lambda(\sigma : x)\, d\sigma + \sum_{k=1}^{K} \int_{t_0}^{t} \ln \lambda_k(\sigma : x) N(d\sigma, k). \tag{4.53}$$

The E-step of the expectation-maximization algorithm yields

$$Q[x \mid x^{(i-1)}] \stackrel{\Delta}{=} E[\mathcal{L}_{\text{CD}}(x) \mid \{N(\sigma) : t_0 \le \sigma < T\}, x^{(i-1)}] \tag{4.54}$$

$$= -\int_{t_0}^{T} \lambda(\sigma : x)\, d\sigma + \sum_{k=1}^{K} \int_{t_0}^{t} \ln \lambda_k(\sigma : x) E[N(d\sigma, k) \mid \{N(\sigma) : t_0 \le \sigma < T\}, x^{(i-1)}].$$

Since, from (4.49),

$$E[N(d\sigma, k) \mid \{N(\sigma) : t_0 \le \sigma < T\}, x^{(i-1)}] = \frac{\lambda_k(\sigma : x^{(i-1)})}{\lambda(\sigma : x^{(i-1)})} N(d\sigma), \tag{4.55}$$

the E-step yields

$$Q[x \mid x^{(i-1)}] = -\int_{t_0}^{T} \lambda(\sigma : x)\, d\sigma + \sum_{k=1}^{K} \int_{t_0}^{t} \ln \lambda_k(\sigma : x) \left[\frac{\lambda_k(\sigma : x^{(i-1)})}{\lambda(\sigma : x^{(i-1)})} \right] N(d\sigma).$$

$$\tag{4.56}$$

Then, for the M-step of the expectation-maximization algorithm, we maximize Q with respect to x to get

$$x^{(i)} = \underset{x}{\operatorname{argmax}}\, Q[x \mid x^{(i-1)}]. \tag{4.57}$$

Whether or not this maximization to get from the (i-1)st step to the ith step of the expectation-maximization iteration can be performed any more easily than the original maximization of the incomplete-data loglikelihood depends on the detailed form of the functions involved and the way they depend upon the parameters to be estimated. Here are two examples where the expectation-maximization method has proven useful.

Example 4.5.2 *Optical Position-Sensing, continued* ———————————
    Consider again the problem of optical position-sensing introduced in Ex's. 4.4.1 and 4.4.2. Suppose, now, that the points measured in the photodetector are modeled as a two-dimensional mixed Poisson-process on $X = \{x: -1 \le \alpha \le 1, -1 \le \beta \le 1\}$ with an intensity given by

$$\lambda(x:m) = I(x:m) + \lambda_0, \tag{4.58}$$

where $I(x:m) = \Lambda \exp[-(x-m)'(x-m)/2\rho^2]$, and $m = [\alpha \quad \beta]'$. Here, λ_0 models spontaneous *dark-current* points created within the detector due to thermal effects, and m is the unknown position of the spot of light. The loglikelihood function for estimating m based on measurements of the locations of each point event is

$$L(m) = -\int_X \lambda(x:m)dx + \int_X \ln \lambda(x:m)N(dx). \tag{4.59}$$

Setting the gradient of $L(m)$ with respect to m to zero yields the following necessary equation for the maximum-likelihood estimate of m

$$-\frac{1}{\rho^2}\int_X (x-m)I(x:m)dx + \frac{1}{\rho^2}\int_X \frac{(x-m)I(x:m)}{I(x:m)+\lambda_0}N(dx) = 0. \tag{4.60}$$

If $\lambda_0 = 0$, solving (4.60) for m shows that the maximum-likelihood estimate of m is the arithmetic average of the measured point-locations

$$\hat{m}_{ML} = \frac{1}{N(X)}\sum_{i=1}^{N(X)} x_i,$$

as found in Ex. 4.4.2. However, if $\lambda_0 > 0$, (4.60) cannot be solved explicitly for the estimate, which must then be produced numerically. One way to accomplish this is with the expectation-maximization algorithm. We imagine that a mark $\gamma \in (s,n)$ accompanies each point indicating whether it is a signal event ($\gamma = s$) or a dark-current, noise event ($\gamma = n$). These marks are assigned independently. The complete data may then be taken to be the number of points observed, their locations and their marks, $\{N(X), x_1, \cdots, x_{N(X)}, \gamma_1, \cdots, \gamma_{N(X)}\}$. The complete-data loglikelihood is

$$\mathcal{L}_{CD}(m) = \mathcal{L}_s(m) + \mathcal{L}_n(m), \tag{4.61}$$

where

$$\mathcal{L}_s(m) = -\int_X I(x:m)\,dx + \int_X \ln[I(x:m)]N_s(dx), \tag{4.62}$$

is the loglikelihood for the signal points, and

$$\mathcal{L}_n(m) = -\int_X \lambda_0\,dx + \int_X \ln \lambda_0 N_n(dx), \tag{4.63}$$

is the loglikelihood for the noise points. In these expressions, $N_s(\cdot)$ and $N_n(\cdot) = N(\cdot) - N_s(\cdot)$ count points with the respective marks. The loglikelihood for the noise (4.63) does not depend on m, so the second term on the right in (4.61) can be disregarded for estimating m. The E-step of the expectation-maximization algorithm then yields

$$Q(m \mid m^{(i-1)}) = E[\mathcal{L}_{CD}(m) \mid N(X), x_1, \cdots, x_{N(X)}, m^{(i-1)}] \tag{4.64}$$

$$= -\int_X I(x:m)\,dx + \int_X w(x:m^{(i-1)})\ln[I(x:m)]N(dx).$$

where the weighting function $w(\cdot)$, determined from (4.55) as

$$w(x:m^{(i-1)}) = \frac{I(x:m^{(i-1)})}{I(x:m^{(i-1)}) + \lambda_0},$$

is the probability that a point located at x is a signal point when the position of the spot is $m^{(i-1)}$. Then, the E-step yields

$$m^{(i)} = \underset{m}{\operatorname{argmax}}\, Q(m \mid m^{(i-1)}). \tag{4.65}$$

There is no appreciable simplification of this in general, but if the size of the detector is large in comparison to the size of the spot of light, so that edge effects can be neglected, (4.65) yields

$$\hat{m}^{(i)} = \frac{\int_x xw(x:\hat{m}^{(i-1)})N(dx)}{\int_x w(x:\hat{m}^{(i-1)})N(dx)}.$$ (4.66)

If $\lambda_0 = 0$, then this equation reduces to the arithmetic average of the measured point-locations, as obtained before, and the algorithm converges in one step. B. Slocumb [10, 11] examines this problem further, and analyzes the estimation performance through computer simulations and the Cramér-Rao bound. ∎

Example 4.5.3 *Multiexponential Models, continued* ────────────
Consider Ex. 4.5.1 again. We noted before that the problem of estimating the parameters x by maximizing the loglikelihood (4.51) results in an analytically intractable problem, so that some numerical method is required. The expectation-maximization algorithm is one such method. For this, we associate a mark with each point to indicate the exponential mode that produced it. The complete data may then be selected as the occurrence time and mark for each point along with the total number of points in each mode. The independence of the modes for this mixed Poisson process implies that the complete-data loglikelihood is given by

$$\mathcal{L}_{CD}(x) = \sum_{k=1}^{K} \mathcal{L}_k(a_k, b_k),$$ (4.67)

where $\mathcal{L}_k$ is given in (4.52). Substitution of (4.50) and (4.52) yields

$$\mathcal{L}_{CD}(x) = \sum_{k=1}^{K} \left\{ \frac{a_k}{b_k}\left[e^{-b_k T} - e^{-b_k t_0}\right] + N_k \ln a_k - b_k s_k \right\},$$ (4.68)

where

$$N_k = \int_{t_0}^{T} N(dt,k) \quad \text{and} \quad s_k = \int_{t_0}^{T} t N(dt,k).$$

The E-step of the expectation-maximization algorithm yields

$$Q(x \mid \hat{x}^{(i-1)}) = \sum_{k=1}^{K} \left\{ \frac{a_k}{b_k}\left[e^{-b_k T} - e^{-b_k t_0}\right] + \hat{N}_k^{(i-1)}\ln a_k - b_k \hat{s}_k^{(i-1)} \right\},$$ (4.69)

where

$$\hat{N}_k^{(i-1)} = E[N_k \mid N(t): t_0 \leq t < T, \hat{x}^{(i-1)}]$$

$$= \int_{t_0}^{T} \frac{\hat{\lambda}_k^{(i-1)}(t)}{\hat{\lambda}^{(i-1)}(t)} N(dt), \tag{4.70}$$

and

$$\hat{s}_k^{(i-1)} = E[s_k \mid N(t): t_0 \leq t < T, \hat{x}^{(i-1)}]$$

$$= \int_{t_0}^{T} \frac{t\,\hat{\lambda}_k^{(i-1)}(t)}{\hat{\lambda}^{(i-1)}(t)} N(dt), \tag{4.71}$$

in which

$$\hat{\lambda}_k^{(i-1)}(t) = \hat{a}_k^{(i-1)} \exp[-\hat{b}_k^{(i-1)} t] \quad \text{and} \quad \hat{\lambda}^{(i-1)}(t) = \sum_{k=1}^{K} \hat{\lambda}_k^{(i-1)}(t).$$

For the M-step, the function Q in (4.69) must be maximized with respect to x to produce $\hat{x}^{(i)}$. This is easier than the maximization of (4.51) because the maximizing parameters for each mode can be determined separately from those of other modes,

$$(\hat{a}_k^{(i)}, \hat{b}_k^{(i)}) = \underset{a_k, b_k}{\mathrm{argmax}} \left\{ \frac{a_k}{b_k} \left[e^{-b_k T} - e^{-b_k t_0} \right] + \hat{N}_k^{(i-1)} \ln a_k - b_k \hat{s}_k^{(i-1)} \right\}. \tag{4.70}$$

This maximization over the two parameters (a_k, b_k) must be performed numerically within each step of the expectation-maximization algorithm. See D. Snyder [12] for an application of these ideas in emission tomography. ∎

4.6 References

1. L. Breiman, *Probability*, Addison-Wesley, Reading MA., 1968.

2. I. I. Gikhman and A. V. Skorokhod, *Introduction to the Theory of Random Processes*, Saunders, Philadelphia, PA., 1969.

3. J. A. Jacques, "Tracer Kinetics" in: *Principles of Nuclear Medicine* (H. N. Wagner, Jr., Ed.), Saunders, Philadelphia, 1968.

4. D. Konig and K. Matthes, "Verallgemeinerungen den Erlangsche Formeln," *I. Math. Nachr.*, Vol. 26, pp. 45-56, 1963.

5.   M. A. Mintun, M. E. Raichle, W. R. W. Martin, and P. Herscovitch, "Brain Oxygen Utilization Measured with ^{15}O Radiotracers and Positron Emission Tomography," *J. Nuclear Medicine*, Vol. 25, No. 2, pp. 177-187, 1984.

6.   K. Matthes, "Stationare Zufallige Puntfolgen," *I. Jber. Deutsch. Math.-Verein*, Vol. 66, pp. 66-79, 1963.

7.   K. Matthes, "Unbeschrankt Teilbare Verteilumsgestze Stationarer Zufalliger Puntfolgen," *Wiss. Z. Hochsch. Electro. Ilmenau*, Vol. 9, pp. 235-238, 1963.

8.   N. Mullani and H. Huang, "Gamma Camera Calibration Procedures and Programs," Ch. 12 in: *Computer Processing of Dynamic Images from an Anger Scintillation Camera*, (K. B. Larson and J. R. Cox, Jr., Ed's.), Proc. Gamma Camera Workshop, Washington University, St. Louis, Mo., Society of Nuclear Medicine, 1971.

9.   D. A. Ross and C. C. Harris, "Measurement of Radioactivity," Ch. V in: *Principles of Nuclear Medicine*, (H. N. Wagner, Ed.), Saunders, Philadelphia, PA., 1968.

10.   B. J. Slocumb, "Position-Sensing Algorithms for Optical Communications," M.S. Thesis, Department of Electrical Engineering, Washington University, St. Louis, 1988.

11.   B. J. Slocumb and D. L. Snyder, "Maximum Likelihood Estimation Applied to Quantum-Limited Optical Position-Sensing," Proc. SPIE Technical Symp. on Optical Engineering and Photonics in Aerospace Sensing, Orlando, FL, April 1990.

12.   D. L. Snyder, "Parameter Estimation for Dynamic Studies in Emission-Tomography Systems Having List-Mode Data," *IEEE Transactions on Nuclear Science*, Vol. NS-31, pp. 925-931, April 1984.

4.7 Problems

4.2.1. Suppose that a source of coherent light incident on a photomultiplier produces primary photoelectrons as a Poisson process with intensity $\{\lambda(t): t \geq t_0\}$. By a cascade process, each primary electron produces secondary electrons that are collected at the output of the photomultiplier. Suppose that the number of secondary electrons produced independently by each primary electron is a Poisson-distributed random-variable with parameter 1. Determine the characteristic function, mean, and variance for the number of electrons appearing at the photomultiplier output during the interval $[s, t)$.

4.2.2 Let $\{x(t): t \geq t_0\}$ be a compound Poisson-process. Assume that points occur with intensity $\{\lambda(t): t \geq t_0\}$ and that the common probability distribution function for the marks is $P_u(U)$. Let the probability distribution function for $x(t)$ be defined by

$$P_{x(t)}(X) = \Pr[x(t) \leq X],$$

where the vector inequality means that each component of $x(t) - X$ is nonpositive. Show that

$$P_{x(t)}(X) = \sum_{n=0}^{\infty} P_u^{*(n)}(X) \frac{1}{n!} \left(\int_{t_0}^{t} \lambda(\sigma) d\sigma \right)^n \exp\left[-\int_{t_0}^{t} \lambda(\sigma) d\sigma \right],$$

where $P_u^{*(n)}(U)$ is the nth iterated convolution of $P_u(U)$ with itself. Here

$$P_u^{*(0)}(U) = \begin{cases} 1, & U \geq 0 \\ 0, & U < 0, \end{cases}$$

$P_u^{*(1)}(U) = P_u(U)$, and $P_u^{*(n)}(U)$, for $n \geq 2$, can be determined from the recursive relation

$$P_u^{*(n)}(U) = \int_U P_u^{*(n-1)}(U - \xi) dP_u(\xi).$$

4.2.3 Establish (4.17). Hint: The following series may be useful

$$\ln(1 - z) = -\sum_{n=1}^{\infty} \frac{z^n}{n} \qquad \text{for } |z| \leq 1 \quad \text{and} \quad z \neq 1.$$

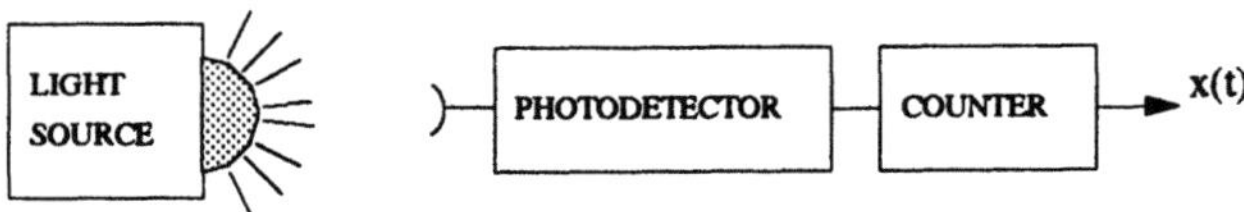

Figure P.4.2.4

4.2.4 The light source of Fig. P.4.2.4 can operate at two intensity levels. The levels occur with equal probability when the source is first energized. Assume electrons are generated at the output of the photodetector as a homogeneous Poisson-process at a rate $\lambda_1 = \lambda$ or $\lambda_2 = 2\lambda$ corresponding to the low- and high-level intensities of the light source. Electrons are counted, but the counter is defective. Suppose an electron fails to be registered with probability p independently of all other electrons. Denote by $x(t)$ the total number of registered electrons in $[0,t)$.

   a. Determine the characteristic functional for $\{x(t){:}0 \leq t < T\}$; that is, evaluate

$$\phi_x(jv) \overset{\Delta}{=} E\left[e^{j\int_0^T v(\sigma)dx(\sigma)} \right].$$

   b. Determine the characteristic function for $x(t)$.
   c. Evaluate the mean and variance of $x(t)$.

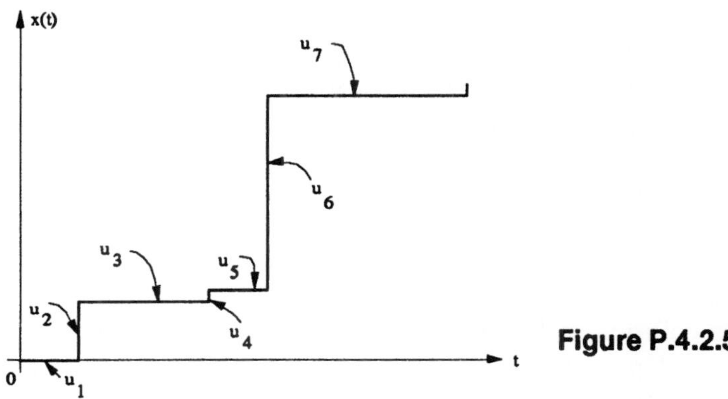

Figure P.4.2.5

4.2.5 The following construction is used to generate a random staircase function of time. Let $\{u_i\}$ be a sequence of independent, identically distributed random variables with the common distribution-function

$$P_u(U) = \Pr[u \leq U] = \begin{cases} 1 - e^{-\lambda U}, & U \geq 0 \\ 0, & U < 0. \end{cases}$$

Let $\{x(t){:}t \geq t_0\}$ be constructed as shown in Fig. P.4.2.5. Odd-indexed variables in the sequence define the length of the treads of the staircase, and even-indexed variables define the heights of the risers. Evaluate either the characteristic function or the probability density for the height $x(t)$ of the staircase at time t. Determine the expected value and variance of the height.

4.2.6 A random process $\{x(t): t \geq t_0\}$ is said to continuous in the mean-square sense if for all $[t_0, t)$

$$\lim_{\delta \to 0} E[|x(t+\delta)-x(t)|^2] = 0.$$

Is a compound Poisson-process continuous in this sense?

4.2.7 Consider the motion of a small particle as a function of time. Its movement is confined to one dimension and is influenced by random collisions with other particles. Let $x(t)$ denote the position of the particle at time t for $[t_0, t)$. Assume:

i.   the particle is initially at the origin, $x(0) = 0$;
ii. collisions occur at random instants corresponding to a homogeneous Poisson-process with intensity λ;
iii. each collision, independently, may be with either a light or heavy particle with probability 1/3 and 2/3, respectively;
iv. a collision with a light particle changes the position of the small particle by an amount of either +1 or -1 unit with equal probability;
v.   a collision with a heavy particle changes the position of the small particle by an amount of +2 or -2 units with equal probability.

a.   Determine the characteristic functional for $\{x(t): 0 \leq t < T\}$.

b.   Determine the characteristic function for $x(t)$.

c.   Determine the mean, variance, and covariance functions for $\{x(t): 0 \leq t < T\}$.

d.   Define

$$x^*(t) = \frac{x(t) - E[x(t)]}{\sqrt{\text{var}[x(t)]}}.$$

Then $x^*(t)$ is the position of the particle normalized to have zero mean and unit variance. Determine the characteristic function for $x^*(t)$ and discuss what happens in the limit as λt tends to infinity.

4.3.1 Rederive equations (4.19), (4.20), and (4.23) for the mean, instantaneous covariance-matrix, and covariance matrix of a compound Poisson-process by starting with the representations (4.26) and (4.31).

4.4.1 Suppose that $\{x(t):t \geq t_0\}$ is the staircase function constructed in Problem 4.2.5. Show that the likelihood function $\mathcal{L}(\lambda)$ for estimating λ in terms of an observed path of $x(\cdot)$ on $[0,T)$ is

$$\mathcal{L}(\lambda) = -\lambda T + (2\ln\lambda)N - \lambda x(T),$$

where N is the total number of treads and risers occurring on $[0,T)$. Derive an expression for the maximum-likelihood estimate of λ in terms of the observed data.

4.4.2 (*Cramér-Rao Bound for Unbiased Estimates*). Let $\{x(t):t \geq t_0\}$ be a compound Poisson-process with mark space $\mathcal{U}$. Assume that points occur with intensity $\{\lambda(t,\theta):t \geq t_0\}$, where θ is a vector of unknown parameters to be estimated. If $\mathcal{U}$ is denumerable, the mark occurrence-probabilities $P(U_k:\theta)$, for $k = 1,2,\cdots$, may also depend on the parameters. If $\mathcal{U}$ is nondenumerable, assume that a mark probability-density $p_u(U:\theta)$ exists and may also depend on the parameters. Let θ^* denote any unbiased estimate of θ in terms of $\{x(\sigma):t_0 \leq \sigma < T\}$. Show that

$$E[(\theta* - \theta)(\theta* - \theta)':\theta] \geq F^{-1}(\theta),$$

where $F(\theta)$ is a Fisher-information matrix given for a denumerable mark-space by

$$F(\theta) = \sum_{k=1}^{\infty} \int_{t_0}^{T} \frac{1}{\lambda(\sigma,\theta)P(U_k:\theta)} \left[\frac{\partial\lambda(\sigma,\theta)P(U_k:\theta)}{\partial\theta} \right]\left[\frac{\partial\lambda(\sigma,\theta)P(U_k:\theta)}{\partial\theta} \right]' d\sigma,$$

and for a nondenumerable mark-space by

$$F(\theta) = \int_{\mathcal{U}} \int_{t_0}^{T} \frac{1}{\lambda(\sigma,\theta)p_u(U:\theta)} \left[\frac{\partial\lambda(\sigma,\theta)p_u(U:\theta)}{\partial\theta} \right]\left[\frac{\partial\lambda(\sigma,\theta)p_u(U:\theta)}{\partial\theta} \right]' d\sigma dU,$$

where $\partial(\cdot)/\partial\theta$ denotes the gradient with respect to θ. Hint: Parallel the proof of Theorem 2.4.1.

4.4.3 Suppose that $\{x(t){:}t \geq t_0\}$ is the staircase function constructed in Problem 4.2.5. Use the result of Problem 4.4.2 to derive a lower bound on the mean square-error for any unbiased estimate of the parameter λ in terms of an observed path of $x(\cdot)$ on $[0, T)$.

CHAPTER FIVE

FILTERED POISSON-PROCESSES

5.1 Introduction

There are numerous physical phenomena that can be modeled as a response to the points of a marked point process. The simplest models occur when the response can be expressed as a superposition of separate responses to each marked point. These models are developed in Sec. 5.2. Later, in Sec. 5.3, we remove the superposition requirement but impose the additional structure of Markov processes.

5.2 Superposition of Point Responses

Let $\{x(t): t \geq t_0\}$ be a marked point process as defined in Sec. 4.1. Denote the nth occurrence time and mark by τ_n and $\mathbf{u}_n$, respectively, and let $\{N(t): t \geq t_0\}$ be a counting process that counts points regardless of their mark. The processes of interest in this section have the form of the following superposition,

$$
y(t) = \begin{cases} 0, & N(t) = 0 \\ \\ \sum_{n=1}^{N(t)} h(t, \tau_n: \mathbf{u}_n), & N(t) \geq 1, \end{cases}
$$

(5.1)

for $[t_0, t)$, where $h(t, \tau_n: \mathbf{u}_n)$ is termed the impulse response or weighting function and is the response at time t to the nth marked point. A process in the form of (5.1) will be referred to as a *filtered Poisson-process* when $\{x(t): t \geq t_0\}$ is a compound Poisson-process. Then, $\{N(t): t \geq t_0\}$ is a Poisson counting-process, and the marks $\{\mathbf{u}_n\}$ are mutually independent, independent of $N(\cdot)$, identically distributed, vector-valued random variables.

We assume throughout the discussion that the weighting function h is causal; that is, $h(t, \tau: \mathbf{u}) = 0$ for $\tau > t$.

Some physical phenomena for which filtered point processes provide a reasonable model are given in the following examples.

Example 5.2.1 *Shot Noise* —————————————————————————

After being emitted from a heated cathode in a vacuum-tube diode, an electron travels to the anode. In doing so, it makes an incremental contribution to the anode current. Denote the anode current at time t by $i(t)$. A model for the current $\{i(t): t \geq t_0\}$ is

$$i(t) = \begin{cases} 0, & N(t) = 0 \\ \sum_{n=1}^{N(t)} h(t - \tau_n), & N(t) \geq 1, \end{cases} \tag{5.2}$$

where t_0 is the time the diode is energized, $N(t)$ is the number of electrons emitted during $[t_0, t)$, $\{\tau_n\}$ are the emission times, and $h(t - \tau_n)$ is the contribution to $i(t)$ due to the *nth* electron emitted. The process $\{i(t): t \geq t_0\}$ is commonly called *shot noise*. The form of $h(\cdot)$ is a function of the cathode-anode geometry, cathode temperature, and anode voltage; thus, the detailed form of $h(\cdot)$ can be difficult to predict. It is of interest, therefore, that only more global aspects of this function enter into certain calculation we make below. Specifically, the area of $h(\cdot)$ is the electronic charge $e \approx 1.6 \times 10^{-19}$ coulombs, and the duration of $h(\cdot)$ is approximately the cathode to anode transit time, which is typically on the order of 10^{-9} seconds. When the electron emission rate is temperature limited so that space-charge effects are negligible, $\{N(t): t \geq t_0\}$ can be reasonably modeled as a Poisson counting-process. W. Davenport and W. Root [4, Ch. 7] give a detailed study of shot noise models. ∎

Example 5.2.2 *Busy Telephone Channels* —————————————————————

Denote the time at which a telephone call is initiated by τ and its duration by u. Then, the number of telephone connections in use at time t in a telephone network energized at time t_0 and having an unlimited number of possible connections can be expressed in the form of (5.1) by setting

$$h(t, \tau_n : u_n) = \begin{cases} 1, & \tau_n \leq t < \tau_n + u_n \\ \\ 0, & \text{otherwise} \end{cases} \tag{5.3}$$

where τ_n and u_n are the initiation time and duration of the *nth* call. Here, $\{y(t): t \geq t_0\}$ is a filtered point process that is not a linear function of the marks. E. Parzen [23, p. 147] gives a detailed study of this situation when $y(\cdot)$ is a homogeneous filtered Poisson process; see Prob. 5.2.3. ∎

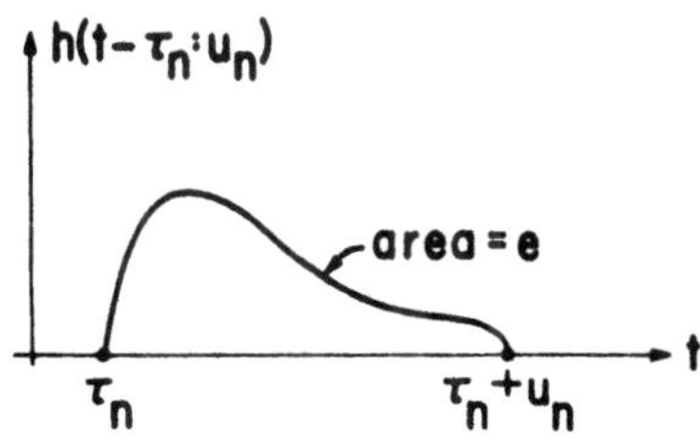

Figure 5.1   Incremental current contribution of a photon absorption.

Example 5.2.3 *Photoconductive Device* ————————————————————
Electron-hole pairs created in a semiconductor by the absorption of optical photons increase the conductivity of the material during the life-time of the pairs. This effect, called photoconductivity, provides a mech-anism for the detection of optical signals. Upon application of a bias voltage, the change in conductivity due to the absorption of a photon results in a measurable change in the current through the device. To a first approximation, the output current $\{i(t): t \geq t_0\}$ can be modeled as

$$i(t) = \begin{cases} 0, & N(t) = 0 \\ \sum_{n=1}^{N(t)} h(t - \tau_n : u_n), & N(T) \geq 1 \end{cases}$$

where $N(t)$ is the number of photon absorptions that occur during $[t_0, t)$, $\{\tau_n\}$ are the times of these absorptions, and u_n is the duration of the in-cremental contribution $h(t - \tau_n : u_n)$ to $i(t)$, as shown in Fig. 5.1. The du-ration u_n of h is a random variable depending on the lifetime of the electron-hole pair created by the photon absorption. Evidence suggests that this lifetime can be taken as exponentially distributed. ∎

Example 5.2.4 *ELF-VLF Atmospheric Noise* ————————————————
Atmospheric radio-noise in the frequency bands below 30 kHz (i.e., the extremely-low- and very-low-frequency bands) is mainly due to lightning discharges. The effect of such noise in the radio receiver can be modeled to a first approximation as a filtered point process of the form

$$y(t) = \begin{cases} 0, & N(t) = 0 \\ \sum_{n=1}^{N(t)} u_n h(t - \tau_n), & N(t) \geq 1 \end{cases} \tag{5.4}$$

where $n(t)$ denotes the number of discharges during $[t_0, t)$ subsequent to time t_0 when the receiver is first energized, $\{\tau_n\}$ denotes their occurrence times, and $h(\cdot)$ is the receiver response to a single discharge. Some sample functions of ELF noise recorded at the output of a narrowband receiver and reported by J. Evans [6] are shown in Fig. 5.2. The amplitude marks $\{u_n\}$

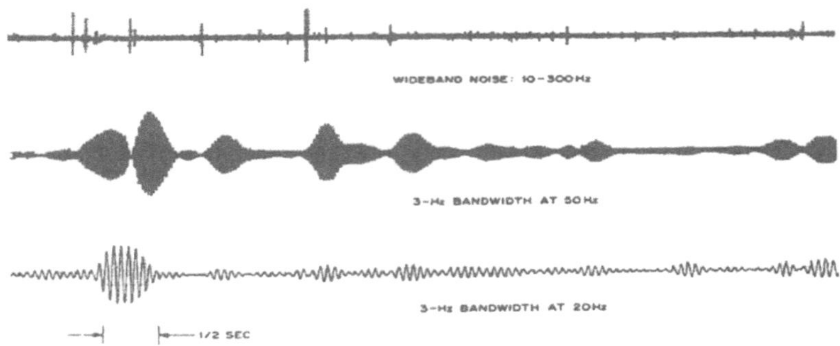

Figure 5.2 Sample functions of ELF Noise. From J. Evans [6].

are included to account for the variability in the amplitude of the responses, which arises because of the varying amounts of energy released by different lightning strokes. ∎

Example 5.2.5 *Radar Clutter, Scatter Communication Channels* ———
Suppose that a signal $s(t)$ is propagated over a radio channel consisting of small scatterers distributed in space. Examples are radar signals reflected from water droplets in rain and communication signals reflected from small, needle-sized dipoles as dispersed, for example, in the Project West Ford experiment [25]. Let $s(t) = e(t)\cos(\omega_c t + \phi(t))$ for $t \geq t_0$. Here, $e(t)$ and $\phi(t)$ are deterministic amplitude and phase modulation, respectively, of a carrier of frequency ω_c. A model described by R. Kennedy [16, Ch. 2] for the signal received $r(\cdot)$ after scattering is

$$r(t) = \begin{cases} 0, & N(t) = 0 \\[2ex] \displaystyle\sum_{n=1}^{N(t)} \rho_n e(t - \tau_n)\cos[\omega_c t + \phi(t - \tau_n) + \theta_n], & N(t) \geq 1 \end{cases} \tag{5.5}$$

where $N(t)$ is the number of scattered signals received during $[t_0, t)$, $\{\tau_n\}$ are their arrival times at the receiver, and $\{\rho_n\}$ and $\{\theta_n\}$ are random attenuation and phase variables associated with each scatterer. Attenuation variables $\{\rho_n\}$ are included because the amount of energy reflected from a scatterer depends on its shape and aspect. The phase variable $\{\theta_n\}$ can usually be taken to be mutually independent, uniformly distributed random variables because of the high frequencies ω_c normally used in radar and communication systems. Thus, $\{r(t): t \geq t_0\}$ is a filtered point process with a two-dimensional mark variable $\mathbf{u} = [\rho_n \quad \theta_n]'$. ∎

It is evident from these examples that filtered point processes defined in (5.1) provide a model for a wide variety of phenomena. We now develop the statistics for a special class of these processes; namely, for filtered Poisson-processes. The most important quantities we develop are the characteristic functional, mean, and covariance function for the filtered Poisson-process. The characteristic functional is important because it provides a complete statistical characterization of the process in the same way we have seen for other processes in earlier chapters. The first moment and covariance function provide important information in applications.

5.2.1 Statistics for a Filtered Poisson-Process

The characteristic functional for a filtered Poisson-process is given in the following theorem. This functional is then used to evaluate various first- and second-order statistics for these processes.

Theorem 5.1 (*Characteristic Functional for a Filtered Poisson Process*). Let $\{y(t): t \geq t_0\}$ be a filtered Poisson process in which points occur with an integrable intensity function $\{\lambda(t): t \geq t_0\}$. Then, the characteristic functional

$$\phi_y(jv) = E\left\{\exp\left[j\int_{t_0}^{T} y(\sigma)\,dv(\sigma)\right]\right\} \tag{5.6}$$

of the random variables $\{y(\sigma): t_0 \leq \sigma < T\}$ has the evaluation

$$\phi_y(jv) = \exp\left\{\int_{t_0}^{T}\lambda(\tau)E\left[\exp\left(j\int_{\tau}^{T} h(\sigma,\tau{:}\mathbf{u})\,dv(\sigma)\right) - 1\right]d\tau\right\}, \tag{5.7}$$

where the expectation is with respect to the distribution of the marks, and $v(\cdot)$ is an arbitrary real-valued function such that

$$\int_{t_0}^{T}\int_{t_0}^{T} f(\alpha,\beta)\,dv(\alpha)\,dv(\beta) < \infty, \tag{5.8}$$

where $f(\cdot,\cdot)$ is defined by

$$f(\alpha,\beta) = \int_{t_0}^{\min(\alpha,\beta)}\lambda(\tau)E[h(\alpha,\tau{:}\mathbf{u})h(\beta,\tau{:}\mathbf{u})]d\tau$$

$$+ \int_{t_0}^{\alpha}\lambda(\tau)E[h(\alpha,\tau{:}\mathbf{u})]d\tau\int_{t_0}^{\beta}\lambda(\tau)E[h(\beta,\tau{:}\mathbf{u})]d\tau.$$

Proof. The proof of Theorem 5.2.1 is accomplished using iterated expectation and the properties of Poisson processes. We need to observe initially that because $N(T) \geq N(\sigma)$ for $\sigma \in [t_0, T)$, and because h is causal, there holds

$$\phi_y(jv) = \Pr(N(T) = 0) \tag{5.9}$$

$$+ \sum_{k=1}^{\infty} \Pr(N(T) = k) E\left\{ \exp\left[j \sum_{n=1}^{k} \int_{t_0}^{T} h(\sigma, T_n : \mathbf{u}_n) dv(\sigma) \right] \Big| N(T) = k \right\}.$$

The summation in the expectation is unchanged by a random reordering of the occurrence times. With this reordering, the occurrence times $\{T_1, T_2, \cdots, T_k\}$ given $N(T) = k$ are independent and identically distributed, their common density being

$$p_{T_i}(T) = \frac{\lambda(T)}{\displaystyle\int_{t_0}^{T} \lambda(\sigma) d\sigma}$$

from (2.34). Since the mark variables are also independent and identically distributed, we obtain

$$E\left\{ \exp\left[j \sum_{n=1}^{k} \int_{t_0}^{T} h(\sigma, T_n : \mathbf{u}_n) dv(\sigma) \right] \Big| N(T) = k \right\}$$

$$= \left\{ \int_{t_0}^{T} \lambda(\sigma) d\sigma \right\}^{-1} \int_{t_0}^{T} \lambda(T) E\left[\exp\left(j \int_{t_0}^{T} h(\sigma, T : \mathbf{u}) dv(\sigma) \right) \right] dT \right\}^{k},$$

where the expectation on the right is with respect to the distribution of the mark random variable $\mathbf{u}$. Substitution of this expression into (5.9) and a routine calculation with the Poisson distribution results in (5.7). The restriction on $v(\cdot)$ in (5.8) insures that $E\left[\left(\int_{t_0}^{T} y(\sigma) dv(\sigma) \right)^2 \right] < \infty$, as developed below. $\square$

We now illustrate the use of the characteristic functional (5.7) with several examples. These follow from particular selections of the function $v(\cdot)$.

Let us first examine the statistics for the instantaneous response $y(t)$ at time t. Take the function $v(\cdot)$ in (5.6) to be

$$v(\sigma) = \begin{cases} 0, & t_0 \leq \sigma < t \\ \alpha, & t \leq \sigma < T, \end{cases}$$

where α is a constant. Then

$$\phi_y(jv) = E[e^{j\alpha y(t)}] \overset{\Delta}{=} M_{y(t)}(j\alpha)$$

is by definition the characteristic function for $y(t)$. We have

$$\int_{\tau}^{T} h(\sigma,\tau{:}\mathbf{u})\,dv(\sigma) = \begin{cases} \alpha h(t,\tau{:}\mathbf{u}), & \tau \leq t \\ \\ 0, & \tau > t. \end{cases}$$

Consequently, from (5.7), the evaluation of the characteristic function for $y(t)$ is

$$M_{y(t)}(j\alpha) = \exp\left\{ \int_{t_0}^{t} \lambda(\tau)E[e^{j\alpha h(t,\tau{:}\mathbf{u})} - 1]d\tau \right\}. \tag{5.10}$$

This expression is generally difficult to invert to determine the probability distribution for $y(t)$. It is useful, however, for evaluating the moments of $y(t)$. For this purpose, let γ_n be the nth cumulant for $y(t)$; the cumulants are defined in Prob. 2.2.3. By using $j^n\gamma_n = \partial^n M_{y(t)}(j\alpha)/\partial\alpha^n$ at $\alpha = 0$, we deduce from (5.10) that

$$\gamma_n = \int_{t_0}^{t} \lambda(\sigma)E[h^n(t,\tau{:}\mathbf{u})]d\tau. \tag{5.11}$$

the mean $E[y(t)]$ and variance $\mathrm{var}[y(t)]$ of the response at time t are easily obtained from (5.11) because $E[y(t)] = \gamma_1$ and $\mathrm{var}[y(t)] = \gamma_2$. Higher-order moments of $y(t)$ can be similarly obtained using the results of Prob. 2.2.3.

Example 5.2.6 *White Noise Property of Poisson Processes* ───────
Suppose that $h(t,\tau{:}\mathbf{u}) = g(t-\tau)$, so the point process generating $y(\cdot)$ can be viewed as unmarked and exciting a linear, time-invariant system with causal impulse response $g(t)$. Further, suppose that $\lambda(t) \equiv \lambda$, a constant for all t, so the excitation process is homogeneous. Set $t_0 = -\infty$. Then

$$\gamma_n = \lambda \int_{-\infty}^{t} g^n(t-\tau)d\tau = \lambda \int_{0}^{\infty} g^n(\alpha)d\alpha. \tag{5.12}$$

Consequently,

$$E[y(t)] = \gamma_1 = \lambda \int_0^\infty g(\alpha)\, d\alpha = \lambda G(0), \tag{5.13}$$

where

$$G(f) = \int_{-\infty}^\infty g(t) e^{-j2\pi f t}\, dt$$

is the frequency response function of the system. Thus, the expected value of the response $y(t)$ is time independent and equals the product of the average event rate λ and the zero-frequency response of the system, $G(0)$. By Parseval's equation, we also have

$$\mathrm{var}[y(t)] = \gamma_2 = \lambda \int_0^\infty g^2(\alpha)\, d\alpha = \lambda \int_{-\infty}^\infty |G(f)|^2\, df. \tag{5.14}$$

It is evident that the variance of the response is the same as if the filter were excited by a stationary process with a power spectral density of constant level λ. This "white noise" property of a homogeneous Poisson process is further developed in Ex. 5.2.7. ∎

The joint statistics for the response at times t_1 and t_2 can be determined by taking the function $v(\cdot)$ to be

$$v(\sigma) = \begin{cases} 0, & t_0 \le \sigma < t_1 \\ \alpha_1, & t_1 \le \sigma < t_2 \\ \alpha_1 + \alpha_2, & t_2 \le \sigma < T \end{cases}$$

where α_1 and α_2 are constants. Then

$$\phi_y(jv) = E\!\left[e^{j\alpha_1 y(t_1) + j\alpha_2 y(t_2)} \right] \overset{\Delta}{=} M_{y(t_1),\,y(t_2)}(j\alpha_1, j\alpha_2)$$

is by definition the joint characteristic function for $y(t_1)$ and $y(t_2)$. Consequently, from (5.7), the evaluation of this joint characteristic function is

$$M_{y(t_1),\,y(t_2)}(\alpha_1, \alpha_2)$$

$$= \exp\left\{ \int_{t_0}^T \lambda(\tau) E\!\left[e^{j\alpha_1 h(t_1,\tau;u) + j\alpha_2 h(t_2,\tau;u)} - 1 \right] d\tau \right\}. \tag{5.15}$$

We use (5.15) to evaluate the covariance function $K_y(t_1, t_2)$, defined by $K_y(t_1, t_2) = E[y(t_1)y(t_2)] - E[y(t_1)]E[y(t_2)]$. Using the relation

$$j^{-2}\frac{\partial^2 M_{y(t_1), y(t_2)}(\alpha_1, \alpha_2)}{\partial\alpha_1\partial\alpha_2}\bigg|_{\alpha_1=0, \alpha_2=0} = E[y(t_1)y(t_2)]$$

and (5.11) with $n = 1$, we conclude that

$$K_y(t_1, t_2) = \int_{t_0}^{\min(t_1, t_2)} \lambda(\tau)E[h(t_1, \tau:\mathbf{u})h(t_2, \tau:\mathbf{u})]d\tau. \qquad (5.16)$$

The function $f(\cdot, \cdot)$ in (5.8) is now seen to be the correlation function $f(t_1, t_2) = K_y(t_1, t_2) + E[y(t_1)]E[y(t_2)]$. Thus, condition (5.8) ensures that $\int_{t_0}^T y(\sigma)dv(\sigma)$ exists in the mean square sense and that

$$E\left[\left(\int_{t_0}^T y(\sigma)dv(\sigma)\right)^2\right] = \int_{t_0}^T \int_{t_0}^T f(\sigma_1, \sigma_2)dv(\sigma_1)dv(\sigma_2) \qquad (5.17)$$

holds and is finite. ∎

Example 5.2.7 *Campbell's Theorem* ——————————————————————
For the situation of Ex. 5.2.6, equation (5.16) becomes

$$K_y(t_1, t_2) = \lambda \int_{-\infty}^{\min(t_1, t_2)} g(t_1 - \tau)g(t_2 - \tau)d\tau = \lambda \int_{-\infty}^{\infty} g(t_1 - \tau)g(t_2 - \tau)d\tau.$$

Let $z = t_2 - \tau$, and change variables of integration to get

$$K_y(t_1, t_2) = \lambda \int_{-\infty}^{\infty} g(t_1 - t_2 + z)g(z)dz.$$

Thus, the covariance function for $y(\cdot)$ depends only on the difference of its arguments and not on each separately; therefore, $y(\cdot)$ is wide-sense stationary. Let $t = t_1 - t_2$, and abbreviate $K_y(t, 0)$ by $K_y(t)$. Then

$$K_y(t) = \lambda \int_{-\infty}^{\infty} g(t + z)g(z)dz \qquad (5.18)$$

and

$$S_y(f) = \int_{-\infty}^{\infty} K_y(t)e^{-j2\pi ft}dt = \lambda |G(f)|^2, \qquad (5.19)$$

where $S_y(f)$ is the power-density spectrum for $y(t) - E[y(t)]$. Thus, the power-density spectrum for $y(\cdot)$ is the same as if the system were excited by a random process with a constant power-density spectrum of intensity λ. ∎

Equations (5.18) and (5.19) are known as *Campbell's theorem*. In this terminology, (5.16) can be viewed as a generalized form of Campbell's theorem. A classic application of Campbell's theorem is the modeling of shot noise, which is discussed in the next example.

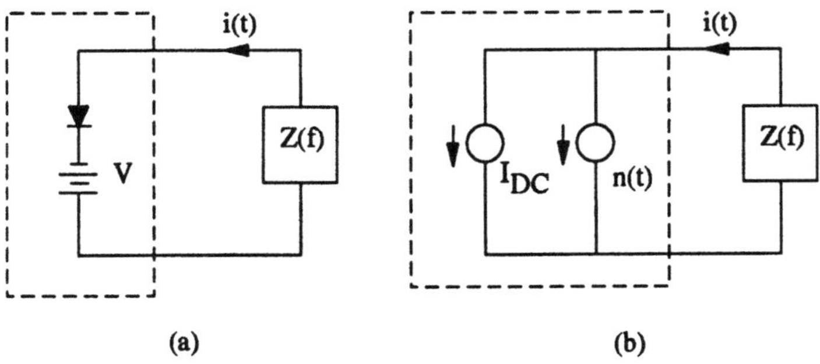

Figure 5.3 Diode circuit and its equivalent.

Example 5.2.8 *Shot Noise* ──────────────────────────

Consider, again, the shot noise model of Ex. 5.2.1. Suppose that a diode is connected to a steady bias voltage V and a load with frequency response function $Z(f)$, as diagrammed in Fig. 5.3(a). An equivalent circuit, Fig. 5.3(b), for the diode and biasing source can be obtained by use of Campbell's theorem as follows. From Ex. 5.2.1, we have

$$i(t) = \sum_{n=1}^{N(t)} h(t - \tau_n).$$

Assume that $\{N(t): t > -\infty\}$ is a homogeneous Poisson process with constant intensity λ representing the rate of electron emissions from the cathode. The incremental contribution $h(\cdot)$ of an electron to the total current $i(\cdot)$ has area e, the electronic charge. Hence, $H(0) = e$, and from (5.13), we have

$$E[i(t)] \overset{\Delta}{=} I_{DC} = \lambda e. \tag{5.20}$$

From elementary considerations, $I_{DC} = V[R_d + Z(0)]^{-1}$, where R_d is the diode resistance at the operating point. Thus, the emission rate λ can be determined from (5.20) when V, R_d, and $Z(0)$ are known. Furthermore, we have from Campbell's theorem (5.19) that the power-density spectrum for $i(t) - I_{DC}$ is given by

$$S_i(f) = \lambda |H(f)|^2. \tag{5.21}$$

That $i(t)$ can be expressed as $I_{DC} + n(t)$, where $n(t) = i(t) - E[i(t)]$, leads to the diode model in Fig. 5.3(b); here, $n(\cdot)$ is a zero mean, wide-sense stationary process with power-density spectrum $\lambda |H(f)|^2$. The power-density spectrum for the load voltage (minus its expectation) is $\lambda |H(f)Z(f)|^2$. Under usual conditions, $H(f)$ is a slowly changing function of f by comparison to $Z(f)$. For instance, the bandwidth of $H(\cdot)$ is of the order of the inverse transit time, or about 10^9 Hz; the bandwidth of $Z(\cdot)$ is ordinarily a small fraction of this. Consequently,

$$\lambda |H(f)Z(f)|^2 \approx \lambda |H(0)Z(f)|^2 = I_{DC} e |Z(f)|^2.$$

In these conditions, $n(\cdot)$ can be modeled as a process with constant power-density spectrum eI_{DC}, which is known as Schottky's formula for shot noise. ∎

Example 5.2.9 *Signal-to-Noise Ratio in Photodetection* ────────────

A model described by E. Hoversten [13] for a wide variety of photodetectors is shown in Fig. 5.4. The output current $\{i(t): t \geq t_0\}$ is the sum of two independent filtered Poisson processes, $\{i_s(t): t \geq t_0\}$ and $\{i_d(t): t \geq t_0\}$, and an independent Gaussian process $i_{th}(t): t \geq t_0\}$ according to $i(t) = i_s(t) + i_d(t) + i_{th}(t)$. For $t \geq t_0$, the "signal" current is

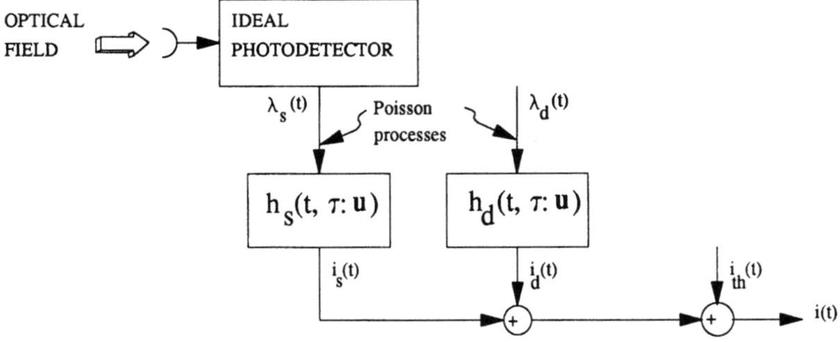

Figure 5.4 Photodetector model.

$$i_s(t) = \sum_{n=1}^{N_s(t)} h_s(t, \tau_n : u_n),$$

where $\{N_s(t): t \geq t_0\}$ is an inhomogeneous Poisson process such that $N_s(t)$ is the number of photoelectrons generated during $[t_0, t)$. The intensity $\lambda_s(t)$ is related to a quasi-monochromatic optical signal-field

$$s(t, \vec{r}) = \sqrt{2}\,\text{Re}[S(t, \vec{r})e^{j2\pi\nu t}],$$

that is incident on the active surface $\mathcal{A}$ of the detector, according to

$$\lambda_s(t) = \frac{\eta}{h\nu} \int_{\mathcal{A}} |S(t, \vec{r})|^2 d^2\vec{r}, \tag{5.22}$$

where η is a constant called the quantum efficiency of the detector, h is Planck's constant, ν is the unmodulated optical-carrier frequency, and $S(\cdot, \cdot)$ is the complex envelope of the signal, which is assumed here to be deterministic. The filter response $h_s(t, \tau:u)$ has a form that depends on the particular type of photoconductor used. Some examples are given in Table 5.1. In the presence of background radiation (for example, scattered sunlight), it is often assumed that $\lambda_s(t)$ is simply replaced by $\lambda_s(t) + \lambda_b$, where

Table 5.1 Photodetector Models

$h_s(t,\tau{:}u)$	Remark	Type of Photodetector Modeled
$e\delta(t-\tau)$	e is the electronic charge and $\delta(t)$ is a Dirac delta function	ideal photodetector (without gain)
$g(t-\tau)$	$g(t)$ is a deterministic impulse response	photodiode
$ug(t-\tau)$	u is a random variable modeling random gain (e.g., number of secondary electrons generated for each primary photoelectron)	photomultiplier, avalanche photodiode
$h(t-\tau{:}u)$	response function shown in Fig. 5.1	photoconductor

$\lambda_b = \eta P_b/h\nu$ and P_b is the background noise power incident on the detector. Conditions for the validity of this assumption are discussed by E. Hoversten [13].

The disturbance current $\{i_d(t){:}\, t \geq t_0\}$ is given by

$$i_d(t) = \sum_{n=1}^{N_d(t)} h_d(t,\tau_n{:}\nu_n)$$

where $\{N_d(t){:}\, t \geq t_0\}$ is an inhomogeneous Poisson counting process. This current is termed "dark current" and accounts for the extraneous electrons generated during $[t_0,t)$. The intensity function $\{\lambda_d(t){:}\, t \geq t_0\}$ is the instantaneous rate at which such spontaneous events occur. The filter responses h_s and h_d may or may not be identical depending on the type of photodetector used.

The disturbance current $\{i_{th}(t){:}\, t \geq t_0\}$ is a zero-mean Gaussian process modeling the combined effects of thermally generated noise in amplifiers or other circuits to which the detector is connected. Usually, $i_{th}(\cdot)$ can be assumed to be stationary and having a constant power-density spectrum of level $2kT/R$, where k is Boltzmann's constant, T is the temperature, and R is the noise resistance. This thermal noise is especially significant for detectors that do not have internal gain.

As an application of this model and our results on the second-moment properties of filtered Poisson processes, we evaluate the signal-to-noise ratio at the output of an amplifier in response to the detector current $i(\cdot)$. For this purpose, assume that the amplifier output $\{y(t): t \geq t_0\}$ is given by

$$y(t) = \sum_{n=1}^{N(t)} h(t - T_n : u_n) + \int_{t_0}^{t} k(t - T) i_{th}(T) dT,$$

where $N(t) = N_s(t) + N_b(t) + N_d(t)$ is the total number of detection events during $[0, t)$ corresponding to signal, background, and dark current, and where $k(t)$ is the impulse response of the amplifier, and $h(t - T : u) = uek(t - T)$. We have assumed that the detector response functions h_s and h_d both equal $ue\delta(t - T)$ corresponding to an ideal photodetector with a random gain u. By using (5.11), we conclude that the signal-to-noise ratio S/N defined by

$$\frac{S}{N} = \frac{E^2[y(t) \mid i_b(t) = i_d(t) = i_{th}(t) = 0]}{\mathrm{var}[y(t)]} \tag{5.23}$$

has the evaluation

$$\frac{S}{N} = \frac{e^2 E^2(u) \left[\int_0^\infty k(t - T) \lambda_s(T) dT \right]^2}{e^2 E(u^2) \int_0^\infty k^2(t - T)[\lambda_s(T) + \lambda_b + \lambda_d(T)] dT + \left(\frac{2kT}{R}\right) \int_0^\infty k^2(t - T) dT}. \tag{5.24}$$

This signal-to-noise ratio can be simplified when the signal field corresponds to an unmodulated optical carrier for then we have $\lambda_s(t) = \eta P_s / h\nu$, where P_s is the total carrier power incident on the detector. Assume further that the dark current rate $\lambda_d(t)$ is a constant λ_d. Then

$$\frac{S}{N} = \frac{\left[\dfrac{\eta P_s}{h\nu}\right]^2}{2B\left[\dfrac{E(u^2)}{E^2(u)}\right]\left[\dfrac{\eta(P_s + P_b)}{h\nu} + \lambda_d\right] + 2B\left[\dfrac{2kT}{e^2 E^2(u)R}\right]}, \tag{5.25}$$

where

$$2B \overset{\Delta}{=} \frac{\displaystyle\int_0^\infty k^2(t-\tau)d\tau}{\left[\displaystyle\int_0^\infty k(t-\tau)d\tau\right]^2}.$$

Here, B is the equivalent rectangular noise bandwidth of the filter $k(\cdot)$. We note, finally, that S/N can also be expressed as

$$\frac{S}{N} = \frac{I_s^2}{2Be\left[\frac{E(u^2)}{E^2(u)}\right](I_s+I_b+I_d)+2B\left[\frac{2kT}{E^2(u)R}\right]} \tag{5.26}$$

in which I_s, I_b, and I_d are the signal, background, and dark currents, respectively. We note that for detectors without gain (that is, $u=1$), the last term in the denominator usually dominates S/N, and the converse holds for photomultipliers and other detectors with gain. ∎

A Central Limit Theorem

A filtered Poisson process is a superposition of independent components. As such, it is not unexpected that under some conditions, a filtered Poisson process can tend to a Gaussian process as certain parameters, such as the intensity, tend to a limit. This fact is the basis for the widespread use of Gaussian models for random phenomena that would at first appear to be better modeled as filtered point processes. Examples of phenomena where Gaussian statistics are frequently used are shot noise, photoelectron conversion noise, atmospheric noise above the VLF band, radar clutter, and communication in a scattering medium. In Theorem 5.2.2, we give conditions such that a normalized version of the response process $y(\cdot)$ converges to a Gaussian random process on $[t_0, T)$.

Theorem 5.2.2 (*Central Limit Theorem for Filtered Poisson Processes*). Let

$$\left\{ y(t) = \sum_{n=1}^{N(t)} h(t,\tau_n{:}\mathbf{u}_n){:} t \ge t_0 \right\}$$

be a filtered Poisson process in which points occur with intensity $\{\lambda(t){:}t \ge t_0\}$. Suppose for $t \in [t_0, T)$ that

$$\gamma_1(t) = E[y(t)] = \int_{t_0}^t \lambda(\tau)E[h(t,\tau{:}\mathbf{u})]d\tau < \infty, \tag{5.27}$$

$$0 < \gamma_2(t) = \text{var}[y(t)] = \int_{t_0}^{t} \lambda(\tau)E[h^2(t,\tau;\mathbf{u})]d\tau < \infty$$

and that

$$\frac{\Gamma(t_1, t_2, t_3)}{\sqrt{\gamma_2(t_1)\gamma_2(t_2)\gamma_2(t_3)}}$$

tends to zero uniformly on $[t_0, T) \times [t_0, T) \times [t_0, T)$ as certain parameters tend to prescribed limits, where

$$\Gamma(t_1, t_2, t_3) = \int_{t_0}^{\min(t_1, t_2, t_3)} \lambda(\tau)E\{|\, h(t_1, \tau;\mathbf{u})h(t_2, \tau;\mathbf{u})h(t_3, \tau;\mathbf{u})|\,\}d\tau.$$

$$(5.28)$$

Then

$$y^*(t) \overset{\Delta}{=} \frac{y(t) - \gamma_1(t)}{\sqrt{\gamma_2(t)}}$$

tends to a Gaussian process on $[t_0, T)$ with zero mean and covariance function $[\gamma_2(t_1)\gamma_2(t_2)]^{-1/2}K_y(t_1, t_2)$, where $K_y(t_1, t_2)$ is given in (5.16).

For the proof of this theorem, we need the following two lemmas.

Lemma 1. For some $\theta(\beta)$ satisfying $|\,\theta(\beta)| \leq 1$ there holds

$$e^{j\beta} - 1 - j\beta + \frac{1}{2}\beta^2 = \frac{1}{6}\beta^3\theta(\beta).$$

$$(5.29)$$

Proof of Lemma 1. Integrate by parts to establish the identity

$$6\left(e^{j\beta} - 1 - j\beta + \frac{1}{2}\beta^2\right)\beta^{-3} = 3j^3 e^{j\beta} \int_0^1 x^2 e^{-j\beta x}dx.$$

The assertion of Lemma 1 follows by defining $\theta(\beta)$ as the right side of this equation. $\square$

Lemma 2. (*Characteristic Functional for a Gaussian Process*). Let $\{z(t): t \geq t_0\}$ be a Gaussian mean-square integrable random process with zero mean and covariance function $\rho(t_1, t_2)$. The characteristic functional for this process on $[t_0, T)$, defined by

$$\phi_z(jv) \overset{\Delta}{=} E\left[\exp\left(j\int_{t_0}^{T} z(\sigma)dv(\sigma)\right)\right],$$

has the evaluation

$$\phi_z(jv) = \exp\left\{-\frac{1}{2}\int_{t_0}^{T}\int_{t_0}^{T}\rho(\sigma_1,\sigma_2)dv(\sigma_1)dv(\sigma_2)\right\}. \tag{5.30}$$

Proof of Lemma 2. Let $x = \int_{t_0}^{T} z(\sigma)dv(\sigma)$. Then, x is a Gaussian random variable with zero mean and variance

$$\sigma_x^2 = \int_{t_0}^{T}\int_{t_0}^{T}\rho(\sigma_1,\sigma_2)dv(\sigma_1)dv(\sigma_2).$$

Equation (5.30) then follows from the observation that $\phi_x(jv) = M_x(j1)$, where $M_x(j\alpha)$ is the characteristic function for x, which is given by $M_x(j\alpha) = \exp(-\alpha^2\sigma_x^2/2)$. $\square$

Proof of Theorem 5.2.2. From the definition of $y^*(\cdot)$ in terms of $y(\cdot)$, note that

$$\phi_y(jv) \overset{\Delta}{=} E\left[\exp\left(j\int_{t_0}^{T} y^*(\sigma)dv(\sigma)\right)\right]$$

$$= E\left[\exp\left(j\int_{t_0}^{T} y(\sigma)\gamma_2^{-1/2}(\sigma)dv(\sigma)\right)\right]\exp\left(-j\int_{t_0}^{T}\gamma_1(\sigma)\gamma_2^{-1/2}(\sigma)dv(\sigma)\right).$$

Thus, by using

$$y(t)\gamma_2^{-1/2}(t) = \sum_{n=1}^{N(t)} \gamma_2^{-1/2}(t)h(t,\tau_n:\mathbf{u}_n)$$

and (5.7), we have

$$\ln\phi_y(jv) = \int_{t_0}^{T}\lambda(\tau)E\left(\exp\left[j\int_{t_0}^{T}\gamma_2^{-1/2}(\sigma)h(\sigma,\tau:\mathbf{u})dv(\sigma)\right] - 1\right)d\tau$$

$$-j\int_{t_0}^{T}\gamma_1(\sigma)\gamma_2^{-1/2}(\sigma)dv(\sigma). \tag{5.31}$$

Suppose, now, that z is a Gaussian process on $[t_0, T)$ with zero mean and covariance function $\rho(t_1, t_2) = [\gamma_2(t_1)\gamma_2(t_2)]^{-1/2}K_y(t_1, t_2)$, where $K_y(\cdot, \cdot)$ is given in (5.16). Then subtracting $\ln \phi_z(jv)$ from both sides of (5.31), and using (5.16) and (5.27) results in

$$\ln\phi_{y^*}(jv) + \frac{1}{2}\int_{t_0}^{T}\int_{t_0}^{T}[\gamma_2(\sigma_1)\gamma_2(\sigma_2)]^{-1/2}K_y(\sigma_1, \sigma_2)dv(\sigma_1)dv(\sigma_2)$$

$$= \int_{t_0}^{T}\lambda(\tau)E\left\{e^{j\beta(\tau:\mathbf{u})} - 1 - j\beta(\tau:\mathbf{u}) + \frac{1}{2}\beta^2(\tau:\mathbf{u})\right\}d\tau$$

$$= \int_{t_0}^{T}\lambda(\tau)E\left\{\frac{1}{6}\beta^3(\tau:\mathbf{u})\theta(\tau:\mathbf{u})\right\}d\tau,$$

where

$$\beta(\tau:\mathbf{u}) \overset{\Delta}{=} \int_{t_0}^{T}\gamma_2^{-1/2}(\sigma)h(\sigma, \tau:\mathbf{u})dv(\sigma),$$

and the last equality follows from Lemma 1, where $|\theta(\tau:\mathbf{u})| \leq 1$. Upon expansion of β^3 and rearrangement of the integrals, the right side becomes

$$\frac{1}{6}\int_{t_0}^{T}\int_{t_0}^{T}\int_{t_0}^{T}\left[\frac{E\left\{\int_{t_0}^{T}\lambda(\tau)h(\sigma_1, \tau:\mathbf{u})h(\sigma_2, \tau:\mathbf{u})h(\sigma_3, \tau:\mathbf{u})\theta(\tau:\mathbf{u})d\tau\right\}}{\sqrt{\gamma_2(\sigma_1)\gamma_2(\sigma_2)\gamma_3(\sigma_3)}}\right]dv(\sigma_1)dv(\sigma_2)dv(\sigma_3).$$

Taking the magnitude of both sides then yields

$$\left|\ln\phi_{y^*}(jv) + \frac{1}{2}\int_{t_0}^{T}\int_{t_0}^{T}[\gamma_2(\sigma_1)\gamma_2(\sigma_2)]^{-1/2}K_y(\sigma_1, \sigma_2)dv(\sigma_1)dv(\sigma_2)\right|$$

$$\leq \frac{1}{6}\int_{t_0}^{T}\int_{t_0}^{T}\int_{t_0}^{T}[\gamma_2(\sigma_1)\gamma_2(\sigma_2)\gamma_2(\sigma_3)]^{-1/2}\Gamma(\sigma_1, \sigma_2, \sigma_3)|\,dv(\sigma_1)dv(\sigma_2)dv(\sigma_3)|,$$

$$\tag{5.32}$$

where Γ is defined in (5.28). The integrand on the right tends uniformly to zero as certain parameters tend to prescribed limits. Consequently, the left side also tends to zero, and in the limit, we conclude that $y^*(\cdot)$ has the characteristic functional of z on $[t_0, T)$. Thus, in the limit, $y^*(\cdot)$ has the same finite distributions as a Gaussian process with zero mean and covariance $\rho(t_1, t_2)$. $\square$

Example 5.2.10 *Homogeneous Point Process* ————————————

Suppose that the point process underlying the generation of $y(\cdot)$ is homogeneous with a constant intensity λ. Then (5.32) becomes

$$\left| \phi_{y^*}(jv) + \frac{1}{2} \int_{t_0}^{T} \int_{t_0}^{T} \rho(\sigma_1, \sigma_2)\, dv(\sigma_1)\, dv(\sigma_2) \right| \tag{5.33}$$

$$\leq \frac{1}{6}\lambda^{-1/2} \int_{t_0}^{T} \int_{t_0}^{T} \int_{t_0}^{T} f(\sigma_1, \sigma_2, \sigma_3)\, dv(\sigma_1)\, dv(\sigma_2)\, dv(\sigma_3),$$

where

$$\rho(\sigma_1, \sigma_2) \overset{\Delta}{=} \frac{\displaystyle\int_{t_0}^{T} E[h(\sigma_1, \tau{:}\mathbf{u})h(\sigma_2, \tau{:}\mathbf{u})]\, d\tau}{\left(\displaystyle\int_{t_0}^{T} E[h^2(\sigma_1, \tau{:}\mathbf{u})]\, d\tau \int_{t_0}^{T} E[h^2(\sigma_2, \tau{:}\mathbf{u})]\, d\tau \right)^{1/2}},$$

and where

$$f(\sigma_1, \sigma_2, \sigma_3)$$

$$\overset{\Delta}{=} \frac{\displaystyle\int_{t_0}^{T} E\{|\, h(\sigma_1, \tau{:}\mathbf{u})h(\sigma_2, \tau{:}\mathbf{u})h(\sigma_3, \tau{:}\mathbf{u})|\, \}\, d\tau}{\left(\displaystyle\int_{t_0}^{E} [h^2(\sigma_1, \tau{:}\mathbf{u})]\, d\tau \int_{t_0}^{E} [h^2(\sigma_2, \tau{:}\mathbf{u})]\, d\tau \int_{t_0}^{E} [h^2(\sigma_3, \tau{:}\mathbf{u})]\, d\tau \right)^{1/2}}.$$

Consequently, for intervals where the triple integral in (5.33) is finite, we conclude that as λ tends to infinity, y^* tends to a Gaussian process with covariance function $\rho(t_1, t_2)$. ∎

Example 5.2.11 *Bounded Response Function* ─────────────
Let the function $v(\cdot)$ in (5.33) be defined by

$$v(\sigma) = \begin{cases} 0, & t_0 \leq \sigma < t \\[2ex] \alpha, & t \leq \sigma < T \end{cases}$$

where α is a constant. Then, $\phi_{y^*}(jv)$ is the characteristic function $M_{y^*(t)}(j\alpha)$ for $y^*(t)$, and (5.33) becomes

$$\left| \ln M_{y^*(t)}(j\alpha) + \frac{1}{2}\alpha^2 \right| \leq \frac{1}{6}\lambda^{-1/2}\alpha^3 \frac{\displaystyle\int_{t_0}^{T} E\{|h^3(t,\tau\!:\!\mathbf{u})|\}d\tau}{\left[\displaystyle\int_{t_0}^{T} E\{h^2(t,\tau\!:\!\mathbf{u})\}d\tau\right]^{3/2}}. \qquad (5.34)$$

Suppose, further, that the response function h has the form $h(t,\tau\!:\!\mathbf{u}) = ug(t,\tau)$, where u is a real-valued mark, and g is bounded as $|g(\cdot,\cdot)| < C$. Then, (5.34) implies

$$\left| \ln M_{y^*(t)}(j\alpha) + \frac{1}{2}\alpha^2 \right| \leq \frac{1}{6}\lambda^{-1/2}\alpha^3 E(u^3)E^{-3/2}(u^2)C\left(\int_{t_0}^{T} g^2(t,\tau)d\tau\right)^{-1/2}.$$

$$(5.35)$$

This is used in the next example when the response function h is time invariant and band limited. ∎

Example 5.2.12 *Time-Invariant, Band-Limited Response Function* ───────
Suppose that $h(t,\tau\!:\!\mathbf{u}) = ug(t-\tau)$ so that the response function is time invariant and linearly dependent on the mark variable u. Further, assume that $t_0 = -\infty$ and that

$$G(f) = \int_0^{\infty} g(t)e^{-j2\pi ft}dt$$

is zero for $|f| > f_c/2$ so that the response is also band limited. Then, by paralleling an argument due to A. Papoulis [22], we can show that $g(\cdot)$ is bounded so that (5.35) can be used. From the Fourier inversion equation and the Schwarz inequality, we have that

$$|g(t)|^2 = \left| \int_{-f_c/2}^{f_c/2} G(f) e^{j2\pi ft} df \right|^2 \le f_c \int_{-f_c/2}^{f_c/2} |G(f)|^2 df \stackrel{\Delta}{=} C^2.$$

We also have from Parseval's equation that

$$\int_{-\infty}^{t} g^2(t - \tau) d\tau = \int_{0}^{\infty} g^2(t) dt = \int_{-f_c/2}^{f_c/2} |G(f)|^2 df.$$

These results in (5.35) imply that

$$\left| M_{y^*(t)}(j\alpha) + \frac{1}{2}\alpha^2 \right| \le \frac{1}{6} E(u^3) E^{-3/2}(u^2) \left[\frac{f_c}{\lambda} \right]^{1/2} \alpha^3. \tag{5.36}$$

Thus, we conclude that under these conditions, $y^*(t)$ approaches a Gaussian random variable as the bandwidth to rate ratio f_c/λ approaches zero.

It is difficult to assess from (5.36) how close the actual distribution for $y^*(t)$ is approximated by a Gaussian distribution for f_c/λ small but non-zero. A. Papoulis [22] gives the following upper bound on the error when $u = 1$:

$$\left| \Pr[y^*(t) \le Y] - \frac{1}{\sqrt{2\pi}} \int_{-\infty}^{Y} e^{-\xi^2/2} d\xi \right| \le \frac{4}{3} \left(\frac{2\pi f_c}{\lambda} \right)^{1/2}. \tag{5.37}$$

∎

Example 5.2.13 *Shot Noise* ────────────────────────────────

For shot noise, $f_c/\lambda \approx 10^9$ Hz, $\lambda^{-1} = e I_{DC}^{-1}$, and for $I_{DC} = 10^{-3}$ A, we have $\sqrt{2\pi f_c/\lambda} \approx 1.4 \times 10^{-3}$. Consequently, the distribution function for $y^*(t)$ is uniformly within ± 0.002 of a zero-mean, unit-variance Gaussian distribution function. This justifies, in part, the common modeling of n in Fig. 5.3(b) as a Gaussian process. ∎

Summary. In this section, we have studied a class of processes that arise as the response to the points of a marked point process. Filtered Poisson processes were emphasized. These are the superposition of the separate responses to each marked point of a compound Poisson process. Several applications where a filtered Poisson process provides a reasonable model were mentioned; additional applications are given by E. Parzen [23, Sec. 4.5]. Statistics for filtered Poisson processes were developed. The main result is the characteristic functional given in Theorem 5.2.1. Second-moment properties of filtered Poisson processes are widely used. Most particularly, Campbell's theorem, which is developed in Ex. 5.2.7, is often cited. The central limit theorem given in Theorem 5.2.2 provides the mathematical justification for the frequent use of Gaussian models for processes that would appear more appropriately modeled by a filtered Poisson process.

We now turn to the development of a model in which the response need not be a superposition of individual responses to each marked point.

5.3 Poisson Driven Markov Processes

The response to a marked point process studied in Sec. 5.2 is comprised of the separate responses to each point. In this section, we develop a model in which the response is generally not a superposition of individual effects. The importance of the model is due to the wide variety of phenomena encountered in physics and engineering for which it provides an accurate mathematical representation.

For reasons that will become evident, we call the process developed in this section a *Poisson driven Markov process*. These processes satisfy an integral equation of the form

$$\mathbf{x}(t) = \mathbf{x}_0 + \int_{t_0}^{t} \mathbf{a}(\sigma, \mathbf{x}(\sigma)) d\sigma + \int_{t_0}^{t} \int_{\mathcal{U}} \mathbf{b}(\sigma, \mathbf{x}(\sigma), \mathbf{U}) M(d\sigma, d\mathbf{U}), \tag{5.38}$$

where M is the time-space Poisson process of Theorem 4.3.2, and $\mathcal{U}$ is the mark space associated with it. The vector-valued functions $\mathbf{a}(t, \mathbf{x}(t))$ and $\mathbf{b}(t, \mathbf{x}(t), \mathbf{U})$ satisfy certain technical conditions to be stated later. As developed below, the last integral in (5.38) has the evaluation

$$\int_{t_0}^{t} \int_{\mathcal{U}} \mathbf{b}(\sigma, \mathbf{x}(\sigma), \mathbf{U}) M(d\sigma, d\mathbf{U}) = \begin{cases} 0, & N(t) = 0 \\ \sum_{n=1}^{N(t)} \mathbf{b}(\tau_n, \mathbf{x}(\tau_n), \mathbf{u}_n), & N(t) \geq 1 \end{cases}$$

where $N(t) = \int_{t_0}^{t} \int_{\mathcal{U}} M(d\sigma, d\mathbf{U})$ is the number of incident points during $[t_0, t)$ regardless of their marks, and τ_n and $\mathbf{u}_n$ are the time of occurrence and mark on the nth point.

We also write the equation for $\{\mathbf{x}(t): t \geq t_0\}$ somewhat more concisely in differential form as

$$d\mathbf{x}(t) = \mathbf{a}(t, \mathbf{x}(t)) dt + \int_{\mathcal{U}} \mathbf{b}(t, \mathbf{x}(t), \mathbf{U}) M(dt, d\mathbf{U}), \quad \mathbf{x}(t_0) = \mathbf{x}_0 \tag{5.39}$$

This stochastic differential equation can be viewed as defining a nonlinear transformation of the Poisson process M into the process $\mathbf{x}$. At $t = t_0$, $\mathbf{x}(t_0) = \mathbf{x}_0$. The differential $d\mathbf{x}(t) = \mathbf{x}(t + dt) - \mathbf{x}(t)$ denotes an infinitesimal increment in $\mathbf{x}$ that occurs during $[t, t + dt)$. If during this interval no points occur in the incident marked point process, $M(dt, d\mathbf{U}) = 0$, and the increment in $\mathbf{x}$ is $\mathbf{a}(t, \mathbf{x}(t))dt$. If a point having a mark $\mathbf{U}$ does occur in $[t, t + dt)$, the increment in $\mathbf{x}$ is $\mathbf{a}(t, \mathbf{x}(t))dt + \mathbf{b}(t, \mathbf{x}(t), \mathbf{U})$, which is dominated by $\mathbf{b}(t, \mathbf{x}(t), \mathbf{U})$. It is evident that $\mathbf{x}$ will have discontinuities at the occurrence times of the incident points and that the size of the discontinuity for a point occurring at time t with mark $\mathbf{U}$ is $\mathbf{b}(t, \mathbf{x}(t), \mathbf{U})$.

We will refer to $\mathbf{x}(t)$ as the *state* of the Poisson driven Markov process at time t. The state takes values in the *state space* X of the process; this is an n-dimensional Euclidean space.

The following examples motivate our study and provide additional interpretation of (5.39)

Example 5.3.1 *Brownian Motion* ━━━━━━━━━━━━━━━━━━━━━━
Consider the motion in one dimension of a small particle of mass m immersed in a fluid. Thermally agitated molecules of the fluid strike the particle causing it to move chaotically. The motion is damped to some extend depending upon the viscosity of the fluid. Let $v(t)$ denote the velocity of the particle at time $t \geq t_0$. Then, in the absence of any external force, such as gravitational force, the equation of motion is

$$m \frac{dv(t)}{dt} = -\alpha v(t) + f(t), \qquad (5.40)$$

where α is a friction coefficient that depends on the viscosity of the fluid, and $f(t)$ represents fluctuation forces due to molecular collisions. This equation is called *Langevin's equation* in honor of P. Langevin [17], who first proposed it as a model for Brownian motion. If it is supposed that the time of contact during a collision is very short, the resulting force may be assumed to be impulsive. In this case, the velocity will not be differentiable and (5.40) must be replaced by

$$mdv(t) = -\alpha v(t)dt + du(t), \qquad (5.41)$$

where $\{u(t): t \geq t_0\}$ is the accumulator process of a marked point process. The occurrence times and marks of this point process correspond to the instants that collisions take place and the force induced on the particle, respectively. If we suppose further that $\{u(t): t \geq t_0\}$ is a compound Poisson process and use the representation for these processes in Theorem 4.3.2, Langevin's equation becomes

$$dv(t) = -m^{-1}\alpha v(t)dt + \int_{u} m^{-1}UM(dt, dU). \qquad (5.42)$$

This is in the form (5.39) with $a(t,v(t)) = -m^{-1}\alpha v(t)$ and $b(t,v(t),U) = m^{-1}U$. It is straightforward to verify that the solution to (5.42) is

$$v(t) = v_0 e^{-\alpha(t-t_0)/m} + \int_{t_0}^{t}\int_{\mathcal{U}} m^{-1}U e^{-\alpha(t-\sigma)/m} M(d\sigma, dU), \qquad (5.43)$$

where v_0 is the velocity of the particle at time t_0. A solution can be obtained in this example due to the linearity of (5.42). Explicit solution of (5.39) is not generally possible when the equation is nonlinear. The last integral in (5.43) is zero if no impacts occur during $[t_0, t)$. Otherwise, its evaluation is

$$\sum_{n=1}^{N(t)} h(t,\tau_n:u_n),$$

where τ_n and u_n are the occurrence time and force induced by the nth impact, and

$$h(t,\tau:u) = \begin{cases} m^{-1}e^{-\alpha(t-\tau)/m}, & t \geq \tau \\ 0, & t < \tau. \end{cases}$$

Thus, if the initial velocity is zero, $\{v(t): t \geq t_0\}$ is a filtered Poisson process of the type studied in Sec. 5.2. ∎

Example 5.3.2 *Linear Dynamical Systems* ─────────────────
    Linear dynamical systems are employed as models in all branches of engineering. These are systems that can be described by ordinary, linear differential equations. Langevin's equation (5.40) is a simple example. More generally, the equations take the form

$$\frac{d\mathbf{x}(t)}{dt} = \mathbf{A}(t)\mathbf{x}(t) + \mathbf{B}(t)\mathbf{f}(t), \quad \mathbf{x}(t_0) = \mathbf{x}_0 \qquad (5.44)$$

$$\mathbf{y}(t) = \mathbf{C}(t)\mathbf{x}(t), \qquad (5.45)$$

where $\mathbf{f}(\cdot)$ is the input function, $\mathbf{x}(\cdot)$ is termed the *state* of the system, and $\mathbf{y}(\cdot)$ is the output function. If $\mathbf{f}(\cdot)$, $\mathbf{x}(\cdot)$, and $\mathbf{y}(\cdot)$ are vectors of dimension p, n, and r, respectively, then $\mathbf{A}(\cdot)$, $\mathbf{B}(\cdot)$, and $\mathbf{C}(\cdot)$ are matrices of dimension $n \times n$, $n \times p$, and $r \times n$, respectively. There are a number of ways for reducing an nth order linear differential equation to the form of (5.44) and (5.45). These are described by L. Zadeh and C. Desoer [29], R.

Brockett [3], and T. Kailath [15]. If the input function $f(\cdot)$ in (5.44) is impulsive with the nth impulse occurring at time τ_n and carrying weight u_n, then $x(\cdot)$ will not be differentiable, and the state equation must be re-written as

$$d\mathbf{x}(t) = \mathbf{A}(t)\mathbf{x}(t)\,dt + \mathbf{B}(t)\,d\mathbf{u}(t), \quad \mathbf{x}(t_0) = \mathbf{x}_0 \qquad (5.46)$$

where $\mathbf{u}(\cdot)$ is a piecewise constant function. Jumps in $\mathbf{u}(\cdot)$ occur at the impulse times of $f(\cdot)$ and have a size equal to the corresponding weight. Suppose that $\{\mathbf{u}(t): t \geq t_0\}$ is a compound Poisson process. Then, by using the representation in Theorem 4.3.2, we can write (5.46) as

$$d\mathbf{x}(t) = \mathbf{A}(t)\mathbf{x}(t)\,dt + \int_{\mathcal{U}} \mathbf{B}(t)\mathbf{U}M(dt, d\mathbf{U}), \quad \mathbf{x}(t_0) = \mathbf{x}_0. \qquad (5.47)$$

This equation is in the form of (5.39) with $\mathbf{a}(t, \mathbf{x}(t)) = \mathbf{A}(t)\mathbf{x}(t)$ and $\mathbf{b}(t, \mathbf{x}(t), \mathbf{U}) = \mathbf{B}(t)\mathbf{U}$. Because the equation is linear, it can be solved readily as

$$\mathbf{x}(t) = \Phi(t, t_0)\mathbf{x}_0 + \int_{t_0}^{t}\int_{\mathcal{U}} \Phi(t, \sigma)\mathbf{B}(\sigma)\mathbf{U}M(d\sigma, d\mathbf{U}), \qquad (5.48)$$

where $\Phi(t, t_0)$ is an $n \times n$ matrix called the *state-transition matrix*, which satisfies

$$\frac{\partial \Phi(t, t_0)}{\partial t} = \mathbf{A}(t)\Phi(t, t_0), \quad \Phi(t_0, t_0) = \mathbf{I}. \qquad (5.49)$$

According to (5.48), the state is a superposition of responses to each marked point. If $\mathbf{x}(t_0) = 0$, then $\{\mathbf{x}(t): t \geq t_0\}$ is a vector-valued, filtered Poisson process of the type studied in Sec. 5.2. ∎

Example 5.3.3 *Phase-Tracking Loop Excited by Shot Noise* ───────
  Suppose that an optical communication system is used to establish a phase reference. For this purpose, a subcarrier $\cos[\omega_r t + \theta_r(t)]$, of reference frequency ω_r and phase $\theta_r(t)$, is used to modulate the intensity of a coherent light source, such as a laser, which is pointed at the receiver. As shown by W. Pratt [24, p. 178], the term $|S(t, \vec{r})|^2$ appearing in (5.22) is of the form $|S(t, \vec{r})|^2 = P\{1 + \alpha\cos[\omega_r t + \theta_r(t)]\}$, where P is a nonnegative constant, and α is the subcarrier modulation index with $\alpha \leq 1$. As shown in Fig. 5.5, the modulated optical field is incident on a photodetector with

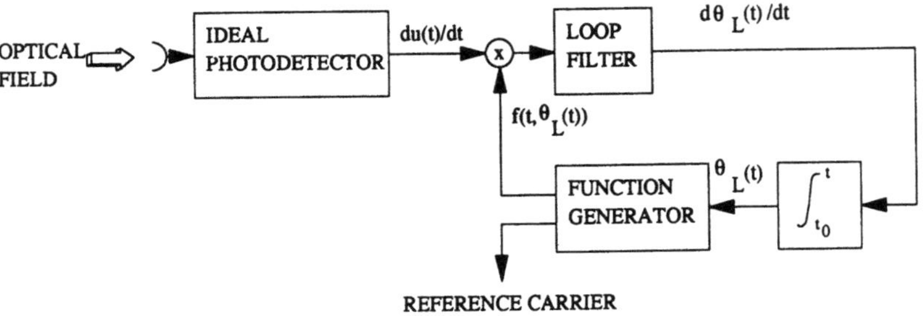

Figure 5.5 Phase-tracking loop.

gain, such as a photomultiplier. We take the detector response to be an impulse $u\,\delta(t-\tau)$, where u is a nonnegative random variable modeling the detector gain. The photodetector output $\{u(t): t \geq t_0\}$ is a compound Poisson process given by

$$u(t) = \begin{cases} 0, & N(t) = 0 \\ \sum_{n=1}^{N(t)} u_n, & N(t) \geq 1, \end{cases}$$

where the intensity of the Poisson process $\{N(t): t \geq t_0\}$ is

$$\lambda(t) = \lambda_s(t) + \lambda_0 = A\{1 + m\cos[\omega_r t + \theta_r(t)]\},$$

where $A \stackrel{\Delta}{=} (\eta\mathcal{A}P/h\nu) + \lambda_0$, $Am \stackrel{\Delta}{=} \eta\mathcal{A}P\alpha/h\nu$, and λ_0 models extraneous counts due to background radiation and dark current. The parameters η, $\mathcal{A}$, h, and ν are as defined in Ex. 5.2.9.

The output of the photodetector is the input to a tracking loop that generates a reference carrier $\cos[\omega_L t + \theta_L(t)]$, where $\theta_L(t)$ satisfies

$$\theta_L(t) = \begin{cases} 0, & N(t) = 0 \\ \displaystyle\sum_{n=1}^{N(t)} u_n f[\tau_n, \theta_L(\tau_n)] h(t, \tau_n), & N(t) \geq 1. \end{cases}$$

Alternatively, $\theta_L(t)$ can be written as

$$\theta_L(t) = \int_{t_0}^{t} \int_{\mathcal{U}} U f[\tau, \theta_L(\tau)] h(t, \tau) M(d\tau, dU). \tag{5.50}$$

Various forms for the function $f(\cdot, \cdot)$ are of interest. The most common is

$$f[t, \theta_L(t)] = \sin[\omega_L t + \theta_L(t)], \tag{5.51}$$

for which the tracking loop is termed a *phase-locked loop*; these are discussed by A. Viterbi [27, Ch's. 2-4] and W. Lindsey and M. Simon [18, Ch. 2]. Another useful form is

$$f[t, \theta_L(t)] = \frac{\sin[\omega_L t + \theta_L(t)]}{1 + m \cos[\omega_L t + \theta_L(t)]},$$

for which the tracking loop is termed a *tanlock loop*; motivation for this loop is given in Ch. 7.

Equation (5.50) cannot be solved explicitly for $\theta_L(\cdot)$ and, therefore, cannot be placed in the form of the superposition (5.1). Therefore, the analysis of the behavior of the loop cannot be accomplished using the procedures of Sec. 5.2. However, by describing $\theta_L(\cdot)$ in the form of (5.39), R. Gagliardi and M. Haney [9] and R. Forrester and D. Snyder [8] have analyzed the loop to some extent; see Ex. 5.3.12 below. For this purpose, we assume that the loop filter $h(\cdot)$ has a finite-dimensional state realization as

$$d\mathbf{z}(t) = \mathbf{A}(t)\mathbf{z}(t)\,dt + \int_{\mathcal{U}} U\mathbf{b}(t) f[t, \theta_L(t)] M(dt, dU),$$

$$\tag{5.52}$$

$$d\theta_L(t) = \mathbf{c}'(t)\mathbf{x}(t) + \int_{\mathcal{U}} U d(t) f[t, \theta_L(t)] M(dt, dU).$$

That is, h is described by the matrix $\mathbf{A}(\cdot)$, the vectors $\mathbf{b}(\cdot)$ and $\mathbf{c}(\cdot)$, the scalar $d(\cdot)$, and the state vector $\mathbf{z}(\cdot)$. To place (5.52) in the form of (5.39), define $\mathbf{x}(t)$ by

$$\mathbf{x}(t) = \begin{bmatrix} \mathbf{z}(t) \\ \theta_L(t) \end{bmatrix}.$$

Then, in (5.39), we have

$$\mathbf{a}(t, \mathbf{x}(t)) = \begin{bmatrix} \mathbf{A}(t) & \mathbf{0} \\ \mathbf{c}'(t) & \mathbf{0} \end{bmatrix}$$

and

$$\mathbf{b}(t, \mathbf{x}(t), U) = U \begin{bmatrix} \mathbf{b}(t) \\ d(t) \end{bmatrix} f[t, \mathbf{h}'\mathbf{x}(t)]$$

where $\mathbf{h}' = [0 \quad 0 \quad \cdot \quad \cdot \quad \cdot \quad 0 \quad 1]$. ∎

In the remainder of this section, we give an introduction to the theory of Poisson driven Markov processes by studying solutions of (5.38) and (5.39). Our first consideration is with a more careful treatment of the last integral in (5.38), which we term a *counting integral.* A counting integral is a special form of integral introduced by K. Itô [14] in his fundamental memoir on stochastic differential equations. The scalar version of the integral is discussed by J. Doob [5, p. 436] and the vector version by A. Skorokhod [26, Ch. 2] and I. Gihman and A. Skorokhod [11, Pt. II, Ch. 2]. Also, see R. Lipster and A. Shiryayev [19, 20]. Properties of the counting integral will be used to establish conditions for the existence and uniqueness of solutions to (5.38). The rule for differentiating a function of a Poisson driven Markov process is also given in Sec 5.3.1. This rule is important here as well as in later chapters where it is used repeatedly. The differential rule we give for Poisson driven Markov processes is analogous to the Itô differential rule for differentiating a function of a Markov diffusion process; see A. Skorokhod [26, p. 24] or E. Wong and B. Hajek [28, p. 148] for a statement of the Itô rule. In Sec. 5.3.2, we investigate the statistics of solutions to (5.38). For the definition of the counting integral adopted in Sec 5.3.1, $\{\mathbf{x}(t): t \geq t_0\}$ is a Markov process, which motivates our terminology. Markov processes will be defined and reviewed briefly, and they are developed further in Ch. 7.

5.3.1 Calculus for Poisson Driven Markov Processes

Integration

We have so far let the last integral in (5.38) pass without detailed comment in order to motivate its occurrence in applications where nonlinear systems are excited by marked point-processes. We now give a definition of the integral and indicate its properties.

The integral of interest has the general form

$$\mathbf{I}(t, A, \omega) = \int_{t_0}^{t} \int_{A} \mathbf{b}(\sigma, \mathbf{U}, \omega) M(d\sigma, d\mathbf{U}, \omega), \qquad (5.53)$$

where $A \subseteq \mathcal{U}$, and $\mathbf{b}(\sigma, \mathbf{U}, \omega)$ is a random function of σ and $\mathbf{U}$. We have included the sample space variable ω of an underlying probability space to emphasize that $\mathbf{b}$ and M are random quantities. In (5.38), this randomness in $\mathbf{b}$ enters through the influence of $\mathbf{x}$, which, in turn, is random because of the influence of M. We will henceforth suppress the explicit dependence on ω and write (5.53) simply as

$$\mathbf{I}(t, A) = \int_{t_0}^{t} \int_{A} \mathbf{b}(\sigma, \mathbf{U}) M(d\sigma, d\mathbf{U}), \qquad (5.54)$$

It is the possible dependence of this integral on M that makes the definition and evaluation of $\mathbf{I}$ somewhat delicate. As an example illustrating the potential difficulties, suppose that $\mathbf{b}$ does not depend on the marks $\mathbf{U}$, so

$$\mathbf{I}(t, \mathcal{U}) = \int_{t_0}^{t} \mathbf{b}(\sigma) dN(\sigma),$$

where $N(t) = \int_{t_0}^{t} \int_{\mathcal{U}} M(d\sigma, d\mathbf{U})$ is the number of points on $[t_0, t)$ regardless of their mark. Suppose, further, that $\mathbf{b}$ is real valued and $b(\sigma) = N(\sigma)$. Then, $I(t) = \int_{t_0}^{t} N(\sigma) dN(\sigma)$. The evaluation of $I(t)$ according to the usual rules of integration would suggest for this example that $I(t) = N^2(t)/2$. However, closer examination indicates, first, that the integral $\int_{t_0}^{t} N(\sigma) dN(\sigma)$ cannot exist in the usual Riemann-Stieltjes sense because of the simultaneous jumps in the integrand $b(\sigma) = N(\sigma)$ and the integrator $N(\sigma)$; see T. Apostol [1, Theorem 9-28]. Secondly, an effort to evaluate $\int_{t_0}^{t} N(\sigma) dN(\sigma)$ from first principles can lead to more than one answer. To see this, consider a sequence of partitions $t_0 = t_{0,n} \leq t_{1,n} \leq t_{2,n} \leq \cdots \leq t_{n,n} = t$ of $[t_0, t)$ such that $\max_{1 \leq i \leq n}(t_{i,n} - t_{i-1,n}) \to 0$ as $n \to \infty$, and examine the sum

$$S_n(\alpha) = \sum_{i=1}^{n} N(t'_{i,n})[N(t_{i,n}) - N(t_{i-1,n})]$$

as $n \to \infty$, where $t'_{i,n} = \alpha t_{i-1,n} + (1-\alpha)t_{i,n}$, for $0 \le \alpha \le 1$, is a point in the subinterval $[t_{i-1}, t_i]$. The choice of α affects the evaluation of this limit, which is not the case for the usual Riemann-Stieltjes integral. First, set $\alpha = 1$ so that $t'_{i,n}$ is the left end $t_{i-1,n}$ of the subinterval. We have

$$\lim_{n \to \infty} S_n(1) = \lim_{n \to \infty} \sum_{i=1}^{n} N(t_{i-1,n})[N(t_{i,n}) - N(t_{i-1,n})]$$

$$= \frac{1}{2} \lim_{n \to \infty} \sum_{i=1}^{n} [N^2(t_{i,n}) - N^2(t_{i-1,n})] - \frac{1}{2} \lim_{n \to \infty} \sum_{i=1}^{n} [N(t_{i,n}) - N(t_{i-1,n})]^2$$

$$= \frac{1}{2} N^2(t) - \frac{1}{2} N(t).$$

Similarly, for $\alpha = 0$, so that $t'_{i,n} = t_{i,n}$ is the right end of the subinterval,

$$\lim_{n \to \infty} S_n(0) = \lim_{n \to \infty} \sum_{i=1}^{n} N(t_{i,n})[N(t_{i,n}) - N(t_{i-1,n})]$$

$$= \frac{1}{2} \lim_{n \to \infty} \sum_{i=1}^{n} [N^2(t_{i,n}) - N^2(t_{i-1,n})] + \frac{1}{2} \lim_{n \to \infty} \sum_{i=1}^{n} [N(t_{i,n}) - N(t_{i-1,n})]^2$$

$$= \frac{1}{2} N^2(t) + \frac{1}{2} N(t).$$

And for $0 < \alpha < 1$,

$$\lim_{n \to \infty} S_n(\alpha) = \lim_{n \to \infty} \sum_{i=1}^{n} N(\alpha t_{i-1,n} + (1-\alpha)t_{i,n})[N(t_{i,n}) - N(t_{i-1,n})]$$

$$= \lim_{n \to \infty} \sum_{i=1}^{n} [N(\alpha t_{i-1,n} + (1-\alpha)t_{i,n})[N(t_{i,n}) - N(\alpha t_{i-1,n} + (1-\alpha)t_{i,n})]$$

$$+ \lim_{n \to \infty} \sum_{i=1}^{n} N(\alpha t_{i-1,n} + (1-\alpha)t_{i,n})[N(\alpha t_{i-1,n} + (1-\alpha)t_{i,n}) - N(t_{i-1,n})]$$

$$= \frac{1}{2} N^2(t). \tag{5.55}$$

Thus, we conclude that

$$\int_{t_0}^{t} N(\sigma)\,dN(\sigma) = \begin{cases} \dfrac{1}{2}N^2(t)+\dfrac{1}{2}N(t), & \alpha=0 \\[2ex] \dfrac{1}{2}N^2(t), & 0<\alpha<1 \\[2ex] \dfrac{1}{2}N^2(t)-\dfrac{1}{2}N(t), & \alpha=1. \end{cases} \tag{5.56}$$

Only for the range $0<\alpha<1$ does the integral $\int_{t_0}^{t} N(\sigma)\,dN(\sigma)$ have the usual evaluation, and the selection of α in this range would appear to be natural. However, there is strong motivation to forgo this familiar evaluation and use $\alpha=1$ so that $t'_{i,n}$ is at the left end of the interval. The reason is that for this choice alone, the term $N(t'_{i,n})$ does not anticipate the increment $N(t_{i,n})-N(t_{i-1,n})$ because $\{N(t): t \geq t_0\}$ has independent increments. Furthermore, with the choice $\alpha=1$, the theory of *Martingales* can be used to study the statistical properties of the integral; see Prob. 5.3.1. While we will not use the Martingale theory explicitly, we will follow Itô's precedent and select $\alpha=1$ in the definition of $I(t,A)$ in (5.54). A consequence of this selection is that the counting integral that results is incompatible with the integration rules of ordinary calculus, as is evident in the above simple example. Thus, some care is needed in the evaluation of counting integrals. We will give an explicit evaluation consistent with that of previous sections where counting integrals have been used formally.

The following conditions will be assumed to hold.

a. M is a Poisson process on $[t_0,t)\times \mathcal{U}$. This is process of Theorem 4.3.2 for representing a compound Poisson process on $[t_0,\infty)$ with mark space $\mathcal{U}$. Here, $\int_a^b \int_A M(dt,d\mathbf{U})$ is the random number of points occurring in the interval $[a,b)$ with marks in $A \subseteq \mathcal{U}$. We take $\mathcal{U}$ to be an m-dimensional Euclidean space.

b. For all finite intervals $[a,b)$ and $A \subseteq \mathcal{U}$, there holds

$$E\left[\int_a^b \int_A M(d\sigma,d\mathbf{U})\right] = \Pr(\mathbf{u} \in A)\int_a^b \lambda(\sigma)\,d\sigma < \infty.$$

This assumption implies that regardless of their marks, the number of points in $[a,b)$ is finite with probability one.

c.   The process $\{\mathbf{b}(t,\mathbf{U}): t \geq t_0, \mathbf{U} \in \mathcal{U}\}$ does not anticipate the underlying point process. We mean by this that $\mathbf{b}(t,\mathbf{U}')$ is statistically independent of $\int_a^b \int_A M(d\sigma,d\mathbf{U})$ for any choice of $\mathbf{U}' \in \mathcal{U}$ and $A \subseteq \mathcal{U}$ and for all triples $t \leq a \leq b$.

d.   Almost every realization of the process $\mathbf{b}(t,\mathbf{U})$ is continuous in $\mathbf{U}$, left continuous in t, and bounded for $(t,\mathbf{U}) \in [a,b) \times A$, where $[a,b)$ is any bounded interval in $[t_0,\infty)$ and A is any bounded set in $\mathcal{U}$.

These four conditions will ensure the existence of the counting integral we introduce next.

Definition 5.3.1 (*Counting Integral*). Let A be a set in $\mathcal{U}$, and consider a sequence of partitions of $[t_0,t) \times A$ obtained as follows. Let $A_{1,p}, A_{2,p}, \cdots$ for $p = 1,2,\cdots$, be a sequence of pairwise disjoint partitions of A such that $A_{k,p} \cap A_{j,p} = \varnothing$ for $k \neq j$, $A = \cup_k A_{k,p}$, and $\max_k \| A_{k,p} \| \to 0$ as $p \to \infty$, where $\| A_{k,p} \|$ denotes the diameter of the set $A_{k,p}$.[1] Similarly, let $t_0 = t_{0,q} \leq t_{1,q} \leq \cdots \leq t_{q,q} = t$, for $q = 1,2,\cdots$, be a sequence of partitions of $[t_0,t)$ such that $\max_{1 \leq j \leq q} |t_{j,q} - t_{j-1,q}| \to 0$ as $q \to \infty$. Define

$$\mathbf{I}_{p,q}(t,A) = \sum_{j=1}^{q} \sum_{k} \mathbf{b}(t_{j-1,q},\mathbf{U}'_{k,p}) \int_{t_{j-1,q}}^{t_{j,q}} \int_{A_{k,p}} M(d\sigma,d\mathbf{U}), \qquad (5.57)$$

where $\mathbf{U}'_{k,p}$ is an arbitrary point in $A_{k,p}$. If independently of the partition used, $\mathbf{I}_{p,q}(t,A)$ converges with probability one to a unique finite limit as p and q tend to infinity, then we shall say that the counting integral of $\mathbf{b}$ with respect to M exists on $[t_0,t) \times A$ and denote the limit by

$$\mathbf{I}(t,A) = \int_{t_0}^{t} \int_A \mathbf{b}(\sigma,\mathbf{U}) M(d\sigma,d\mathbf{U}). \qquad (5.58)$$

The following theorem shows that this integral exists under the conditions (a)-(d) given above. The evaluation of the integral and some of its moment properties are also given.

---

[1] The diameter of a set is the maximum distance between any two points in the set.

Theorem 5.3.1 (*Existence and Some Properties of Counting Integrals*).
Under the conditions (a)-(d), the counting integral (5.58) exists and has the
evaluation

$$\mathbf{I}(t,A) = \int_{t_0}^{t} \int_{A} \mathbf{b}(\sigma,\mathbf{U}) M(d\sigma, d\mathbf{U})$$

$$= \begin{cases} 0, & N(t)=0 \\ \sum_{n=1}^{N(t)} \mathbf{b}(\tau_n, \mathbf{u}_n), & N(t) \geq 1, \end{cases} \qquad (5.59)$$

where $N(t) = \int_{t_0}^{t} \int_{A} M(d\sigma, d\mathbf{U})$ is the total number of points in $[t_0, t)$ with
mark in A, and τ_n and $\mathbf{u}_n$ are the occurrence time and mark for the n*th*
point. Furthermore, if

$$\tilde{\mathbf{I}}(t,A) = \mathbf{I}(t,A) - \int_{t_0}^{t} \int_{A} \mathbf{b}(\sigma,\mathbf{U}) \lambda(\sigma) d\sigma dP_{\mathbf{u}}(\mathbf{U})$$

then

$$E[\tilde{\mathbf{I}}(t,A)] = E[\mathbf{I}(t,A)] - \int_{t_0}^{t} \int_{A} E[\mathbf{b}(\sigma,\mathbf{U})] \lambda(\sigma) d\sigma dP_{\mathbf{u}}(\mathbf{U}) = \mathbf{0}, \qquad (5.60)$$

and

$$E[|\tilde{\mathbf{I}}(t,A)|^2] = \int_{t_0}^{t} \int_{A} E[|\mathbf{b}(\sigma,\mathbf{U})|^2] \lambda(\sigma) d\sigma dP_{\mathbf{u}}(\mathbf{U}), \qquad (5.61)$$

where $P_{\mathbf{u}}(\mathbf{U})$ is the probability distribution function for the mark random
variable, and $|\mathbf{v}|$ is the length of $\mathbf{v}$.

Proof. $\mathbf{I}(t,A)$ exists, and (5.59) holds trivially if $N(t)=0$. Therefore,
suppose that $N(t) \geq 1$, and choose p and q sufficiently large in (5.57) that
each cell in the partition $[t_{j-1,q}, t_{j,q}) \times A_{k,p}$ of $[t_0, t) \times A$ contains no more
than one point. This can be done for almost every realization of the
marked point process because it is uniformly orderly, and the total number
of points in $[t_0, t) \times A$ is finite with probability one. Then,

$$\mathbf{b}(t_{j-1,q}, \mathbf{U}'_{k,p}) \int_{t_{j-1,q}}^{t_{j,q}} \int_{A_{k,p}} M(d\sigma, d\mathbf{U}) = \mathbf{b}(t_{j-1,q}, \mathbf{U}'_{k,p}) \qquad (5.62)$$

if $(\tau_n, \mathbf{u}_n) \in [t_{j-1,q}, t_{j,q}) \times A_{k,p}$ for some n, $1 \leq n \leq N(t)$. Otherwise,

$$\mathbf{b}(t_{j-1,q}, \mathbf{U}'_{k,p}) \int_{t_{j-1,q}}^{t_{j,q}} \int_{A_{k,p}} M(d\sigma, d\mathbf{U}) = 0.$$

Hence, the sum on k in (5.57) is well defined because only finitely many terms are nonzero with probability one. Denote the $N(t)$ cells in the partition of $[t_0, t) \times A$ for which (5.62) holds by $[t_{j_n-1,q}, t_{j_n,q}) \times A_{k_n,p}$ for $n = 1, 2, \cdots, N(t)$, where these are ordered such that $(\tau_n, \mathbf{u}_n) \in [t_{j_n-1,q}, t_{j_n,q}) \times A_{k_n,p}$. These $N(t)$ cells are bounded for p sufficiently large and any q. Hence, from condition (d), (5.57) becomes

$$\mathbf{I}_{p,q}(t,A) = \sum_{n=1}^{N(t)} \mathbf{b}\left(t_{t_{j_n-1,q}}, \mathbf{U}'_{k_n,p}\right).$$

Consequently,

$$\left| \mathbf{I}_{p,q}(t,A) - \sum_{n=1}^{N(t)} \mathbf{b}(\tau_n, \mathbf{u}_n) \right| \le \sum_{n=1}^{N(t)} \left| \mathbf{b}\left(t_{j_n-1,q}, \mathbf{U}'_{k_n,p}\right) - \mathbf{b}(\tau_n, \mathbf{u}_n) \right|.$$

By selecting p and q sufficiently large, the right side can be made arbitrarily small with probability one due to conditions (b) and (d). These conditions also imply that $\sum_{n=1}^{N(t)} \mathbf{b}(\tau_n, \mathbf{u}_n)$ is finite with probability one. Thus, $\mathbf{I}_{p,q}(t,A)$ converges with probability one to a finite limit as p and q tend to infinity, and the limit $\mathbf{I}(t,A)$ has the evaluation (5.59).

Let us next establish the moment relation in (5.60). Due to conditions (a) and (c),

$$E[\mathbf{I}_{p,q}(t,A)] = \sum_{j=1}^{q} \sum_{k=1}^{p} E[\mathbf{b}(t_{j-1,q}, \mathbf{U}'_{k,p})] \int_{t_{j-1,q}}^{t_{j,q}} \int_{A_{k,p}} \lambda(\sigma) d\sigma dP_\mathbf{u}(\mathbf{U}).$$

Thus,

$$E[\mathbf{I}(t,A)] = E\left[\lim_{p,q \to \infty} \mathbf{I}_{p,q}(t,A) \right] = \lim_{p,q \to \infty} E[\mathbf{I}_{p,q}(t,A)]$$

$$= \int_{t_0}^{t} \int_{A} E[\mathbf{b}(\sigma, \mathbf{U})] \lambda(\sigma) d\sigma dP_\mathbf{u}(\mathbf{U}),$$

which establishes (5.60). The interchange of the limit and expectation may be justified by the finiteness of $\mathbf{I}_{p,q}$ for all p and q, the finiteness of $\mathbf{I}(t)$, and the bounded convergence theorem, Theorem 1.4.1. The second moment in (5.61) is evaluated in a similar fashion; see Problem 5.3.2. $\square$

Examination of the proof of Theorem 5.3.1 shows that the existence and evaluation of the counting integral do not depend upon condition (a). The underlying marked point process does not need to be a compound Poisson process for the validity of (5.59). What is required is that the point process be orderly and, with probability one, have a finite number of points in a finite interval. This observation will be important in subsequent chapters when we study point processes that do not have Poisson statistics. Of course, the moment relations (5.60) and (5.61) do require condition (a) for their validity.

The continuity and boundedness restrictions of condition (a) can be relaxed for alternative definitions of the counting integral. Under the condition

$$\int_{t_0}^{t} \int_{A} E[|\, \mathbf{b}(\sigma, \mathbf{U})|^2]\lambda(\sigma)\, d\sigma dP_u(\mathbf{U}) < \infty \tag{5.63}$$

the sequence $\mathbf{I}_{p,q}(t,A)$ in (5.57) is shown by I. Gihman and A. Skorokhod [11, p. 254] to converge in the mean square sense to a limit that is then taken as the definition for $\mathbf{I}(t,A)$. Properties (5.59), (5.60), and (5.61) will hold. A. Skorokhod [26, p. 34] discusses counting integrals under still weaker conditions.

A special case of Theorem 5.3.1 will appear repeatedly in later chapters. This occurs when the mark space reduces to a single point, $\mathcal{U} = \{1\}$, so all marks are unity. Equation (5.59) becomes

$$\mathbf{I}(t) = \int_{t_0}^{t} \mathbf{b}(\sigma)dN(\sigma) = \begin{cases} 0, & N(t) = 0 \\ \displaystyle\sum_{n=1}^{N(t)} \mathbf{b}(\tau_n), & N(t) \geq 1, \end{cases} \tag{5.64}$$

where $N(t)$ is the number of point that occur during $[t_0, t)$.

Example 5.3.4 ──

Consider, again, the integral $I(t) = \int_{t_0}^{t} N(\sigma)dN(\sigma)$ appearing in (5.56). According to (5.64), its evaluation is

$$\mathbf{I}(t) = \begin{cases} 0, & N(t) = 0 \\ \displaystyle\sum_{n=1}^{N(t)} N(\tau_n), & N(t) \geq 1. \end{cases}$$

Thus,

$$I(t) = \begin{cases} 0, & N(t) = 0 \\ \sum_{n=1}^{N(t)} [n-1], & N(t) \geq 1, \end{cases} \qquad (5.65)$$

which follows because $N(\cdot)$ is left continuous, so $N(\tau_n)$ is the number of points in $[t_0, \tau_n)$ and does not include the point at τ_n. Consequently, by evaluating the sum in (5.65), we conclude that

$$I(t) = \int_{t_0}^{t} N(\sigma) dN(\sigma) = \frac{1}{2} N^2(t) - \frac{1}{2} N(t) \qquad (5.66)$$

in correspondence with the choice $\alpha = 1$ in (5.56). ∎

Differentiation

Differentiation of random processes defined by counting integrals does not obey the rules of ordinary calculus. As an example, consider the process $\{\phi(t): t \geq t_0\}$ defined by

$$\phi(t) = \frac{1}{2} N(t) + \int_{t_0}^{t} N(\sigma) dN(\sigma).$$

the differential of this process is $d\phi(t) = (1/2 + N(t))N(t)$. However, from (5.66), we see that $\phi(t) = N^2(t)/2$, of which the ordinary differential is $N(t) dN(t)$. Consequently, the differential rules of ordinary calculus do not apply. The explanation for this departure may be seen roughly as follows. As $\phi(t) = N^2(t)/2$, we have

$$d\phi(t) = \phi(t+dt) - \phi(t) = \frac{1}{2} [N^2(t+dt) - N^2(t)]$$

$$= \frac{1}{2} [\{dN(t) + N(t)\}^2 - N^2(t)]$$

$$= \frac{1}{2} [dN(t)]^2 + N(t) dN(t).$$

In ordinary calculus, the term $[dN(t)]^2/2$ is of second order compared to $N(t) dN(t)$ and, therefore, it is neglected. Here, however, $[dN(t)]^2 = dN(t)$ because $dN(t)$ is, with high probability, one or zero corresponding to whether or not a point occurs in $[t, t+dt)$. Replacing $[dN(t)]^2$ by $dN(t)$ results in $d\phi(t) = [1/2 + N(t)] dN(t)$. It follows from this simple example that some care must be exercised when differentiating random processes defined by counting integrals. A general differential rule is given in the following theorem.

Theorem 5.3.2 (*Differential Rule*). Let $\{\zeta(t): t \geq t_0\}$ be a random process defined in terms of counting integrals by

$$\zeta(t) = \zeta_0 + \int_{t_0}^{t} \mathbf{a}(\sigma)d\sigma + \int_{t_0}^{t}\int_{\mathcal{U}} \mathbf{b}(\sigma, \mathbf{U})M(d\sigma, d\mathbf{U}), \qquad (5.67)$$

where the initial condition ζ_0 is a random variable, and the integral $\int_{t_0}^{t}\mathbf{a}(\sigma)d\sigma$ exists as an ordinary integral for almost every realization of $\{\mathbf{a}(t): t \geq t_0\}$. The vectors $\zeta(\cdot)$, ζ_0, $\mathbf{a}(\cdot)$, and $\mathbf{b}(\cdot, \cdot)$ are all of dimension n. Assume that conditions (a)-(d) hold, so the counting integral exists according to Theorem 5.3.1. Let the function $\phi(t, \mathbf{X})$ be bounded for t and $\mathbf{X}$ finite and have continuous first derivatives. Then, with probability one, the random process $\{\phi(t, \zeta(t)): t \geq t_0\}$ satisfies

$$\phi(t, \zeta(t)) = \phi_0 + \int_{t_0}^{t}\left[\frac{\partial\phi(\sigma, \zeta(\sigma))}{\partial\sigma} + <\mathbf{a}(\sigma), \frac{\partial\phi(\sigma, \zeta(\sigma))}{\partial\zeta(\sigma)}> \right]d\sigma \qquad (5.68)$$

$$+ \int_{t_0}^{t}\int_{\mathcal{U}} [\phi(\sigma, \zeta(\sigma) + \mathbf{b}(\sigma, \mathbf{U})) - \phi(\sigma, \zeta(\sigma))]M(d\sigma, d\mathbf{U}),$$

where $\phi_0 = \phi(t_0, \zeta_0)$, $\partial\phi(\sigma, \zeta(\sigma))/\partial\zeta(\sigma)$ denotes the gradient of $\phi(\sigma, \zeta(\sigma))$ with respect to $\zeta(\sigma)$, and $<\mathbf{a}(\sigma), \partial\phi(\sigma, \zeta(\sigma))/\partial\zeta(\sigma)>$ denotes the vector inner product,

$$<\mathbf{a}(\sigma), \frac{\partial\phi(\sigma, \zeta(\sigma))}{\partial\zeta(\sigma)}> = \sum_{i=1}^{n} a_i(\sigma)\frac{\partial\phi(\sigma, \zeta(\sigma))}{\partial\zeta_i(\sigma)}.$$

In differential form, (5.67) and (5.68) imply that if

$$d\zeta(t) = \mathbf{a}(t)dt + \int_{\mathcal{U}} \mathbf{b}(t, \mathbf{U})M(dt, d\mathbf{U}), \qquad (5.69)$$

then

$$d\phi(t,\zeta(t)) = \left[\frac{\partial\phi(t,\zeta(t))}{\partial t} + <\mathbf{a}(t), \frac{\partial\phi(t,\zeta(t))}{\partial\zeta(t)} > \right] dt$$

$$+ \int_{\mathcal{U}} \phi(t,\zeta(t)+\mathbf{b}(t,\mathbf{U})) - \phi(t,\zeta(t)))]M(dt,d\mathbf{U}). \qquad (5.70)$$

Proof. (I. Gihman and A. Dorogovcev [10], I. Gihman and A. Skorokhod [11, p. 263], P. Fishman [7]). We first note that as $\zeta(t)$ defined by (5.67) is a left-continuous function of t, so is $\phi(t,\zeta(t))$ defined by (5.68). Similarly, $\phi(t,\zeta(t)+\mathbf{b}(t,\mathbf{U}))$ is a left-continuous function of t and a continuous function of $\mathbf{U}$. Therefore, the counting integral in (5.68) exists according to Theorem 5.3.1. Let $t_0 = t_{0,q} < t_{1,q} < \cdots < t_{q,q} = t$ be a sequence of partitions of $[t_0,t)$ such that $\max_{1 \le j \le q} |t_{j,q} - t_{j-1,q}|$ tends to zero as q tends to infinity. We have

$$\phi(t,\zeta(t)) - \phi_0 = \sum_{j=1}^{q} [\phi(t_{j,q}, \Delta\zeta(t_{j,q}) + \zeta(t_{j-1,q})) - \phi(t_{j-1,q}, \zeta(t_{j-1,q}))],$$

$$(5.71)$$

where from (5.67) $\Delta\zeta(t_{j,q}) = \zeta(t_{j,q}) - \zeta(t_{j-1,q}) = \Delta\alpha_{j,q} + \Delta\beta_{j,q}$, in which we define

$$\Delta\alpha_{j,q} = \int_{t_{j-1,q}}^{t_{j,q}} \mathbf{a}(\sigma)d\sigma \qquad \text{and} \qquad \Delta\beta_{j,q} = \int_{t_{j-1,q}}^{t_{j,q}} \int_{\mathcal{U}} \mathbf{b}(\sigma,\mathbf{U})M(d\sigma,d\mathbf{U}).$$

The sum (5.71) can be rewritten as

$$\phi(t,\zeta(t)) - \phi_0 = \Sigma_{1,q} + \Sigma_{2,q} + \Sigma_{3,q},$$

where

$$\Sigma_{1,q} = \sum_{j=1}^{q} [\phi(t_{j,q}, \Delta\zeta(t_{j,q}) + \zeta(t_{j-1,q})) - \phi(t_{j-1,q}, \Delta\zeta(t_{j,q}) + \zeta(t_{j-1,q}))]$$

$$\Sigma_{2,q} = \sum_{j=1}^{q} [\phi(t_{j-1,q}, \Delta\zeta(t_{j,q}) + \zeta(t_{j-1,q})) - \phi(t_{j-1,q}, \zeta(t_{j-1,q}) + \Delta\beta_{j,q}))]$$

and

$$\Sigma_{3,q} = \sum_{j=1}^{q} [\phi(t_{j-1,q}, \zeta(t_{j-1,q}) + \Delta\beta_{j,q}) - \phi(t_{j-1,q}, \zeta(t_{j-1,q}))].$$

Now, by the mean-value theorem, we can write

$$\Sigma_{1,q} = \sum_{j=1}^{q} \dot{\phi}(t_{j,q}, \zeta(t_{j,q}))[t_{j,q} - t_{j-1,q}],$$

where $\dot{\phi}(t, \zeta(t)) = \partial\phi(t, \zeta(t))/\partial t$ and $t_{j-1,q} \le t'_{j,q} \le t_{j,q}$ for $j = 1, 2, \cdots, q$. Thus, in the limit as q tends to infinity, $\Sigma_{1,q}$ converges to the Riemann integral $\int_{t_0}^{t} [\partial\phi(\sigma, \zeta(\sigma)/\partial\sigma] d\sigma$. Similarly, by the mean-value theorem, $\Sigma_{2,q}$ converges to $\int_{t_0}^{t} < \mathbf{a}(\sigma), \partial\phi(\sigma, \zeta(\sigma))/\partial\zeta(\sigma) > d\sigma$. By an argument identical to that in the proof of Theorem 5.3.1, the quantities $\Delta\beta_{j,q}$ in $\Sigma_{3,q}$ are either nonzero or zero for q sufficiently large in correspondence with whether or not a point occurs in $[t_{j-1,q}, t_{j,q})$. Denote the $N(t)$ intervals where $\Delta\beta_{j,q}$ is nonzero by $[t_{j_n-1,q}, t_{j_n,q})$ for $n = 1, 2, \cdots, N(t)$, where these intervals are ordered such that $(\tau_n, \mathbf{u}_n) \in [t_{j_n-1,q}, t_{j_n,q}) \times \mathcal{U}$. Here, $(\tau_n, \mathbf{u}_n)$ are the occurrence time and mark for the nth point in the underlying point process. Then, for q sufficiently large

$$\Sigma_{3,q} = \begin{cases} 0, & N(t) = 0 \\ \sum_{n=1}^{N(t)} \left[\phi\left(t_{j_n-1,q}, \zeta(t_{j_n-1,q}) + \Delta\beta_{j_n,q}\right) - \phi\left(t_{j_n-1,q}, \zeta(t_{j_n-1,q})\right) \right], & N(t) \ge 1. \end{cases}$$

In the limit $q \to \infty$, this sum then becomes

$$\Sigma_{3,q} = \begin{cases} 0, & N(t) = 0 \\ \sum_{n=1}^{N(t)} [\phi(\tau_n, \zeta(\tau_n) + \mathbf{b}(\tau_n, \mathbf{u}_n)) - \phi(\tau_n, \zeta(\tau_n))], & N(t) \ge 1, \end{cases}$$

where we have used the continuity of ϕ with respect to t and ζ and the left continuity of ζ. According to Theorem 5.3.1, we therefore have

$$\Sigma_{3,q} = \int_{t_0}^{t} \int_{\mathcal{U}} [\phi(\sigma, \zeta(\sigma) + \mathbf{b}(\sigma, \mathbf{U})) - \phi(\sigma, \zeta(\sigma))] M(d\sigma, d\mathbf{U}).$$

Combining these expressions for $\Sigma_{1,q}$, $\Sigma_{2,q}$, and $\Sigma_{3,q}$ yields (5.68), and the proof of Theorem 5.3.2 is complete. $\square$

Examination of the proof of Theorem 5.3.2 shows that it does not depend upon condition (a). The underlying point process does not need to be a compound Poisson process for the validity of (5.68) and (5.70). What is required is that the point process be uniformly orderly and have a finite number of points in a finite interval. This observation will be important in later chapters when we study point processes that do not have Poisson statistics.

A special case of Theorem 5.3.2 will appear repeatedly in later chapters. This occurs when all the marks are unity, $\mathcal{U} = \{1\}$. Then, (5.69) and (5.70) become

$$d\zeta(t) = \mathbf{a}(t)dt + \mathbf{b}(t)dN(t) \tag{5.72}$$

and

$$d\phi(t,\zeta(t)) = \left[\frac{\partial\phi(t,\zeta(t))}{\partial t} + < \mathbf{a}(t), \frac{\partial\phi(t,\zeta(t))}{\partial\zeta(t)} > \right] dt$$

$$+ [\phi(t,\zeta(t)+\mathbf{b}(t)) - \phi(t,\zeta(t))]dN(t), \tag{5.73}$$

respectively, where $N(t)$ is the number of points occurring in $[t_0, t)$.

The following are some examples illustrating the use of the differential rule. These will be useful in later calculations.

Example 5.3.5 *Power Law* ───────────────────────────
Let $\{\zeta(t): t \geq t_0\}$ be a real-valued process satisfying

$$d\zeta(t) = a(t)dt + \int_{\mathcal{U}} b(t,\mathbf{U})M(dt,d\mathbf{U}), \qquad \zeta_0 = 0.$$

Suppose that $\phi(\zeta(t)) = \zeta^\nu(t)$. Then

$$d\zeta^\nu(t) = \nu\zeta^{\nu-1}(t)a(t)dt + \int_{\mathcal{U}} [(\zeta(t)+b(t,\mathbf{U}))^\nu - \zeta^\nu(t)]M(dt,d\mathbf{U}). \tag{5.74}$$

Alternatively,

$$\zeta^\nu(t) = \int_{t_0}^{t} \nu\zeta^{\nu-1}(\sigma)a(\sigma)d\sigma + \int_{t_0}^{t}\int_{\mathcal{U}} [(\zeta(\sigma)+b(\sigma,\mathbf{U}))^\nu - \zeta^\nu(\sigma)]M(d\sigma,d\mathbf{U}).$$

$$\tag{5.75}$$

In particular, for $\nu = 2$, $a(t) = 0$, and $b(t,\mathbf{U}) = 1$, we have $\zeta(t) = N(t)$ and $dN^2(t) = [(N(t)+1)^2 - N^2(t)]dN(t)$. From this, we conclude that

$$\int_{t_0}^{t} N(\sigma)\,dN(\sigma) = \frac{1}{2}N(t)[N(t)-1].$$

This reemphasizes the care with which counting integrals must be treated as their evaluation differs from ordinary integrals. ∎

Example 5.3.6 *Component Products* ────────────────────
Let $\{\zeta(t): t \geq t_0\}$ be an n-dimensional process satisfying

$$d\zeta(t) = \mathbf{a}(t)\,dt + \int_{\mathcal{U}} \mathbf{b}(t,\mathbf{U})M(dt,d\mathbf{U}). \tag{5.76}$$

Denote by $\mathbf{e}_i$ an n-dimensional unit vector with ith element equal to one and all other elements equal to zero. Define $\phi_{i,j}(\zeta(t)) = \mathbf{e}'_i\zeta(t)\zeta'(t)\mathbf{e}_j$. This function is the product of the ith and jth elements of $\zeta(t)$. From (5.70), we have

$$d\phi_{i,j}(\zeta(t)) = \langle \mathbf{a}(t),(\mathbf{e}_i\mathbf{e}'_j + \mathbf{e}_j\mathbf{e}'_i)\zeta(t)\rangle\,dt \tag{5.77}$$

$$+ \mathbf{e}'_i \int_{\mathcal{U}}[\mathbf{b}(t,\mathbf{U})\zeta'(t) + \zeta(t)\mathbf{b}'(t,\mathbf{U}) + \mathbf{b}(t,\mathbf{U})\mathbf{b}'(t,\mathbf{U})]M(dt,d\mathbf{U})\mathbf{e}_j.$$

This expression will be useful when we examine the second-moment statistics for a Poisson driven Markov process. ∎

Example 5.3.7 *Exponential* ──────────────────────────
Suppose that $\zeta(t)$ satisfies (5.76), and let $\phi(\zeta(t)) = \exp[j\langle \mathbf{v},\zeta(t)\rangle]$, where $\mathbf{v}$ is a constant n-dimensional vector. Then, $\partial\phi(\zeta(t))/\partial\zeta(t) = j\mathbf{v}\exp[j\langle \mathbf{v},\zeta(t)\rangle]$, and we have from (5.70) that

$$de^{j\langle \mathbf{v},\zeta(t)\rangle} = e^{j\langle \mathbf{v},\zeta(t)\rangle}\left\{ j\langle \mathbf{v},\mathbf{a}(t)\rangle\,dt + \int_{\mathcal{U}}[e^{j\langle \mathbf{v},\mathbf{b}(t,\mathbf{U})\rangle} - 1]M(dt,d\mathbf{U})\right\}.$$

$$\tag{5.78}$$

This expression will be useful below when we examine the characteristic function for a Poisson driven Markov process. ∎

Example 5.3.8 *Logarithm* ────────────────────────────
Suppose that $\zeta(t)$ satisfies (5.76), and let $\phi(\zeta(t)) = \ln[\langle \mathbf{v},\zeta(t)\rangle]$, where $\mathbf{v}$ is a constant n-dimensional vector. Then, we have from (5.70) that

$$d \ln[<\mathbf{v}, \zeta(t)>] = \frac{<\mathbf{v}, \mathbf{a}(t)>}{<\mathbf{v}, \zeta(t)>} dt + \ln\left[1 + \frac{<\mathbf{v}, \mathbf{b}(t)>}{<\mathbf{v}, \zeta(t)>}\right] dN(t), \quad (5.79)$$

∎

Existence and Uniqueness of Poisson Driven Markov Processes

Equation (5.38) is a nonlinear integral equation defining the Poisson driven Markov process $\{\mathbf{x}(t): t \geq t_0\}$. With Theorem 5.3.1, we have developed an interpretation for the last integral on the right in (5.38), and with Theorem 5.3.2, we have developed a rule for differentiating functions of $\mathbf{x}(t)$. These results will be useful below when we study the statistics of $\{\mathbf{x}(t): t \geq t_0\}$. However, we have not yet given conditions on $\mathbf{a}(t, \mathbf{x}(t))$ and $\mathbf{b}(t, \mathbf{x}(t), \mathbf{U})$ so that a solution to (5.38) exists and is unique. These technical considerations are covered in the following theorem.

Theorem 5.3.3 (*Conditions for the Existence and Uniqueness of a Poisson Driven Markov Process*). Let $\{M([t_1, t_2), A: [t_1, t_2) \subseteq [t_0, \infty), A \subseteq \mathcal{U}\}$ be the time-space Poisson process defined in Theorem 4.3.2 for representing a compound Poisson process on $[t_0, \infty)$ with mark space $\mathcal{U}$. Let $[t_0, T)$ be a finite interval, and assume that

$$\sup_{t_0 \leq t \leq T} \lambda(t) < \infty. \qquad (5.80)$$

Denote n-dimensional Euclidean space by $\mathcal{R}^n$. Suppose that $\mathbf{x}_0$ and the functions $\mathbf{a}(t, \mathbf{X})$ and $\mathbf{b}(t, \mathbf{X}, \mathbf{U})$, for $t \in [t_0, T)$, $\mathbf{X} \in \mathcal{R}^n$, and $\mathbf{U} \in \mathcal{U}$, satisfy the following four conditions:

i. $(T - t_0)E(|\mathbf{x}_0|^2) < \infty$

ii. $\mathbf{b}(t, \mathbf{X}, \mathbf{U})$ is a continuous function of $\mathbf{X}$ and $\mathbf{U}$ and a left-continuous function of t;

iii. for all $\mathbf{X} \in \mathcal{R}^n$ and $t \in [t_0, T)$, there exist finite nonnegative constants K_1 and K_2 such that

$$(T - t_0)|\mathbf{a}(t, \mathbf{X}) + \lambda(t)E\{\mathbf{b}(t, \mathbf{X}, \mathbf{u})\}|^2 \leq K_1(1 + |\mathbf{X}|^2)$$

and

$$E\{|\mathbf{b}(t, \mathbf{X}, \mathbf{u})|^2\} \leq K_2(1 + |\mathbf{X}|^2);$$

iv. for all $\mathbf{X} \in \mathcal{R}^n$, $\mathbf{Y} \in \mathcal{R}^n$, and $t \in [t_0, T)$, there exist finite nonnegative constants L_1 and L_2 such that

$$(T - t_0)|\,\mathbf{a}(t, \mathbf{X}) - \mathbf{a}(t, \mathbf{Y}) + \lambda(t)E\{\mathbf{b}(t, \mathbf{X}, \mathbf{u}) - \mathbf{b}(t, \mathbf{Y}, \mathbf{u})\}|^2 \le L_1|\,\mathbf{X} - \mathbf{Y}|^2$$

and

$$E\{|\,\mathbf{b}(t, \mathbf{X}, \mathbf{u}) - \mathbf{b}(t, \mathbf{Y}, \mathbf{u})|^2\} \le L_2|\,\mathbf{X} - \mathbf{Y}|^2.$$

Under these four conditions, there exists with probability one a solution $\{\mathbf{x}(t): t \ge t_0\}$ to the stochastic integral equation

$$\mathbf{x}(t) = \mathbf{x}_0 + \int_{t_0}^{t} \mathbf{a}(\sigma, \mathbf{x}(\sigma))d\sigma + \int_{t_0}^{t} \int_{\mathcal{U}} \mathbf{b}(\sigma, \mathbf{x}(\sigma), \mathbf{U})M(d\sigma, d\mathbf{U}). \tag{5.81}$$

Furthermore, the solution satisfies $\int_{t_0}^{T} E\{|\,\mathbf{x}(t)|^2\}dt < \infty$ and is unique in the sense that if $\mathbf{x}$ and $\mathbf{y}$ are solutions on $[t_0, T)$, then $\Pr[\sup_{t_0 \le t < T} |\,\mathbf{x}(t) - \mathbf{y}(t)| > 0] = 0$.

The expectation in condition (i) is with respect to the distribution of $\mathbf{x}_0$, and those in (iii) and (iv) are with respect to the mark distribution. Condition (iii) is a growth condition; we will show in the proof of Theorem 5.3.3 that this implies a boundedness condition on $\mathbf{b}$. Condition (iv) is termed a *uniform Lipschitz condition*.

Proof. (A. Skorokhod [26, p. 47], A. Bharucha-Reid [2, p. 230]) Let Ξ be the complete, normed, linear space of left-continuous, $\mathcal{R}^n$-valued random function $\{\xi(t): t_0 \le t < T\}$ such that

$$\sup_{t_0 \le t < T} E(|\,\xi(t)|^2) < \infty, \tag{5.82}$$

and take the norm on Ξ to be

$$\|\xi\| = \left(\sup_{t_0 \le t < T} E(|\,\xi(t)|^2) \right)^{1/2}.$$

We demonstrate, first, that the counting integral

$$\int_{t_0}^{t} \int_{\mathcal{U}} \mathbf{b}(\sigma, \xi(\sigma), \mathbf{U})M(d\sigma, d\mathbf{U}) \tag{5.83}$$

exists for all $t \in [t_0, T)$, where $\xi(\cdot) \in \Xi$. It follows from condition (ii) and the left continuity of functions in Ξ that $\mathbf{b}(t, \xi(t), \mathbf{U})$ is left continuous in t and continuous in $\mathbf{U}$. Assumption (5.80) implies that $E\{M([t_0, T), \mathcal{U})\} = \int_{t_0}^{T} \lambda(\sigma) d\sigma < \infty$. To use Theorem 5.3.1, we must also show that almost every realization of $\mathbf{b}(t, \xi(t), \mathbf{u})$ is bounded. This follows from condition (iii) because

$$E(| \mathbf{b}(t, \xi(t), \mathbf{u})|^2) \le K_2[1 + E(| \xi(t)|^2]. \tag{5.84}$$

Thus, $\sup_{t_0 \le t < T} E\{\mathbf{b}(t, \xi(t), \mathbf{u})|^2\} < \infty$, which implies that $\Pr(\sup_{t_0 \le t < T} | \mathbf{b}(t, \xi(t), \mathbf{u})| < \infty) = 1$. It then follows from Theorem 5.3.1 that the counting integral (5.83) exists.

As the integral (5.83) exists, we can define a mapping S of $\mathcal{R}^n$-valued functions from Ξ into $\mathcal{R}^n$-valued functions by

$$(S\xi)(t) = \mathbf{x}_0 + \int_{t_0}^{t} \mathbf{a}(\sigma, \xi(\sigma)) d\sigma + \int_{t_0}^{t} \int_{\mathcal{U}} \mathbf{b}(\sigma, \xi(\sigma), \mathbf{U}) M(d\sigma, d\mathbf{U}). \tag{5.85}$$

We will rewrite S in terms of the zero-mean process defined by

$$\tilde{M}([t_1, t_2), A) = M([t_1, t_2), A) - \int_{t_1}^{t_2} \lambda(\sigma) d\sigma dP_A(\mathbf{U})$$

as

$$(S\xi)(t) = \mathbf{x}_0 + \int_{t_0}^{t} \left\{ \mathbf{a}(\sigma, \xi(\sigma)) + \lambda(\sigma) \int_{\mathcal{U}} \mathbf{b}(\sigma, \xi(\sigma), \mathbf{U}) dP_u(\mathbf{U}) \right\} d\sigma$$

$$+ \int_{t_0}^{t} \int_{\mathbf{U}} \mathbf{b}(\sigma, \xi(\sigma), \mathbf{U}) \tilde{M}(d\sigma, d\mathbf{U}). \tag{5.86}$$

The map S has the property that it maps functions of Ξ into functions of Ξ. This can be seen by noting first that

$$E\{|(S\xi)(t)|^2\} \le 3E\{|\mathbf{x}_0|^2\}$$

$$+3(t-t_0)\int_{t_0}^{t} E\left\{\left|\mathbf{a}(\sigma,\xi(\sigma))+\lambda(\sigma)\int_{\mathcal{U}}\mathbf{b}(\sigma,\xi(\sigma),\mathbf{U})dP_{\mathbf{u}}(\mathbf{U})\right|^2\right\}d\sigma$$

$$+3\int_{t_0}^{t}\int_{\mathcal{U}} E\{|\mathbf{b}(\sigma,\xi(\sigma),\mathbf{U})|^2\}\lambda(\sigma)d\sigma dP_{\mathbf{u}}(\mathbf{U}), \tag{5.87}$$

where we have used (5.61) and the Cauchy-Schwarz inequalities

$$\left|\sum_{i=1}^{k}\alpha_i\right|^2 \le k\sum_{i=1}^{k}|\alpha_i|^2 \quad \text{and} \quad \left|\int_{t_0}^{t}\alpha(\sigma)d\sigma\right|^2 \le (t-t_0)\int_{t_0}^{t}|\alpha(\sigma)|^2 d\sigma.$$

Now, from (5.87) and condition (iii), we obtain

$$E\{|(S\xi)(t)|^2\} \le 3E\{|\mathbf{x}_0|^2\}+3K_1\int_{t_0}^{t}E\{1+|\xi(\sigma)|^2\}d\sigma$$

$$+3K_2\int_{t_0}^{t}E\{1+|\xi(\sigma)|^2\}\lambda(\sigma)d\sigma.$$

Hence,

$$\|S\xi\| \le 3E\{|\mathbf{x}_0|^2\}+3K(1+\|\xi\|)\int_{t_0}^{T}[1+\lambda(\sigma)]d\sigma < \infty,$$

where $K = \max(K_1, K_2)$. Moreover, $(S\xi)(t)$ is a left-continuous function of t because of its definition (5.85) and the properties of counting integrals. Thus, $S\xi \in \Xi$ if $\xi \in \Xi$. From this property of S, it is easily seen that for the iterated map $S^k\xi$, defined by $S^k\xi = S(S^{k-1}\xi)$ for $k \ge 1$ with $S^0\xi = \xi$, we also have $S^k\xi \in \Xi$ if $\xi \in \Xi$.

We show next that the map S^k is a contraction for sufficiently large k. For this purpose, let $\xi \in \Xi$ and $\zeta \in \Xi$, and note that

$$E\{|(S\xi)(t)-(S\zeta)(t)|^2\} \le 2(t-t_0)\int_{t_0}^{t} E\{|\mathbf{a}(\sigma,\xi(\sigma))-\mathbf{a}(\sigma,\zeta(\sigma))$$

$$+\lambda(\sigma)\int_{\mathcal{U}}[\mathbf{b}(\sigma,\xi(\sigma),U)-\mathbf{b}(\sigma,\zeta(\sigma),U)]dP_u(U)|^2\}d\sigma$$

$$+2\int_{t_0}^{t}\int_{\mathcal{U}} E\{|\mathbf{b}(\sigma,\xi(\sigma),U)-\mathbf{b}(\sigma,\zeta(\sigma),U)|^2\}\lambda(\sigma)d\sigma dP_u(U).$$

Then, using condition (iv), we obtain the inequality

$$E\{|(S\xi)(t)-(S\zeta)(t)|^2\} \le L\int_{t_0}^{t} E\{|\xi(\sigma)-\zeta(\sigma)|^2\}d\sigma,$$

where we define $L = 2[L_1 + L_2\sup_{t_0 \le t < T}\lambda(t)]$. Iteration of this inequality yields

$$E\{|(S^k\xi)(t)-(S^k\zeta)(t)|^2\}$$

$$\le L^k\int_{t_0}^{t}\int_{t_0}^{\sigma_1}\cdots\int_{t_0}^{\sigma_{k-2}}\int_{t_0}^{\sigma_{k-1}} E\{|\xi(\sigma_k)-\zeta(\sigma_k)|^2\}d\sigma_k d\sigma_{k-1}\cdots d\sigma_2 d\sigma_1$$

$$\le \frac{L^k(t-t_0)^{k-1}}{(k-1)!}\int_{t_0}^{T} E\{|\xi(\sigma_k)-\zeta(\sigma_k)|^2\}d\sigma_k$$

$$\le \frac{L^k(T-t_0)^k}{(k-1)!}\|\xi-\zeta\|.$$

Therefore, for every $\xi \in \Xi$ and $\zeta \in \Xi$, there holds

$$\|S^k\xi - S^k\zeta\| \le \frac{L^k(T-t_0)^k}{(k-1)!}\|\xi-\zeta\|. \tag{5.88}$$

Thus, S^k is a contraction mapping in the space Ξ for k sufficiently large that the constant coefficient $L^k(T-t_0)^k/(k-1)!$ is less than unity.

As S^k is a contraction mapping in the space Ξ for k sufficiently large, there exists a unique function, say $\mathbf{x}(\cdot) \in \Xi$ such that $S^k\mathbf{x} = \mathbf{x}$ for k sufficiently large because of the contraction mapping theorem (see L. Liusternik and V. Sobolev [21, p. 27].) We next show that for this $\mathbf{x}$, there holds

$S\mathbf{x} = \mathbf{x}$, which establishes the existence of a solution to (5.81). For $k_0 \geq 1$, and k sufficiently large for S^k to be a contraction, we have $S^{kk_0}\mathbf{x} = \mathbf{x}$, and from the inequality (5.88), we then have

$$\| \mathbf{x} - S\mathbf{x}\| = \left\| S^{kk_0}\mathbf{x} - S^{kk_0+1}\mathbf{x} \right\| \leq \frac{L^{kk_0}(T-t_0)^{kk_0}}{(kk_0-1)!} \| \mathbf{x} - S\mathbf{x}\| . \tag{5.89}$$

However, for $k_0 \geq 1$, we have $L^{kk_0}(T-t_0)^{kk_0}/(kk_0-1)! < 1$, which with (5.89) implies $S\mathbf{x} = \mathbf{x}$ in the sense that $\| S\mathbf{x} - \mathbf{x}\| = 0$. Consequently, $\Pr[\sup_{t_0 \leq t < T} |\mathbf{x}(t) - (S\mathbf{x})(t)| > 0] = 0$.

Finally, $\mathbf{x}$ is a unique solution to $S\mathbf{x} = \mathbf{x}$ because any such solution is also a solution to $S^k\mathbf{x} = \mathbf{x}$, which has a unique solution because S^k is a contraction. This uniqueness is in the sense that if $S\mathbf{x} = \mathbf{x}$ and $S\mathbf{y} = \mathbf{y}$, then $\| \mathbf{x} - \mathbf{y}\| = 0$, which implies $\Pr[\sup_{t_0 \leq t < T} |\mathbf{x}(t) - \mathbf{y}(t)| > 0] = 0$. $\square$

A. Skorokhod [26, Ch. 3] and I. Gihman and A. Skorokhod [11, Pt. II, Ch. 2] may be consulted for further results on the existence and uniqueness of solutions to the stochastic integral equation (5.81). In particular, these authors relax the continuity condition (ii) and establish the existence and uniqueness of solutions to (5.81) in a mean-square sense. Furthermore, they consider stochastic integral equations that have an additional term on the right side of (5.81) that is derived from a Wiener process, which is a zero-mean random process with independent, normally distributed increments. The form of the more general equation they study is

$$\mathbf{x}(t) = \mathbf{x}_0 + \int_{t_0}^t \mathbf{a}(\sigma, \xi(\sigma))d\sigma + \int_{t_0}^t \mathbf{B}(\sigma, \mathbf{x}(\sigma))d\mathbf{w}(\sigma) \tag{5.90}$$

$$+ \int_{t_0}^t \int_{u} \mathbf{b}(\sigma, \xi(\sigma), \mathbf{U})M(d\sigma, d\mathbf{U}),$$

where $\{\mathbf{w}(t): t \geq t_0\}$ is a vector of independent Wiener processes normalized so that $E\{\mathbf{w}(s)\mathbf{w}'(t)\} = \mathbf{I}\min(s,t)$, where $\mathbf{I}$ is the identity matrix. Several authors develop the theory for the second integral on the right side of (5.90); see, for example, E. Wong and B. Hajek [28, Ch. 4], A. Bharucha-Reid [2, Ch. 7], J. Doob [5, Ch. 6], and I. Gikhman and A. Skorokhod [12, Ch. 8]. We mention, finally, that the differential rule of Theorem 5.3.2 can be generalized for processes satisfying (5.90). I. Gihman and A. Dorogovcev [10] and Gihman and Skorokhod [11] show that if $\{\mathbf{x}(t): t \geq t_0\}$ satisfies (5.90), then $\{\phi(t, \mathbf{x}(t)): t \geq t_0\}$ satisfies

$$\phi(t,\mathbf{x}(t)) = \phi_0 + \int_{t_0}^{t} \left\{ \frac{\partial\phi(\sigma,\mathbf{x}(\sigma))}{\partial\sigma} + \left< \mathbf{a}(\sigma,\xi(\sigma)), \frac{\partial\phi(\sigma),\mathbf{x}(\sigma))}{\partial\mathbf{x}(\sigma)} \right> \right.$$

$$\left. + \frac{1}{2}\mathrm{tr}\left[\mathbf{B}(\sigma,\mathbf{x}(\sigma))\mathbf{B}'(\sigma,\mathbf{x}(\sigma))\frac{\partial^2\phi(\sigma,\mathbf{x}(\sigma))}{\partial\mathbf{x}(\sigma)^2} \right] \right\} d\sigma$$

$$+ \int_{t_0}^{t} \left< \frac{\partial\phi(\sigma,\mathbf{x}(\sigma))}{\partial\mathbf{x}(\sigma)}, \mathbf{B}(\sigma,\mathbf{x}(\sigma))d\mathbf{w}(\sigma) \right> \tag{5.91}$$

$$+ \int_{t_0}^{t} \int_{u} [\phi((\sigma,\mathbf{x}(\sigma)+\mathbf{b}(\sigma,x(\sigma),\mathrm{U})) - \phi(\sigma,\mathbf{x}(\sigma))]M(d\sigma,d\mathrm{U}).$$

where $\phi_0 = \phi(t_0,\mathbf{x}(t_0))$, $\mathrm{tr}(\cdot)$ is the trace operator, and $\partial^2\phi(\mathbf{x})/\partial\mathbf{x}^2$ is the Hessian matrix; the i,j*th* element is $\partial^2\phi(\mathbf{x})/\partial x_i \partial x_j$.

We now examine the statistics of the process defined by (5.81).

5.3.2 Statistics of Poisson Driven Markov Processes

We assume throughout this subsection that the process $\{\mathbf{x}(t): t \geq t_0\}$ is defined by

$$\mathbf{x}(t) = \mathbf{x}_0 + \int_{t_0}^{t} \mathbf{a}(\sigma,\mathbf{x}(\sigma))d\sigma + \int_{t_0}^{t} \int_{u} \mathbf{b}(\sigma,\mathbf{x}(\sigma),\mathrm{U})M(d\sigma,\mathrm{U}). \tag{5.92}$$

Under the conditions of Theorem 5.3.3, there exists a unique solution to this nonlinear integral equation. These conditions are assumed to hold for (5.92). In addition, we assume that the initial condition $\mathbf{x}_0$ does not anticipate the underlying point process on $[t_0,\infty)$, by which we mean that $\mathbf{x}_0$ is statistically independent of $\int_{t_1}^{t_2}\int_{u}M(d\sigma,\mathrm{U})$ for all t_1 and t_2 such that $t_0 \leq t_1 \leq t_2$.

The statistics of $\{\mathbf{x}(t): t \geq t_0\}$ are studied in this subsection. The principal conclusion is that $\{\mathbf{x}(t): t \geq t_0\}$ is a Markov process. The importance of this is that $\{\mathbf{x}(t): t \geq t_0\}$ can be completely characterized statistically by two quantities, the distribution for $\mathbf{x}_0$ and the conditional distribution for $\mathbf{x}(t)$ given $\mathbf{x}(s)$ for $t_0 \leq s \leq t$. The distribution for $\mathbf{x}_0$ is assumed to be known, and we shall investigate the conditional distribution of $\mathbf{x}(t)$ given $\mathbf{x}(s)$.

Let us recall, first, that a random process $\{\xi(t): t \geq t_0\}$ is by definition a *Markov process* if for any time s such that $t_0 \leq s \leq t$, the conditional distribution of $\xi(t)$ given the random variables $\{\xi(\sigma): t_0 \leq \sigma \leq s\}$ is the same as the conditional distribution of $\xi(t)$ given $\xi(s)$ alone. In other words, the process is a Markov process if $\xi(t)$ is conditionally independent of $\{\xi(\sigma): t_0 \leq \sigma < s\}$ given $\xi(s)$.

We have assumed that the initial condition $\mathbf{x}_0$ in (5.92) is statistically independent of $\int_{t_1}^{t_2} \int_A M(d\sigma, \mathbf{U})$ for any choice of t_1 and t_2 such that $t_0 \leq t_1 \leq t_2$ and any $A \subseteq \mathcal{U}$. This nonanticipatory property of the initial condition reproduces itself in time. To verify this, let s split $[t_0, t)$ into two subintervals $[t_0, s)$ and $[s, t)$. It is evident from (5.92) that $\mathbf{x}(t)$ can be written in terms of $\mathbf{x}(s)$ as

$$\mathbf{x}(t) = \mathbf{x}(s) + \int_s^t \mathbf{a}(\sigma, \mathbf{x}(\sigma))d\sigma + \int_s^t \int_{\mathcal{U}} \mathbf{b}(\sigma, \mathbf{x}(\sigma), \mathbf{U})M(d\sigma, d\mathbf{U}), \qquad (5.93)$$

where

$$\mathbf{x}(s) = \mathbf{x}_0 + \int_{t_0}^s \mathbf{a}(\sigma, \mathbf{x}(\sigma))d\sigma + \int_{t_0}^s \int_{\mathcal{U}} \mathbf{b}(\sigma, \mathbf{x}(\sigma), \mathbf{U})M(d\sigma, d\mathbf{U}). \qquad (5.94)$$

Equation (5.93) can be regarded as a stochastic integral equation for $\{\mathbf{x}(\sigma): s \leq \sigma < t\}$ with $\mathbf{x}(s)$ as the initial condition. But from (5.94), this initial condition is a function only of $\mathbf{x}_0$ and the underlying compound Poisson process on $[t_0, s)$. As the compound Poisson process has independent increments, $\mathbf{x}(s)$ is, therefore, statistically independent of $\int_{t_1}^{t_2} \int_A M(d\sigma, \mathbf{U})$ for all t_1 and t_2 such that $s \leq t_1 \leq t_2$ and any $A \subseteq \mathcal{U}$. In other words, $\mathbf{x}(s)$ does not anticipate the point process on $[s, \infty)$. Furthermore, according to (5.93), $\mathbf{x}(t)$ is a function only of $\mathbf{x}(s)$ and the underlying Poisson process on $[s, t)$. We conclude that because $\mathbf{x}(s)$ is determined by $\{\mathbf{x}(\sigma): t_0 \leq \sigma < s\}$ and because the point process on $[s, t)$ is independent of $\{\mathbf{x}(\sigma): t_0 \leq \sigma < s\}$ by virtue of its independent increments, the conditional distribution for $\mathbf{x}(t)$ given $\{\mathbf{x}(\sigma): t_0 \leq \sigma \leq s\}$ is the same as the conditional distribution for $\mathbf{x}(t)$ given $\mathbf{x}(s)$ alone. Hence, $\{\mathbf{x}(t): t \geq t_0\}$ is a Markov process.

An immediate consequence of $\{x(t): t \geq t_0\}$ being a Markov process is that it can be completely characterized by two quantities, the distribution of x_0 and the conditional distribution of $x(t)$ given $x(s)$ for any pair of times t and s such that $t_0 \leq s \leq t$. Recall that $\{x(t): t \geq t_0\}$ is completely characterized statistically when the joint probability

$$\Pr[x(t_1) \in B_1, x(t_2) \in B_2, \cdots, x(t_K) \in B_K] \tag{5.95}$$

is known for any countable collection of ordered times $t_1, t_2, \cdots, t_K$ in $[t_0, \infty)$ and any choice of subsets $B_1, B_2, \cdots, B_K$ in $\mathcal{R}^2$ for which the probability is defined. But by virtue of $\{x(t): t \geq t_0\}$ being a Markov process, (5.95) can be written as

$$\Pr[x(t_1) \in B_1, x(t_2) \in B_2, \cdots, x(t_K) \in B_K]$$

$$= \Pr[x(t_1) \in B_1] \prod_{k=2}^{K} \Pr[x(t_k) \in B_k | x(t_{k-1}) \in B_{k-1}]. \tag{5.96}$$

Thus, the probability (5.95) can be determined knowing only the distribution of $x(t)$ for any $t \geq t_0$ and the conditional distribution of $x(t)$ given $x(s)$ for any pair of times s and t, $t_0 \leq s \leq t$. This conditional distribution is called the *transition distribution*. The distribution of $x(t)$ can, in turn, be determined from the distribution of x_0 and the transition distribution by use of the relation

$$\Pr[x(t_1) \in B_1] = \int_{\mathcal{R}^2} \Pr[x(t_1) \in B_1 | x_0 = X] dP_{x_0}(X). \tag{5.97}$$

Thus, the probability (5.95) can be determined from the initial distribution, which we have assumed to be known, and the transition distribution, which we characterize in the following theorem.

Theorem 5.3.4 (*Transition Statistics for a Poisson Driven Markov Process*). Let $\{x(t): t \geq t_0\}$ be a Poisson driven Markov process satisfying (5.92) and the conditions of Theorem 5.3.3. Define the *transition characteristic function*

$$M_{x(t)|x(s)}(j v | W) = E(e^{j<v, x(t)>} | x(s) = W), \tag{5.98}$$

for $t_0 \leq s \leq t$. This satisfies

$$dM_{\mathbf{x}(t)|\mathbf{x}(s)}(j\mathbf{v}\mid\mathbf{W})=E\{e^{j<\mathbf{v},\mathbf{x}(t)>}\Psi_t(\mathbf{v}\mid\mathbf{x}(t))\mid\mathbf{x}(s)=\mathbf{W}\}dt$$

$$(5.99)$$

$$M_{\mathbf{x}(s)|\mathbf{x}(s)}(j\mathbf{v}\mid\mathbf{W})=e^{j<\mathbf{v},\mathbf{W}>},$$

where Ψ_t is given by

$$\Psi_t(\mathbf{v}\mid\mathbf{x}(t))=j<\mathbf{v},\mathbf{a}(t,\mathbf{x}(t))>+\lambda(t)\int_{\mathcal{U}}\{e^{j<\mathbf{v},\mathbf{b}(t,\mathbf{x}(t),\mathbf{U})>}-1\}dP_{\mathbf{u}}(\mathbf{U}).$$

$$(5.100)$$

Proof. The proof is straightforward using (5.78), which for (5.92) becomes

$$d\,e^{j<\mathbf{v},\mathbf{x}(t)>}$$

$$=e^{j<\mathbf{v},\mathbf{x}(t)>}\left\{j<\mathbf{v},\mathbf{a}(t,\mathbf{x}(t))>dt+\int_{\mathcal{U}}\{e^{j<\mathbf{v},\mathbf{b}(t,\mathbf{x}(t),\mathbf{U})>}-1\}M(dt,d\mathbf{U})\right\}.$$

Now take the conditional expectation of this expression given $\mathbf{x}(s)=\mathbf{W}$ to get

$$dM_{\mathbf{x}(t)|\mathbf{x}(s)}(j\mathbf{v}\mid\mathbf{W})=E\{e^{j<\mathbf{v},\mathbf{x}(t)>}E(j<\mathbf{v},\mathbf{a}(t,\mathbf{x}(t))>dt$$

$$+\int_{\mathcal{U}}\{e^{j<\mathbf{v},\mathbf{b}(t,\mathbf{x}(t),\mathbf{U})>}-1\}M(dt,d\mathbf{U})\mid\mathbf{x}(t),\mathbf{x}(s)=\mathbf{W}\Big)\mid\mathbf{x}(s)=\mathbf{W}\}.$$

The interchange of differentiation and expectation can be justified using the bounded convergence theorem, Theorem 1.4.1. $\square$

The function Ψ appearing in Theorem 5.3.4 can be related qualitatively to the infinitesimal transition statistics of the process $\{\mathbf{x}(t): t\geq t_0\}$ of (5.92). For this purpose, note that the increment $\Delta\mathbf{x}(t)=\mathbf{x}(t+dt)-\mathbf{x}(t)$ has either the value $\Delta\mathbf{x}(t)=\mathbf{a}(t,\mathbf{x}(t))\Delta t+o(\Delta t)$ with probability $1-\lambda(t)\Delta t+o(\Delta t)$, where $o(\Delta t)/\Delta t\to 0$ as $\Delta t\to 0$, or it has the value $\Delta\mathbf{x}(t)=\mathbf{a}(t,\mathbf{x}(t))\Delta t+\mathbf{b}(t,\mathbf{x}(t),\mathbf{U})+o(\Delta t)$ with probability $\lambda(t)\Delta t\Delta P_{\mathbf{u}}(\mathbf{U})+o(\Delta t)$. Thus,

$$E\{e^{j<v,\Delta x(t)>}-1\,|\,x(t)\} = [e^{j<v,a(t,x(t))>\Delta t}-1][1-\lambda(t)\Delta t]$$

$$+\lambda(t)\Delta t\int_{\mathcal{U}}[e^{j<v,a(t,x(t))\Delta t+b(t,x(t),U)>}-1]dP_{u}(U)+o(\Delta t).$$

The limit as Δt tends to zero of the right side of this equation divided by Δt exists and equals the function Ψ in (5.100). Consequently, the limit exists also on the left side, and we obtain

$$\Psi_t(v\,|\,x(t)) = \lim_{\Delta t\to0}\frac{1}{\Delta t}E\{e^{j<v,\Delta x(t)>}-1\,|\,x(t)\}. \qquad (5.101)$$

Because of this relation, Ψ is termed the *characteristic form for the differential generator* of the Markov process $\{x(t): t\ge t_0\}$.

The following examples illustrate the use of (5.99).

Example 5.3.9 *Poisson Counting Process* ——————————————————

As a simple example illustrating the use of (5.99), suppose that $\{x(t): t\ge t_0\}$ is a real-valued process satisfying (5.92) with $a=0$ and $b=1$. Then, $x(t)=N(t)$, where $N(t)$ is the number of points on $[t_0,t)$ regardless of their marks. We easily have that $\Psi_t(v\,|\,N(t))=\lambda(t)[\exp(jv)-1]$, and (5.99) becomes

$$dM_{N(t)|N(s)}(jv\,|\,k)=E\{e^{jvN(t)}\lambda(t)(e^{jv}-1)\,|\,N(s)=k\}dt$$

$$=M_{N(t)|N(s)}(jv\,|\,k)\lambda(t)(e^{jv}-1)dt$$

$$M_{N(s)|N(s)}(jv\,|\,k)=e^{jvk}.$$

The solution is

$$M_{N(t)|N(s)}(jv\,|\,k)=e^{jvk}\exp\left\{(e^{jv}-1)\int_{s}^{t}\lambda(\sigma)d\sigma\right\}. \qquad (5.102)$$

Compare this to the expression in (2.16) for the characteristic function for the increment $N(t)-N(s)$. ∎

Example 5.3.10 *Linear Dynamical Systems* ——————————————————

Linear dynamical systems were introduced in Ex. 5.3.2. These are Poisson driven Markov processes satisfying the linear differential equation (5.47). We write this equation as

$$d\mathbf{x}(t) = \mathbf{A}(t)\mathbf{x}(t)\,dt + \int_{\mathcal{U}} \mathbf{b}(t,\mathbf{U})M(dt,d\mathbf{U}), \quad \mathbf{x}(t_0) = \mathbf{x}_0 \tag{5.103}$$

and we set $\mathbf{b}(t,\mathbf{U}) = \mathbf{B}(t)\mathbf{U}$ because there is no need to assume that $\mathbf{b}(t,\mathbf{U})$ is a linear function of $\mathbf{U}$ in the present consideration. From (5.45), the output process is

$$\mathbf{y}(t) = \mathbf{C}(t)\mathbf{x}(t). \tag{5.104}$$

We will need the following two properties of the state transition matrix $\Phi(t,t_0)$ defined in (5.49): (1), the inverse $\Phi^{-1}(t,t_0)$ exists and satisfies $\Phi^{-1}(t,t_0) = \Phi(t_0,t)$; and (2), $\Phi(t,t_0)\Phi(t_0,\tau) = \Phi(t,\tau)$. These are established by R. Brockett [3, p. 27] and T. Kailath [15].

We want to determine the transition characteristic function for $\mathbf{x}(t)$ given that $\mathbf{x}(s) = \mathbf{W}$. As an intermediate step, it is convenient to evaluate the transition characteristic function for the process $\mathbf{z}(t) = \Phi(t_0,t)\mathbf{x}(t)$ given that $\mathbf{z}(s) = \Phi(t_0,s)\mathbf{W}$; the desired characteristic functions will follow from this because $\mathbf{x}(t) = \Phi(t,t_0)\mathbf{z}(t)$.

By using (5.48) and the definition of $\mathbf{z}(t)$, we have that $\{\mathbf{z}(t): t \geq t_0\}$ satisfies

$$d\mathbf{z}(t) = \int_{\mathcal{U}} \Phi(t_0,t)\mathbf{b}(t,\mathbf{U})M(d\sigma,d\mathbf{U}), \qquad \mathbf{z}(t_0) = \mathbf{x}_0. \tag{5.105}$$

The characteristic form for the differential generator of this process is

$$\Psi_t(\alpha \mid \mathbf{z}(t)) = \lambda(t)\int_{\mathcal{U}}\left\{ e^{j<\alpha,\,\Phi(t_0,t)\mathbf{b}(t,\mathbf{U})>} - 1\right\} dP_{\mathbf{u}}(\mathbf{U}),$$

and by Theorem 5.3.4 we have after an elementary calculation that

$$M_{\mathbf{z}(t)\mid\mathbf{z}(s)}(j\alpha \mid \Phi(t_0,s)\mathbf{W}) \tag{5.106}$$

$$= e^{j<\alpha,\,\Phi(t_0,s)\mathbf{W}>}\exp\left[\int_s^t\int_{\mathcal{U}}\lambda(\sigma)\left\{ e^{j<\alpha,\,\Phi(t_0,\sigma)\mathbf{b}(\sigma,\mathbf{U})>} - 1\right\} d\sigma\,dP_{\mathbf{u}}(\mathbf{U})\right].$$

Observing that $<\mathbf{v},\mathbf{x}(t)> = <\Phi(t,t_0)\mathbf{v},\mathbf{z}(t)>$, we then deduce the desired result that the transition characteristic function for $\mathbf{x}(t)$ given that $\mathbf{x}(s) = \mathbf{W}$ is given by

$$M_{\mathbf{x}(t)|\mathbf{x}(s)}(j\mathbf{v}\mid \mathbf{W}) \tag{5.107}$$

$$= e^{j<\mathbf{v},\Phi(t,s)\mathbf{W}>}\exp\left[\int_s^t\int_{\mathcal{U}}\lambda(\sigma)\{e^{j<\mathbf{v},\Phi(t,\sigma)\mathbf{b}(\sigma,\mathbf{U})>}-1\}d\sigma dP_{\mathbf{u}}(\mathbf{U})\right].$$

The conditional characteristic function for $\mathbf{y}(t)$, the response (5.104), given that $\mathbf{y}(s)=\mathbf{C}(s)\mathbf{W}$ can be similarly obtained from (5.106) by noting that $<\mathbf{v},\mathbf{y}(t)> \,=\, <\Phi'(t,t_0)\mathbf{C}'(t)\mathbf{v},\mathbf{z}(t)>$. Thus,

$$M_{\mathbf{y}(t)|\mathbf{y}(s)}(j\mathbf{v}\mid \mathbf{C}(s)\mathbf{W}) \tag{5.108}$$

$$= e^{j<\mathbf{v},\mathbf{C}(t)\Phi(t,s)\mathbf{W}>}\exp\left[\int_s^t\int_{\mathcal{U}}\lambda(\sigma)\{e^{j<\mathbf{v},\mathbf{h}(t,\sigma:\mathbf{U})>}-1\}d\sigma dP_{\mathbf{u}}(\mathbf{U})\right],$$

where we define $\mathbf{h}(t,\sigma:\mathbf{U})=\mathbf{C}(t)\Phi(t,\sigma)\mathbf{b}(\sigma,\mathbf{U})$, which is the response at time t to a point at time σ having mark $\mathbf{U}$. ∎

Example 5.3.11 *Covariance for a Linear Dynamical System* ——————
    The covariance of the state of a linear dynamical system can be determined by using (5.107) and the moment generating properties of characteristic functions. An alternative approach is to use the result of Ex. 5.3.6. Thus, suppose that $\{\mathbf{x}(t): t \geq t_0\}$ is a Poisson driven Markov process satisfying (5.103) and that the initial state $\mathbf{x}_0$ has mean $E(\mathbf{x}_0)$ and co var i ance $\Lambda_0 = E[\mathbf{x}_0-E(\mathbf{x}_0))(\mathbf{x}_0-E(\mathbf{x}_0)']$. It is convenient to form a zero-mean version of $\mathbf{x}(\cdot)$ by defining $\tilde{\mathbf{x}}(t)=\mathbf{x}(t)-E\{\mathbf{x}(t)\}$ for $t \geq t_0$. Then, from (5.103)

$$d\tilde{\mathbf{x}}(t)=[\mathbf{A}(t)\tilde{\mathbf{x}}(t)-\lambda(t)\bar{\mathbf{b}}(t)]dt+\int_{\mathcal{U}}\mathbf{b}(t,\mathbf{U})M(dt,d\mathbf{U}),$$

$$\tilde{\mathbf{x}}(t_0)=\mathbf{x}_0-E(\mathbf{x}_0), \tag{5.109}$$

where $\bar{\mathbf{b}}(t)=E[\mathbf{b}(t,\mathbf{u})]$. Furthermore, in parallel with (5.48), there holds

$$\tilde{\mathbf{x}}(t)=\Phi(t,\tau)\tilde{\mathbf{x}}(\tau)-\int_\tau^t\Phi(t,\sigma)\lambda(\sigma)\bar{\mathbf{b}}(\sigma)d\sigma$$

$$+\int_\tau^t\int_{\mathcal{U}}\Phi(t,\sigma)\mathbf{b}(\sigma,\mathbf{U})M(d\sigma,d\mathbf{U}). \tag{5.110}$$

First, we want to derive an equation for the instantaneous covariance matrix for $\mathbf{x}(t)$, which is defined by $\operatorname{cov}[\mathbf{x}(t),\mathbf{x}(t)] = \Lambda(t) = E\{\tilde{\mathbf{x}}(t)\tilde{\mathbf{x}}'(t)\}$. As in Ex. 5.3.6, let $\mathbf{e}_i$ be an n-dimensional unit vector with the ith element equal to one and all others equal to zero. Also, let $\phi_{i,j}(\mathbf{X}) = \mathbf{e}'_i\mathbf{X}\mathbf{X}'\mathbf{e}_j$. Then, from (5.109) and (5.77), we have

$$d\phi_{i,j}(\tilde{\mathbf{x}}(t)) = < (\mathbf{e}_i\mathbf{e}'_j + \mathbf{e}_j\mathbf{e}'_i)\tilde{\mathbf{x}}(t), \mathbf{A}(t)\tilde{\mathbf{x}}(t) - \lambda(t)\overline{\mathbf{b}}(t) > dt \qquad (5.111)$$

$$+\mathbf{e}'_i\int_{\mathcal{U}}[\mathbf{b}(t,\mathbf{U})\tilde{\mathbf{x}}'(t) + \tilde{\mathbf{x}}(t)\mathbf{b}'(t,\mathbf{U}) + \mathbf{b}(t,\mathbf{U})\mathbf{b}'(t,\mathbf{U})]M(dt,d\mathbf{U})\mathbf{e}_j.$$

Noting that the (i,j)-element, $\Lambda_{i,j}(t)$, of $\Lambda(t)$ equals $E[\phi_{i,j}(\tilde{\mathbf{x}}(t))]$ and taking the expectation of both sides of (5.111), we have that

$$d\Lambda_{i,j}(t)$$

$$= \sum_{k=1}^{n}[A_{i,k}(t)\Lambda_{k,j}(t) + \Lambda_{i,k}(t)A_{j,k}(t)]dt + \lambda(t)\mathbf{e}'_i E[\mathbf{b}(t,\mathbf{u})\mathbf{b}'(t,\mathbf{u})]\mathbf{e}_j dt.$$

Consequently, the instantaneous covariance for $\mathbf{x}(t)$ satisfies the linear, matrix differential-equation

$$\frac{d\Lambda(t)}{dt} = \mathbf{A}(t)\Lambda(t) + \Lambda(t)\mathbf{A}'(t) + \lambda(t)E[\mathbf{b}(t,\mathbf{u})\mathbf{b}'(t,\mathbf{u})]$$

$$\Lambda(t_0) = \Lambda_0. \qquad (5.112)$$

Let us next determine the covariance matrix $\mathbf{K}_x(t,\tau)$, which by definition equals $E[\tilde{\mathbf{x}}(t)\tilde{\mathbf{x}}(\tau)]$. For $t \geq \tau$, we have from (5.110) that $\mathbf{K}_x(t,\tau) = \Phi(t,\tau)\Lambda(\tau)$ because

$$E\left\{\left[-\int_{\tau}^{t}\Phi(t,\sigma)\lambda(\sigma)\overline{\mathbf{b}}(\sigma)d\sigma + \int_{\tau}^{t}\int_{\mathcal{U}}\Phi(t,\sigma)\mathbf{b}(\sigma,\mathbf{U})M(d\sigma,d\mathbf{U})\right]\tilde{\mathbf{x}}'(\tau)\right\}$$

$$= E\left\{E\left[-\int_{\tau}^{t}\Phi(t,\sigma)\lambda(\sigma)\overline{\mathbf{b}}(\sigma)d\sigma + \int_{\tau}^{t}\int_{\mathcal{U}}\Phi(t,\sigma)\mathbf{b}(\sigma,\mathbf{U})M(d\sigma,d\mathbf{U})|\mathbf{x}(\tau)\right]\mathbf{x}'(\tau)\right\}$$

$$= 0.$$

A similar procedure for $t \leq \tau$ yields

$$\mathbf{K_x}(t,\tau) = \begin{cases} \Phi(t,\tau)\Lambda(\tau), & t \geq \tau \\[2ex] \Lambda(t)\Phi'(\tau,t), & t \leq \tau. \end{cases} \tag{5.113}$$

Finally, let us note that if $\mathbf{y}(t) = \mathbf{C}(t)\mathbf{x}(t)$, then the covariance matrix for $\mathbf{y}(\cdot)$ is

$$\mathbf{K_y}(t,\tau) = \mathbf{C}(t)\mathbf{K_x}(t,\tau)\mathbf{C}'(\tau). \tag{5.114}$$

By using the fact that the solution to (5.112) can be expressed in terms of $\Phi(t,\tau)$ as

$$\Lambda(t) = \Phi(t,t_0)\Lambda_0\Phi'(t,t_0)$$

$$+ \int_{t_0}^{t} \Phi(t,\tau)\lambda(\tau)E[\mathbf{b}(\tau,\mathbf{u})\mathbf{b}'(\tau,\mathbf{u})]\Phi'(t,\tau)d\tau, \tag{5.115}$$

it can be verified (5.114) and (5.16) are equivalent when $\Lambda_0 = 0$ and the dimension of $\mathbf{y}(\cdot)$ is unity. ∎

The stochastic integral equations associated with the above examples are *linear*. It is for this reason that an explicit expression for the transition characteristic function of the state can be found by solving (5.99). More generally, when the integral equation is not linear, the solution of (5.99) poses an insurmountable problem analytically. Even so, (5.99) still provides a useful point of departure for the *approximate* statistical analysis of the influence of a compound Poisson process on a nonlinear system when an exact analysis is infeasible. The appropriate approximations will, of course, depend strongly on the functions involved, so we cannot give any general rules to follow. As one instance illustrating that (5.99) can be useful in engineering analysis even when it cannot be solved explicitly, we continue the investigation begun in Ex. 5.3.3.

Example 5.3.12 *Phase-Tracking Loop Excited by Shot Noise* ———————
Consider, again, the model in Ex. 5.3.3, and make the following simplifying assumptions: $\omega_L = \omega_R = \omega$; $\theta_R(t) = 0$; $h(t,\tau) = g\delta(t-\tau)$; and $U = e$, where e is the electronic charge. Since $\theta_R(t)$ is zero, the loop phase $\theta_L(t)$ is ideally zero also, and any departure from zero is the phase tracking error. Our interest is in the statistics of the phase error process $\phi(t) = \theta_L(t) - \theta_R(t) = \theta_L(t)$. The loop filter h corresponds to an amplifier of gain g. The phase-tracking loop for this choice of filter is termed *first order*. For these assumptions, (5.50) becomes

$$d\phi(t) = f[\psi(t)]dN(t), \quad \phi(t_0) = \phi_0 \tag{5.116}$$

where $\psi(t) = \omega t + \phi(t)$, and the function $f[\cdot]$ is defined for a phase-lock loop by

$$f[\psi(t)] = ge \sin(\psi(t)) \tag{5.117}$$

and for a tanlock loop by

$$f[\psi(t)] = \frac{ge}{1 + m \cos(\omega t)} \sin(\psi(t)). \tag{5.118}$$

The underlying Poisson process has intensity $\lambda(t) = A[1 + m \cos(\omega t)]$ for $t \geq t_0$. Let $M_{\phi(t)}(jv) = E[\exp(jv\phi(t))]$ be the characteristic function for the phase tracking error $\phi(t)$ at time $t \geq t_0$. As there holds $M_{\phi(t)}(jv) = E\{E[\exp(jv\phi(t))|\phi_0]\}$, we have from (5.99) that

$$\frac{dM_{\phi(t)}(jv)}{dt} = A[1 + m \cos(\omega t)]E\{e^{jv\phi(t)}(e^{jv\,f[\psi(t)]} - 1)\}. \tag{5.119}$$

No closed form solution to this differential equation is evident. Consequently, we resort to an approximate analysis based on the additional assumption that the phase-error process $\phi(\cdot)$ approaches a steady state as t tends to infinity in the sense that the distribution for $\phi(t)$ becomes independent of t. Then for t large, (5.119) becomes

$$0 \doteq A[1 + m \cos(\omega t)]E\{e^{jv\phi(t)}[e^{jv\,f[\psi(t)]} - 1]\}. \tag{5.120}$$

where the symbol $\doteq$ is used to indicate that equality holds in the usual sense for t large when all terms on the right that depend explicitly on t are isolated and removed; these terms presumably average to zero as t approaches infinity for a steady-state distribution to exist. Now, it appears to be a reasonable approximation to replace $\exp\{jv f[\psi(t)]\}$ by the first few terms in its Taylor series if $|f[\psi(t)]|$ is sufficiently small, say less than some prescribed number δ, which for three terms might be on the order of 0.1. It is easy to verify from (5.117) and (5.118) that for a phase-lock loop, $\max_{\psi(t)}|f[\psi(t)]| = |ge|$, and for a tanlock loop $\max_{\psi(t)}|f[\psi(t)]| = |ge|/(1 - m^2)$. Thus, if for a phase-lock loop, the condition $|ge| < \delta$ is satisfied or if for a tanlock loop, the condition $|ge|/(1 - m^2) < \delta$ is satisfied, we will use the approximation

$$e^{jvf[\psi(t)]} - 1 \approx jvf[\psi(t)] + \frac{1}{2}(jv)^2 f^2[\psi(t)]. \tag{5.121}$$

The right side of this equation is a periodic function of $\psi(t)$; the two terms can be expanded in a Fourier series to yield

$$e^{jvf[\psi(t)]} - 1 \approx a_0 + a_1 \cos \psi(t) + b_1 \sin \psi(t) + a_2 \cos 2\psi(t) + \cdots \tag{5.122}$$

in which $a_0, a_1, \cdots$ are the Fourier coefficients. Upon multiplication of (5.122) by $\lambda(t)$, we conclude that

$$\lambda(t)e^{jvf[\psi(t)]} \approx A a_0 + \frac{1}{2}Ama_1 \cos \phi(t) + \frac{1}{2}Amb_1 \sin \phi(t), \tag{5.123}$$

where all terms that depend explicitly on t have been dropped. The three coefficients that are required are

$$a_0 = \frac{1}{4}(jv)^2(ge)^2, \quad a_1 = 0, \quad \text{and} \quad b_1 = jv(ge), \tag{5.124}$$

for a phase-lock loop, and

$$a_0 = \frac{1}{2}(jv)^2[(1-m^2)^{-1/2} - 1]m^{-2}(ge)^2$$

$$a_1 = (jv)^2[1 - (1-m^2)^{1/2}][1 - (1-m^2)^{-1/2}]m^{-3}(ge)^2 \tag{5.125}$$

$$b_1 = (jv)2[1 - (1-m^2)^{1/2}]m^{-2}(ge)$$

for a tanlock loop.

By introducing (5.123) and (5.124) into (5.120), we see for a phase-lock loop that

$$0 = E\left\{ [(jv)^2 + (jv)\alpha \sin(\phi_\infty)]e^{jv\phi_\infty} \right\}, \tag{5.126}$$

where $\phi_\infty = \lim_{t \to \infty} \phi(t)$ and

$$\alpha = 2m/ge. \tag{5.127}$$

Inverse Fourier transformation of (5.126) yields

$$0 = \frac{d^2 p_{\phi_\infty}(\Phi)}{d\Phi^2} + \frac{d\,\alpha\sin(\Phi)p_{\phi_\infty}(\Phi)}{d\Phi}. \tag{5.128}$$

This second-order differential equation also arises in the study of phase-lock loops operating in a wideband Gaussian noise; see A. Viterbi [27, p. 89]. It must be solved on the interval $(-\pi, \pi)$ and subject to the boundary condition $p_{\phi_\infty}(\pi) = p_{\phi_\infty}(-\pi)$ and the normalization $\int_{-\pi}^{\pi} p_{\phi_\infty}(\Phi)\,d\Phi = 1$. The solution is

$$p_{\phi_\infty}(\Phi) = \frac{e^{\alpha\cos(\Phi)}}{2\pi I_0(\alpha)}, \qquad -\pi \le \Phi \le \pi \tag{5.129}$$

where $I_0(\alpha)$ is a zero-order Bessel function of the first kind. Equation (5.129) was first obtained by R. Gagliardi and M. Haney [9] for phase-lock loops operating in shot noise. The single parameter α of (5.127) is seen to determine the steady-state phase tracking performance under the condition $|ge| < \delta \approx 0.1$. This parameter can be interpreted qualitatively as follows. The constant $B_L = Amge/4$ is shown by A. Viterbi [27, p. 36] to be the "linearized-loop noise bandwidth" of a phase-lock loop with forward gain g and input signal $Ame\cos(\omega t)$. The signal $Ame\cos(\omega t)$ is the instantaneous average current associated with $\lambda(t)$. As it is usually desired to operate the loop with a specified bandwidth, it is necessary to adjust the gain so that $g = 4B_L/Ame$, in which case $\alpha = Am^2/2B_L$. Hence, multiplication of the numerator and denominator of (5.127) by Ae^2 results in

$$\alpha = \frac{(Ame)^2/2}{Ae^2 B_L}. \tag{5.130}$$

The constant $(Ame)^2/2$ is the average power in the synchronization signal $Ame\cos(\omega t)$, and from Ex. 5.2.8, the constant Ae^2 is the power spectral density of the input noise signal, which is the average current of associated with $\lambda(t)$. The constant $Ae^2 B_L$ is, therefore, the noise power in the noise bandwidth of the loop. Thus, α is the signal-to-shot-noise power ratio in the noise bandwidth of the loop.

Similar results can be obtained for a tanlock loop; see Prob. 5.3.8. ∎

Summary. Several important ideas have been developed in this section. Our first concern was with the calculus for integral and differential equations excited by a marked point process. The most important ideas about this calculus are:

1. *Counting Integrals.* Counting integrals do not obey the usual rules of integration. Their evaluation is given in Theorem 5.3.1.

2. *Differentiation.* Differentiation of functions of processes defined by counting integrals does not obey the usual rules of differentiation. A special differential rule is given in Theorem 5.3.2.

Poisson driven Markov processes are defined by counting integrals in which the counting process is derived from a compound Poisson process. These processes may be regarded as the response to a compound Poisson process of a system that can be described by an ordinary differential equation. Two important ideas about the statistics for Poisson driven Markov processes are:

1. Under certain conditions given in Theorem 5.3.3, the solution to a nonlinear integral or differential equation is a Markov process.

2. The transition characteristic function satisfies a differential equation given in Theorem 5.3.4. While this equation can be solved only in certain special circumstances, it still provides a useful point of departure for the analysis through approximations of a nonlinear system excited by a compound Poisson process.

5.4 References

1. T. M. Apostol, *Mathematical Analysis*, Addison-Wesley, Reading MA., 1957.

2. A. T. Bharucha-Reid, *Random Integral Equations*, Wiley, New York, 1970.

3. R. W. Brockett, *Finite Dimensional Linear Systems*, Wiley, New York, 1970.

4. W. B. Davenport, Jr. and W. L. Root, *An Introduction to the Theory of Random Signals and Noise*, McGraw-Hill, New York, 1958.

5. J. L. Doob, *Stochastic Processes*, Wiley, New York, 1953.

6. J. Evans, "Preliminary Analysis of ELF Noise," Tech. Note 1969-18, M.I.T. Lincoln Laboratory, Lexington, MA, March 1969.

7. P. M. Fishman, "Statistical Inference for Space-Time Point Processes," D.Sc. Thesis, Sever Institute, Washington University, St. Louis, MO, June 1974.

8. R. H. Forrester and D. L. Snyder, "Phase-Tracking Performance of Direct-Detection Optical Receivers," *IEEE Trans. on Communications*, Vol. COM-21, pp. 1037-1039, September 1973.

9. R. Gagliardi and M. Haney, "Optical Synchronization - Phase Locking with Shot Noise Processes," Tech. Rpt. USCEE - 396, Dep't. of Electrical Engineering, Univ. of Southern California, Los Angeles, CA, August 1970.

10. I. I. Gihman and A. Ja. Dorogovcev, "On Stability of Solutions of Stochastic Differential Equations," *Ukrain. Mat. Z.*, Vol. 17, pp. 229-250, 1965.

11. I. I. Gihman and A. V. Skorokhod, *Stochastic Differential Equations*, Springer-Verlag, New York, 1972.

12. I. Gikhman and A. Skorokhod, *Introduction to the Theory of Random Processes*, Saunders, Philadelphia, PA, 1969.

13. E. V. Hoversten, "Optical Communication Theory," in: *Laser Handbook* (F. T. Arecchi and E. O. Schultz, Eds.), North Holland, Amsterdam, pp. 1805-1862, 1972.

14. K. Itô, "On Stochastic Differential equations," *Mem. Am. Math. Soc.*, No. 4, pp. 1-51, 1951.

15. T. Kailath, *Linear Systems*, Prentice-Hall, Englewood Cliffs, NJ, 1980.

16. R. S. Kennedy, *Fading Dispersive Communication Channels*, Wiley, New York, 1969.

17. P. Langevin, "Sur La Theorie Du Mouvement Brownian," *C. R. Acad. Sci.*, Paris, Vol. 146, pp. 530-533, 1908.

18. W. C. Lindsey and M. K. Simon, *Telecommunication Systems Engineering*, Prentice-Hall, Englewood Cliffs, NJ, 1973.

19. R. S. Lipster and A. N. Shiryayev, *Statistics of Random Processes I: General Theory*, Springer-Verlag, New York, 1977.

20. R. S. Lipster and A. N. Shiryayev, *Statistics of Random Processes I: Applications*, Springer-Verlag, New York, 1978.

21. L. Liusternik and V. Sobolev, *Elements of Functional Analysis*, Ungar, New York, 1961.

22. A. Papoulis, "High Density Shot Noise and Gaussianity," *J. Appl. Prob.*, Vol. 8, pp. 118-127, March 1971.

23. E. Parzen, *Stochastic Processes*, Holden-Day, San Francisco, CA, 1962.

24. W. K. Pratt, *Laser Communications Systems*, Wiley, New York, 1969.

25. "Project West Ford," *Proc. IEEE*, Vol. 52, May 1964.

26. A. V. Skorokhod, *Studies in the Theory of Random Processes*, Addison-Wesley, Reading, MA, 1964.

27. A. J. Viterbi, *Principles of Coherent Communication*, McGraw-Hill, New York, 1966.

28. E. Wong and B. Hajek, *Stochastic Processes in Engineering Systems*, Springer-Verlag, New York, 1985.

29. L. A. Zadeh and C. A. Desoer, *Linear System Theory*, McGraw-Hill, New York, 1963.

5.5 Problems

5.2.1. Suppose that electrons are emitted from a cathode as a homogeneous Poisson process with intensity λ electrons per second. The incremental contribution of an electron emitted at time τ to the anode current at time t is

$$h(t,\tau) = \begin{cases} \dfrac{2e(t-\tau)}{T^2}, & \tau \leq t \leq T+\tau \\ 0, & \text{otherwise} \end{cases}$$

where e is the electronic charge, and T is the time for the electron to traverse from the cathode to the anode.

   a. Evaluate the nth cumulant of the current $i(t)$, where

$$i(t) = \begin{cases} 0, & N(t)=0 \\ \sum_{n=1}^{N(t)} h(t,\tau_n), & N(t) \geq 1. \end{cases}$$

   b. Determine the mean and variance of $i(t)$.

   c. Determine the covariance function for $i(t)$.

5.2.2. Suppose that a sinusoidal bias voltage is applied so that the electron emissions described in Prob. 5.2.1 are emitted as an inhomogeneous Poisson process with intensity $\lambda(t) = \lambda_0[1 + m \sin(\omega t)]$. Repeat parts (a)-(c) of Prob. 5.2.1.

5.2.3. Let

$$y(t) = \begin{cases} 0, & N(t)=0 \\ \sum_{n=1}^{N(t)} h(t,\tau_n{:}u_n), & N(t) \geq 1 \end{cases}$$

be a filtered homogeneous Poisson process of rate λ for $t \geq 0$. Assume that the weighting function h has the form

$$h(t,\tau_n{:}u_n) = \begin{cases} 1, & \tau_n \leq t \leq \tau_n + u_n \\ 0, & \text{otherwise} \end{cases}$$

Here, $y(t)$ can be interpreted as the number of busy telephone lines when τ_n is the time the nth call of duration u_n is placed, where $u_n \geq 0$.

a. Verify that the characteristic function for $y(t)$ is

$$M_{y(t)}(j\alpha) = E(e^{j\alpha y(t)}) = \exp\left\{\lambda(e^{j\alpha}-1)\int_0^t [1-P_u(U)]dU\right\},$$

where $P_u(U)$ is the probability distribution function for u. Thus, conclude that $y(t)$ is Poisson distributed with parameter $\lambda\int_0^t[1-P_u(U)]dU$. Is $\{y(t): t \geq 0\}$ a Poisson process?

b. Verify that $E(u) = \int_0^\infty [1-P_u(U)]dU$. Hence, conclude for large t that $y(t)$ is Poisson distributed with parameter $\lambda E(u)$.

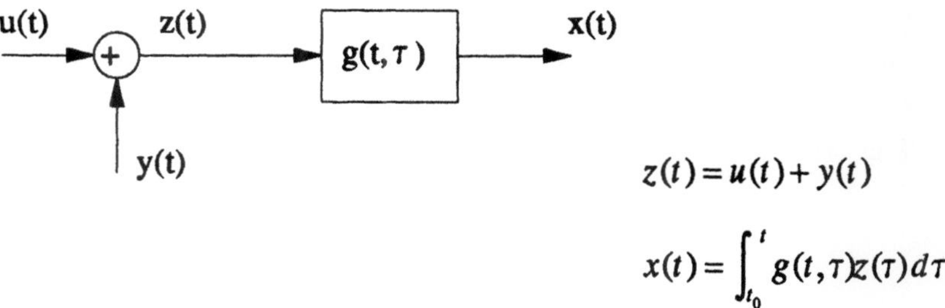

Figure P.5.2.4

5.2.4. For the system shown in Fig. P.5.2.4, $\{u(t): t \geq t_0\}$ is a deterministic control signal applied to a plant with impulse response $g(t,\tau)$, and $\{y(t): t \geq t_0\}$ is a filtered Poisson process defined by

$$y(t) = \begin{cases} 0, & N(t)=0 \\ \sum_{n=1}^{N(t)} u_n h(t,\tau_n), & N(t) \geq 1 \end{cases}$$

You can view $y(\cdot)$ as a model for lightning induced VLF atmospheric noise that additively disturbs the control signal. Then, $N(t)$ is the number of lightning strokes at time t, $\{\tau_n\}$ are the stroke occurrence times, and $\{u_n\}$ are energy marks. Assume that strokes occur with rate $\{\lambda(t): t \geq t_0\}$.

Determine the characteristic functional of the output $x(\cdot)$ on the interval $[t_0, T)$. What are the mean value function $m(t) = E[x(t)]$ and covariance function $K(t, \tau) = [\{x(t) - m(t)\}\{x(\tau) - m(\tau)\}]$ for the output process?

5.2.5. The random process $\{y(t): t \geq t_0\}$ is described by the following assumptions:

a. The probability of an event occurring in a particular interval is independent of the probability of an event occurring in any disjoint interval.

b.
$$\lim_{\Delta t \downarrow 0}\left\{\frac{\Pr[1 \quad \text{event in} \quad [t, t + \Delta t)]}{\Delta t}\right\} = \lambda,$$

$$\lim_{\Delta t \downarrow 0}\left\{\frac{\Pr[2 \quad \text{or more events in} \quad [t, t + \Delta t)]}{\Pr[1 \quad \text{event in} \quad [t, t + \Delta t)]}\right\} = 0.$$

c. If an event occurs at t, the amplitude of the process increases by one unit immediately and then increases by another unit Δ seconds later. Fig. P.5.2.5 shows a typical sample function.

d. The process starts at 0 at $t = 0$.

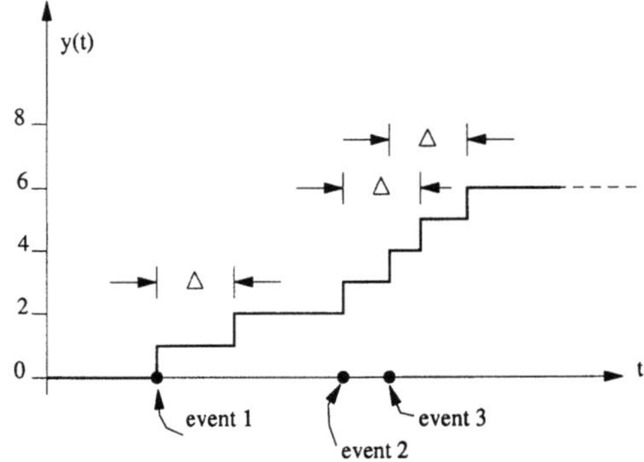

Figure P.5.2.5

i.   Find either $P_{y(t)}(Y)$, the probability distribution for $y(t)$ for $t \geq 0$, or $M_{y(t)}(jv)$, the corresponding characteristic function.

ii.   Is $\{y(t): t \geq 0\}$ an independent increment process? Explain.

5.2.6 Let $\{y(t): t > -\infty\}$ be a filtered Poisson process defined by

$$y(t) = \sum_{n=-\infty}^{\infty} h(t - T_n),$$

where $\{\tau_n\}$ are the occurrence times of a homogeneous Poisson process with constant intensity λ. Assume that $h(t) = 1/t$ for $-\infty < t < \infty$. Show that $y(t)$ has a Cauchy distribution.

5.2.7 (Random Response Function). For $t \geq t_0$, let

$$y(t) = \begin{cases} 0, & N(t) = 0 \\ \displaystyle\sum_{n=1}^{N(t)} h_n(t, \tau_n), & N(t) \geq 0, \end{cases}$$

where

i. $\{N(t): t \geq t_0\}$ is a Poisson counting process with intensity function $\{\lambda(t): t \geq t_0\}$ and occurrence times $\{\tau_n\}$.

ii. $\{h_n(t, \tau): t \geq t_0, \tau > t_0\}$ for $n = 1, 2, \cdots$ are mutually independent identically distributed random processes all having the same distribution as the process $\{h(t, \tau): t \geq t_0, \tau > t_0\}$. Also, these processes are independent of $\{N(t): t \geq t_0\}$.

iii. The known characteristic functional for $h(t, \tau)$ for $t \in [t_0, T)$ and τ fixed is defined by

$$\phi_h(jv; \tau) \overset{\Delta}{=} E\left[\exp\left\{ j \int_{t_0}^{T} h(\sigma, \tau) dv(\sigma) \right\} \right].$$

iv. Assume that $h_n(t, \tau) = 0$ for $t < \tau$.

a. Show that the characteristic functional for $y(\cdot)$ on the interval $[t_0, T)$ is given by

$$\phi_y \overset{\Delta}{=} E\left[\exp\left\{ \int_{t_0}^{T} y(\sigma) dv(\sigma) \right\} \right] = \exp\left\{ \int_{t_0}^{T} \lambda(\sigma) [\phi_h(jv; \tau) - 1] d\tau \right\}.$$

b. Determine the mean and variance for $y(t)$, $t \in [t_0, T)$.

c. Determine the covariance function for $y(t)$.

5.2.8 Let $\{y(t): t \geq t_0\}$ be a process of the type defined in Prob. 5.2.7. Assume that $\lambda(t) \equiv \lambda$, a constant. Suppose, now, that $h(t,\tau)$ is an independent increment process in t for each fixed τ, $h(t,\tau) = 0$ for $t < \tau$, $h(\tau,\tau) = 1$, and

$$\Pr[h(t,\tau) = k] = \frac{[\gamma(t-\tau)]^{k-1}}{(k-1)!} e^{-\gamma(t-\tau)}, \qquad \gamma > 0$$

for $t \geq \tau$ and $k = 1, 2, \cdots$. Thus, h is a homogeneous Poisson counting process that starts at time τ with unit count. The process $y(\cdot)$ is a model for a photomultiplier in which $N(t)$ and $\{\tau_n\}$ represent the number and occurrence times of primary photoelectrons, and $h_n(t,\tau_n)$ is the number of secondary electrons out of the photomultiplier due to the nth photoelectron; typically, $\gamma \approx 10^6$. Determine the characteristic function for $y(t)$, $M_{y(t)} = E\{\exp[j\alpha y(t)]\}$.

5.2.9 Consider a process in which events happen in clusters rather than singly. For $k = 1, 2, \cdots$, suppose the event that k particles are released occurs in accordance with a Poisson counting process $N_k(t)$ with constant intensity λ_k. At each event of this process, k particles are released, so the total number of particles released in $[0, T)$ due to this process is $kN_k(t)$. Assume that the processes $\{N_k(t)\}$ are mutually independent. Let $N(t)$ be the total number of particles produced in $[0, T)$,

$$N(t) = N_1(t) + 2N_2(t) + \cdots + kN_k(t) + \cdots$$

a. Show that $E[N(t)] = \mu_1 t$ and $\text{var}[N(t)] = \mu_2 t$, where

$$\mu_i = \sum_{k=1}^{\infty} k^i \lambda_k, \qquad i = 1, 2.$$

b. Show that $\{N(t): t \geq 0\}$ is a compound Poisson process.

c. Suppose that the particles communicate equal charges to an electrometer whose deflection at time t is

$$y(t) = \begin{cases} 0, & N(t) = 0 \\ \displaystyle\sum_{n=1}^{N(t)} h(t - \tau_n), & N(t) \geq 1, \end{cases}$$

where $\{\tau_n\}$ are the times at which the charges are observed, and $h(\cdot)$ is the impulse response of the electrometer. Show that

$$E[y(t)] = \mu_1 \int_0^t h(\sigma)d\sigma \quad \text{and} \quad \text{var}[y(t)] = \mu_2 \int_0^t h^2(\sigma)d\sigma.$$

5.2.10 Suppose that $\{y(t): t \geq 0\}$ has the form

$$y(t) = \begin{cases} 0, & N(t) = 0 \\ \displaystyle\sum_{n=1}^{N(t)} u_n e^{-(t-\tau_n)/RC}, & N(t) \geq 1 \end{cases}$$

in which $\{\tau_n\}$ are the times of occurrence of points of a homogeneous Poisson process with constant intensity λ, $\{u_n\}$ are independent and identically distributed random variables with finite second moments, and R and C are positive constants.

   a. Evaluate the mean-value function $m(t) = E[y(t)]$ for $t \geq 0$. What happens as $t \to \infty$?

   b. Evaluate the covariance function $K(t,v) = E[y(t)y(v)] - m(t)m(v)$ for $t > 0, v > 0$.

   c. Determine the characteristic function $M(j\alpha) = E\{\exp[j\alpha y(t)]\}$ for $t > 0$.

   d. Show that $y^*(t) = [y(t) - m(t)]/\sqrt{\text{var}[y(t)]}$ converges in distribution to a zero-mean, unit variance Gaussian random variable as $RC \to \infty$.

5.2.11 Let $\{y(t): t \geq t_0\}$ be the process defined in Prob. 5.2.7. Define

$$\gamma_1(t) = \int_{t_0}^t \lambda(\tau)E[h(t,\tau)]d\tau, \qquad \gamma_2(t) = \int_{t_0}^t \lambda(\tau)E[h^2(t,\tau)]d\tau,$$

$$\Gamma(t_1,t_2,t_3) = \int_{t_0}^{\min(t_1,t_2,t_3)} \lambda(\tau)E[|\,h(t_1,\tau)h(t_2,\tau)h(t_3,\tau)\,|\,]d\tau.$$

Show that if $[\gamma_2(t_1)\gamma_2(t_2)\gamma_2(t_3)]^{-1/2}\Gamma(t_1,t_2,t_3)$ tends uniformly to zero on $[t_0,T) \times [t_0,T) \times [t_0,T)$ as certain parameters tend to prescribed limits, then the normalized process

$$y^*(t) = \frac{y(t) - \gamma_1(t)}{\sqrt{\gamma_2(t)}}$$

tends to a Gaussian random process on $[t_0, T)$. Determine the mean and covariance for $\{y^*(t): t \geq t_0\}$.

5.2.12 A typical sample function for a *nonlinearly* filtered Poisson process is shown in Fig. P.5.2.12. Starting at $t = 0$, $y(t)$ decreases exponentially

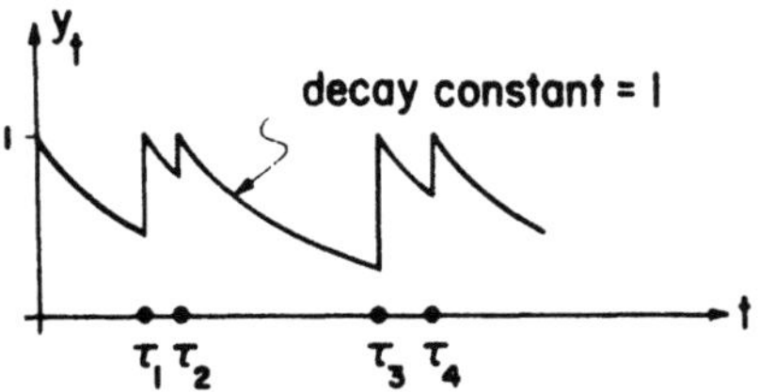

Figure P.5.2.12

from value 1 until time τ_1, the first arrival time. At $t = \tau_1$, y returns to 1 and then decreases exponentially with the same decay constant until τ_2, the second arrival time. At $t = \tau_2$, y returns to 1, and so on. Assume that points arrive as a homogeneous Poisson process with constant intensity λ. Evaluate the mean and variance of $y(t)$ at any $t \geq 0$.

5.3.1 Let $\{N(t): t \geq t_0\}$ be a Poisson counting process with intensity $\{\lambda(t): t \geq t_0\}$. Define the zero-mean process $\{\tilde{N}(t): t \geq t_0\}$ by $\tilde{N}(t) = N(t) - \int_{t_0}^{t} \lambda(\sigma) d\sigma$. Consider the integral

$$\tilde{I}_\alpha(t) = \int_{t_0}^{t} N(\sigma) d\tilde{N}(\sigma) = \lim_{n \to \infty} \sum_{i=1}^{n} N(t'_{i,n})\{\tilde{N}(t_{i,n}) - \tilde{N}(t_{i-1,n})\},$$

where $t_0 = t_{0,n} < t_{1,n} < \cdots < t_{n,n} = t$ is a sequence of partitions of $[t_0, t)$ such that $\max_{1 \leq i \leq n}|t_{i,n} - t_{i-1,n}|$ tends to zero as n tends to infinity, and where $t'_{i,n} = \alpha t_{i-1,n} + (1-\alpha)t_{i,n}$ for $0 \leq \alpha \leq 1$.

a. Evaluate $\tilde{I}_\alpha(t)$ as a function of α.

b. Evaluate the mean and variance of $\tilde{I}_\alpha(t)$ as a function of α.

c. Evaluate $E[\tilde{I}_\alpha(t)|N(\sigma): t_0 \leq \sigma \leq s]$ as a function of α, where $s \leq t$.

d. A random process $\{m(t): t \geq t_0\}$ is termed a *Martingale, submartingale,* or *supermartingale* adapted to $\{N(t): t \geq t_0\}$ if

$E[m(t)|N(\sigma): t_0 \le \sigma \le s]$ is equal, greater, or less than $m(s)$, respectively. As a function of α, classify $\{\tilde{I}_\alpha(t): t \ge t_0\}$ as a Martingale, submartingale, or supermartingale.

5.3.2 [Derivation of (5.61)]. Let $I_{p,q}(t,A): t \ge t_0\}$ be as defined in (5.57), and suppose that

$$\tilde{I}_{p,q}(t,A) = I_{p,q}(t,A) - \sum_{j=1}^{q}\sum_{k=1}^{p} b(t_{j-1,q}, U'_{k,p}) \int_{t_{i-1,q}}^{t_{j,q}} \int_{A_{k,p}} \lambda(\sigma)\,d\sigma dP_u(U).$$

a. Evaluate $E[|\tilde{I}_{p,q}(t,A)|^2]$.

b. Deduce (5.61) from the result of (a) as p and q tend to infinity.

5.3.3 Suppose that $\{\zeta(t): t \ge t_0\}$ satisfies

$$\zeta(t) = 1 + \int_{t_0}^{t} \zeta(\sigma)f(\sigma)\,dN(\sigma),$$

where $f(t)$ is a deterministic function of t for $t \ge t_0$, and $\{N(t): t \ge t_0\}$ is a Poisson process with intensity $\{\lambda(t): t \ge t_0\}$.

a. Show that $\zeta(t) = \exp\left\{ \int_{t_0}^{t} \ln[1 + f(\sigma)]dN(\sigma)\right\}$. Hint: let $\phi(\zeta(t)) = \ln[\zeta(t)]$, and use the result in Ex. 5.3.8.

b. Show that $E[\zeta(t)] = \exp\left\{ \int_{t_0}^{t} f(\sigma)\lambda(\sigma)\,d\sigma\right\}$.

c. Show that $\zeta^2(t) = 1 + \int_{t_0}^{t} \zeta^2(\sigma)f(\sigma)[2 + f(\sigma)]dN(\sigma)$.

d. Evaluate the variance of $\zeta(t)$.

5.3.4 (Formula for Integration by Parts). The common formula for integration by parts is

$$\int_{t_0}^{t} \Phi_1(\sigma)\,d\Phi_2(\sigma) = \int_{t_0}^{t} d[\Phi_1(\sigma)\Phi_2(\sigma)] - \int_{t_0}^{t} \Phi_2(\sigma)\,d\Phi_1(\sigma),$$

where $\Phi_1(t)$ and $\Phi_2(t)$ are deterministic functions of t. The analog of this when $\Phi_1(t)$ and $\Phi_2(t)$ are defined by counting integrals is the following. Suppose that

$$\Phi_i(t) = \int_{t_0}^{t} \int_{\mathcal{u}} \phi_i(\sigma, U) M(d\sigma, dU), \qquad i = 1,2$$

where $\{\phi_i(t, U): t \geq t_0, U \in \mathcal{U}\}$, $i = 1,2$, satisfy the conditions of Theorems 5.3.1 and 5.3.3. Show that

$$\int_{t_0}^{t} \Phi_1(\sigma) d\Phi_2(\sigma) = \int_{t_0}^{t} d[\Phi_1(\sigma)\Phi_2(\sigma)] - \int_{t_0}^{t} \Phi_2(\sigma) d\Phi_1(\sigma)$$

$$- \int_{t_0}^{t} \int_{\mathcal{u}} \phi_1(\sigma, U)\phi_2(\sigma, U) M(d\sigma, dU).$$

5.3.5 Let $\{\zeta(t): t \geq t_0\}$ be defined by

$$\zeta(t) = -\int_{t_0}^{t} \lambda(\sigma) d\sigma + \int_{t_0}^{t} \ln \lambda(\sigma) dN(\sigma),$$

where $\{N(t): t \geq t_0\}$ is a Poisson counting process with intensity function $\{\lambda(t): t \geq t_0\}$. The quantity $\phi(\zeta(t)) = \exp[\zeta(t)]$ is the sample-function density for $N(\cdot)$ on the interval $[t_0, t)$.

a. Show that

$$\phi^k(\zeta(t)) = 1 - k \int_{t_0}^{t} \phi^k(\zeta(\sigma))\lambda(\sigma) d\sigma + \int_{t_0}^{t} \phi^k(\zeta(\sigma))[\lambda^k(\sigma) - 1] dN(\sigma).$$

b. Use the result in (a) to evaluate the $k{th}$ moment of the sample-function density, $E[\phi^k(\zeta(t))]$.

5.3.6 Let $\{\zeta(t): t \geq t_0\}$ be defined by

$$\zeta(t) = -\int_{t_0}^{t} \lambda(\sigma) d\sigma + \int_{t_0}^{t} \int_{\mathcal{u}} \ln[p_u(U)\lambda(\sigma)] M(d\sigma, dU),$$

where M is the time-space Poisson process of the representation of a compound Poisson process in Theorem 4.3.2, and $p_u(U)$ is the probability density of the mark random variable u. Recall from (4.45) that $\phi(\zeta(t)) = \exp[\zeta(t)]$ is the sample-function density for the compound Poisson process on $[t_0, t)$.

a. Derive an integral equation for $\phi^k(\zeta(t))$.

b. Evaluate $E[\phi^k(\zeta(t))]$.

5.3.7 (I. Bar-David's Expectation Formula)

a. Let $\{N(t): t \geq t_0\}$ be a Poisson counting process with intensity $\{\lambda(t): t \geq t_0\}$. Denote occurrence times by τ_1, τ_2, $\cdots$, and let $\{f(t): t \geq t_0\}$ be a deterministic function. Define the process $\{\zeta(t): t \geq t_0\}$ by $\zeta(t) = \int_{t_0}^{t} \ln f(\sigma) dN(\sigma)$, and let

$$\phi(\zeta(t)) = e^{\zeta(t)} = \begin{cases} 1, & N(t) = 0 \\ \prod_{n=1}^{N(t)} f(\tau_n), & N(t) \geq 1 \end{cases}$$

Show that

$$E[\phi(\zeta(t))] = \exp\left[\int_{t_0}^{t} \{f(\sigma) - 1\}\lambda(\sigma) d\sigma \right].$$

Hint: establish and use the fact that

$$\phi(\zeta(t)) = 1 + \int_{t_0}^{t} \phi(\zeta(\sigma))[f(\sigma) - 1] dN(\sigma).$$

b. As an extension, consider a compound Poisson process on $[t_0, t)$ with occurrence times $\{\tau_n\}$ and marks $\{\mathbf{u}_n\}$. Let $f(t, \mathbf{U})$ be a deterministic function of t and $\mathbf{U}$. Define

$$\zeta(t) = \int_{t_0}^{t} \int_{u} \ln f(\sigma, \mathbf{U}) M(d\sigma, d\mathbf{U}).$$

Show that

$$E[\phi(\zeta(t))] = \exp\left\{ \int_{t_0}^{t} \int_{u} [f(\sigma, \mathbf{U}) - 1]\lambda(\sigma) d\sigma dP_u(\mathbf{U}) \right\},$$

where

$$\phi(\zeta(t)) = e^{\zeta(t)} = \begin{cases} 1, & N(t) = 0 \\ \prod_{n=1}^{N(t)} f(\tau_n, \mathbf{u}_n), & N(t) \geq 1. \end{cases}$$

5.3.8 (Error Density for a Tanlock Loop [8]).

a. By following the procedure used to derive (5.128), show for a tan-lock loop that

$$\frac{d^2[1+\Gamma\cos(\Phi)]p_{\phi_\infty}(\Phi)}{d\Phi^2}+\frac{d\Lambda\sin(\Phi)p_{\phi_\infty}(\Phi)}{d\Phi}=0,$$

where $\Gamma=\sqrt{1-m^2}-1$ and $\Lambda=(2m/ge)\sqrt{1-m^2}$.

b. Show that

$$p_{\phi_\infty}(\Phi)=\frac{[1+\Gamma\cos(\Phi)]^{(\Lambda/\Gamma)-1}}{\displaystyle\int_{-\pi}^{\pi}[1+\Gamma\cos(\Phi)]^{(\Lambda/\Gamma)-1}d\Phi},\qquad -\pi\le\Phi\le\infty.$$

CHAPTER SIX

SELF-EXCITING POINT PROCESSES

6.1 Introduction

A temporal point process must be both orderly and without aftereffects to be a Poisson process. The orderliness restriction that points be isolated from one another is relaxed in Ch. 4 where a generalized Poisson process is defined and studied. The restriction that the process be without aftereffects is removed in this chapter about *self-exciting point processes*.

We first relax completely the restriction that the point process evolve without aftereffects. Thus, in the most general self-exciting point processes, the entire past - that is, the number $N(t)$ and occurrence times w_1, $w_2, \cdots, w_{N(t)}$ for all points realized before some time t - can influence the number and occurrence times for all points subsequent to t. This influence of the past on the future is characterized by the influence on only the microscopic future: the foremost quantity required in the development is defined in terms of the conditional probability of a single point occurring in the infinitesimal interval $[t, t + \Delta t)$ given the number and time of occurrence for all points realized up to time t. A general self-exciting point process can be conceived of as a modified inhomogeneous Poisson process in which the intensity is not only a function of time but also the entire past of the point process.

An example of a phenomenon that may be modeled as a self-exciting point process occurs in the operation of an electronic vacuum tube under space-charge limited conditions. Electron emissions from the cathode are suppressed by electrons emitted previously and still in transit to the anode. In fact, the cloud of transiting electrons can have a total charge of sufficient magnitude that further electron emissions are nearly extinguished. Thus, emission times form a self-exciting point process. This phenomenon is investigated in detail by W. Davenport and W. Root [7, Ch. 7] and S. Srinivasan and R. Vasudevan [36, Ch. 5].

General self-exciting point processes are examined in Sec. 6.2. Expressions are derived for the counting and occurrence-time statistics. Perhaps the most important quantity developed for applications is the sample-function density for self-exciting point processes. This plays the same role as in previous chapters. Moreover, the sample-function density will prove to be useful in the next chapter where we study the doubly-stochastic Poisson process, which is a form of self-exciting point process.

For many applications, the point process does not depend on its entire past but, rather, only on some small fraction of it such as the total number of accumulated counts or the most recent occurrence time. The most notable of these processes are *Markov birth processes* and *renewal processes*, both of which have been extensively studied. These self-exciting point processes with limited memory are investigated in Sec. 6.3.

It is perhaps worth alerting the reader in advance that many of the results obtained in this chapter are characterizations rather than explicit formulas. Thus, for example, we derive but do not solve a differential equation for the counting probability $\Pr[N(t) = n]$ of the number of points in the interval $[t_0, t)$ for a general self-exciting point process. Indeed, it is only in very special circumstances that explicit expressions for this probability can be obtained. The renewal process of Sec. 6.3 is in many ways the simplest non-Poisson self-exciting point process we will encounter, and an explicit expression for its counting probability is generally unavailable. However, the characterizations we develop do provide considerable insight into the nature of point processes that evolve with aftereffects.

The approach that we have adopted is largely due to I. Rubin [31, 32]. We have preferred this approach because it is based on a natural and intuitively appealing modification of the Poisson process. Moreover, it turns out that important expressions, such as for the sample-function density, bear a remarkable similarity to those of a Poisson process. As a consequence, qualitative interpretations we have already developed for the Poisson process often carry over unchanged to self-exciting point processes.

We refer to J. Moyal [23], H. Wold [38], J. McFadden [21], O. Macchi [18, 19], A. Ramakrishnan [27, 28], and D. Daley and D. Vere-Jones [6] for details and additional references on alternative approaches for general self-exciting point processes. One characterization used first by S. Rice [29] in his 1944 studies of the zero crossings of a random process and then developed thoroughly by A. Ramakrishnan [27, 28] beginning in 1950 and P. Kuznetsov and R. Stratonovich [17] in 1956 is outlined in Problems 6.2.6 and 6.2.7. This characterization appears especially well suited for the study of filtered, self-exciting point processes. There are numerous treatments of Markov birth processes and renewal processes, both of which are special forms of self-exciting point processes with limited memory. A. Bharucha-Reid [1] and W. Smith [34] treat these important models. Martingale methods also encompass general self-exciting point processes. These are pursued by P. Bremaud [3] and A. Karr [16].

6.2 General Self-Exciting Point Processes

We first introduce some assumptions, notation, and terminology that will be used in this chapter. Let $\{N(t): t \geq t_0\}$ be a counting process, and assume that

a. $\{N(t): t \geq t_0\}$ is conditionally orderly;

b.   the limit of the following functions $a(\Delta t, N(t))$ exist as Δt tends to zero for almost every realization of $\{N(t): t \geq t_0\}$

$a(\Delta t, N(t))$

$$= \begin{cases} \dfrac{1}{\Delta t} \Pr[N(t, t+\Delta t) = 1 \mid N(t)], & N(t) = 0 \\[2ex] \dfrac{1}{\Delta t} \Pr[N(t, t+\Delta t) = 1 \mid N(t); w_1, w_2, \cdots, w_{N(t)}], & N(t) \geq 1; \end{cases} \tag{6.1}$$

c. $\Pr[N(t_0) = 0] = 1$.

The property of conditional orderliness in (a) is defined in Ch. 2. Qualitatively, it means that the conditional probability of two or more points in $[t, t+\Delta t)$ can be made an arbitrarily small fraction of there being only one point by choosing Δt sufficiently small. We denote the limits in (b) by $\mu(t, 0)$ and $\mu(t, N(t); w_1, w_2, \cdots, w_{N(t)})$, respectively,

$$\mu(t, 0) = \lim_{\Delta t \downarrow 0} \frac{1}{\Delta t} \Pr[N(t, t+\Delta t) = 1 \mid N(t)], \quad \text{for } N(t) = 0 \tag{6.2}$$

and

$$\mu(t, N(t); w_1, w_2, \cdots, w_{N(t)}) \tag{6.3}$$

$$= \lim_{\Delta t \downarrow 0} \frac{1}{\Delta t} \Pr[N(t, t+\Delta t) = 1 \mid N(t); w_1, w_2, \cdots, w_{N(t)}], \quad \text{for } N(t) \geq 1.$$

These limiting functions characterize the infinitesimal behavior of the point process conditioned on its previous global behavior. It is convenient to introduce a random process $\{\lambda(t): t \geq t_0\}$ defined by

$$\lambda(t) \overset{\Delta}{=} \begin{cases} \mu(t, 0), & t_0 \leq t \leq w_1 \\[2ex] \mu(t, N(t); w_1, w_2, \cdots, w_{N(t)}), & w_{N(t)} < t \leq w_{N(t)+1}. \end{cases} \tag{6.4}$$

A typical sample function for this process is shown in Fig. 6.1. It is continuous from the left, and we assume for $t \geq t_0$ that $E[\lambda(t)] < \infty$.

We term $N(\cdot)$ a *self-exciting counting process* if $\lambda(\cdot)$ depends on more than t. In correspondence with the terminology in Ch. 2 for Poisson pro-

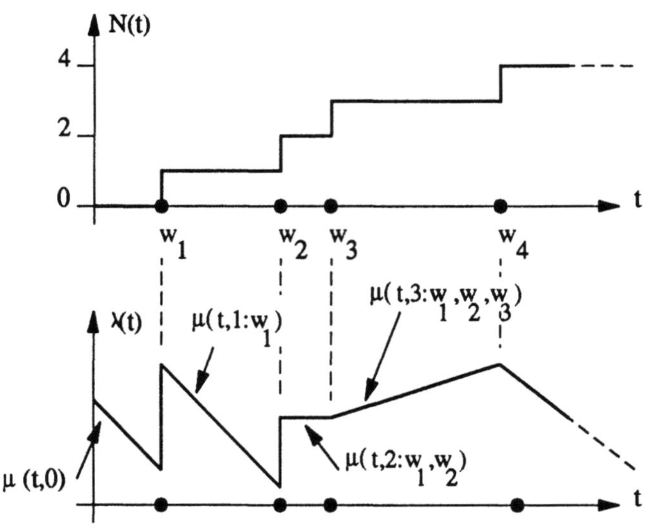

Figure 6.1 Typical sample function of an intensity process $\{\lambda(t): t \geq t_0\}$.

cesses, $\lambda(\cdot)$ will be called the intensity function or more appropriately here, the *intensity process*, the latter because $\lambda(\cdot)$ is itself a random process whose sample path depends on the realization of $N(\cdot)$.

It is evident by examination of (2.8) that if $\lambda(\cdot)$ does not depend in any way on $N(\cdot)$, the point process is not self exciting, and $\{N(t): t \geq t_0\}$ is an inhomogeneous Poisson process. If $\lambda(\cdot)$ depends on the number of points in $[t_0, t)$ but not on their occurrence times, $N(\cdot)$ is the birth process described in Problem 2.2.6. More generally, $\lambda(\cdot)$ can depend on one or more of the occurrence times as well as the number of points.

The most immediate interpretation of $\lambda(\cdot)$ is that the probability of the occurrence of one point in the infinitesimal interval $[t, t + \Delta t)$ given that the number of points in $[t_0, t)$ is n and that their times of occurrence are $W_1, W_2, \cdots, W_n$ is approximated to order Δt by $\lambda(t)\Delta t$, where $\lambda(t)$ is the value at time t of that sample function of the intensity process corresponding to the realized count n and occurrence times $W_1, W_2, \cdots, W_n$ on $[t_0, t)$. Because $N(\cdot)$ is conditionally orderly, the conditional probability of there being no point in $[t, t + \Delta t)$ is approximated by $1 - \lambda(t)\Delta t$. Further interpretation is provided in the following comment.

Suppose that $N(s) = 0$, and consider the conditional probability of there being no points in $[s, t + \Delta t)$, where $t_0 \leq s \leq t < t + \Delta t$. It follows from (6.2) that

$$\Pr[N(s, t + \Delta t) = 0 | N(s) = 0]$$

$$= \Pr[N(s, t) = 0 | N(s) = 0]\Pr[N(t, t + \Delta t) = 0 | N(t) = 0]$$

$$= \Pr[N(s, t) = 0 | N(s) = 0][1 - \mu(t, 0)\Delta t + o(\Delta t)].$$

We deduce from this that

$$\frac{d\Pr[N(s, t) = 0 | N(s) = 0]}{dt} = -\mu(t, 0)\Pr[N(s, t) = 0 | N(s) = 0],$$

with the initial condition $\Pr[N(s, s) = 0 | N(s) = 0] = 1$. Hence,

$$\Pr[N(s, t) = 0 | N(s) = 0] = \exp\left[-\int_s^t \mu(\sigma, 0) d\sigma \right]. \tag{6.5}$$

Similarly, for $N(t) \geq 1$, we obtain

$$\Pr[N(s, t) = 0 | N(s); w_1, w_2, \cdots, w_{N(s)}]$$

$$= \exp\left[-\int_s^t \mu(\sigma, N(s); w_1, w_2, \cdots, w_{N(s)}) d\sigma \right]. \tag{6.6}$$

Thus, the functions $\mu(\cdot)$ defined by conditional occurrence probabilities for infinitesimal intervals in (6.2) and (6.3) specify and are specified by conditional probabilities of there being no points in the global interval $[s, t)$ in (6.5) and (6.6).

6.2.1 Counting Statistics for Self-Exciting Point Processes

In this subsection, we characterize the counting statistics for a self-exciting point process by studying the distribution of the number of points in $[t_0, t)$. A central role is played by the conditional expectation $E[\lambda(t) | N(t)]$ of $\lambda(t)$ given the number of points but not their locations. This conditional expectation is by definition a function of the random variable $N(t)$, which we denote by

$$\hat{\lambda}(t, N(t)) = E[\lambda(t) | N(t)]. \tag{6.7}$$

We term $\hat{\lambda}(t,N(t))$ the *count-conditional intensity*. The principal conclusion we draw is that $N(t)$ has the same distribution as the pure birth process described in Problem 2.2.6, having a birthrate equal to the count-conditional intensity. This assertion is established in Theorem 6.2.1 by deriving a differential-difference equation for the probability of the number of points in $[t_0, t)$.

The following lemma motivates the interpretation of $\hat{\lambda}(t,N(t))$ as an intensity and provides a result needed in the proof of Theorem 6.2.1.

Lemma 6.2.1. Let $\{N(t): t \geq t_0\}$ be a self-exciting point process with intensity process $\{\lambda(t): t \geq t_0\}$ defined in (6.3). Assume that $\{N(t): t \geq t_0\}$ is conditionally orderly and that $E[\lambda(t)] < \infty$ for $t \geq t_0$. Then, the function $\hat{\lambda}(t,N(t))$ satisfies

$$\hat{\lambda}(t,N(t)) = \lim_{\Delta t \downarrow 0} \frac{1}{\Delta t} \Pr[N(t,t+\Delta t) \geq 1 | N(t)] \tag{6.8}$$

and

$$\hat{\lambda}(t,N(t)) = \lim_{\Delta t \downarrow 0} \frac{1}{\Delta t} \Pr[N(t,t+\Delta t) = 1 | N(t)]. \tag{6.9}$$

Proof.[1] Suppose first that $N(t) = 0$. From (6.5), we see that

$$\lim_{\Delta t \downarrow 0} \frac{1}{\Delta t} \Pr[N(t,t+\Delta t) \geq 1 | N(t) = 0] = \mu(t,0).$$

Thus, (6.8) holds for $N(t) = 0$ because $\mu(t,0) = E[\lambda(t) | N(t) = 0]$ from (6.3). Equation (6.9) is obtained from (6.8) using the conditional orderliness of $\{N(t): t \geq t_0\}$, which implies that for any $\varepsilon > 0$, there is a sufficiently small Δt that $\Pr[N(t,t+\Delta t) > 1 | N(t) = 0] \leq \varepsilon \Pr[N(t,t+\Delta t) = 1 | N(t) = 0]$. For this choice of ε and Δt, there holds

$$\frac{1}{(1+\varepsilon)\Delta t} \Pr[N(t,t+\Delta t) \geq 1 | N(t) = 0]$$

$$\leq \frac{1}{\Delta t} \Pr[N(t,t+\Delta t) = 1 | N(t) = 0] \leq \frac{1}{\Delta t} \Pr[N(t,t+\Delta t) \geq 1 | N(t) = 0].$$

Hence, for any $\varepsilon > 0$,

---

[1] This proof is largely due to P. Fishman [9]

$$\frac{1}{1+\varepsilon}\mu(t,0) \le \lim_{\Delta t \downarrow 0}\frac{1}{\Delta t}\Pr[N(t,t+\Delta t)=1\,|\,N(t)=0] \le \mu(t,0).$$

Equation (6.9) follows from this for $N(t)=0$ by virtue of the arbitrariness of ε.

Suppose next that $N(t) \ge 1$. Let $\{\rho_n\}$ be a sequence of positive numbers converging to zero, and define

$$f_n = \frac{1}{\rho_n}\Pr[N(t,t+\rho_n)\ge 1\,|\,N(t);w_1,w_2,\cdots,w_{N(t)}]$$

and

$$g_n = \frac{1}{\rho_n}\int_t^{t+\rho_n}\mu(\sigma,N(t);w_1,w_2,\cdots,w_{N(t)})d\sigma.$$

The sequence $\{g_n\}$ converges with probability one to $\mu(t,N(t);w_1,w_2,\cdots,w_{N(t)})$ and $E(g_n)<\infty$ for each n because $E[\mu(t,N(t);w_1,w_2,\cdots,w_{N(t)})]<\infty$ by assumption. From (6.6), the sequence $\{f_n\}$ also converges with probability one to $\mu(t,N(t);\ w_1,w_2,\cdots,w_{N(t)})$, and

$$f_n = \frac{1}{\rho_n}\left\{1-\exp\left[-\int_t^{t+\rho_n}\mu(\sigma,N(t);w_1,w_2,\cdots,w_{N(t)})d\sigma\right]\right\}$$

$$\le \frac{1}{\rho_n}\int_t^{t+\rho_n}\mu(\sigma,N(t);w_1,w_2,\cdots,w_{N(t)})d\sigma = g_n.$$

Since

$$E[\mu(t,N(t);w_1,w_2,\cdots,w_{N(t)})]$$

$$= \lim_{n\to\infty}\frac{1}{\rho_n}\int_t^{t+\rho_n}E[\mu(\sigma,N(t);w_1,w_2,\cdots,w_{N(t)})]d\sigma = \lim_{n\to\infty}E(g_n),$$

it follows from the generalized bounded convergence theorem, Theorem 1.4.1, that

$$E[\mu(t,N(t);w_1,w_2,\cdots,w_{N(t)})\,|\,N(t)]$$

$$= \lim_{n\to\infty}E[f_n\,|\,N(t)] = \lim_{n\to\infty}\frac{1}{\rho_n}\Pr[N(t,t+\rho_n)\ge 1\,|\,N(t)],$$

which proves (6.8) for $N(t) \geq 1$. The proof of (6.9) for $N(t) \geq 1$ parallels the proof when $N(t) = 0$ and will not be repeated. $\square$

It is evident from Lemma 6.2.1 that to order of Δt, the conditional probabilities of zero, one, and more than on point in $[t, t + \Delta t)$ given the number of points in $[t_0, t)$, and not their occurrence times, are $1 - \hat{\lambda}(t, N(t))\Delta t$, $\hat{\lambda}(t, N(t))\Delta t$, and zero, respectively. As a consequence, we have the following theorem.

Theorem 6.2.1 (*Counting Statistics for a Self-Exciting Point Process*). Let $\{N(t): t \geq t_0\}$ be a self-exciting point process with intensity process $\{\lambda(t): t \geq t_0\}$ defined in (6.3). Assume that $\{N(t): t \geq t_0\}$ is conditionally orderly, $E[\lambda(t)] < \infty$, and $\Pr[N(t_0) = 0] = 1$. Define the count-conditional intensity process $\{\hat{\lambda}(t, N(t)): t \geq t_0\}$ as in (6.7). Then, the probability $\Pr[N(t) = n]$ that the number of points in $[t_0, t)$ is n satisfies

$$\frac{\partial \Pr[N(t) = 0]}{\partial t} = -\hat{\lambda}(t, 0)\Pr[N(t) = 0] \tag{6.10}$$

for $n = 0$ and

$$\frac{\partial \Pr[N(t) = n]}{\partial t} = -\hat{\lambda}(t, n)\Pr[N(t) = n] + \hat{\lambda}(t, n - 1)\Pr[N(t) = n - 1]$$

$$\tag{6.11}$$

for $n \geq 1$, with the initial condition $\Pr[N(t_0) = 0] = 1$.

Proof. Equation (6.10) is a consequence of (6.5) because $\hat{\lambda}(t, 0) = \mu(t, 0)$. We leave the derivation of (6.11) as an exercise as it closely parallels that for Theorem 2.2.1.; see Problem 6.2.1. $\square$

Comparison of (6.10) and (6.11) with the result of Problem 2.2.6(a) shows that the counting statistics for a self-exciting point process are identical with those of a pure birth process with a population dependent birth rate $\hat{\lambda}(t, n)$ that is the count-conditional expectation of the intensity of the self-exciting point process. Equation (6.11) can in principle be solved sequentially for the counting probabilities $\Pr[N(t) = 1]$, $\Pr[N(t) = 2]$, $\cdots$. For this purpose, note from (6.10) and (6.5) that

$$\Pr[N(t) = 0] = \exp\left[-\int_{t_0}^{t} \mu(\sigma, 0)d\sigma \right] \tag{6.12}$$

and from (6.11) that, for $n = 1, 2, \cdots$,

$$\Pr[N(t) = n] = \int_{t_0}^{t} \hat{\lambda}(\tau, n-1) \Pr[N(\tau) = n-1] \exp\left[-\int_{\tau}^{t} \hat{\lambda}(\sigma, n) d\sigma \right] d\tau.$$

$$(6.13)$$

Example 6.2.1 ―――――――――――――――――――――――――――――――――――――

Let $\{N(t): t \geq t_0\}$ be a self-exciting counting process with conditional infinitesimal occurrence probabilities defined according to $\mu(t, 0) = \lambda_2$ and

$$\mu(t, N(t); w_1, w_2, \cdots, w_{N(t)}) = \begin{cases} \lambda_1, & N(t) = 1, 3, 4, \cdots \quad \text{(odd)} \\ \lambda_2, & N(t) = 2, 4, 6, \cdots \quad \text{(even)}. \end{cases}$$

where $\lambda_1 > \lambda_2$. Thus, $\{\lambda(t): t \geq t_0\}$ is a two-state process as indicated in Fig. 6.2. From (6.7), the count-conditional intensity is given by

$$\hat{\lambda}(t, n) = \begin{cases} \lambda_1, & n = 1, 3, 5, \cdots \\ \lambda_2, & n = 2, 4, 6, \cdots. \end{cases}$$

The probability that the number of points at time t, $t \geq t_0$, is zero is from (6.12)

$$\Pr[N(t) = 0] = e^{\lambda_2 t}.$$

By substituting this expression into (6.13) and integrating, we obtain

$$\Pr[N(t) = 1] = \frac{\lambda_2}{\lambda_1 - \lambda_2}\left(e^{-\lambda_2 t} - e^{-\lambda_1 t} \right).$$

Similarly, by substituting this expression into (6.13) and integrating, we obtain

$$\Pr[N(t) = 2] = \frac{\lambda_1 \lambda_2}{\lambda_1 - \lambda_2} t e^{-\lambda_2 t} - \frac{\lambda_1 \lambda_2}{(\lambda_1 - \lambda_2)^2}\left(e^{-\lambda_2 t} - e^{-\lambda_1 t} \right).$$

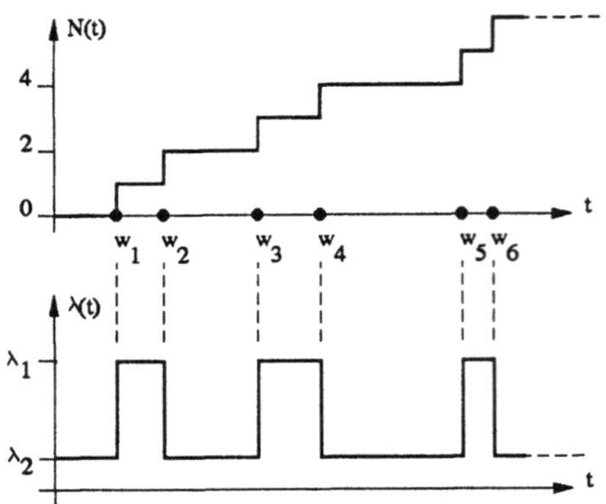

Figure 6.2 Typical sample functions for $\{N(t): t \geq t_0\}$ and $\{\lambda(t): t \geq t_0\}$.

This procedure can be continued to determine $\Pr[N(t) = n]$ for any $n \geq 0$. These counting probabilities are identical to those of a birth process in which the birth rate depends on whether n is even or odd in correspondence with $\hat{\lambda}(t, n)$. ∎

6.2.2 Time Statistics for Self-Exciting Point Processes

In this subsection, we derive various expressions associated with the statistics of the occurrence times of a self-exciting point process. The quantities of interest are the *survival probability*, the *joint occurrence density*, and the *sample-function density*. The last of these will be the most important.

Let $w_1, w_2, \cdots$ denote the occurrence times of a self-exciting point process with intensity process $\{\lambda(t): t \geq t_0\}$. Denote by

$$\mathcal{P}_{w_{n+1} | w_1, w_2, \cdots, w_n}(t \mid W_1, W_2, \cdots, W_n)$$

the conditional probability that $w_{n+1} \geq t$ given that $w_i = W_i$ for $i = 1, 2, \cdots, n$. If the occurrence times correspond to some anomalous behavior in a machine, then $\mathcal{P}$ is the conditional probability that the machine survives beyond time t before the (n+1)*st* anomaly occurs, given that previous anomalies occurred at times $W_1, W_2, \cdots, W_n$. Motivated by this, $\mathcal{P}$ is called the *conditional survival probability* for the (n+1)*st* point.

Because of the identity of the events $\{N(t) = 0\}$ and $\{w_1 \geq t\}$, the survival probability for the first occurrence time is easily determined from (6.12) to be

$$\mathcal{P}_{w_1}(t) = \Pr(w_1 \geq t) = \exp\left[-\int_{t_0}^{t} \mu(\sigma, 0) d\sigma \right]. \tag{6.14}$$

More generally, the expression for the conditional survival probability for the (n+1)st point is given for $t \geq W_n$ and $n \geq 1$ by

$$\mathcal{P}_{w_{n+1}|w_1,w_2,\cdots,w_n}(t \mid W_1, W_2, \cdots, W_n)$$

$$= \exp\left[-\int_{W_n}^{t} \mu(\sigma,n;W_1,W_2,\cdots,W_n)d\sigma\right]. \tag{6.15}$$

This equation can be established by straightforward manipulation as follows. Write the conditional probability of the (n+1)st point being in the interval $[t, t+\Delta t)$ in two ways. First,

$$\Pr(t \leq w_{n+1} < t + \Delta t \mid N(t) = n; w_1 = W_1, \cdots, w_n = W_n)$$

$$= \Pr[N(t,t+\Delta t) = 1 \mid N(t) = n; w_1 = W_1, \cdots, w_n = W_n]$$

$$= \mu(t,n;W_1,\cdots,W_n)\Delta t + o(\Delta t).$$

Secondly, for $t \geq W_n$,

$$\Pr(t \leq w_{n+1} < t + \Delta t \mid N(t) = n; w_1 = W_1, \cdots, w_n = W_n)$$

$$= \frac{\Pr(t \leq w_{n+1} < t + \Delta t \mid w_1 = W_1, \cdots, w_n = W_n)}{\Pr(w_{n+1} \geq t \mid w_1 = W_1, \cdots, w_n = W_n)},$$

where we have used the identity between the events $\{N(t) = n\}$ and $\{w_{n+1} \geq t\}$ for $t \geq W_n$. Equating these expressions, dividing by Δt, and taking the limit as Δt tends to zero results in

$$\mu(t,n;W_1,\cdots,W_n) = \frac{p_{w_{n+1}|w_1,\cdots,w_n}(t \mid W_1,\cdots,W_n)}{\mathcal{P}_{w_{n+1}|w_1,\cdots,w_n}(t \mid W_1,\cdots,W_n)}, \tag{6.16}$$

where the numerator on the right side is the conditional density for the (n+1)st occurrence time. Thus, we deduce the following relation between the survival probability and the intensity

$$\mu(t,n;W_1,\cdots,W_n) = -\frac{\partial \ln \mathcal{P}_{w_{n+1}|w_1,\cdots,w_n}(t \mid W_1,\cdots,W_n)}{\partial t}, \tag{6.17}$$

from which (6.15) follows.

Example 6.2.2 *Dead Time in Photon Detectors* ————————————
Practical devices used for detecting gamma photons from a radioactive source have the undesirable property that they become inoperative for a brief period of time, which can be deterministic or random, following each detected photon. During this *dead time*, any incident photons go undetected. If incident photons arriving during a dead time further extend the inoperative period, the detector is said to be *paralyzable*; otherwise it is *nonparalyzable*.

Consider a nonparalyzable detector in which the successive dead times $\tau_1, \tau_2, \cdots$ are independent and identically distributed as a nonnegative random variable τ having a probability distribution function $P_\tau(T) = \Pr(\tau \le T)$. With reference to Fig. 6.3, let the incident point process be an inhomogeneous Poisson process with intensity $\{v(t): t \ge t_0\}$. Let $N(t)$ be the number of detected points on $[t_0, t)$, and let $w_1, w_2, \cdots, w_{N(t)}$ be their occurrence times.

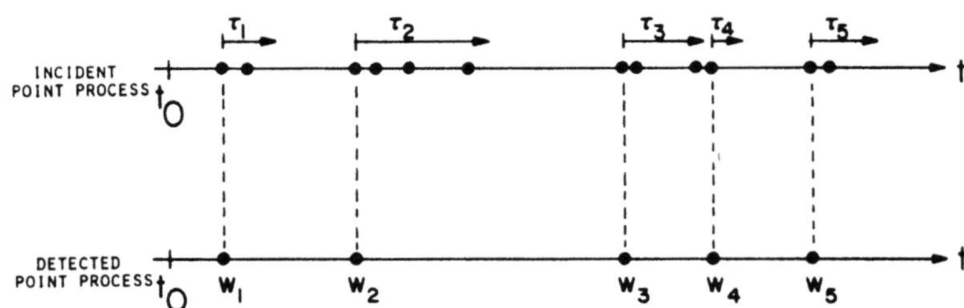

Figure 6.3 Typical realization for an incident and detected point process in a nonparalyzable detector.

Clearly, $\{N(t): t \ge t_0\}$ is a self-exciting counting process. Its intensity process $\{\lambda(t): t \ge t_0\}$ can be determined from the conditional probability of there being a detected point in the infinitesimal interval $[t, t + \Delta t)$ given the point process on $[t_0, t)$. If the detector is initially not in a dead time and, therefore, free to detect an incident point, we obtain $\mu(t, 0) = v(t)$. Similarly, for $n \ge 1$, we obtain

$$\mu(t, n; W_1, W_2, \cdots, W_n) = v(t)P_\tau(t - W_n). \tag{6.18}$$

The intensity process is then determined using (6.3), (6.17), and (6.18). In particular, for a detector with a fixed dead time t_d,

$$P_\tau(T) = \begin{cases} 0, & T < t_d \\ 1, & T \ge t_d \end{cases}$$

and

$$\mu(t,n;W_1,W_2,\cdots,W_n)=\begin{cases} 0, & W_n \leq t < W_n+t_d \\ \nu(t), & W_n+t_d \leq t. \end{cases} \qquad (6.19)$$

The intensity process $\{\lambda(t):t \geq t_0\}$, therefore, switches between 0 and the intensity of the incident process $\nu(t)$ depending on whether or not the detector is in a dead time or not, respectively.

From (6.14) and (6.15), the survival probability for the nth point is

$$\mathcal{P}_{w_1}(t)=\exp\left(-\int_{t_0}^{t}\nu(\sigma)d\sigma\right) \qquad (6.20)$$

for $n=1$ and

$$\mathcal{P}_{w_{n+1}|w_1,\cdots,w_n}(t\mid W_1,\cdots,W_n)=\exp\left[-\int_{W_n}^{t}\nu(\sigma)P_\tau(\sigma-W_n)d\sigma\right] \qquad (6.21)$$

for $n \geq 2$ and $t \geq W_n$. $\blacksquare$

Denote by $p_w^{(n)}(\mathbf{W})$ the joint probability density for the first n occurrence times. This *joint occurrence density* for a self-exciting point process is given by

$$p_{w_1}^{(1)}(W_1)=\mu(W_1,0)\exp\left[-\int_{t_0}^{W_1}\mu(\sigma,0)d\sigma\right] \qquad (6.22)$$

for $n=1$ and

$$p_w^{(n)}(\mathbf{W})=\mu(W_1,0)\left\{\prod_{i=2}^{n}\mu(W_i,i-1;W_1,\cdots,W_{i-1})\right\} \qquad (6.23)$$

$$\times\exp\left[-\int_{t_0}^{W_1}\mu(\sigma,0)d\sigma-\sum_{i=2}^{n}\int_{W_{i-1}}^{W_i}\mu(\sigma,i-1;W_1,\cdots,W_{i-1})d\sigma\right]$$

for $n \geq 2$ and $t_0 \leq W_1 < \cdots = W_n$. For the proof of (6.22), simply differentiate (6.14) with respect to t. For the proof of (6.23), use the relation

$$p_w^{(n)}(\mathbf{W})=p_{w_1}(W_1)\prod_{i=2}^{n}p_{w_i|w_1,\cdots,w_{i-1}}(W_i\mid W_1,\cdots,W_{i-1}).$$

The densities on the right side can be determined by differentiation of (6.14) and (6.15) with respect to t , and then (6.23) follows.

The *sample-function density* for a self-exciting point process plays a central role in decision and estimation problems in which the entire realization of the process is observed as data. We now derive and give some applications for this important quantity.

The sample-function density for a self-exciting point process is the joint probability of the number of points in $[t_0, t)$ and the density of their occurrence times. It is denoted by

$$p[\{N(t): t_0 \le \sigma < T\}] = \begin{cases} \Pr[N(T) = 0], & N(T) = 0 \\ p_{\mathbf{w}}[\mathbf{W}, N(T) = n], & N(T) = n \ge 1 \end{cases} \tag{6.24}$$

where

$$p_{\mathbf{w}}[\mathbf{W}, N(T) = n] = \Pr[N(T) = n \mid w_1 = W_1, \cdots, w_n = W_n] p_{\mathbf{w}}^{(n)}(\mathbf{W}).$$

The counting probability on the right side can be expressed in terms of the conditional survival probability of the $(n+1)st$ point as

$$\Pr[N(T) = n \mid w_1 = W_1, \cdots, w_n = W_n] = \mathcal{P}_{w_{n+1} \mid w_1, \cdots, w_n}(T \mid W_1, \cdots, W_n)$$

for $n \ge 1$. Thus, on combining (6.15), (6.22), and (6.23) in (6.24), we obtain

$$p[\{N(\sigma): t_0 \le \sigma < T\}]$$

$$= \begin{cases} \exp\left[-\int_{t_0}^{T} \mu(\sigma, 0) d\sigma \right], & N(T) = 0 \\ \mu(W_1, 0) \exp\left[-\int_{t_0}^{W_1} \mu(\sigma, 0) d\sigma - \int_{W_1}^{T} \mu(\sigma, W_1) d\sigma \right], & N(T) = 1 \end{cases}$$

$$\tag{6.25}$$

for $N(T) = 0$ and $N(T) = 1$ and by

$$p[\{N(\sigma): t_0 \leq \sigma < T\}]$$

$$= \mu(W_1, 0)\left\{\prod_{i=2}^{n} \mu(W_i, i-1; W_1, \cdots, W_{i-1})\right\}$$

$$\times \exp\left[-\int_{t_0}^{W_1} \mu(\sigma, 0)d\sigma - \sum_{i=2}^{n} \int_{W_{i-1}}^{W_i} \mu(\sigma, i-1; W_1, \cdots, W_{i-1})d\sigma\right.$$

$$\left. - \int_{W_n}^{T} \mu(\sigma, n; W_1, \cdots, W_n)d\sigma\right] \tag{6.26}$$

for $n \geq 2$ and $t_0 \leq W_1 < W_2 < \cdots < W_n < T$. This result can be placed in a more compact and suggestive form by incorporating into (6.25) and (6.26) the intensity process $\{\lambda(t): t \geq t_0\}$ defined in (6.3) and the counting integrals of Ch. 5. Note first that

$$\int_{t_0}^{T} \lambda(\sigma)d\sigma = \begin{cases} \displaystyle\int_{t_0}^{T} \mu(\sigma, 0)d\sigma, & N(T) = 0 \\[3em] \displaystyle\int_{t_0}^{w_1} \mu(\sigma, 0)d\sigma + \int_{w_1}^{T} \mu(\sigma, 1; w_1)d\sigma, & N(T) = 1 \end{cases} \tag{6.27a}$$

and

$$\int_{t_0}^{T} \lambda(\sigma)d\sigma = \int_{t_0}^{w_1} \mu(\sigma, 0)d\sigma$$

$$+ \sum_{i=2}^{N(T)} \int_{w_{i-1}}^{w_i} \mu(\sigma, i-1; w_1, \cdots, w_{i-1})d\sigma \tag{6.27b}$$

$$+ \int_{w_{N(T)}}^{T} \mu(\sigma, N(T); w_1, \cdots, w_{N(T)})d\sigma, \qquad N(T) \geq 2.$$

Similarly, for the counting integrals defined in Sec. 5.3.1,

$$\int_{t_0}^{T} \ln \lambda(\sigma) \, dN(\sigma)$$

$$= \begin{cases} 0, & N(T) = 0 \\ \ln \mu(w_1, 0) & N(T) = 1 \\ \ln \mu(w_1, 0) + \sum_{i=2}^{N(T)} \ln \mu(w_i, i-1; w_1, \cdots, w_{i-1}), & N(T) \geq 2. \end{cases}$$

$$(6.28)$$

By combining (6.25) to (6.28), we conclude that the sample-function density can be expressed in terms of the counting process $\{N(t): t \geq t_0\}$ and its intensity process $\{\lambda(t): t \geq t_0\}$ as in the following theorem.

Theorem 6.2.2 (*Sample-Function Density for a Self-Exciting Point Process*). Let $\{N(t): t \geq t_0\}$ be a self-exciting counting process with an intensity process $\{\lambda(t): t \geq t_0\}$. The sample-function density for $\{N(\sigma): t_0 \leq \sigma < T\}$ is

$$p[\{N(\sigma): t_0 \leq \sigma < T\}] = \exp\left[-\int_{t_0}^{T} \lambda(\sigma) \, d\sigma + \int_{t_0}^{T} \ln \lambda(\sigma) \, dN(\sigma)\right]. \quad (6.29)$$

When $\lambda(\cdot)$ is a deterministic function, $\{N(t): t \geq t_0\}$ is a Poisson counting process. In this case, (6.29) and (2.32) are identical, as expected. More generally, it is interesting to note that when $\lambda(\cdot)$ depends on the counting path, so that $N(\cdot)$ is self exciting, the form of the sample-function density remains unchanged. The use of (6.29) in estimation and decision problems involving observed self-exciting point processes is precisely the same as the use of (2.32) for observed Poisson processes. This is outlined in the following examples.

Example 6.2.3 *Parameter Estimation* ────────────────
The problem of estimating parameters arises frequently when self-exciting point processes are used as models for observed data. Let $\{N(t): t_0 \leq t < T\}$ be a self-exciting point process with an intensity process $\{\lambda(t, \mathbf{X}): t \geq t_0\}$ that is a function of a set of parameters $\mathbf{X} \in \mathcal{X}$. The log-likelihood function $\mathcal{L}(\mathbf{X})$ is the logarithm of the sample-function density in (6.29),

$$\mathcal{L}(\mathbf{X}) = -\int_{t_0}^{T} \lambda(\sigma, \mathbf{X}) \, d\sigma + \int_{t_0}^{T} \ln \lambda(\sigma, \mathbf{X}) \, dN(\sigma), \quad (6.30)$$

where $\{\lambda(t, \mathbf{X}): t \geq t_0\}$ is that sample function of the intensity process corresponding to the observed data $\{N(t): t_0 \leq t < T\}$. A maximum-likelihood estimate $\hat{\mathbf{X}}_{\mathrm{ML}}$ of $\mathbf{X}$ in terms of the data is a value of $\mathbf{X} \in X$ that maximizes $\mathcal{L}(\mathbf{X})$. A particular application of this principle is in the next example. $\blacksquare$

Example 6.2.4 *Correction for Dead Time* ————————————

Suppose that a homogeneous Poisson process having a constant intensity ν is incident on a detector having a fixed dead time t_d. Here, ν is an unknown, nonnegative parameter to be estimated in terms of the point process observed at the output of the detector on the interval $[0, T)$. Let $\{N(t): t \geq 0\}$ be the counting process associated with the output point process. This is a self-exciting point process for which $\mu(t, 0) = \nu$ and, from (6.19),

$$\mu(t, n; W_1, \cdots, W_n) = \begin{cases} 0, & W_n \leq t < W_n + t_d \\ \nu, & W_n + t_d \leq t. \end{cases}$$

The intensity process $\{\lambda(t, \nu): t \geq 0\}$ is a two-state process having an instantaneous jump from ν to zero following each detected point and then returning instantaneously to ν after t_d seconds. From (6.30), the loglikelihood function $\mathcal{L}(\nu)$ for estimating ν in terms of an observed counting path with $N(T) = n$, $w_1 = W_1, \cdots, w_n = W_n$ is given by

$$\mathcal{L}(\nu) = \begin{cases} -\nu[W_n - (n-1)t_d] + n \ln \nu, & W_n \leq T < W_n + t_d \\ -\nu[T - nt_d] + n \ln \nu, & W_n + t_d \leq T. \end{cases} \tag{6.31}$$

By setting $\partial \mathcal{L}(\nu)/\partial \nu$ equal to zero, we determine that

$$\hat{\nu}_{\mathrm{ML}} = \begin{cases} \bar{n}\left[\dfrac{W_n + t_d}{T} - \bar{n}t_d\right]^{-1}, & W_n \leq T < W_n + t_d \\ \bar{n}[1 - \bar{n}t_d]^{-1}, & W_n + t_d \leq T, \end{cases} \tag{6.32}$$

where $\bar{n} = n/T$ is the time-average rate at which points are observed. If $\bar{n}T \gg 1$, the end effect in (6.32) can be neglected so that

$$\hat{\nu}_{\mathrm{ML}} = \frac{\bar{n}}{1 - \bar{n}t_d}. \tag{6.33}$$

The factor $[1 - \overline{n}t_d]^{-1}$ can be regarded as the correction to be applied to the average rate at which points are observed in order to obtain the true maximum-likelihood estimate of the average rate of incident points. We note, finally, that the entire counting path is not needed to form this estimate; only the total number of detected points n is required. If end effects cannot be neglected, then the last occurrence time W_n is also needed so that (6.32) can be used. ∎

6.2.3 A Limit Theorem for Sums of Self-Exciting Point Processes

In this subsection, we briefly examine the situation indicated in Fig. 6.4 in which a point process is obtained by pooling the points of many independent point processes. The independent components may, in general, be self-exciting and inhomogeneous. A theorem is given indicating that if the component processes are sufficiently sparse so that no one of them dominates the contributions to the pooled process, then the pooled process is approximately a Poisson process. This result explains, in part, the relative frequency with which the Poisson process is found to be a reasonable model in applications. The theorem may be viewed as the analog for point processes of the central limit theorem for sums of random variables.

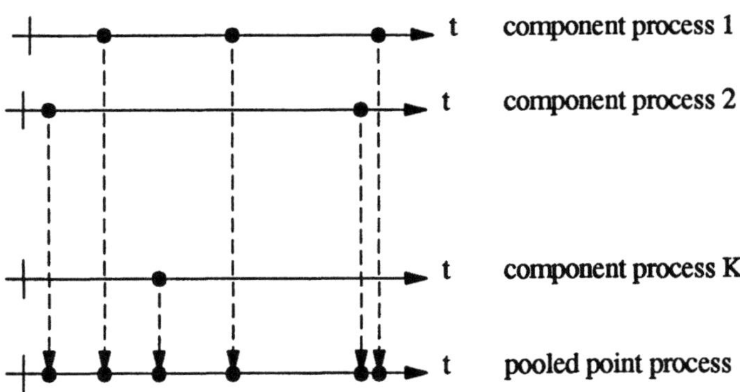

Figure 6.4 A pooled point-process.

Let $\{N_{i,k}(t): t \geq t_0\}$ be the counting process associated with the ith component process, for $i = 1, 2, \cdots, k$. Denote by $\{M_k(t): t \geq t_0\}$ the counting process of the pooled point process,

$$M_k(t) = \sum_{i=1}^{k} N_{i,k}(t). \tag{6.34}$$

We will say that the component processes are *uniformly sparse* if for t finite there holds

$$\lim_{k \to \infty} \sup_{1 \le i \le k} \Pr[N_{i,k}(t) \ge 1] = 0.$$

The interpretation of this expression is that the probability of any component contributing a point to the pooled process tends to zero on any finite interval as the number of components tends to infinity.

Under relatively weak additional conditions, $\{M_k(t): t \ge t_0\}$ tends to a Poisson process as k tends to infinity. The principal result is contained in the following theorem, due to P. Franken [10] and B. Grigelinois [14], which we state without proof.

Theorem 6.2.3 (*Limit Theorem for Pooled Point Processes*). Suppose that the component counting processes $\{N_{i,k}(t): t \ge t_0\}$ for $i = 1, 2, \cdots, k$ are mutually independent and uniformly sparse. Then, the pooled counting process $\{M_k(t): t \ge t_0\}$ converges in distribution to a Poisson process with intensity $\{\lambda(t): t \ge t_0\}$ if and only if both

$$\lim_{k \to \infty} \sum_{i=1}^{k} \Pr[N_{i,k}(t) > 1] = 0 \quad \text{and} \quad \lim_{k \to \infty} \sum_{i=1}^{k} \Pr[N_{i,k}(t) = 1] = \int_{t_0}^{t} \lambda(\sigma) d\sigma$$

for $t_0 \le t < \infty$.

This theorem is proven and discussed in a comprehensive review of limit theorems for pooled point processes by E. Çinlar [4].

Summary. Statistical characterizations for general self-exciting point processes were developed in this section. The following important observations were made:

1. The counting statistics for a self-exciting point process are identical to those of a pure birth process having a birth rate equal to the count-conditional intensity of the self-exciting point process.

2. The sample-function density for a self-exciting point process has the same form as that for an inhomogeneous Poisson process.

3. A point process obtained by pooling the points of independent and sparse self-exciting point processes converges under weak conditions to a Poisson process.

6.3 Self-Exciting Point Processes with Limited Memory

The future evolution of the self-exciting point processes introduced in Sec. 6.2 do, in general, depend on the occurrence times of all past points as well as their number. This is overly complicated for many applications because it often happens that only a few of the most recent points influence the future evolution. In this section, a class of self-exciting point processes with limited memory is introduced. These *m-memory self-exciting point processes* are defined as follows.

Definition (*m-memory self-exciting point processes*). A self-exciting point process is *0-memory* if for all $N(t) > 1$, the function $\mu(t, N(t);$ $w_1, w_2, \cdots, w_{N(t)})$ of (6.2) is not a function of any of the occurrence times $w_1, w_2, \cdots, w_{N(t)}$. It is *m-memory*, for $m = 1, 2, \cdots$, if for all $N(t) \geq 1$, $\mu(t, N(t); w_1, w_2, \cdots, w_{N(t)})$ depends only on t, $N(t)$, and the m most recent occurrence times $w_{N(t)-m+1}, \cdots, w_{N(t)-1}, w_{N(t)}$ and is not a function of the $N(t) - m$ earlier times.

Characterizations due to I. Rubin [31, 32] are given in the following subsections for these m-memory self-exciting point processes.

6.3.1 Characterization of m-Memory Point Processes for $m = 0$

Zero-memory self-exciting point processes are characterized in the following theorem.

Theorem 6.3.1 (*Characterization of 0-Memory Self-Exciting Point Processes*) Let $\{N(t): t \geq t_0\}$ be a self-exciting counting process satisfying the conditions of Theorem 6.2.1. Then, the underlying point process is 0-memory if and only if $\{N(t): t \geq t_0\}$ is a Markov process.

Proof. Suppose, first, that $\{N(t): t \geq t_0\}$ is a Markov process. As the events generated by the random variables $\{N(t); w_1, \cdots, w_{N(t)}\}$ are the same as those generated by the random variables $\{N(\sigma): t_0 \leq \sigma < t\}$, we have

$$\Pr[N(t, t+\Delta t) = 1 | N(t); w_1, \cdots, w_{N(t)}] = \Pr[N(t, t+\Delta t) = 1 | N(\sigma): t_0 \leq \sigma < t]$$

$$= \Pr[N(t, t+\Delta t) = 1 | N(t)], \tag{6.35}$$

where the second equality follows because of the Markov hypothesis for $N(\cdot)$. Thus,

$$\lim_{\Delta t \downarrow 0} \frac{1}{\Delta t} \Pr[N(t,t+\Delta t)=1 \mid N(t); w_1, \cdots, w_{N(t)}]$$

$$= \lim_{\Delta t \downarrow 0} \frac{1}{\Delta t} \Pr[N(t,t+\Delta t)=1 \mid N(t)] = \mu(t,N(t)), \tag{6.36}$$

and the point process is 0-memory. Conversely, suppose that the point process is 0-memory so that (6.36) holds. We need to show that for any $s \in [t_0,t)$ there holds

$$\Pr[N(t)=n \mid N(\sigma): t_0 \leq \sigma < s] = \Pr[N(t)=n \mid N(s)]. \tag{6.37}$$

We establish this equation by showing first that the left and right sides of it satisfy the same linear differential-difference equation and then appealing to the uniqueness of the solution for such equations. For the left side, note for $t_0 \leq s \leq t$ that

$$\lim_{\Delta t \downarrow 0} \frac{1}{\Delta t} \Pr[N(t,t+\Delta t) \geq 1 \mid N(t), N(\sigma): t_0 \leq \sigma < s]$$

$$= \lim_{\Delta t \downarrow 0} \frac{1}{\Delta t} E\{\Pr[N(t,t+\Delta t) \geq 1 \mid N(\sigma): t_0 \leq \sigma < t] \mid N(t), N(\sigma): t_0 \leq \sigma < s\}$$

$$= E\left\{ \lim_{\Delta t \downarrow 0} \frac{1}{\Delta t} \Pr[N(t,t+\Delta t) \geq 1 \mid N(\sigma): t_0 \leq \sigma < t] \mid N(t), N(\sigma): t_0 \leq \sigma < s \right\}$$

$$= E\{\mu(t,N(t)) \mid N(t), N(\sigma): t_0 \leq \sigma < s\} = \mu(t,N(t)), \tag{6.38}$$

where the interchange of the limit and expectation can be justified using the generalized bounded convergence theorem, Theorem 1.4.1, in the same manner as in the proof of Lemma 6.2.1. It also follows from (6.38) and the conditional orderliness of $\{N(t): t \geq t_0\}$ that

$$\lim_{\Delta t \downarrow 0} \frac{1}{\Delta t} \Pr[N(t,t+\Delta t)=1 \mid N(t), N(\sigma): t_0 \leq \sigma < s] = \mu(t,N(t)). \tag{6.39}$$

Thus, we conclude that

$$\Pr[N(t,t+\Delta t)=0 \mid N(t)=n, N(\sigma): t_0 \leq \sigma < s] = 1 - \mu(t,n)\Delta t + o(\Delta t),$$

$$\tag{6.40}$$

$$\Pr[N(t,t+\Delta t)=1|N(t)=n,N(\sigma):t_0\le\sigma<s]=\mu(t,n)\Delta t+o(\Delta t),$$

$$(6.41)$$

and

$$\Pr[N(t,t+\Delta t)>1|N(t)=n,N(\sigma):t_0\le\sigma<s]=o(\Delta t). \qquad (6.42)$$

These conditional, incremental occurrence probabilities can be used in the manner of deriving (6.10) and (6.11) to obtain

$$\frac{\partial\Pr[N(t)=0|N(\sigma):t_0\le\sigma<s]}{\partial t}=-\mu(t,0)\Pr[N(t)=0|N(\sigma):t_0\le\sigma<s]$$

$$(6.43)$$

for $n=0$ and

$$\frac{\partial\Pr[N(t)=n|N(\sigma):t_0\le\sigma<s]}{\partial t}=-\mu(t,n)\Pr[N(t)=n|N(\sigma):t_0\le\sigma<s]$$

$$+\mu(t,n-1)\Pr[N(t)=n-1|N(\sigma):t_0\le\sigma<s]$$

$$(6.44)$$

for $n\ge1$. Now, consider the right side of (6.37), and note that

$$\Pr[N(t)=n|N(s)]=E\{\Pr[N(t)=n|N(\sigma):t_0\le\sigma<s]|N(s)\}. \qquad (6.45)$$

Thus, by taking the conditional expectation of both sides of (6.43) and (6.44) given $N(s)$, we see that

$$\frac{\partial\Pr[N(t)=0|N(s)]}{\partial t}=-\mu(t,0)\Pr[N(t)=0|N(s)] \qquad (6.46)$$

and, for $n\ge1$,

$$\frac{\partial\Pr[N(t)=n|N(s)]}{\partial t} \qquad (6.47)$$

$$=-\mu(t,n)\Pr[N(t)=n|N(s)]+\mu(t,n-1)\Pr[N(t)=n-1|N(s)].$$

At $t = s$, we have the boundary condition $\Pr[N(s) = n \mid N(\sigma): t_0 \le \sigma < s] = \Pr[N(s) = n \mid N(s)]$. Therefore, the left and right sides of (6.37) satisfy the same linear differential-difference equation and boundary condition at $t = s$ for each n. On the basis of the uniqueness of the solution to such equations, we conclude that (6.37) holds. $\square$

It is evident for a 0-memory self-exciting point process that the count-conditional intensity $\hat{\lambda}(t, n)$ of (6.7) equals $\mu(t, n)$ of (6.36) for $n = 0, 1, 2, \cdots$. Examination of Theorems 6.2.1 and 6.3.1 indicates that the associated counting process $\{N(t): t \ge t_0\}$ is a Markov birth process with birth rate $\mu(t, n)$. It is possible for $N(\cdot)$ to be unstable for certain birthrates that depend nonlinearly on the population size. The instability manifests itself in the form of a population explosion in which $N(t)$ becomes infinite for t finite. This phenomenon occurs in the following example.

Example 6.3.1 *A Population Explosion* ———————————————————
Let $\{N(t): t \ge t_0\}$ be a Markov birth process with a birth rate $\{\mu(t, N(t)): t \ge t_0\}$ that depends quadratically on the population as $\mu(t, N(t)) = [N(t) + 1]^2$ for $N(t) = 1, 2, \cdots$. Let $P_n(t) = \Pr[N(t) = n]$ denote the probability that the population is of size n at time t, and assume that $P_n(0) = 1$ for $n = 0$ and $P_n(0) = 0$ for $n \ge 1$. From (6.10) and (6.11), $P_n(t)$ satisfies

$$\frac{\partial P_0(t)}{\partial t} = -P_0(t), \quad P_0(0) = 1 \tag{6.48}$$

for $n = 0$ and

$$\frac{\partial P_n(t)}{\partial t} = -(n+1)^2 P_n(t) + n^2 P_{n-1}(t), \quad P_n(0) = 0 \tag{6.49}$$

for $n \ge 1$. These equations can be solved, for example, by using Laplace transforms. For this, let $\tilde{P}_n(s)$ be the Laplace transform of $P_n(t)$. Transforming (6.48) and (6.49) yields

$$s\tilde{P}_0(s) - 1 = -\tilde{P}_0(s) \tag{6.50}$$

and

$$s\tilde{P}_n(s) = -(n+1)^2 \tilde{P}_n(s) + n^2 \tilde{P}_{n-1}(s), \tag{6.51}$$

respectively. It follows that

$$\tilde{P}_n(s) = \frac{1}{(n+1)^2} \prod_{m=1}^{n+1} \frac{m^2}{s+m^2} \tag{6.52}$$

for $n = 0, 1, 2, \cdots$. Define $S_K(t)$ as the probability that the population is at most of size K at time t,

$$S_K(t) = \Pr[N(t) < K+1] = \sum_{n=0}^{K} P_n(t). \tag{6.53}$$

By summing (6.48) and (6.49), it is seen that

$$\frac{\partial S_K(t)}{\partial t} = -(K+1)^2 P_K(t), \quad S_K(0) = 1. \tag{6.54}$$

Thus, the Laplace transform $\tilde{S}_K(s)$ of $S_K(t)$ is given by

$$\tilde{S}_K(s) = \frac{1}{s} - \frac{1}{s}(K+1)^2 \tilde{P}_K(s) = \frac{1}{s} - \frac{1}{s}\prod_{m=1}^{K+1} \frac{m^2}{s+m^2}, \tag{6.55}$$

where we used (6.52) to obtain the second equality. By expanding the product in a partial fraction expansion, we have the alternative expression

$$\tilde{S}_K(s) = \sum_{i=1}^{K+1} c_i(K) \frac{1}{s+i^2}, \tag{6.56}$$

where

$$c_i(K) = \lim_{s \to -i^2} (s+i^2) \frac{1}{s} \prod_{m=1}^{K+1} \frac{m^2}{s+m^2} = \prod_{\substack{m=1 \\ m \neq i}} \frac{m^2}{m^2 - i^2}. \tag{6.57}$$

In the limit $K \to \infty$, it can be verified that $c_i(\infty) = 2(-1)^{i+1}$. Thus, we conclude upon inverse Laplace transforming this equation and letting $K \to \infty$ that

$$\Pr[N(t) < \infty] = -2 \sum_{i=1}^{\infty} (-1)^i e^{-i^2 t}. \tag{6.58}$$

This expression is related to the theta function, which is tabulated. A graph of this probability as a function of t is shown in Fig. 6.5. The probability that the population is of finite size rapidly approaches zero indicating a population explosion. ∎

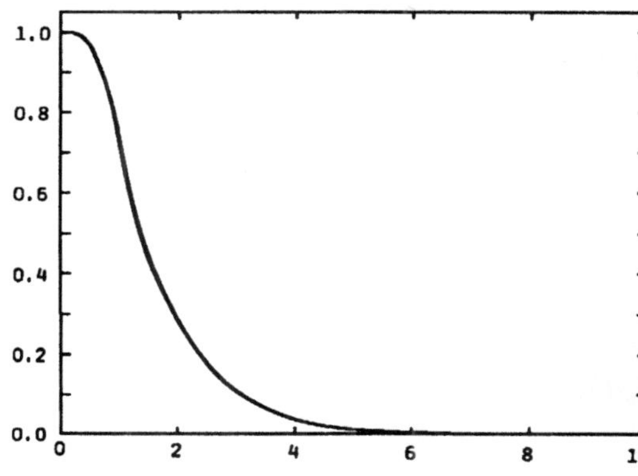

Figure 6.5 Probability of a population of finite size, $\Pr[N(t) < \infty]$ versus t.

The population explosion seen in Ex. 6.3.1 can be prevented by suitably restricting the birth rate according to the following theorem.

Theorem 6.3.2 (*Conditions for Nonexploding 0-Memory Point Processes*) Let $\{N(t): t \geq t_0\}$ be a Markov birth process with rate $\{\mu(t, N(t): t \geq t_0\}$. If the sum

$$\sum_{n=0}^{\infty} \left[\sup_{t_0 \leq \sigma < t} \mu(\sigma, n) \right]^{-1} \tag{6.59}$$

is infinite, then for all finite $t \geq t_0$ there holds

$$\Pr[N(t) < \infty] = \sum_{n=0}^{\infty} \Pr[N(t) = n] = 1, \tag{6.60}$$

and there is no population explosion with probability one.

Proof. We parallel the proof of W. Feller [8, p. 452] for the homogeneous situation when $\mu(t, n)$ is not a function of t. As in Ex. 6.3.1, let $P_n(t) = \Pr[N(t) = n]$, and set

$$S_K(t) = \Pr[N(t) < K + 1] = \sum_{n=0}^{K} P_n(t). \tag{6.61}$$

Clearly, $1 - S_k(t)$ is a bounded, nonincreasing function of K, so the limit

$$\xi(t) = \lim_{K \to \infty} [1 - S_k(t)] \tag{6.62}$$

exists. Summing (6.10) and (6.11) over n shows that

$$\frac{\partial S_K(t)}{\partial t} = -\mu(t,K)P_K(t), \quad S_K(t_0) = 1. \tag{6.63}$$

Hence,

$$1 - S_K(t) = \int_{t_0}^{t} \mu(\sigma,K)P_K(\sigma)d\sigma \tag{6.64}$$

for all $K \geq 0$. As $\xi(t) \leq 1 - S_n(t)$, we have

$$\xi(t) \leq \int_{t_0}^{t} \mu(\sigma,n)P_n(\sigma)d\sigma \leq \left[\sup_{t_0 \leq \sigma < t} \mu(\sigma,n) \right] \int_{t_0}^{t} P_n(\sigma)d\sigma. \tag{6.65}$$

After multiplying by $[\sup \mu(\sigma,n)]^{-1}$, summing on n, and using $S_K(\sigma) \leq 1$ for $t_0 \leq \sigma < t$, we have

$$\xi(t) \sum_{n=0}^{K} \left[\sup_{t_0 \leq \sigma < t} \mu(\sigma,n) \right]^{-1} \leq \int_{t_0}^{t} S_K(\sigma)d\sigma = t - t_0. \tag{6.66}$$

Consequently, if the sum (6.59) diverges, then $\xi(t) = 0$ for all finite $t \geq t_0$, and in view of (6.62), this implies (6.60). $\square$

   Two additional remarks about Theorem 6.3.2 can be made. The first is that the condition (6.59) is also necessary as well as sufficient for there to be no explosions when the birth rate is not a function of time; see Problem 6.3.1. As the number of point in $[t_0, t)$ for a general self-exciting point process has the same distribution as if it were a Markov birth process with the birth rate equal to the count conditional intensity $\hat{\lambda}(n, N(t))$, it follows from Theorem 6.3.2 that a sufficient condition for the nonexistence of explosions in a general self-exciting point process is that the count-conditional intensity (6.7) be such that

$$\sum_{n=0}^{\infty} \left[\sup_{t_0 \leq \sigma < t} \hat{\lambda}(\sigma,n) \right]^{-1} \tag{6.67}$$

is infinite for all finite $t \geq t_0$.

Further information about Markov birth processes is given by A. Bharucha-Reid [1, Ch. 2], W. Feller [8, p. 448], and I. Gikhman and A. Skorokhod [12, Ch. 7].

6.3.2 Characterization of *m*-Memory Point Processes for $m > 0$

Self-exciting point processes having memory greater than zero are characterized in the next theorem.

Theorem 6.3.3 (*Characterization of m-Memory Self-Exciting Point Processes*). A self-exciting point process is m-memory for $m = 1, 2, \cdots$ if and only if its occurrence times $\{w_n : n = 1, 2, \cdots\}$ form an m*th*-order Markov sequence.

Proof. Suppose, first, that the occurrence times do form an m*th*-order Markov sequence. Then, the survival probability for the (n+1)*st* point, where $n \geq m$, satisfies

$$\mathcal{P}_{w_{n+1}|w_1,\cdots,w_n}(t \mid W_1,\cdots,W_n) = \mathcal{P}_{w_{n+1}|w_{n-m+1},\cdots,w_n}(t \mid W_{n-m+1},\cdots,W_n). \qquad (6.68)$$

It follows from (6.17) that μ of (6.3) depends only on the m most recent occurrence times, and, therefore, the self-exciting point process is *m*-memory. Conversely, suppose that the point process is *m*-memory. We have by differentiating (6.15) that the conditional probability density for the (n+1)*st* waiting time is given in general by

$$p_{w_{n+1}|w_1,\cdots,w_n}(t \mid W_1,\cdots,W_n) \qquad (6.69)$$

$$= \mu(t,n:W_1,\cdots,W_n)\exp\left[-\int_{W_n}^{t} \mu(\sigma,n:W_1,\cdots,W_n)d\sigma\right].$$

Thus, for an *m*-memory process, we have for $n \geq m$

$$p_{w_{n+1}|w_1,\cdots,w_n}(t \mid W_1,\cdots,W_n) \qquad (6.70)$$

$$= \mu(t,n:W_{n-m+1},\cdots,W_n)\exp\left[-\int_{W_n}^{t} \mu(\sigma,n:W_{n-m+1},\cdots,W_n)d\sigma\right].$$

In view of this and the relation

$$p_{w_{n+1}|w_{n-m+1},\cdots,w_n}(t \mid W_{n-m+1}, \cdots, W_n)$$

$$= \int \cdots \int [p_{w_{n+1}|w_1,\cdots,w_n}(t \mid W_1, \cdots, W_n)$$

$$\times p_{w_1,\cdots,w_{n-m}|w_{n-m+1},\cdots,w_n}(W_1, \cdots, W_{n-m}\mid W_{n-m+1}, \cdots, W_n)]dW_1\cdots dW_{n-m},$$

it is clear that

$$p_{w_{n+1}|w_1,\cdots,w_n}(t\mid W_1,\cdots,W_n) = p_{w_{n+1}|w_{n-m+1},\cdots,w_n}(t\mid W_{n-m+1},\cdots,W_n)$$

for $n \geq m$. Consequently, the occurrence times $\{w_n\}$ form an mth-order Markov sequence. $\square$

The remainder of our discussion of m-memory processes will be for $m = 1$ and $t_0 = 0$. The future evolution of a 1-memory point process after a time t depends at most on t, the number of points $N(t)$ up to time t, and the most recent occurrence time $w_{N(t)}$. It is often hypothesized in applications that this dependence is not on t and $w_{N(t)}$ separately but only on their difference $t - w_{N(t)}$, so it is only the elapsed time since a point occurs that matters. These homogeneous, 1-memory processes are characterized in the following theorem and its corollary.

Theorem 6.3.4 (*Homogeneous 1-Memory Point Processes*). A self-exciting point process is 1-memory with

$$\mu(t,N(t);w_{N(t)}) = f(N(t),t - w_{N(t)}) \tag{6.71}$$

for some function $f(\cdot)$ if and only if the interarrival times $t_1 = w_1$, $t_2 = w_2 - w_1$, $\cdots$, $t_n = w_n - w_{n-1}$, $\cdots$ form a sequence of statistically independent random variables.

Proof. Suppose, first, that (6.71) holds. The exceedence probability for the nth interarrival time can be expressed in terms of the survival probability for the nth point as

$$\Pr[t_n \geq T_n \mid t_1 = T_1, \cdots, t_{n-1} = T_{n-1}] \tag{6.72}$$

$$= \mathcal{P}_{w_n|w_1,\cdots,w_{n-1}}(T_1 + \cdots + T_n\mid T_1, T_1 + T_2, \cdots, T_1 + T_2 + \cdots + T_{n-1})$$

But (6.71) implies that the process is 1-memory and, consequently, the occurrence times form a first-order Markov sequence by Theorem 6.3.3. Thus,

$$\Pr[t_n \geq T_n | t_1 = T_1, \cdots, t_{n-1} = T_{n-1}]$$

$$= \mathcal{P}_{w_n | w_{n-1}}(T_1 + \cdots + T_n | T_1 + \cdots + T_{n-1})$$

$$= \exp\left[- \int_{T_1 + \cdots + T_{n-1}}^{T_1 + \cdots + T_n} \mu(\sigma, n-1 : T_1 + \cdots + T_{n-1}) d\sigma \right], \qquad (6.73)$$

where the last equality follows from (6.70). By incorporating (6.71), we obtain

$$\Pr[t_n \geq T_n | t_1 = T_1, \cdots, t_{n-1} = T_{n-1}] = \exp\left[- \int_0^{T_n} f(n-1, \sigma) d\sigma \right]. \qquad (6.74)$$

It is clear from this and the relation

$$\Pr[t_n \geq T_n] = \int \cdots \int \Pr[t_n \geq T_n | t_1 = T_1, \cdots, t_{n-1} = T_{n-1}]$$

$$\times p_{t_1, \cdots, t_{n-1}}(T_1, \cdots, T_{n-1}) dT_1 \cdots dT_{n-1}$$

that

$$\Pr[t_n \geq T_n | t_1 = T_1, \cdots, t_{n-1} = T_{n-1}] = \Pr[t_n \geq T_n]. \qquad (6.75)$$

Thus, (6.71) implies that the interarrival times form an independent sequence. Suppose next that the interarrival times do form an independent sequence. We need to show that (6.71) holds. Equation (6.72) and the independence of the interarrival times imply that the conditional survival probability for the $(n+1)st$ point satisfies

$$\mathcal{P}_{w_{n+1} | w_1, \cdots, w_n}(t \mid W_1, \cdots, W_n) = \Pr[t_{n+1} \geq t - W_n]. \qquad (6.76)$$

By (6.17), we then have

$$\mu(t, n; W_1, \cdots, W_n) = - \frac{\partial \ln \Pr[t_{n+1} \geq t - W_n]}{\partial t}. \qquad (6.77)$$

The right side depends only on the nth occurrence time W_n and, therefore, the process in 1-memory. Furthermore, the dependence on t and W_n is only through their difference and, therefore, (6.71) holds with

$$f(n, t - W_n) = -\frac{\partial \ln \Pr[t_{n+1} \geq t - W_n]}{\partial t}. \tag{6.78}$$

This ends the proof of Theorem 6.3.4. $\square$

Renewal processes form a special but important class of homogeneous 1-memory self-exciting point processes. These are defined as follows.

Renewal Processes. A self-exciting point process with independent, identically distributed interarrivals is termed an *ordinary renewal process*. If all interarrivals but the first are identically distributed, it is termed a *modified renewal process*. A modified renewal process in which the first interarrival has the probability density

$$p_{t_1}(T) = \frac{1 - P_{t_n}(T)}{E(t_n)}, \tag{6.79}$$

where $P_{t_n}(T)$ is the common probability distribution function for the nth interarrival time, for $n > 1$, is termed an *equilibrium renewal process*.

Renewal processes are characterized in the following corollary to Theorem 6.3.4.

Corollary to Theorem 6.3.4. A self-exciting point process is a modified renewal process if and only if the function $\mu(\cdot)$ in (6.2) has the form

$$\mu(t, N(t); w_1, \cdots, w_{N(t)}) = h(t - w_{N(t)}) \tag{6.80}$$

for some function $h(\cdot)$. It is an ordinary renewal process if and only if in addition to (6.80), $\mu(t, 0) = h(t)$.

Proof. If (6.80) holds, the process is 1-memory, and by Theorem 6.3.4 it has independent interarrivals. By (6.14), the exceedence probability for the first interarrival, $t_1 = w_1$, is given by

$$\Pr[t_1 \geq T] = \exp\left(-\int_0^T \mu(\sigma, 0) d\sigma\right).$$

By (6.74) and (6.75), the exceedence probability for the nth interarrival t_n, for $n > 1$, is given by

$$\Pr(t_n \geq T) = \exp\left(-\int_0^T h(\sigma)d\sigma\right). \qquad (6.81)$$

Thus, interarrivals for $n > 1$ are identically distributed, and (6.80) implies that the process is a modified renewal process. It is an ordinary renewal process if in addition $\mu(t,0) = h(t)$. Conversely, suppose that the point process is a modified renewal process with the first interarrival t_1 having a probability density function $p_{t_1}(T)$ and all subsequent interarrival times identically distributed as a random variable t having a probability density function $p_t(T)$. The survival probability for the first point is

$$\mathcal{P}_{w_1}(T) = \Pr[t_1 \geq T] = \int_T^\infty p_{t_1}(\alpha)d\alpha.$$

Hence, from (6.14)

$$\mu(T,0) = \frac{p_{t_1}(T)}{1 - P_{t_1}(T)}, \qquad (6.82)$$

where $P_{t_1}(T) = \int_0^T p_{t_1}(\alpha)d\alpha$ is the probability distribution function for t_1. Similarly, the survival probability for the $(n+1)st$ point, $n > 1$, is given by

$$\mathcal{P}_{w_{n+1}|w_1,\cdots,w_n}(T \mid W_1,\cdots,W_n) = \Pr[t_{n+1} \geq T - W_n] = \int_{T-W_n}^\infty p_t(\alpha)d\alpha.$$

Hence, from (6.17)

$$\mu(T,n{:}W_1,\cdots,W_n) = \frac{p_t(T-W_n)}{1 - P_t(T-W_n)}, \qquad (6.83)$$

where $P_t(T)$ is the distribution function for t. Thus, $\mu(T,n{:}W_1,\cdots,W_n)$ has the form in (6.80) with

$$h(T) = \frac{p_t(T)}{1 - P_t(T)}. \qquad (6.84)$$

Furthermore, $\mu(T,0) = h(T)$ if $p_{t_1}(T) = p_t(T)$. $\quad\square$

The function $h(T)$ in (6.84) is called the *hazard function*. From (6.80) and the definition of μ in (6.2) that, to order Δt, the quantity $h(t-\tau)\Delta t$ is the conditional probability that the *nth* point occurs in the infinitesimal interval $[t, t+\Delta t)$ given that the (n-1)*st* point occurs at time τ. An easy calculation shows that the hazard function for a homogeneous Poisson process is a constant independent of T. Hazard functions are often estimated in applications to make a judgement about the appropriateness of a Poisson model; significant departures from a constant hazard may preclude the use of a homogeneous Poisson model and suggest employing a more complicated renewal model. The estimated hazard function can be used with (6.84) to form an estimate of the interarrival density needed in this renewal model.

The following observations may be useful in problems of parameter estimation for renewal processes. An ordinary renewal process is characterized by the common distribution of its interarrival times. It is not surprising, therefore, that the expressions required for parameter estimation for observed renewal processes can be put in a form involving only this distribution. To see how, suppose that $\{N(t): t \geq 0\}$ is a counting process associated with an ordinary renewal process having interarrival times with the probability density and distribution functions $p_t(T)$ and $P_t(T) = \int_0^T p_t(\sigma)d\sigma$, respectively. The hazard function is then

$$h(T) = \frac{p_t(T)}{1 - P_t(T)}. \qquad (6.85)$$

Note that

$$\int_\alpha^\beta h(\sigma)d\sigma = \ln\left\{\frac{1 - P_t(\alpha)}{1 - P_t(\beta)}\right\}. \qquad (6.86)$$

The intensity process defined in (6.3) has the form of the hazard function following each point and lasting until the next point. By using (6.29), (6.86), and some manipulations, it is seen that the sample-function density for an ordinary renewal process is given by

$$p[\{N(\sigma): t_0 \leq \sigma < T\}]$$

$$= \begin{cases} 1 - P_t(T), & N(T) = 0 \\ [1 - P_t(T)]h(W_1), & N(T) = 1 \\ [1 - P_t(T)]h(W_1)h(W_2 - W_1)\cdots h(W_n - W_{n-1}), & N(T) = n \geq 2 \end{cases}$$

$$(6.87)$$

As $h(T)$ can be expressed in terms of $P_t(T)$, so can the sample function density. If the interarrival distribution depends on a set of parameters $\mathbf{X}$ to be estimated from an observed path of $N(\cdot)$ on $[0, T)$, then the maximum-likelihood estimate of $\mathbf{X}$ maximizes the loglikelihood function

$$\mathcal{L}(\mathbf{X}) = \begin{cases} \ln\{1 - P_t(T{:}\mathbf{X})\}, & N(T) = 0 \\ \ln\{[1 - P_t(T{:}\mathbf{X})]h(W_1{:}\mathbf{X})\}, & N(T) = 1 \\ \ln\{[1 - P_t(T{:}\mathbf{X})]h(W_1{:}\mathbf{X})\cdots h(W_n - W_{n-1}{:}\mathbf{X})\} & N(T) = n \geq 2 \end{cases} \qquad (6.88)$$

subject to constraints on the possible values of $\mathbf{X}$.

Estimating a Hazard Function[1]

   While models for electrical discharges in the auditory nerve are most certainly of the point-process type, as described in the Ex's. 1.1.2 and 2.4.4, it is well established that a Poisson process is not an adequate description of actual neural firing patterns. Fundamental to all neural responses are effects due to the refractory properties of neurons as predicted by the basic Hodgkin-Huxley model for biological membranes. Immediately following a discharge, the probability of another discharge is zero because the membrane activation voltage is infinite; as the elapsed time following a discharge increases, the probability of another discharge increases due to a monotonic decrease in activation potential. There has been a substantial effort in auditory-neuroscience to develop models of auditory-nerve discharges that generalize the Poisson model with the goal of accounting for these refractory effects. Beginning with the work of W. Siebert and P. Gray [33] and P. Gray [13], and more recently R. Gaumond, C. Molnar, and D. Kim [11], there now exists a model of auditory-nerve discharges that accounts for these phenomena. In particular, the Siebert/Gaumond model describes an intensity process for auditory-nerve discharges as the product of two functions, a stimulus related function, denoted as $s(t)$, and a refractory related function, denoted as $r(\tau)$, and given according to

$$\begin{aligned} \mu(t, 0) &= s(t), & 0 \leq t \leq w_1 \\ \mu(t, N(t); w_0, w_1, \cdots, w_{N(t)}) &= s(t)r(t - w_{N(t)}), & w_{N(t)} < t \leq w_{N(t)+1}. \end{aligned} \qquad (6.89)$$

With this model, the conditional intensity of discharge at a time t is seen to be a function of the time $w_{N(t)}$ of the most recent discharge prior to t, so this is a 1-memory point process. In general, both the $s(\cdot)$ and $r(\cdot)$ are

---

[1] We appreciate the help of K. Mark in preparing this section based on his dissertation [20].

unknown, nonnegative functions to be estimated from data acquired over a measurement interval $[0, T)$, with loglikelihood functional given from (6.3), (6.30), and (6.89) by

$$\mathcal{L}(s,r) = -\int_0^T s(\sigma) r(\sigma - w_{N(\sigma)}) d\sigma + \int_0^T \ln[s(\sigma) r(\sigma - w_{N(\sigma)})] dN(\sigma). \quad (6.90)$$

The maximum-likelihood estimates of $s(\cdot)$ and $r(\cdot)$ satisfy coupled non-linear equations. M. Miller [22] and N. Karamanos and M. Miller [15] derive expectation-maximization algorithms for generating the maximum-likelihood estimates of both of these functions under fairly general stimulus conditions. Here, we focus on just the estimation of the recovery function, which is precisely the hazard function $h(\cdot)$ described above. Assume the stimulus function is the known function $s(t) = 1$. Then, the loglikelihood functional (6.90) for the hazard function $r(\cdot)$ simplifies to

$$\mathcal{L}(r) = -\int_0^T r(\sigma - w_{N(\sigma)}) d\sigma + \ln[r(w_1)] + \sum_{i=2}^{N(T)} \ln[r(w_i - w_{i-1})], \quad (6.91)$$

and the process corresponds to a stationary renewal process with independent interarrivals $\{t_i = w_i - w_{i-1}; i = 1, \cdots, N(T); w_0 = 0\}$.

An important aspect of the problem of estimating the recovery function is that it reflects a physical process that monotonically increases as a function of time following a discharge, which forces a particular structure into the estimation problem. As first shown by M. Boswell [2] for estimating monotonic intensities of Poisson processes, the function maximizing the likelihood is piecewise constant, with jumps exactly at the points $\tau \in [0, \infty)$ for which there exist interarrival times t_i. To see this for a renewal process having a monotonic hazard function, simply examine (6.91), and notice that the first term is made smaller by allowing $r(\tau)$ to increase between interarrival times, whereas the second two terms are unaffected. Therefore, the estimation of a monotonically increasing hazard $r(\cdot)$ becomes the estimation of the jump values of a piecewise constant function with jumps occurring at no more than $N(T)$ instants, given $N(T)$ occurrence times $w = \{w_1, \cdots, w_{N(T)}\}$. In this way, the infinite dimensional problem of estimating the function $r(\cdot)$ becomes a finite-dimensional one of estimating at most $N(T)$ parameters. The difficulty that remains is that $N(T)$ occurrence times will, in general, not support the accurate estimation of a monotonic hazard-function containing $N(T)$ free parameters without exhibiting excessive instabilities. This is precisely the same problem as encountered in Ch. 3 for which sieve constraints and random field texture-models were introduced to remove the variability inherent to the problem of estimating image intensities. Because of the finiteness of the problem induced by the monotonicity constraint, we use a Bayesian approach for choosing the number of jumps.

First, we establish a family of models and their associated maximum-likelihood solutions, where each member of the family is indexed by the number of jumps k present in the hazard function estimate. Thus, select a set of times $\tau_{k0} = 0, \tau_{k1}, \tau_{k2}, \cdots, \tau_{k,k-1}, \tau_{kk} = T$ that partition the measurement interval $[0, T)$ into disjoint subintervals, and define functions

$$r_k(\tau) = \begin{cases} r_k[1], & \tau \in [0, \tau_{k1}) \\ r_k[2], & \tau \in [\tau_{k1}, \tau_{k2}) \\ \vdots & \vdots \\ r_k[k], & \tau \in [\tau_{k,k-1}, T) \end{cases} \tag{6.92}$$

that are piecewise constant over these intervals. Then, the model of order k is a piecewise-constant hazard function with k jumps of arbitrary non-negative values $r_k[1], r_k[2], \cdots, r_k[k]$ occurring, respectively, at the times $0, \tau_{k1}, \cdots, \tau_{k,k-1}$. These functions form a sieve, which we denote by S, in the space of monotonic hazard functions.

The loglikelihood (6.91) for the model of order k becomes

$$\mathcal{L}(r_k) = -\int_0^T r_k(\sigma - w_{N(\sigma)}) d\sigma + \ln[r_k(w_1)] + \sum_{i=2}^{N(T)} \ln[r_k(w_i - w_{i-1})]. \tag{6.93}$$

This loglikelihood can be expressed in terms of a set indicator function $I_A(x)$, which equals one if $x \in A$ and zero otherwise, as follows

$$\mathcal{L}(r_k) = -\int_0^T \sum_{i=1}^{k} r_k[i] I_{[\tau_{k,i-1}, \tau_{ki})}(\sigma - w_{N(\sigma)}) d\sigma + \sum_{i=1}^{k} \ln\{r_k[i]\} \sum_{j=1}^{N(T)} I_{[\tau_{k,i-1}, \tau_{ki})}(t_j)$$

$$= -\sum_{i=1}^{k} r_k[i] \int_0^T I_{[\tau_{k,i-1}, \tau_{ki})}(\sigma - w_{N(\sigma)}) d\sigma + \sum_{i=1}^{k} \ln\{r_k[i]\} \sum_{j=1}^{N(T)} I_{[\tau_{k,i-1}, \tau_{ki})}(t_j).$$

$$\tag{6.94}$$

By maximizing this expression with respect to the parameters $r_k[i]$, the maximum-likelihood estimate under each model is straightforwardly derived to be

$$
\hat{r}_k[i] = \begin{cases} \dfrac{\displaystyle\sum_{j=1}^{N(T)} I_{[\tau_{k,i-1},\tau_{ki})}(t_j)}{\displaystyle\int_0^T I_{[\tau_{k,i-1},\tau_{ki})}(\sigma - w_{N(\sigma)})\,d\sigma}, & \text{for } \displaystyle\int_0^T I_{[\tau_{k,i-1},\tau_{ki})}(\sigma - w_{N(\sigma)})\,d\sigma > 0 \\[2em] 0, & \text{for } \displaystyle\int_0^T I_{[\tau_{k,i-1},\tau_{ki})}(\sigma - w_{N(\sigma)})\,d\sigma = 0, \end{cases}
$$

$$(6.95)$$

for $i = 1, 2, \cdots, k$.

By selecting the times $\{\tau_{ki}\}$ so that the partitions of $[0, T)$ are nested for increasing model orders k, it is clear that, as the number of parameters k increases, the likelihood attained will increase. This suggests that the unconstrained maximum-likelihood solution will achieve it's highest likelihood by allowing a model of order $k = N(T)$, the number of data points.

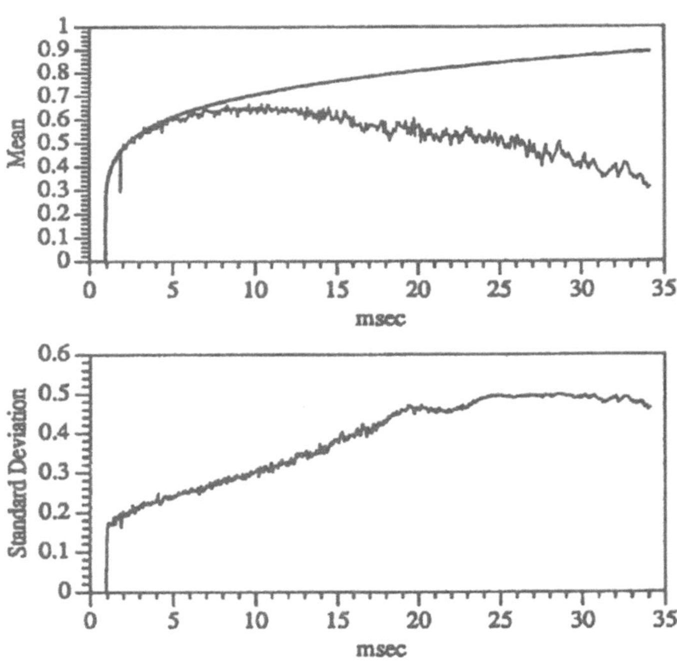

Figure 6.6   Performance of the unconstrained maximum-likelihood estimate of a hazard function.

A demonstration of the fundamental difficulty with this solution is shown in Fig. 6.6, which is the result of a computer simulation of a renewal process with hazard function given in accordance with (6.89) with $s(\cdot) = 1$. The hazard function used was patterned after a typical recovery function exhibited by the auditory nerve, with a 1 ms absolute refractory interval followed by a steep recovery in the next millisecond and then a gradual ascent to full recovery. A measurement interval of duration $T = 360$ ms and containing approximately 1000 events was used. For each of 200 independent datasets, the maximum-likelihood estimate (6.95) was generated, with the number of allowed jumps in the recovery function estimate equal to the number of interarrivals $N(T)$. The performance of the maximum likelihood estimate is shown in Fig. 6.6. The top panel shows the true recovery function (bold line) and the sample mean from the maximum-likelihood algorithm (solid line). The algorithm shows a bias that steadily increases with the interarrival time τ. The bottom graph shows the standard deviation of the maximum-likelihood estimate, which reaches a value of nearly 50 percent. We next describe a Bayesian method for selecting the number of jumps, or model order, k. A lower standard deviation for the resulting estimate will be demonstrated for the same simulated data used in producing the results of Fig. 6.6.

The Bayesian hypothesis-testing method for selecting k with a uniform cost for each selection begins with the assumption that a prior probability P_k for k is given [35, 37]. Then, k is selected to maximize the loglikelihood according to

$$\hat{k} = \underset{k \,\in\, \{1,2,\cdots,N(T)\}}{\operatorname{argmax}} \left\{ \ln E\!\left[e^{\mathcal{L}(r_k)} \,\middle|\, k, w_1, \cdots, w_{N(T)} \right] + \ln P_k \right\}, \tag{6.96}$$

where $\mathcal{L}(r_k)$ is the loglikelihood in (6.93) for a given $r_k(\cdot)$, and the conditional expectation of the likelihood $\exp[\mathcal{L}(r_k)]$ is with respect to the distribution of the random jumps $r_k[1], r_k[2], \cdots, r_k[k]$, given both k and the $N(T)$ occurrence times $w = \{w_1, \cdots, w_{N(T)}\}$. The difficulty in using the Bayes' criterion for selecting the model order k is the evaluation of the conditional expectation in (6.96). A solution is stated in the following theorem.

First define $n_k(i)$ according to

$$n_k(i) = \sum_{j=1}^{N(T)} I_{[\tau_{k,i-1}, \tau_{ki})}(t_j), \tag{6.97}$$

where $t_j = w_j - w_{j-1}$ is the jth interarrival time. Thus, $n_k(i)$ is the number of interarrival times having values in $[\tau_{k,i-1}, \tau_{ki})$.

Theorem 6.3.5 If a uniform prior distribution on the parameters $r_k[i]$ is assumed, then

$$\operatorname*{argmax}_{k \in \{1,2,\cdots,N(T)\}} \left\{ \ln E\left[e^{L(r_k)} \mid k, w_1, \cdots, w_{N(T)} \right] + \ln P_k \right\} \tag{6.98}$$

$$= \operatorname*{argmax}_{k \in \{1,2,\cdots,N(T)\}} \{ L(\hat{r}_k) - C(k) + \ln P_k \},$$

where $L(\hat{r}_k)$ is the maximum value of the loglikelihood (6.93) for functions $r_k(\cdot)$ in the sieve set S, and $C(k)$ is the *complexity of the model* as given by

$$C(k) = -\sum_{i=1}^{k} \left\{ n_k(i) - \ln\left[\frac{n_k(i)^{n_k(i)}}{\hat{r}_k[i](n_k(i)-1)!} \right] \right.$$

$$\left. + \ln\left(1 - \exp\left[-\frac{n_k(i)}{\hat{r}_k[i]} \right] \sum_{j=0}^{n_k(i)} \frac{1}{j!}\left[\frac{n_k(i)}{\hat{r}_k[i]} \right]^j \right) \right\}. \tag{6.99}$$

Proof. The proof of this theorem is an immediate consequence of the following two lemmas. $\square$

Lemma 6.3.1 The loglikelihood functional $L(r_k)$ for the kth order model is expressed in terms of the maximum likelihood $L(\hat{r}_k)$ and the maximum likelihood estimate $\hat{r}_k$ by the following equality:

$$L(r_k) = L(\hat{r}_k) + \sum_{i=1}^{k} n_k(i)\left\{ 1 - \frac{r_k[i]}{\hat{r}_k[i]} + \ln\frac{r_k[i]}{\hat{r}_k[i]} \right\}. \tag{6.100}$$

Proof of Lemma 6.3.1. Reordering and grouping terms of the loglikelihood functional (6.94) yields the following alternative expression:

$$L(r_k) = -\sum_{i=1}^{k} r_k[i] \int_0^T I_{[\tau_{k,i-1},\tau_{ki})}(\sigma - w_{N(o)})\,d\sigma + \sum_{i=1}^{k} n_k(i)\ln r_k[i]$$

$$= \sum_{i=1}^{k} g(r_k[i]), \tag{6.101}$$

where

$$g(r_k[i]) = -r_k[i] \int_0^T I_{[\tau_{k,i-1}, \tau_{ki})}(\sigma - w_{N(\sigma)}) \, d\sigma + n_k(i) \ln r_k[i]. \quad (6.102)$$

Since an expression for $\mathcal{L}(r_k)$ in terms of the maximum likelihood is desired, a Taylor series expansion of $g(r_k[i])$ about the maximum likelihood estimate $\hat{r}_k[i]$ is performed. This expansion yields

$$g(r_k[i]) = g(\hat{r}_k[i]) + g'(\hat{r}_k[i])(r_k[i] - \hat{r}_k[i]) + \frac{1}{2} g''(\hat{r}_k[i])(r_k[i] - \hat{r}_k[i])^2 + R_3$$

$$= g(\hat{r}_k[i]) - \frac{n_k(i)}{2\hat{r}_k^2[i]}(r_k[i] - \hat{r}_k[i])^2 + R_3, \quad (6.103)$$

where

$$R_3 = \frac{1}{2} \int_{\hat{r}_k[i]}^{r_k[i]} (r_k[i] - t)^2 g'''(t) \, dt$$

$$= \frac{1}{2} \int_{\hat{r}_k[i]}^{r_k[i]} (r_k[i] - t)^2 \frac{2n_k(i)}{t^3} \, dt$$

$$= \frac{n_k(i)}{2\hat{r}_k^2[i]}(r_k[i] - 2\hat{r}_k[i])^2 - \frac{n_k(i)}{2} + n_k(i) \ln \frac{r_k[i]}{\hat{r}_k[i]}. \quad (6.104)$$

Substituting (6.104) into (6.103) and simplifying yields

$$g(r_k[i]) = g(\hat{r}_k[i]) - n_k(i) \frac{r_k[i]}{\hat{r}_k[i]} + n_k(i) + n_k(i) \ln \frac{r_k[i]}{\hat{r}_k[i]}. \quad (6.105)$$

Replacing $g(r_k[i])$ in (6.101) with this expression then yields

$$\mathcal{L}(r_k) = \sum_{i=1}^{k} \left\{ g(\hat{r}_k[i]) - n_k(i) \frac{r_k[i]}{\hat{r}_k[i]} + n_k(i) + n_k(i) \ln \frac{r_k[i]}{\hat{r}_k[i]} \right\}$$

$$= \mathcal{L}(\hat{r}_k) + \sum_{i=1}^{k} n_k(i) \left\{ 1 - \frac{r_k[i]}{\hat{r}_k[i]} + \ln \frac{r_k[i]}{\hat{r}_k[i]} \right\}, \quad (6.106)$$

which is (6.100). $\square$

Using Lemma 6.3.1, the expectation required in (6.96) can be performed to yield the following lemma giving the Bayesian solution in Thm. 6.3.5 for model-order selection.

Lemma 6.3.2 If a uniform prior on the parameters $r_k[i]$ is assumed, then the loglikelihood for model-order k is given in terms of the maximum likelihood $L(\hat{r}_k)$, the maximum likelihood estimate $\hat{r}_k$, and the complexity $C(k)$ according to

$$\ln E\left[e^{L(\hat{r}_k)}| k, w_1, \cdots, w_{N(T)}\right] = L(\hat{r}_k) - C(k), \tag{6.107}$$

with $C(k)$ given in (6.99).

Proof of Lemma 6.3.2. Using Lemma 6.3.1 to substitute for $L(r_k)$ in (6.96) yields

$$E\left[e^{L(\hat{r}_k)}| k, w\right] = e^{L(\hat{r}_k)} E\left[\prod_{i=1}^{k} \exp\left(n_k(i)\left\{1 - \frac{r_k[i]}{\hat{r}_k[i]} + \ln\frac{r_k[i]}{\hat{r}_k[i]}\right\}\right)| k, w\right], \tag{6.108}$$

where $w = \{w_1, \cdots, w_{N(T)}\}$. Assuming a uniform conditional-distribution on $r_k[i]$

$$p_{r_k}(x \mid k, w) = \begin{cases} 1, & 0 \le x \le 1 \\ 0, & \text{otherwise}, \end{cases}$$

and assuming that these variables are conditionally independent, allows (6.108) to be written as

$$E\left[e^{L(\hat{r}_k)}| k, w\right] = e^{L(\hat{r}_k)} \prod_{i=1}^{k} \int_0^1 \exp\left(n_k(i)\left\{1 - \frac{x}{\hat{r}_k[i]} + \ln\frac{x}{\hat{r}_k[i]}\right\}\right) dx. \tag{6.109}$$

Straightforward evaluation of the integral in this expression yields (6.107) and completes the proof of Lemma 6.3.2. □

There is a worthwhile relationship to note between selecting the model order k by maximizing the penalized likelihood, as we have described in (6.98), and Rissanen's concept of a minimum description-length [30]. The intensity (6.89) indicates that the corresponding point process is a renewal process having $N(T)$ independent, identically distributed inter-arrival times. Then, Rissanen's method for estimating the k parameters associated with the specification of the kth-order hazard function $r_k(\cdot)$ would introduce into the maximum-likelihood procedure a penalty of size

$(k/2)\ln N(T)$ with k the number of real parameters and $N(T)$ the sample size. K. Mark [20] has proven the following asymptotic connection between the complexity term $C(k)$ in (6.98) and Rissanen's penalty $(k/2)\ln N(T)$. First, define $P(i)$, $i = 1, \cdots, k$ to be the probability that an interarrival has a value in interval $[\tau_{k,i-1}, \tau_{ki})$.

Theorem 6.3.6 If $P(i)$ is uniform across the k intervals, $P(i) = 1/k$, then, with probability 1,

$$\lim_{N(T) \to \infty} \left[\frac{L(\hat{r}_k) - \ln E\{e^{L(\hat{r}_k)} | k, w\}}{(k/2)\ln N(T)} \right] = 1. \qquad (6.110)$$

proof. See K. Mark [20].

Thm. 6.3.6 suggests a simple criterion for choosing a model order k that is asymptotically equivalent to the optimal value selected according to (6.98) for large values of $N(T)$. This is exactly Rissanen's minimum description length criterion:

$$\min_k \left[-L(\hat{r}_k) + \frac{k}{2}\ln N(T) \right]. \qquad (6.111)$$

The interpretation of this equation is that k is chosen to minimize Rissanen's description length, where the description length is defined to be the code length for encoding the observed data relative to a class of probabilistic models. In this discussion, the models under consideration are renewal processes with piecewise constant, nondecreasing hazard functions, where the jump sites are determined by an exponential distribution of intervals. Each model in this class is indexed by k, the number of jump sites to be estimated in the recovery function. This criterion provides an optimal way of both estimating the parameters of the model and choosing the correct model order. The first term in the minimum description length criterion is interpreted as the code length for describing the data given the model. The second term is a bound on describing the model itself. Without the second term, the criterion is equivalent to maximizing likelihood and this leads to an intuitive understanding of the second term. The second term balances maximizing likelihood which tends to result in 'rough' estimates with a cost for describing the model. We emphasize that the minimum description length procedure is valid asymptotically with sample size, whereas that described by Thm. 6.3.5 is precise for all sample sizes.

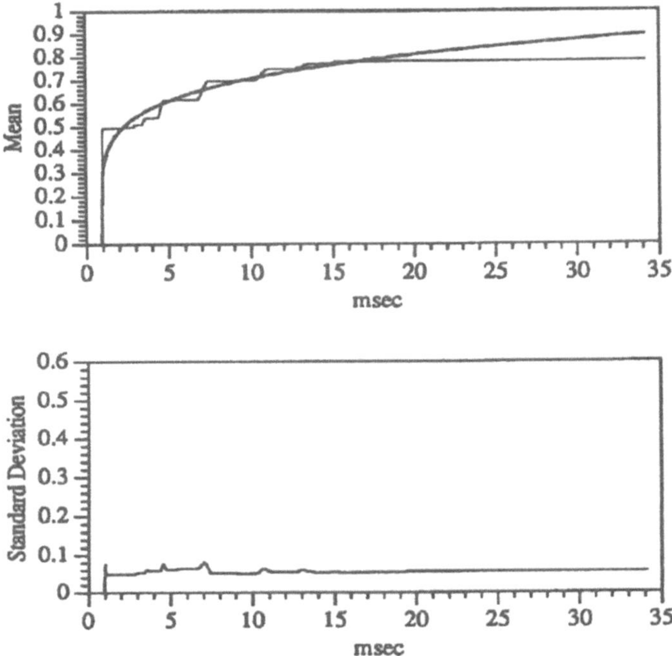

Figure 6.7 Performance of the maximum-likelihood estimate of a hazard function subject to a minimum description length constraint.

The performance of the minimum description length algorithm was evaluated via the same computer simulation used to produce Fig. 6.6. The Bayesian estimates were generated from the same data by running the maximum-likelihood algorithm for different model orders k and then choosing the highest one in Bayes' probability according to Thm. 6.3.6. In this approach, each experiment of the 200 trials would have its own highest probability number of jumps, with that number used each time to generate the estimates. Bias and standard deviation statistics were calculated, and are shown plotted in Fig. 6.7. Shown in the top panel are the true recovery function (bold line) and the sample mean from the Bayesian solution including the complexity term (solid line). Over the approximately 20 ms interval where there are interarrival data, the estimator is unbiased and has standard deviations that are about 25 percent of those of the unconstrained maximum-likelihood estimates shown in Fig. 6.6.

Further discussion of renewal processes is given by W. Smith [34], E. Parzen [26, Ch. 5], and D. Cox and H. Miller [5, Ch. 9].

Summary. Self-exciting point process in which the future evolution depends on only a limited number of past occurrences have been characterized in this section. An m-memory process depends at most on the m most recent occurrence times. The important conclusions of the section are:

1. A necessary and sufficient condition for a self-exciting point process to be a 0-memory process is that its associated counting process be a Markov birth process.

2. A necessary and sufficient condition for a self-exciting point process to be an m-memory process for $m \geq 1$ is that its occurrence time sequence form an mth-order Markov sequence.

3. A renewal process is a particular type of 1-memory process that is temporally homogeneous. Thus, 1-memory self-exciting point processes may be regarded as a generalization of the renewal process without the restriction of temporal homogeneity.

6.4 References

1. A. T. Bharucha-Reid, *Elements of Theory of Markov Processes and Their Applications*, McGraw-Hill, New York, 1960

2. M. T. Boswell, "Estimating and Testing Trend in a Stochastic Process of Poisson Type," *Ann. Math. Stat.*, Vol. 37, pp. 1564-1573, 1966.

3. P. Bremaud, *Point Processes and Queues: Martingale Dynamics*, Springer-Verlag, Berlin, 1981.

4. E. Çinlar, "Superposition of Point Processes," in *Stochastic Point Processes Statistical Analysis, Theory, and Applications* (P. A. W. Lewis, Ed.), Wiley, New York, 1972.

5. D. R. Cox and H. D. Miller, *The Theory of Stochastic Processes*, Wiley, New York, 1965.

6. D. J. Daley and D. Vere-Jones, "A Summary of the Theory of Point Processes," in *Stochastic Point Processes Statistical Analysis, Theory, and Applications* (P. A. W. Lewis, Ed.), Wiley, New York, 1972.

7. W. B. Davenport, Jr. and W. L. Root, *An Introduction to the Theory of Random Signals and Noise*, McGraw-Hill, New York, 1958.

8. W. Feller, *An Introduction to Probability Theory and Its Applications: Vol. 1*, Wiley, New York, 1950.

9. P. M. Fishman, "Statistical Inference for Space-Time Point Processes," D.Sc. Thesis, Sever Institute of Technology, Washington University, St. Louis, MO, June 1974.

10. P. Franken, "A Refinement of the Limit Theorem for the Superposition of Independent Renewal Processes," *Teor. Veroyatnost. i Primen.*, Vol. 8, pp. 320-328, 1963.

11.   R. P. Gaumond, C. E. Molnar, and D. O. Kim, "Stimulus and Recovery Dependence of Cat Cochlear Nerve Fiber Spike Discharge Probability," *J. Neurophysiology*, Vol. 48, pp. 856-873, 1982.

12.   I. I. Gikhman and A. V. Skorokhod, *Introduction to the Theory of Random Processes*, Saunders, Philadelphia, PA, 1965.

13.   P. R. Gray, "Conditional Probability Analysis of the Spike Activity of Single Neurons," *Biophysics J.*, Vol. 7, pp. 759-777, 1967.

14.   B. Grigelionis, "On the Convergence of Sums of Random Step Processes to a Poisson Process," *Teor. Veroyatnost. i Primen.*, Vol. 8, pp. 189-194, 1963.

15.   N. Karamanos and M. I. Miller, "A New Method for Estimating Stimulus and Refractory Related Functions from Auditory-Nerve Discharges," in: *Basic Issues in Hearing, Proc. 8th Internat. Symp. on Hearing*, (H. Duifhuis, J. Horst, and H. Wit, Ed's.), Academic Press, pp. 185-195, 1988.

16.   A. F. Karr, *Point Processes and Their Statistical Inference*, Marcel-Dekker, New York, 1986.

17.   P. I. Kuznetsov and R. L. Stratonovich, "A Note on the Mathematical Theory of Correlated Random Points," *Izv. Akad. Nauk SSSR, Ser. Mat.*, Vol. 20, p. 167, 1956. Also in *Nonlinear Transformations of Random Processes*, Pergamon, New York, 1965.

18.   O. Macchi, "Stochastic Point Processes and Multicoincidences," *IEEE Trans. on Information Theory*, Vol. IT-17, pp. 2-7, January 1971.

19.   O. Macchi, "Processus Ponctuels et Coincidences," D. Sc. These, Centre D'Orsay, Univ. de Paris Sud, January 1972.

20.   K. E. Mark, "Minimum Description Length Estimation in the Analysis of Auditory-Nerve Discharge Patterns," M.S. Thesis, Department of Electrical Engineering, Washington University, St. Louis, MO, December 1990.

21.   J. A. McFadden, "On the Lengths of Intervals in a Stationary Point Process," *J. Roy. Stat. Soc. B*, Vol. 24, pp. 364-382, 1962.

22.   M. I. Miller, "Algorithms for Removing Recovery Related Distortion from Auditory-Nerve Discharge Patterns," *J. Acoustical Society of America*, Vol. 77, pp. 1452-1464, 1985.

23.   J. E. Moyal, "The General Theory of Stochastic Population Processes," *Acta Math.*, Vol. 108, pp. 1-31, 1962.

24.   Y. Ogata, "On Lewis' Simulation Method for Point Processes," *IEEE Trans. on Information Theory*, Vol. IT-27, pp. 23-31, January 1981.

25.   T. Ozaki, "Maximum Likelihood Estimation of Hawkes' Self-Exciting Point Processes," *Annals Institute Stat. Math*, Vol. 30, pp. 145-155, 1978.

26.   E. Parzen, *Stochastic Processes*, Holden-Day, San Francisco, CA, 1962.

27.   A. Ramakrishnan, "Stochastic Processes Relating to Particles Distributed in Continuous Infinity of States," *Proc. Camb. Phil. Soc.*, Vol. 46, pp. 595-602, 1950.

28.   A. Ramakrishnan, "Stochastic Theory of Evolutionary Processes (1937-71)," in *Stochastic Point Processes Statistical Analysis, Theory, and Applications* (P. A. W. Lewis, Ed.), Wiley, New York, 1972.

29. S. O. Rice, "Mathematical Analysis of Random Noise," *Bell System Tech. J.*, Vol's. 23-24, pp. 1-162, 1944.

30. J. Rissanen, "Stochastic Complexity," *J. Royal Statistical Society B*, Vol. 49, pp. 223-239, 1987.

31. I. Rubin, "Reduced Memory Information Processors for Point Processes," Proc. Tenth Annual Allerton Conf. on Circuit and System Theory, Univ. of Illinois, Urbana, IL, pp. 303-312, 1972.

32. I. Rubin, "Regular Point Processes and Their Detection," *IEEE Trans. on Information Theory*, Vol. IT-18, pp. 547-557, September 1972.

33. W. M. Siebert and P. R. Gray, "Random Process Model for the Firing Pattern of Single Auditory Neurons," in: Research Laboratory of Electronics Quarterly Progress Rpt. 71, Massachusetts Institute of Technology, pp. 241-246, 1963.

34. W. L. Smith, "Renewal Theory and Its Ramifications," *J. Royal Stat. Soc. B*, Vol. 20, pp. 243-302, 1958.

35. D. L. Snyder, *Random Point Processes*, Wiley, New York, Ch. 2, 1975.

36. S. K. Srinivasan and R. Vasudevan, *Introduction to Random Differential Equations and Their Applications*, Elsevier, New York, 1971.

37. H. L. Van Trees, *Detection, Estimation, and Modulation Theory: Part 1*, Wiley, New York, Ch. 2, 1968.

38. H. Wold, "Sur les Processus Stationnaires Ponctuels," in *Le Calcul des Probabilities et Applications*, CNRS, Paris, pp. 75-86, 1949.

6.5 Problems

6.2.1 Complete the proof of Theorem 6.2.1.

6.2.2 Let $\{N(t): t \geq t_0\}$ be a self-exciting point process satisfying the hypotheses of Theorem 6.2.1. Establish the following expressions for the conditional probability of there being no points in the interval $[s,t\}$ given the number and occurrence times for those on $[t_0, s)$:

$$\Pr[N(s,t) = 0| N(s)] = \exp\left[-\int_s^t \mu(\sigma,0)d\sigma \right]$$

for $N(s) = 0$ and

$$\Pr[N(s,t) = 0| N(s); w_1, \cdots, w_{N(s)}] = \exp\left[-\int_s^t \hat{\mu}(\sigma; w_1, \cdots, w_{N(s)})d\sigma \right]$$

for $N(s) \geq 1$, where we define

$$\hat{\mu}(\sigma, N(s); w_1, \cdots, w_{N(s)}) = E\{\mu(\sigma, N(\sigma); w_1, \cdots, w_{N(\sigma)}) | N(s); w_1, \cdots, w_{N(s)}\}$$

for $s \le \sigma < t$.

6.2.3 (*Counting Moments for a Self-Exciting Point Process*). Show that on the interval $[t_0, t)$, the k*th* moment of the number of points for a self-exciting point process satisfies

$$\frac{dE\{N^k(t)\}}{dt} = \sum_{j=1}^{k} \binom{k}{j} E\{N^{k-j}(t)\hat{\lambda}(t, N(t))\}, \quad E\{N^k(t_0)\} = 0$$

for $k = 1, 2, \cdots$, where $\hat{\lambda}(t, N(t))$ is the count-conditional intensity of the process as given in (6.7). Deduce, therefore, that the mean number of points is given by $E\{N(t)\} = \int_{t_0}^{t} E\{\lambda(\sigma)\}d\sigma$. [Hint: Multiply (6.11) by n^k and sum on n.

6.2.4 Deduce from Prob. 6.2.3 that

$$E\{N^2(t)\} = \int_{t_0}^{t} E\{\lambda(\sigma)\}d\sigma + 2\int_{t_0}^{t} E\{N(\sigma)\lambda(\sigma)\}d\sigma.$$

Hence, conclude that if $\int_{t_0}^{t} E\{(N(\sigma)+1)\lambda(\sigma)\}d\sigma < \infty$, then $E\{N^2(t)\} < \infty$. Show for a Poisson process that this condition is equivalent to $\int_{t_0}^{t} \lambda(\sigma)d\sigma < \infty$.

6.2.5 Let $\{N(t): t \ge t_0\}$ be a self-exciting counting process such that $\mu(t, 0) = \alpha(t)$ for $N(t) = 0$, and $\mu(t, N(t); w_1, \cdots, w_{N(t)}) = (N(t)+1)\alpha(t)$ for $N(t) \ge 1$, so that $\{N(t): t \ge t_0\}$ is an inhomogeneous, linear birth-process.

   a. Show that $\Pr[N(t) = n] = \Pr[N(t) = 0]\Pr^n[N(t) > 0]$.   [Hint: Use (6.12), (6.13), and induction.]

   b. Let $m(t) = E\{N(t)\}$ be the expected number of points in $[t_0, t)$. Determine $m(t)$ by using the result of Prob. 6.2.3, and verify your answer using the result in part (a).

6.2.6 (*Product Densities*). Product densities are quantities that completely characterize a self-exciting point process. Let $\{N(t): t \ge t_0\}$ be the counting process associated with a self-exciting point process having $\{\lambda(t): t \ge t_0\}$ as its intensity process. The product density of degree n is defined by

$$f_n(t_1, \cdots, t_n) = \lim_{\max(\Delta t_i) \downarrow 0} E\left\{ \prod_{i=1}^{n} \frac{\Delta N(t_i)}{\Delta t} \right\}$$

for $t_0 \le t_1 < \cdots < t_n$ and $n = 1, 2, \cdots$, where $\Delta N(t_i) = N(t_i + \Delta t_i) - N(t_i)$. Because $N(\cdot)$ is orderly, an entirely equivalent definition for $f(\cdot)$ is

$$f_n(t_1, \cdots, t_n) = \lim_{\max(\Delta t_i) \downarrow 0} \left\{ \frac{\Pr[\Delta N(t_1) = 1, \cdots, \Delta N(t_n) = 1]}{\Delta t_1 \cdots \Delta t_n} \right\}.$$

Thus, $f_n(t_1, \cdots, t_n) dt_1 \cdots dt_n$ is the n-fold correlation or n-fold probability between points occurring on the infinitesimal intervals $[t_i, t_i + \Delta t_i)$, $i = 1, \cdots, n$, regardless of all other point occurrences.

a. Establish the following relations between the product densities and the intensity process:

$$f_n(t_1, \cdots, t_n) = \begin{cases} E\{\lambda(t_1)\}, & n = 1 \\ E\{\lambda(t_1)\} \prod_{i=2}^{n} E\{\lambda(t_i) \mid t_1, \cdots, t_{i-1}\}, & n \ge 2 \end{cases}$$

where $E\{\lambda(t_i) \mid t_1, \cdots, t_{i-1}\}$ is the expectation of $\lambda(t)$ at $t = t_i$ given the occurrence times $t_1, t_2, \cdots$, and t_{i-1}. Hint: Use

$$\Pr[\Delta N(t_i) = 1 \mid \Delta N(t_1) = 1, \cdots, \Delta N(t_{i-1}) = 1]$$

$$= E(\Pr[\Delta N(t_i) = 1 \mid \{N(\sigma): t_0 \le \sigma < t_i\}] \mid \Delta N(t_1) = 1, \cdots, \Delta N(t_{i-1}) = 1).$$

b. The probability of a single occurrence point in the infinitesimal interval $[t, t + dt)$ is $f_1(t) dt$ to order dt. In view of this, explain why the integral $\int_{t_0}^{t} f_1(\sigma) d\sigma$ does not equal the probability of one occurrence in $[t_0, T)$ but, rather, equals the expected total number of occurrences in this interval.

c. Determine the product densities for an inhomogeneous Poisson process with intensity function $\{\lambda(t): t \ge t_0\}$.

6.2.7 (*Product Densities, Continued*). The sample-function density for a self-exciting point process can be expressed in terms of all its product densities. The following steps are used by P. Kuznetsov and R. Stratonovich [17] to establish this.

Consider a partition of $[t_0, T)$ as $t_0 \le t_1 < t_2 < \cdots < t_k = T$ and a realization $\{N(\sigma): t_0 \le \sigma < T\}$ of $N(\cdot)$ on this interval. If $N(T) = n \ge 1$, denote the n occurrence times in the interval by $W_1, \cdots, W_n$, and take the partition sufficiently fine that these times fall in different intervals:

$$t_{k_1} \le W_1 < t_{k_1+1}, t_{k_2} \le W_2 < t_{k_2+1}, \cdots, t_{k_n} \le W_n < t_{k_n+1}.$$

There are no occurrence times in $[t_0, T)$ if $N(T) = n = 0$.

Let $\xi_k = N(t_{k+1}) - N(t_k)$. Then, for $n \ge 1$, $\xi_{k_i} = 1$ for $i = 1, 2, \cdots, n$ and $\xi_k = 0$ for $k = 0, 1, \cdots, K-1$ $(k \ne k_1, k_2, \cdots, k_n)$. For $n = 0$ and $k = 0, 1, \cdots, K-1$, $\xi_k = 0$. Denote the $M = K - n$ ξ_k's that are zero by $\gamma_1, \cdots, \gamma_M$. The sample function density for the counting path $\{N(\sigma): t_0 \le \sigma < T\}$ is given by

$$p[\{N(\sigma): t_0 \le \sigma < T\}] = \begin{cases} \lim \mathcal{P}_0, & n = 0 \\ \lim\left[\prod_{i=1}^{n}\left(\Delta t_{k_i}\right)^{-1}\right]\mathcal{P}_n, & n \ge 1 \end{cases}$$

where the limit is taken as $K \to \infty$ and $\max(\Delta t_i) \downarrow 0$, $\Delta t_i = t_{i+1} - t_i$, and where we define $\mathcal{P}_n = \Pr[A_n, \gamma_1 = 0, \cdots, \gamma_M = 0]$ in which A_n is the event

$$A_n = \begin{cases} \Omega, & n = 0 \\ \bigcap_{i=1}^{n}\{\xi_{k_i} = 1\}, & n \ge 1 \end{cases}$$

where Ω is the certain event. The objective of what follows is to express $\mathcal{P}_n$ in terms only of A_n and events of the form $\{\gamma_j = 1\}$.

a. From the orderliness of $N(\cdot)$, we have $\Pr[A_n, \gamma_1 = 0] = \Pr[A_n] - \Pr[A_n, \gamma_1 = 1]$. Show that

$$\Pr[A_n, \gamma_1 = 0, \gamma_2 = 0]$$

$$= \Pr[A_n] - \sum_{i=1}^{2} \Pr[A_n, \gamma_i = 1] + \Pr[A_n, \gamma_1 = 1, \gamma_2 = 1]$$

and that

$$\Pr[A_n, \gamma_1 = 0, \gamma_2 = 0, \gamma_3 = 0]$$

$$= \Pr[A_n] - \sum_{i=1}^{3} \Pr[A_n, \gamma_i = 1] + \sum\sum_{1 \le i_1 < i_2 \le 3} \Pr\left[A_n, \gamma_{i_1} = 1, \gamma_{i_2} = 1\right]$$

$$- \Pr[A_n, \gamma_1 = 1, \gamma_2 = 1, \gamma_3 = 1].$$

Deduce in general that

$$\mathcal{P}_n = \Pr[A_n]$$

$$+ \sum_{m=1}^{M} (-1)^m \left\{ \sum\sum_{i \le i_1 < \cdots < i_m \le M} \cdots \sum \Pr\left[A_n, \gamma_{i_1} = 1, \gamma_{i_2} = 1, \cdots, \gamma_{i_m} = 1\right]\right\}.$$

b. By using the expression for $\mathcal{P}_n$ in part (a), introducing the product densities, and passing to the limit, show that for $N(T) = n = 0$,

$$p[\{N(\sigma): t_0 \le \sigma < T\}]$$

$$= 1 + \sum_{m=1}^{\infty} (-1)^m \iint_{t_0 \le \sigma_1 < \sigma_2 < \cdots < \sigma_m \le T} \cdots \int f_m(\sigma_1, \sigma_2, \cdots, \sigma_m) d\sigma_1 d\sigma_2 \cdots d\sigma_m.$$

And for $N(T) = n \ge 1$, show that

$$p[\{N(\sigma): t_0 \le \sigma < T\}]$$

$$= f_n(W_1, \cdots, W_n) + \sum_{m=1}^{\infty} (-1)^m \iint_{t_0 \le \sigma_1 < \sigma_2 < \cdots < \sigma_m \le T} \cdots \int f_{n+m}(\alpha_1, \cdots, \alpha_{n+m}) d\sigma_1 d\sigma_2 \cdots d\sigma_m,$$

where α_i is the ith term in the ordering of $W_1, W_2, \cdots, W_n, \sigma_1, \sigma_2, \cdots, \sigma_m$; that is

$$\alpha_1 = \min(W_1, \cdots, W_n, \sigma_1, \cdots, \sigma_m)$$

$$\alpha_2 = \min_{\text{excluded } \alpha_1}(W_1, \cdots, W_n, \sigma_1, \cdots, \sigma_m)$$

$$\alpha_{n+m} = \max(W_1, \cdots, W_n, \sigma_1, \cdots, \sigma_m).$$

c. Use the result of part (b) to derive the sample function density for an inhomogeneous Poisson process with intensity function $\{\lambda(t): t \geq t_0\}$.

6.2.8 Let $\{N(t): t \geq t_0\}$ be a self-exciting counting process with intensity process $\{\lambda(t): t \geq t_0\}$. Let $\{b(t): t \geq t_0\}$ be a random process that is at most causally dependent on $N(\cdot)$, so $b(t)$ is independent of any future increment $N(u) - N(t)$ for $u \geq t$. Suppose, further, that for almost every realization of $N(\cdot)$ and for all $\sigma \in [t_0, t)$ that $b(\sigma)$ is bounded and piecewise continuous. Then, from Theorem 6.3.1, the counting integral

$$I(t) = \int_{t_0}^{t} b(\sigma) dN(\sigma)$$

exists and has the evaluation

$$I(t) = \sum_{i=1}^{N(t)} b(w_i).$$

By paralleling the derivation of (6.60), show that

$$E[I(t)] = \int_{t_0}^{t} E[b(\sigma)\lambda(\sigma)] d\sigma.$$

6.2.9 The entropy $H(t_0, T)$ associated with the paths of a self-exciting point process on $[t_0, T)$ is defined by

$$H(t_0, T) = -E(\ln p[\{N(\sigma): t_0 \leq \sigma < T\}]),$$

where $p(\cdot)$ is the sample-function density. Use the result of Prob. 6.2.8 to show that if $\ln \lambda$ is bounded on $[t_0, T)$, then

$$H(t_0, T) = \int_{t_0}^{T} E[\lambda(\sigma)\{1 - \ln \lambda(\sigma)\}] d\sigma.$$

6.2.10 (*Moments of the Sample-Function Density*). Let $(\zeta(t): t \geq t_0\}$ be defined by

$$\zeta(t) = -\int_{t_0}^{t} \lambda(\sigma)d\sigma + \int_{t_0}^{t} \ln\lambda(\sigma)dN(\sigma),$$

where $\{N(t): t \geq t_0\}$ is a self-exciting point process with intensity process $\{\lambda(t): t \geq t_0\}$. Recall from (6.29) that $\phi[\zeta(t)] = \exp[\zeta(t)]$ is the sample-function density for $N(\cdot)$ on $[t_0, t)$.

a. Show that

$$\phi^k[\zeta(t)] = 1 - k\int_{t_0}^{t} \phi^k[\zeta(\sigma)]\lambda(\sigma)d\sigma + \int_{t_0}^{t} \phi^k[\zeta(\sigma)]\{\lambda^k(\sigma) - 1\}dN(\sigma).$$

b. Conclude from (a) and Prob. 6.2.8 that

$$E\{\phi^k[\zeta(t)]\} = 1 - \int_{t_0}^{t} E\{\phi^k[\zeta(\sigma)](\lambda^{k+1}(\sigma) - (k+1)\lambda(\sigma))\}d\sigma.$$

6.2.11 Let u be a random variable that is uniformly distributed on $[0, 1]$, and define w as the solution to

$$u = \exp\left[-\int_{W_n}^{w} \mu(\sigma, n; W_1, \cdots, W_n)d\sigma\right],$$

where $\mu(\cdot)$ is the function defined in (6.3) for a self-exciting point process; w can be determined numerically, for example with Newton's method, when given u, $\mu(\cdot)$, and $W_1, \cdots, W_n$.

a. Determine $p_{w|w_1,\cdots,w_n}(W | W_1, \cdots, W_n)$, the conditional probability density of w given $w_1 = W_1, \cdots, w_n = W_n$.

b. Discuss how the result in (a) can be used to simulate a self-exciting point process with a given intensity process. Hint: consider (6.15). Also, see T. Ozaki [25] and Y. Ogata [24].

6.3.1 Show that if the birth rate $\mu(t, n)$ of a Markov birth process is such that the sum

$$\sum_{n=0}^{\infty}\left[\inf_{t_0 \leq \sigma < \infty} \mu(\sigma, n)\right]^{-1}$$

converges, then $\Pr[N(t)<\infty]<1$ and a population explosion can occur with nonzero probability. Use this result to show that the condition of Theorem 6.3.2 is necessary as well as sufficient for there to be no explosion when the birth rate is not a function of time.

6.3.2 Prove that an m-memory self-exciting point process has the function $\mu(t,N(t);\ w_1,\cdots,w_{N(t)})$ of the form

$$\mu(t,N(t);w_1,\cdots,w_{N(t)})=f(t;w_{N(t)-m+1},\cdots,w_{N(t)})$$

for all $N(t)\geq m$ if and only if the sequence of occurrence times $\{w_i\}$ forms an m*th*-order Markov sequence. Here, $f(\cdot)$ is not a function of $N(t)$ and the $N(t)-m$ earlier occurrence times.

6.3.3 Let $\{N(t):t\geq 0\}$ be an ordinary renewal counting process, $m(t)=E\{N(t)\}$, and $p_t(T)$ the common interarrival-time probability density.

a. Show that

$$m(t)=\sum_{n=1}^{\infty}P_{w_n}(t),$$

where $P_{w_n}(t)$ is the probability distribution function for the n*th* occurrence time.

b. Let

$$M(s)=\int_0^{\infty}m(t)e^{-st}dt\quad\text{and}\quad\Phi(s)=\int_0^{\infty}p_t(T)e^{-sT}dT$$

be the Laplace transforms of $m(t)$ and $p_t(T)$. Show that

$$sM(s)=\frac{\Phi(s)}{1-\Phi(s)}.$$

Conclude from this that the mean-value function $m(t)$ of a renewal process provides a complete statistical characterization because the interarrival density does.

6.3.4 Suppose that the process $N(\cdot)$ in Prob. 6.3.3 is a modified renewal process.

a. Show that

$$sM(s) = \frac{\Phi_1(s)}{1 - \Phi(s)},$$

where $\Phi_1(s)$ is the Laplace transform for the first interarrival density, and $\Phi(s)$ is the transform for the subsequent interarrival times.

b. Use the final value theorem of Laplace transforms to show for a modified renewal process that

$$\lim_{t \to \infty}\left[m(t) - \frac{t}{\mu}\right] = \frac{E(t_1^2)}{2\mu^2} - \frac{E(t_1)}{\mu},$$

where $\mu = E(t_n)$ is the mean interarrival time for $n \geq 2$. Thus, conclude to this process that

$$m(t) = \frac{t}{\mu} + \frac{E(t_1^2)}{2\mu^2} - \frac{E(t_1)}{\mu} + O(t^{-1}),$$

where $O(t^{-1}) \to 0$ as $t \to \infty$; for large t, $m(t) \approx t/\mu$.

c. Show for an equilibrium renewal process that $sM(s) = (s\mu)^{-1}$. Conclude from this that for all $t \geq 0$, $m(t) = t/\mu$.

6.3.5 Pulses arrive at a scintillation detector as a homogeneous Poisson process with intensity λ. The detector is nonparalyzable and has a fixed dead time of duration τ.

a. Show that the output is either a modified or ordinary renewal process corresponding to whether the detector is initially unlocked or locked in a dead time.

b. Determine the hazard function for the output point process if the detector is initially locked. Sketch the intensity process for a typical output counting path.

c. Suppose that an output counting path is observed on the interval $[0, T)$. Use (6.88) to determine an expression for the maximum Likelihood estimate for λ in terms of these data, and compare your answer to that obtained in Ex. 6.2.4.

6.3.6 (*Paralyzable Counters*). Suppose that events arrive at a counter as a homogeneous Poisson process with a constant intensity λ. Each incident event causes the counter to be inoperative for a period τ. Thus, an incident event arriving during an inoperative period extends the inoperative period. Let $\{t_n\}$ be the sequence of interarrival times for the *registered* events. Here, t_1 is exponentially distributed with parameter λ. Denote the probability density for t_n, $n \geq 2$, by $p_{t_n}(T)$ and the Laplace transform of this density by $\Phi(s) = E\{\exp(-st)\}$.

a. Let $\{N(t): t \geq 0\}$ and $\{M(t): t \geq 0\}$ be the counting processes for the incident and registered events, respectively, and let $m(t) = E[M(t)]$. By using

$$E[M(t+\Delta t) - M(t)] = \Pr[M(t, t+\Delta t) = 1] + o(\Delta t)$$
$$= \Pr[N(t-\tau, t) = 0, N(t, t+\Delta t) = 1] + o(\Delta t),$$

show that

$$\frac{dm(t)}{dt} = \begin{cases} \lambda e^{-\lambda \tau}, & t \geq \tau \\ \lambda e^{-\lambda t}, & t < \tau. \end{cases}$$

b. Using this result and part (a) of Prob. 6.3.4, show that

$$\Phi(s) = \frac{\lambda e^{-\tau(s+\lambda)}}{s + \lambda e^{-\tau(s+\lambda)}}.$$

c. Evaluate the mean $E(t_n)$ and variance $\mathrm{var}(t_n)$ for the nth interarrival time, for $n \geq 2$.

6.3.7 Let $m_k(t) = E\{N^k(t)\}$, where $\{N(t): t \geq 0\}$ is an ordinary renewal process, and define $\dot{m}_1(t) = dm_1(t)/dt$. Show that

$$m_2(t) = m_1(t) + 2\int_0^t \dot{m}_1(t-\sigma)\dot{m}_1(\sigma)d\sigma.$$

CHAPTER SEVEN

DOUBLY STOCHASTIC POISSON-PROCESSES

7.1 Introduction

The self-exciting point processes of Ch. 6 were obtained from the Poisson process by allowing its intensity to become causally dependent on the point process itself. In this way, the intensity is transformed into a random process having paths that are known exactly given the point process. Point processes in this chapter are also obtained by randomizing the intensity of the Poisson process. Here, however, this randomization is not through self excitation but, rather, by an "outside" process so that the paths of the resulting intensity process are not known given only the point process.

The point processes of this chapter are conditionally Poisson processes given the intensity process. They are called *doubly stochastic Poisson processes*, a terminology introduced by D. Cox [7].[1] To be specific, let $\{\mathbf{x}(t): t \geq t_0\}$ be a left-continuous, vector-valued stochastic process that is the "outside" process influencing the evolution of a point process whose associated counting process is $\{N(t): t \geq t_0\}$. A doubly stochastic Poisson-process in time is defined as follows.

Definition (*Doubly Stochastic Poisson-Process*). $\{N(t): t \geq t_0\}$ is a doubly stochastic Poisson-process with intensity process $\{\lambda(t, \mathbf{x}(t)): t \geq t_0\}$ if for almost every given path of the process $\{\mathbf{x}(t): t \geq t_0\}$, $N(\cdot)$ is a Poisson process with intensity function $\{\lambda(t, \mathbf{x}(t)): t \geq t_0\}$. In other words, $\{N(t): t \geq t_0\}$ is conditionally a Poisson process with intensity function $\{\lambda(t, \mathbf{x}(t)): t \geq t_0\}$ given $\{\mathbf{x}(t): t \geq t_0\}$.

Doubly stochastic Poisson-processes that produce points on multidimensional spaces are also of much interest and are similarly defined as conditional Poisson processes.

---

[1] This terminology is not used universally; *conditional*, *compound*, and *mixed* Poisson-processes are also popular, the latter especially when the intensity process is a random variable.

The process $\{x(t): t \geq t_0\}$ arises in many applications as the quantity that conveys useful information, and for this reason we call it the *information process*. The simplest examples occur when the information process is a finite collection of random variables independent of time. An information process of this type is encountered in such diverse applications as nuclear medicine, optical detection, and auditory electrophysiology when the parameters of interest are modeled as random variables in contrast to Examples 2.4.2, 2.4.3, and 2.4.4 where they are unknown but not random; the new aspect in this chapter is that the parameters are assumed to have a known probability distribution. More generally, there are only relatively weak restrictions on the time dependence and statistics of the information process; to get our most explicit results, we will assume it to be either a vector-valued Gaussian process or a vector-valued Markov process.

There are numerous physical situations for which doubly stochastic Poisson-processes are reasonable mathematical models. Several specific illustrations are developed in this chapter. Some of these are in the area of optical detection, where the applicability of the model is motivated in the following two examples.

Example 7.1.1 *Optical Detection* ───────────────────────────────

If the optical field incident on the photodetector of Ex. 2.4.3 changes randomly in time, the resulting photoelectron stream forms a doubly stochastic Poisson-process to an accurate approximation because it is well modeled by a Poisson process conditional on the field. The optical field may be randomly varying because of fluctuations in the intensity of an incoherent light source, as with a stellar object, scattered solar radiation, or an incandescent light bulb, or because of fluctuations in the properties of the medium in which the optical wave propagates, as when atmospheric turbulence causes random changes in the refractive index of the atmosphere. ∎

Example 7.1.2 *Optical Phase Tracking* ─────────────────────────

If the reference phase $\theta_r(t)$ in Ex. 5.3.3 changes randomly in time, the resulting photoelectron stream is a doubly stochastic Poisson-process. Randomness in the phase may be intentional, as when phase or frequency modulation are used to communicate an analog information signal, or unintentional, which occurs when the reference oscillator is unstable and has a randomly drifting phase. ∎

The chapter is divided into two broad problem areas. The first is what we term the *forward problem*. This is concerned with the determination of the statistics of a doubly stochastic Poisson-process. Thus, in Sec's. 7.2 and 7.3, we study counting and time statistics, respectively. A fundamental observation in Sec. 7.2 is that a doubly stochastic Poisson-process is a particular form of the self-exciting point processes of Ch. 6. Certain statistical properties of a doubly stochastic Poisson-process are an immediate consequence of this observation. Most importantly, it leads in Sec. 7.3 to a characterization of the sample-function density of a doubly stochastic Poisson-process. The form of this density is remarkably similar to that of a Poisson process, and its use in estimation problems for observed doubly stochastic Poisson-processes parallels that in earlier chapters. The second broad problem area addressed in this chapter is termed the *inverse problem*. This is concerned with the determination of the conditional statistics of the information process given observations of a doubly stochastic Poisson-process it influences. The inverse problems that are studied are important when a doubly stochastic Poisson process is used to model some measured data. It is often not this point process itself that is of primary interest. Rather, what is desired is an underlying mechanism, here the information process, that can only be observed indirectly through its influence on the point process. The conditional statistics for the information process are characterized in Sec. 7.4, and this characterization is used to develop estimates of the information process in terms of the available observations. The mean-square performance of estimates when the information process is normally distributed is also developed. Sections 7.2 to 7.4 examine doubly stochastic Poisson-processes that evolve in time. Multidimensional doubly stochastic Poisson-processes are studied in Sec. 7.5. These include multichannel, time-space, and translated doubly stochastic Poisson-processes.

More information on the theory and application of doubly stochastic Poisson processes is given by D. Cox [7], D. Cox and P. Lewis [8], D. Daley and D. Vere-Jones [9], and J. Grandell [15, 16, 17]. Doubly stochastic Poisson processes can also be addressed using Martingale theory. P. Brémaud [4], A. Karr [23], and R. Lipster and A. Shiryayev [27] may be consulted for this alternative approach.

7.2 Counting Statistics

Let $\{N(t): t \geq t_0\}$ be a doubly stochastic Poisson-process with intensity process $\{\lambda(t, \mathbf{x}(t)): t \geq t_0\}$, where $\{\mathbf{x}(t): t \geq t_0\}$ is the underlying information process. The statistics for $N(t)$, the number of points in $[t_0, t)$ are developed in this section. The statistics of interest include various moments for $N(t)$ as well as the probability that $N(t)$ equals a particular integer n. Two methods for determining these statistics are described, the *method of conditioning*, and the *method of self-exciting point processes*. The first uses the property that $\{N(t): t \geq t_0\}$ is conditionally a Poisson process given the information process, and the second uses the property, yet to be established, that $\{N(t): t \geq t_0\}$ is an example of the self-exciting point processes of Ch. 6. Two observations are worthy of mention before these methods are described. The first is that explicit, closed-form expressions for the statistics of $N(t)$ are difficult to obtain even though the problem to be solved is straightforward conceptually. The second observation is that the conditional expectation

$$\hat{\lambda}(t) = E[\lambda(t,\mathbf{x}(t)) \mid N(\sigma): t_0 \le \sigma < t]$$

will be seen to play a frequent and important role in this and subsequent sections; this conditional expectation of the intensity process will be studied in Sec. 7.4.

7.2.1 Counting Statistics by the Method of Conditioning

The counting statistics for a doubly stochastic Poisson-process can be determined in principle by conditioning on the information process, using the property that thus conditioned, $\{N(t): t \ge t_0\}$ is a Poisson counting process, and then removing the conditioning by taking the expectation over the paths of the information process. Thus, for example, the probability that the number of point occurring in $[t_0, t)$ is n is given by

$$\Pr[N(t) = n] = E\{\Pr[N(t) = n \mid \mathbf{x}(\sigma): t_0 \le \sigma < t]\}$$

$$= E\left\{ \frac{1}{n!} \left(\int_{t_0}^{t} \lambda(\sigma, \mathbf{x}(\sigma)) d\sigma \right)^n \exp\left(-\int_{t_0}^{t} \lambda(\sigma, \mathbf{x}(\sigma)) \right) \right\}. \qquad (7.1)$$

The expectation on the right is generally difficult, if not impossible, to evaluate. Some situations where the evaluation is possible are given in the next two examples and in the problems.

Example 7.2.1 *Randomly Scaled Poisson Process* ———————————
Suppose that the information process is simply a nonnegative random variable x and that for $t \ge t_0$, $\lambda(t,x) = xv(t)$, where $v(t)$ is a deterministic function of t. Then, $\{N(t): t \ge t_0\}$ can be viewed as a Poisson process with an intensity $v(t)$ that is scaled randomly by x. From (7.1),

$$\Pr[N(t) = n] = \frac{1}{n!} \int_0^\infty X^n \exp\left(-X \int_{t_0}^{t} v(\sigma) d\sigma \right) dP_x(X), \qquad (7.2)$$

where $P_x(X) = \Pr[x \le X]$ is the probability distribution function for x. As an example, suppose that x is a gamma distributed random variable having the density

$$p_x(X) = \frac{dP_x(X)}{dx} = \begin{cases} \dfrac{\alpha^{\beta+1}}{\Gamma(\beta+1)} X^\beta e^{-\alpha X}, & X \ge 0 \\ 0, & X < 0 \end{cases} \qquad (7.3)$$

where $\Gamma(\cdot)$ denotes the gamma function. Evaluation of (7.2) yields

$$\Pr[N(t)=n]=\frac{\Gamma(n+\beta+1)}{\Gamma(n+1)\Gamma(\beta+1)}\left[\frac{\alpha}{\alpha+\int_{t_0}^{t}\nu(\sigma)d\sigma}\right]^{\beta+1}\left[\frac{\int_{t_0}^{t}\nu(\sigma)d\sigma}{\alpha+\int_{t_0}^{t}\nu(\sigma)d\sigma}\right]^{n}$$

$$(7.4)$$

for $n=1,2,\cdots$. This is the negative binomial distribution, and the associated process $\{N(t):t\geq t_0\}$ is termed an *inhomogeneous Polya process*. Equation (7.2) can, of course, be evaluated for other distributions, and some examples are given in the problems. If it exists, the moment generating-function for x is given by

$$M(s)=E(e^{sx})=\int_{0}^{\infty}e^{sX}dP_x(X),\qquad(7.5)$$

and we have

$$M_x^{(n)}(s)=\frac{d^nM_x(s)}{ds^n}=\int_{0}^{\infty}X^ne^{sX}dP_x(X).\qquad(7.6)$$

Thus, if the moment generating-function exists, we have from (7.2) that

$$\Pr[N(t)=n]=\frac{1}{n!}\left(\int_{t_0}^{t}\nu(\sigma)d\sigma\right)^{n}M_x^{(n)}\left(-\int_{t_0}^{t}\nu(\sigma)d\sigma\right).\qquad(7.7)\blacksquare$$

Example 7.2.2 *Photocount Statistics for Gaussian Light* ————————

In this example, we examine the statistics at the output of an ideal photodetector when the optical field incident on it is a narrowband Gaussian process. Such an incident field occurs in optical communications when a coherent-signal field is observed in the presence of a background radiation such as scattered sunlight; see E. Hoversten, R. Harger, and S. Halme [19]. It can also occur when the signal field propagates through a scattering medium such as a cloud or fog or when an optical radar ranging pulse is reflected from a rough surface; see R. Kennedy [24]. The model of this example will be used throughout the chapter. As in Ex. 5.2.9 and (5.22), let $S(t,\vec{r})$ be the complex envelope of a scalar, narrowband, optical field incident on an ideal photodetector having an active area $\mathcal{A}$. We assume here that $\mathcal{A}$ is sufficiently small that any spatial variations in $S(t,\vec{r})$

can be neglected; this restriction can easily be removed (see J. Clark and E. Hoversten [5]). Thus, we write $S(t,\vec{r})$ simply as $S(t,\vec{r}) = X(t)$, and (5.22) becomes

$$\lambda(t,X(t)) = \alpha|X(t)|^2, \tag{7.8}$$

where $\alpha = \eta \mathcal{A}/h\nu$. Here, $\{X(t): t \geq t_0\}$ is the complex envelope of a narrowband field, so the real field is $x(t) = \sqrt{2}\,\mathrm{Re}[X(t)e^{j2\pi\nu t}]$, where ν is the mean frequency of the optical field.

From Ex. 5.2.9, the output of the photodetector can be modeled as a Poisson process $\{N(t): t \geq t_0\}$ given $\{X(t): t \geq t_0\}$. We take the real signal $\{x(t): t \geq t_0\}$ to be a narrowband Gaussian process; the mean $m(t) = E[x(t)]$ corresponds to a deterministic signal field, and $b(t) = x(t) - m(t)$ represents a zero-mean Gaussian field due to background radiation. In correspondence with $x(t) = m(t) + b(t)$, we write the complex envelope as $X(t) = M(t) + B(t)$. Then, $\{N(t): t \geq t_0\}$ is a doubly stochastic Poisson-process with intensity process $\{\lambda(t,X(t)): t \geq t_0\}$ given by (7.8). It is convenient to expand $\{X(\sigma): t_0 \leq \sigma < t\}$ in a Karhunen-Loève expansion as

$$X(\sigma) = \sum_{i=1}^{\infty} X_i \phi_i(\sigma), \quad t_0 \leq \sigma < t, \tag{7.9}$$

where the coefficients X_i are given by

$$X_i = \int_{t_0}^{t} X(\sigma)\phi_i{}^*(\sigma)d\sigma. \tag{7.10}$$

The symbol "*" denotes complex conjugation. The coefficients $\{X_i\}$ form a sequence of independent, complex-valued Gaussian random variables if the orthonormal basis functions $\{\phi_i(\sigma)\}$ are the eigenfunctions of the integral equation

$$\lambda_i \phi_i(\sigma) = \int_{t_0}^{t} K(\sigma,u)\phi_i(u)du, \quad t_0 \leq \sigma < t \tag{7.11}$$

in which $K(\sigma,u) = E[X(\sigma)X^*(u)] - M(\sigma)M^*(u)$ is the covariance function for $X(\cdot)$. The eigenvalue λ_i is the variance of the ith coefficient X_i. The real and imaginary parts of X_i are statistically independent Gaussian random variables with mean $\mathrm{Re}(M_i)$ and $\mathrm{Im}(M_i)$, respectively, and each has variance $\lambda_i/2$, where

$$M_i = E(X_i) = \int_{t_0}^{t} M(\sigma)\phi_i^*(\sigma)\,d\sigma.$$

These and other properties of the expansion (7.9) are discussed by M. Rosenblatt [36, pp. 185-195] and H. L. Van Trees [47, pp. 584-589]. Because of the orthonormality of the eigenfunctions on $[t_0, t)$, substitution of (7.9) into (7.8) shows that

$$\int_{t_0}^{t} \lambda(\sigma, X(\sigma))\,dx = \alpha \sum_{i=1}^{\infty} |X_i|^2 \tag{7.12}$$

We now examine the counting statistics following the procedure of S. Karp and J. Clark [21] and S. Karp, E. O'Neill, and R. Gagliardi [22]; an historical discussion of this important problem is given by C. Mehta [32]. Let $M_{N(t)}(ju) = E[e^{juN(t)}]$ be the characteristic function for the number of detected photons at time t. Using (2.16) and the method of conditioning, we have

$$M_{N(t)}(ju) = E\left\{ \exp\left[(e^{ju} - 1) \int_{t_0}^{t} \lambda(\sigma, X(\sigma))\,d\sigma \right] \right\}$$

$$= \prod_{i=1}^{\infty} E\{\exp[\alpha(e^{ju} - 1)|X_i|^2]\}, \tag{7.13}$$

where we have used (7.12) and the independence of the coefficients $\{X_i\}$. After evaluating the expectation in (7.13), we have

$$M_{N(t)}(ju) = \prod_{i=1}^{\infty} \frac{1}{1 - \alpha\lambda_i(e^{ju} - 1)} \exp\left\{ \frac{\alpha|M_i|^2(e^{ju} - 1)}{1 - \alpha\lambda_i(e^{ju} - 1)} \right\}. \tag{7.14}$$

Each factor in the product is the characteristic function of a Laguerre-distributed random variable; call this nonnegative integer-valued variable $N_i(t)$. By inverse Fourier transformation (see B. Lachs [26]), there results

$$\Pr[N_i(t) = n_i] = \frac{(\alpha\lambda_i)^{n_i}}{(1 + \alpha\lambda_i)^{n_i+1}} \exp\left(-\frac{\alpha|M_i|^2}{1 + \alpha\lambda_i} \right) L_{n_i}\left[\frac{|M_i|^2}{\lambda_i(1 + \alpha\lambda_i)} \right], \tag{7.15}$$

where $L_n(x)$ is the Laguerre polynomial

$$L_n(x) = \frac{e^x}{n!} \frac{d^n[x^n e^{-x}]}{dx^n} = \sum_{k=0}^{n} \binom{n}{k} \frac{(-x)^k}{k!}.$$

It follows that the total number of photoelectrons on $[t_0, t)$ can be represented as the sum $N(t) = \sum_{i=0}^{\infty} N_i(t)$ of independent Laguerre random variables. Two limiting situations are of interest. As $\lambda_i \to 0$ for all i, we have from (7.14) that

$$M_{N(t)}(ju) = \exp\left\{ (e^{ju} - 1)\alpha \sum_{i=1}^{\infty} |M_i|^2 \right\}. \tag{7.16}$$

By Parseval's theorem,

$$\sum_{i=1}^{\infty} |M_i|^2 = \int_{t_0}^{t} |M(\sigma)|^2 d\sigma.$$

Consequently, in the limit as the background radiation, $B_i = X_i - M_i$, tends to zero, the photocount $N(t)$ tends to a Poisson distributed random variable with parameter $\alpha \int_{t_0}^{t} |M(\sigma)|^2 d\sigma$. On the other hand, for any component with $M_i = 0$, we have

$$\Pr[N_i(t) = n_i] = \frac{(\alpha\lambda_i)^{n_i}}{(1 + \alpha\lambda_i)^{n_i + 1}}, \tag{7.17}$$

which is the Bose-Einstein distribution; thus, any coordinate in the series (7.9) that has no deterministic component contributes a Bose-Einstein variate to the total photocount. $\blacksquare$

7.2.2 Counting Statistics by the Method of Self-Exciting Point Processes

Doubly stochastic Poisson-processes are self-exciting point processes. This important characterization is established in the following theorem and will be used in the rest of the chapter. The implication is that all the results of Ch. 6 apply in the study of doubly stochastic Poisson-processes.

Theorem 7.2.1 (*Characterization of a Doubly Stochastic Poisson-Process as a Self-Exciting Point Process*). Let $\{N(t): t \geq t_0\}$ be a doubly stochastic

Poisson-process with intensity process $\{\lambda(t, \mathbf{x}(t)): t \geq t_0\}$, and suppose that $\lambda(\cdot)$ is integrable. Then, $\{N(t): t \geq t_0\}$ is a self-exciting counting process with intensity process $\{\hat{\lambda}(t): t \geq t_0\}$, where

$$\hat{\lambda}(t) = E[\lambda(t, \mathbf{x}(t)) | N(\sigma): t_0 \leq \sigma < t]$$

$$= \begin{cases} E[\lambda(t, \mathbf{x}(t)) | N(t)], & N(t) = 0 \\ E[\lambda(t, \mathbf{x}(t)) | N(t); w_1, \cdots, w_{N(t)}], & N(t) \geq 1. \end{cases} \tag{7.18}$$

Proof. We need to evaluate the functions $\mu(t, 0)$ and $\mu(t, N(t); w_1, \cdots, w_{N(t)})$ of (6.2). Suppose, first, that $N(t) \geq 1$. Let $\{\rho_n\}$ be a sequence of positive numbers converging to zero as n tends to infinity, and define

$$f_n = \frac{1}{\rho_n} \Pr[N(t, t + \rho_n) = 1 | N(t); w_1, \cdots, w_{N(t)}, \mathbf{x}(\sigma); \sigma \geq t_0].$$

Then $E(f_n) < \infty$. Since $\{N(t): t \geq t_0\}$ is conditionally a Poisson process given $\{\mathbf{x}(t): t \geq t_0\}$, we have

$$f_n = \frac{1}{\rho_n} \int_t^{t+\rho_n} \lambda(\sigma, \mathbf{x}(\sigma)) d\sigma \exp\left[-\int_t^{t+\rho_n} \lambda(\sigma, \mathbf{x}(\sigma)) d\sigma \right]. \tag{7.19}$$

Consequently, the sequence $\{f_n\}$ converges to $\lambda(t, \mathbf{x}(t))$ with probability one as n tends to infinity. Also, define

$$g_n = \frac{1}{\rho_n} \int_t^{t+\rho_n} \lambda(\sigma, \mathbf{x}(\sigma)) d\sigma.$$

Then $E(g_n) < \infty$ because $\lambda(\cdot)$ is integrable by assumption. The sequence $\{g_n\}$ converges to $\lambda(t, \mathbf{x}(t))$ with probability one as n tends to infinity, $f_n \leq g_n$ for each n, and $E[\lambda(t, \mathbf{x}(t))] = \lim_{n \to \infty} E(g_n)$. Thus, the sequences $\{f_n\}$ and $\{g_n\}$ satisfy the conditions of the generalized bounded convergence theorem, Theorem 1.4.1, and we have that

$$E[\lambda(t,\mathbf{x}(t))|N(t);w_1,\cdots,w_{N(t)}]$$

$$= \lim_{n\to\infty} E(f_n|N(t);w_1,\cdots,w_{N(t)}) \tag{7.20}$$

$$= \lim_{n\to\infty}\frac{1}{\rho_n}\Pr[N(t,t+\rho_n)=1|N(t);w_1,\cdots,w_{N(t)}].$$

The limit on the right side of (7.20) is by definition $\mu(t,N(t);w_1,\cdots,w_{N(t)})$. Consequently, (7.18) holds for $N(t)\geq 1$. The proof for $N(t)=0$ is analogous when f_n is defined to be

$$f_n = \rho_n^{-1}\Pr[N(t,t+\rho_n)=1|N(t)=0,\mathbf{x}(\sigma);\sigma\geq t_0]. \qquad \square$$

Theorem 7.2.1 indicates the basic role of the conditional expectation $\hat{\lambda}(t)$ of the intensity $\lambda(t,\mathbf{x}(t))$ of the doubly stochastic Poisson-process given the entire past $\{N(\sigma): t_0 \leq \sigma < t\}$ of the point process up to time t. We will encounter this conditional expectation repeatedly throughout the rest of the chapter. It is characterized and studied in detail in Sec. 7.4. In particular, we show that $\hat{\lambda}(t)$ may be interpreted as an estimate of $\lambda(t,\mathbf{x}(t))$ in terms of $\{N(\sigma): t_0 \leq \sigma < t\}$ that has the smallest mean square-error $E\{[\lambda(t,\mathbf{x}(t))-\hat{\lambda}(t)]^2\}$ of any other function of $\{N(\sigma): t_0 \leq \sigma < t\}$.

It is straightforward to extend the doubly stochastic Poisson-process model to that of a doubly stochastic self-exciting point process by incorporating a possible dependence in $\lambda(t,\mathbf{x}(t))$ on the count $N(t)$ and the occurrence times $w_1,\cdots,w_{N(t)}$. In this way, doubly stochastic renewal processes, Markov birth processes, and more general self-exciting processes can be accommodated. In most of our results of the chapter, this extension is accomplished by replacing the estimate $\hat{\lambda}(t)$ by the estimate $\hat{\lambda}(N(t);w_1,\cdots,w_{N(t)})$, where

$$\hat{\lambda}(t,N(t);w_1,\cdots,w_{N(t)})$$

$$= E\{\lambda(t,\mathbf{x}(t),N(t);w_1,\cdots,w_{N(t)})|N(t);w_1,\cdots,w_{N(t)}\}. \tag{7.21}$$

As a consequence of Theorems 7.2.1 and 6.2.1, the counting statistics for a doubly stochastic Poisson-process satisfy the following differential difference equations:

$$\frac{\partial\Pr[N(t)=0]}{\partial t} = -\hat{\lambda}(t,0)\Pr[N(t)=0], \quad \Pr[N(t_0)=0]=1 \tag{7.22}$$

and, for $n \geq 1$,

$$\frac{\partial \Pr[N(t)=n]}{\partial t} = -\hat{\lambda}(t,n)\Pr[N(t)=n] + \hat{\lambda}(t,n-1)\Pr[N(t)=n-1],$$

$$\Pr[N(t_0)=n] = 0, \tag{7.23}$$

where $\hat{\lambda}(t,n)$ is the count-conditional intensity

$$\hat{\lambda}(t,n) = E[\hat{\lambda}(t)|N(t)=n] = E[\lambda(t,\mathbf{x}(t))|N(t)=n]. \tag{7.24}$$

This indicates that the counting statistics for a doubly stochastic Poisson-process are identical to those of a pure Markov birth process with a population-dependent birthrate equal for each n to the estimate $\hat{\lambda}(t,n)$. Equations (7.22) and (7.23) can be solved recursively for the counting probabilities $\Pr[N(t)=0]$, $\Pr[N(t)=1]$, $\Pr[N(t)=2]$, $\cdots$. For this purpose, we note from (7.22) that

$$\Pr[N(t)=0] = \exp\left(-\int_{t_0}^{t} \hat{\lambda}(\sigma)d\sigma\right) \tag{7.25}$$

and, from (7.23),

$$\Pr[N(t)=n] = \int_{t_0}^{t} \hat{\lambda}(\tau,n-1)\Pr[N(\tau)=n-1]\exp\left(-\int_{t_0}^{t} \hat{\lambda}(\sigma,n)d\sigma\right)d\tau,$$

$$\tag{7.26}$$

for $n = 1,2,\cdots$.

The estimate $\hat{\lambda}(t,n)$ is, in general, difficult if not impossible to determine, so (7.22)-(7.26) must be viewed as characterizations rather than formulas for the counting statistics. The following is an example where $\hat{\lambda}(t,n)$ can be evaluated explicitly and the counting statistics determined.

Example 7.2.3 *Randomly Scaled Poisson Process* ———————————
Consider again the randomly scaled inhomogeneous Poisson process of Ex. 7.2.1. To use (7.25) and (7.26), we need to determine the estimate $\hat{\lambda}(t,N(t)) = E[x\nu(t)|N(t)]$. By using Bayes' rule, we obtain

$$\hat{\lambda}(t,N(t)) = \frac{\nu(t)\int_0^\infty X^{N(t)+1}\exp\left(-X\int_{t_0}^t \nu(\sigma)d\sigma\right)dP_x(X)}{\int_0^\infty X^{N(t)}\exp\left(-X\int_{t_0}^t \nu(\sigma)d\sigma\right)dP_x(X)}. \qquad (7.27)$$

This determines $\hat{\lambda}(\cdot)$ for any given distribution of x, and the counting statistics can be determined from (7.25) and (7.26). If the moment generating function for x exists, (7.27) can be expressed alternatively by using (7.6) as

$$\hat{\lambda}(t,N(t)) = \frac{\nu(t)M_x^{(N(t)+1)}\left(-\int_{t_0}^t \nu(\sigma)d\sigma\right)}{M_x^{(N(t))}\left(-\int_{t_0}^t \nu(\sigma)d\sigma\right)}. \qquad (7.28)$$

For a specific example, suppose that x has the gamma distribution in (7.3), so that $\{N(t):t \geq t_0\}$ is an inhomogeneous Polya process. Then, an easy calculation shows that $M_x(s) = [\alpha/(\alpha-s)]^{\beta+1}$. It then follows from (7.28) that

$$\hat{\lambda}(t,N(t)) = \frac{N(t)+\beta+1}{\alpha+\int_{t_0}^t \nu(\sigma)d\sigma}. \qquad (7.29)$$

Thus, the count-conditional intensity for an inhomogeneous Polya process is linearly dependent on the count. We leave it as an exercise for the reader to derive the counting statistics in (7.4) for this process from (7.25), (7.26), and (7.29). ∎

7.2.3 Limit Theorems for Counting Statistics

Because of the analytical intractability in evaluating the expectations in (7.1) or (7.24), it is in general impossible to determine an explicit expression for the counting probability $\Pr[N(t)=n]$ for t finite. It is of some importance, therefore, that this probability can be more completely characterized asymptotically for large t. These characterizations can be summarized briefly as follows. Let $\Lambda(t)=\int_{t_0}^t \lambda(\sigma,x(\sigma))d\sigma$, and define centered and normalized versions of $N(t)$ and $\Lambda(t)$ according to

$$N_{cn}(t) = \frac{N(t)-m_N(t)}{\sigma_N(t)} \quad \text{and} \quad \Lambda_{cn}(t) = \frac{\Lambda(t)-m_\Lambda(t)}{\sigma_\Lambda(t)}, \qquad (7.30)$$

where $m_N(t)$ and $m_\Lambda(t)$ are the mean-value functions for $N(t)$ and $\Lambda(t)$, and $\sigma_N^2(t)$ and $\sigma_\Lambda^2(t)$ are the corresponding variances. Suppose that $N_{cn}(t)$ and $\Lambda_{cn}(t)$ approach random variables η and ξ in distribution, respectively, as t approaches infinity, and let g be a zero-mean, unit-variance Gaussian random variable independent of ξ. Then, depending on the value of $c = \lim_{t\to\infty}\sigma_\Lambda^2(t)/m_\Lambda(t)$, η can be characterized as $\eta = g$ for $c = 0$, $\eta = g/\sqrt{1+c}+\xi/\sqrt{1+c^{-1}}$ for $0 < c < \infty$, and $\eta = \xi$ for $c = \infty$. These results are established in the following theorems due to J. Grandell [15] and R. Serfozo [40]. The proofs we use parallel those of Grandell for the situation when $\lambda(\cdot)$ is a stationary process.

Theorem 7.2.2 (*Limit Theorem for Counting Statistics*). Suppose that $m_\Lambda(t) \to \infty$ and $\sigma_\Lambda^2(t)/m_\Lambda(t) \to c$ as $t \to \infty$, where c is a constant such that $0 < c < \infty$. If $\Lambda_{cn}(t)$ converges in distribution to ξ as $t \to \infty$, then $N_{cn}(t)$ converges in distribution to $g(1+c)^{-1/2}+\xi(1+c^{-1})^{-1/2}$ as $t \to \infty$, where g is a zero-mean, unit-variance Gaussian random variable independent of ξ.

The condition that $m_\Lambda(t)$ tends to infinity as t tends to infinity implies that the expected number of points in the infinite interval $[t_0, \infty)$ is infinite. This condition precludes the possibility that $\lambda(t, \mathbf{x}(t))$ be zero for all $t \geq T$ for some finite T.

Proof. Throughout the proof, we use the relations established in Prob. 7.2.5 that

$$m_N(t) = m_\Lambda(t) \tag{7.31}$$

$$\sigma_N^2(t) = \sigma_\Lambda^2(t) + m_\Lambda(t). \tag{7.32}$$

We have to prove that

$$\lim_{t\to\infty} E[\exp(juN_{cn}(t))] = M_\xi\!\left[\frac{ju}{\sqrt{1+c^{-1}}}\right]\exp\!\left[-\frac{u^2}{2(1+c)}\right], \tag{7.33}$$

where $M_\xi(ju) = E(e^{ju\xi})$ is the characteristic function for ξ. By the method of conditioning, and with some manipulations, we have

$$E[\exp(juN_{cn}(t)]$$

$$= \exp\left[-\frac{ju\,m_N(t)}{\sigma_N(t)}\right]E\left\{\exp\left[\Lambda(t)\left(e^{ju/\sigma_N(t)}-1\right)\right]\right\} \tag{7.34}$$

$$= \exp\left[-\frac{ju\,m_N(t)}{\sigma_N(t)}+m_\Lambda(t)\left(e^{ju/\sigma_N(t)}-1\right)\right]E\left\{\exp\left[\Lambda_{cn}(t)\sigma_\Lambda(t)\left(e^{ju/\sigma_N(t)}-1\right)\right]\right\}.$$

Using (7.31) and (7.32), we have

$$\lim_{t\to\infty}\exp\left[-\frac{ju\,m_N(t)}{\sigma_N(t)}+m_\Lambda(t)\left(e^{ju/\sigma_N(t)}-1\right)\right]=\exp\left[-\frac{u^2}{2(1+c)}\right]. \tag{7.35}$$

Provided that the interchange of the limit and expectation is justified, we also have

$$\lim_{t\to\infty}E\left\{\exp\left[\Lambda_{cn}(t)\sigma_\Lambda(t)\left(e^{ju/\sigma_N(t)}-1\right)\right]\right\} \tag{7.36}$$

$$=E\left\{\lim_{t\to\infty}\exp\left[\Lambda_{cn}(t)\sigma_\Lambda(t)\left(e^{ju/\sigma_N(t)}-1\right)\right]\right\}=E\left\{\exp\left[ju\frac{\xi}{\sqrt{1+c^{-1}}}\right]\right\}.$$

Thus, the limit of (7.34) as $t\to\infty$ yields (7.33) and establishes the theorem. The interchange of the limit and the expectation is justified by the bounded convergence theorem, Theorem 1.4.1, if we demonstrate that $\exp\left[\Lambda_{cn}(t)\sigma_\Lambda(t)(e^{ju/\sigma_N(t)}-1)\right]$ is bounded in magnitude independently of t. This we do as follows. Since $\Lambda(t)\ge 0$ and $\cos(u/\sigma_N(t))-1\le 0$, we have

$$\left|\exp\left\{\Lambda_{cn}(t)\sigma_\Lambda(t)\left(e^{ju/\sigma_N(t)}-1\right)\right\}\right| \tag{7.37}$$

$$\le \exp\left\{-m_\Lambda(t)\left[\cos\left(\frac{u}{\sigma_\Lambda(t)}\right)-1\right]\right\}\le \exp\left\{\frac{u^2 m_\Lambda(t)}{2\sigma_N^2(t)}\right\}\le \exp\left\{\frac{1}{2}u^2\right\}<\infty,$$

where we have used the inequalities $1-\cos x\le\frac{1}{2}x^2$ and, from (7.32), $m_\Lambda(t)/\sigma_N^2(t)\le 1$. $\quad\square$

Theorem 7.2.3 (*Limit Theorem for Counting Statistics*). Suppose that $m_\Lambda(t)\to\infty$ and $\sigma_\Lambda^2(t)/m_\Lambda(t)\to 0$ as $t\to\infty$. Then $N_{cn}(t)$ converges in distribution to g as $t\to\infty$, where g is a zero-mean, unit-variance Gaussian random variable.

Proof. The proof is the same as that for Theorem 7.2.2 except that the limits in (7.35) and (7.36) are replaced by

$$\lim_{t \to \infty} \exp\left[-ju \frac{m_N(t)}{\sigma_N(t)} + m_\Lambda(t)\left(e^{ju/\sigma_N(t)} - 1\right)\right] = e^{-\frac{1}{2}u^2} \qquad (7.38a)$$

and

$$\lim_{t \to \infty} E\left\{ \exp\left[\Lambda_{cn}(t)\sigma_\Lambda(t)\left(e^{ju/\sigma_N(t)} - 1\right)\right]\right\} = 1, \qquad (7.38b)$$

respectively. ❑

Theorem 7.2.4 (*Limit Theorem for Counting Statistics*). Suppose that $m_\Lambda(t) \to \infty$ and $\sigma_\Lambda^2(t)/m_\Lambda(t) \to \infty$ as $t \to \infty$. If $\lambda_{cn}(t)$ converges in distribution to ξ as $t \to \infty$, then $N_{cn}(t)$ also converges in distribution to ξ as $t \to \infty$.

Proof. The proof is the same as that for Theorem 7.2.2 except that the limits in (7.35) and (7.36) are replaced by

$$\lim_{t \to \infty} \exp\left[-ju \frac{m_N(t)}{\sigma_N(t)} + m_\Lambda(t)\left(e^{ju/\sigma_N(t)} - 1\right)\right] = 1 \qquad (7.39a)$$

and

$$\lim_{t \to \infty} E\left\{ \exp\left[\Lambda_{cn}(t)\sigma_\Lambda(t)\left(e^{ju/\sigma_N(t)} - 1\right)\right]\right\} = E(e^{ju\xi}), \qquad (7.39b)$$

respectively. ❑

Example 7.2.4 *Poisson Process* ———————————————————
Suppose that $\{N(t) : t \geq t_0\}$ is a Poisson process with intensity $\{v(t) : t \geq t_0\}$. Then, $m_\Lambda(t) = \Lambda(t) = \int_{t_0}^{t} v(\sigma)d\sigma$, and $\sigma_\Lambda^2(t) = 0$ for all $t \geq t_0$. If $\Lambda(\infty) = \infty$, then according to Theorem 7.2.3, there holds

$$\lim_{t \to \infty} \Pr\left[\frac{N(t) - \Lambda(t)}{\sqrt{\Lambda(t)}} \leq Y\right] = \Phi(Y), \qquad (7.40)$$

where

Table 7.1

			$\Pr[N(t) \leq t^{1/2}Y + t]$			$\Phi(Y)$
Y	$t=1$	$t=4$	$t=9$	$t=16$	$t=25$	$t=\infty$
0	0.736	0.629	0.587	0.566	0.553	0.500
1	0.920	0.890	0.876	0.868	0.863	0.841

$$\Phi(Y) = \int_{-\infty}^{Y} \frac{1}{\sqrt{2\pi}} e^{-\frac{1}{2}u^2} du.$$

Table 7.1 contains values of $\Pr[\Lambda^{-1/2}(t)/(N(t) - \Lambda(t)) \leq Y]$ and of $\Phi(Y)$ for a homogeneous Poisson process with $\nu(t) = 1$ and for selected values of Y and t. It is seen for $Y = 0$ and $Y = 1$ that $\Pr[\Lambda^{-1/2}(t)/(N(t) - \Lambda(t)) \leq Y]$ - $\Phi(Y)$ tends to zero as $t \to \infty$. The implication of Theorem 7.2.3 is that trend occurs for any choice of Y. ∎

Example 7.2.5 *A Doubly Stochastic Poisson-Process* ————
Suppose that $\{N(t): t \geq t_0\}$ is a doubly stochastic Poisson-process with an intensity process $\{\lambda(t, \mathbf{x}(t)): t \geq t_0\}$ such that for each $t \geq t_0$, $\Lambda(t) = \int_{t_0}^{t} \lambda(\sigma, \mathbf{x}(\sigma)) d\sigma$ is a gamma distributed random variable,

$$p_{\Lambda(t)}(X) = \frac{\alpha^{\beta t + 1}}{\Gamma(\beta t + 1)} X^{\beta t} e^{-\alpha X}, \quad X \geq 0. \tag{7.41}$$

Then, $m_\Lambda(t) = \alpha^{-1}(\beta t + 1)$. Hence, the constant c in Theorem 7.2.2 equals α^{-1}. It is straightforward to verify that $\Lambda_{cn}(t)$ converges in distribution to zero as $t \to \infty$. If $0 < \alpha < \infty$, then according to Theorem 7.2.2, $N_{cn}(t)$ converges in distribution to a normal random variable with zero mean and variance $\alpha(1 + \alpha)^{-1}$. By using the method of conditioning, we find that

$$\Pr[N(t) = n] = \frac{\Gamma(\beta t + n + 1)}{\Gamma(n + 1)\Gamma(\beta t + 1)(1 + \alpha)^n} \left(\frac{\alpha}{1 + \alpha} \right)^{\beta t + 1} \tag{7.42}$$

for $n = 0, 1, 2, \cdots$. Thus,

Table 7.2

			$\Pr[N(t) \leq \sqrt{2(t+1)}\,Y + (t+1)]$			$\Phi(\sqrt{2}\,Y)$
Y	$t=1$	$t=4$	$t=9$	$t=16$	$t=25$	$t=\infty$
0	0.688	0.656	0.623	0.593	0.568	0.500

$$\Pr\left[N(t) \leq \frac{1}{\alpha}\sqrt{(1+\alpha)(\beta t+1)}\,Y + \frac{1}{\alpha}(\beta t+1)\right] = \Phi\left(\sqrt{\frac{1+\alpha}{\alpha}}\,Y\right). \qquad (7.43)$$

where $\Phi(Y)$ is the standard Gaussian distribution function in (7.40). Table 7.2 contains values of the probability on the left side of (7.43) and its limit on the right side for $\alpha = \beta = 1$, $Y = 0$, and selected values of t. The approach of the probability to its limit as t increases is evident. Theorem 7.2.2 implies that this trend transpires for all values of α, β and Y. ∎

Summary. Counting statistics for a doubly stochastic Poisson-process can be determined by the methods of conditioning and of self-exciting point processes. The method of conditioning uses the property that $N(\cdot)$ is conditionally a Poisson process given the information process that influences its intensity. The unconditional statistics of $N(\cdot)$ are then determined by averaging the Poisson statistics with respect to the distribution of the information process. The method of self-exciting point processes uses the property that a doubly stochastic Poisson-process is a particular form of self-exciting process. Recursion equations developed in Ch. 6 may then be applied to determine the counting probability provided that the count-conditional intensity can be determined. Theorems 7.2.2 to 7.2.4 give conditions under which the counting distribution can be approximated.

7.3 Time Statistics

Various expressions associated with the statistics of the occurrence times of a doubly stochastic Poisson-process are derived in this section. The quantities of interest are the *survival probability*, the *joint-occurrence density*, and the *sample-function density*. Of these, the sample-function density will be the most important in the rest of the chapter.

7.3.1 Time Statistics by the Method of Conditioning

The joint-occurrence density and the sample-function density for a doubly stochastic Poisson-process $\{N(t): t \geq t_0\}$ can be determined in principle by conditioning on the information process $\{\mathbf{x}(t): t \geq t_0\}$, using the property that thus conditioned $\{N(t): t \geq t_0\}$ is a Poisson process, and then removing the conditioning by evaluating an expectation with respect to the distribution of the information process. The last step is where the method of conditioning often fails because of analytic intractability. Even so, there are important models for which the expectation can be accomplished, and two examples are given below.

Let $p_{\mathbf{w}}^{(n)}(\mathbf{W})$ be the joint-occurrence density for the first n occurrence times $w_1, \cdots, w_n$, and let $p[\{N(\sigma): t_0 \leq \sigma < t\}]$ be the sample-function density. Then, using (2.20) and (2.32), and the first two steps of the method of conditioning, we have for the joint-occurrence density that

$$p_{\mathbf{w}}^{(n)}(\mathbf{W}) = E\left\{\prod_{i=1}^{n} \lambda(W_i, \mathbf{x}(W_i)) \exp\left[-\int_{t_0}^{W_n} \lambda(\sigma, \mathbf{x}(\sigma)) d\sigma\right]\right\} \qquad (7.44)$$

and for the sample-function density that

$$p[\{N(\sigma): t_0 \leq \sigma < t\}]$$

$$= E\left\{\exp\left[-\int_{t_0}^{T} \lambda(\sigma, \mathbf{x}(\sigma)) d\sigma + \int_{t_0}^{T} \ln \lambda(\sigma, \mathbf{x}(\sigma)) dN(\sigma)\right]\right\}. \qquad (7.45)$$

The expectation in (7.45) is with respect to the statistics of the information process with the occurrence times given.

Example 7.3.1 *Randomly Scaled Poisson Process* ——————————

As in Ex's. 7.2.1 and 7.2.3, let $\{N(t): t \geq t_0\}$ be an inhomogeneous Poisson-process having an intensity $\{v(t): t \geq t_0\}$ that is randomly scaled by a nonnegative random variable x. The result is a doubly stochastic Poisson-process with intensity process $\{xv(t): t \geq t_0\}$. From (7.45), the sample-function density is

$$p[\{N(\sigma): t_0 \leq \sigma < T\}]$$

$$= E\left\{x^{N(T)} \exp\left(-\int_{t_0}^{T} v(\sigma) d\sigma\right)\right\} \exp\left(\int_{t_0}^{T} \ln v(\sigma) dN(\sigma)\right). \qquad (7.46)$$

The expectation is with respect to the distribution of x. If the moment generating function $M_x(s) = E(e^{sx})$ exists, (7.46) can be written in terms of it as

$$p[\{N(\sigma): t_0 \leq \sigma < T\}]$$

$$= M_x^{(N(t))}\left(-\int_{t_0}^{T} v(\sigma)d\sigma\right)\exp\left(\int_{t_0}^{T} \ln v(\sigma)dN(\sigma)\right), \tag{7.47}$$

where $M_x^{(n)}(s) = d^n M_x(s)/ds^n$. When x has the gamma distribution in (7.3) so that $\{N(t): t \geq t_0\}$ is an inhomogeneous Polya process, we have from (7.46) that

$$p[\{N(\sigma): t_0 \leq \sigma < T\}] = \frac{\Gamma[N(T)] + \beta + 1}{\Gamma(\beta + 1)}\left[\frac{\alpha}{\alpha + \int_{t_0}^{T} v(\sigma)d\sigma}\right]^{\beta+1} \tag{7.48}$$

$$\times \left[\frac{1}{\alpha + \int_{t_0}^{T} v(\sigma)d\sigma}\right]^{N(T)} \exp\left(\int_{t_0}^{T} \ln v(\sigma)d\sigma\right).$$

The particular choice of $\beta = 0$ corresponds to an exponential scaling of the intensity; that is, $p_x(X) = \alpha\exp(-\alpha X)$ for $X \geq 0$. From (7.48), the sample-function density for this exponential scaling is

$$p[\{N(\sigma): t_0 \leq \sigma < T\}]$$

$$= \frac{\alpha\Gamma[N(T) + 1]}{\left[\alpha + \int_{t_0}^{T} v(\sigma)d\sigma\right]^{N(T)+1}}\exp\left(\int_{t_0}^{T} \ln v(\sigma)d\sigma\right). \tag{7.49}$$

By setting the derivative of (7.49) with respect to α equal to zero, it is seen that the maximum-likelihood estimate $\hat{\alpha}_{ML} = \hat{\alpha}_{ML}(N(\sigma): t_0 \leq \sigma < T)$ of α, in terms of an observed counting path is,

$$\hat{\alpha}_{ML} = \frac{1}{N(T)}\int_{t_0}^{T} v(\sigma)d\sigma. \tag{7.50}\blacksquare$$

Example 7.3.2 *Sample-Function Density for Gaussian Light* ———
In Ex. 7.2.2, we examined the photocounting statistics for an ideal photodetector in response to an optical field modeled as a narrowband Gaussian process. As this model is used in the theories of optical coherence, optical communication, and optical radar, it is of some interest that its sample-function density can be determined thus characterizing the time

statistics of the photoelectron process. The procedure is straightforward conceptually using the method of conditioning, but there are some tedious manipulations. Let $\{N(t): t \geq t_0\}$ be a doubly stochastic Poisson-process with intensity process $\{\lambda(t, X(t)) = \alpha|X(t)|^2 : t \geq t_0\}$, where $\{X(t): t \geq t_0\}$ is the complex envelope of a narrowband Gaussian process $\{x(t): t \geq t_0\}$ representing incoherent light. Assume that $X(\cdot)$ has zero mean. Recall from Ex. 7.2.2 that $X(t)$ can be represented as

$$X(t) = \sum_{i=1}^{\infty} X_i \phi_i(t), \quad t_0 \leq t < T, \tag{7.51}$$

where $\{\phi_i(t)\}$ are the eigenfunctions of the covariance function $K(t, u) = E[X(t)X^*(u)]$, and where the coefficients $\{X_i\}$ are independent, complex-valued Gaussian random variables such that $\text{Re}(X_i)$ and $\text{Im}(X_i)$ are independent, zero-mean Gaussian random variables with equal variance $\lambda_i/2$. The eigenvalues and eigenfunctions satisfy (7.11). Define $D_{\mathcal{F}}(\alpha) = \prod_{i=1}^{\infty}(1 + \alpha\lambda_i)$; this function of the eigenvalues is called the *Fredholm determinant* of $K(t, u)$. Finally, define $h(t, u)$ according to

$$h(t, u) = \sum_{i=1}^{\infty} \frac{\lambda_i}{1 + \alpha\lambda_i} \phi_i(t)\phi_i^*(u), \quad (t, u) \in [t_0, T) \times [t_0, T). \tag{7.52}$$

An equivalent specification for $h(t, u)$ is that it should be the solution to the integral equation

$$h(t, u) + \alpha \int_{t_0}^{T} K(t, v)h(v, u)dv = K(t, u), \quad (t, u) \in [t_0, T) \times [t_0, T). \tag{7.53}$$

This integral equation has been studied extensively in the context of estimation theory for Gaussian processes; see T. Kailath [25], A. Baggeroer [3, Ch. 6], and H. L. Van Trees [46, Ch. 6]. By using the method of conditioning, O. Macchi [29, 30, 31] has established that the sample-function density associated with a counting path $\{N(\sigma): t_0 \leq \sigma < T\}$ with $N(T) = n$ and occurrence times $W_1, \cdots, W_n$ is given by

$$p[\{N(\sigma): t_0 \leq \sigma < T\}] \tag{7.54}$$

$$= \begin{cases} D_{\mathcal{F}}^{-1}(\alpha), & N(T) = n = 0 \\ \alpha^n D_{\mathcal{F}}^{-1}(\alpha) \sum_{\eta \in \mathscr{P}_n} h\big(W_1, W_{\eta_1}\big) \cdots h\big(W_n, W_{\eta_n}\big), & N(T) = n \geq 1 \end{cases}$$

where $\mathcal{P}_n$ is the set of all the $n!$ permutations of the integers $1, 2, \cdots, n$. Simply to illustrate the notation, suppose that $n = 2$. Then $\mathcal{P}_n = \{(1,2),(2,1)\}$, and (7.54) becomes

$$p[\{N(\sigma): t_0 \leq \sigma < T\}] \tag{7.55}$$

$$= \alpha^2 D_{\mathcal{F}}^{-1}(\alpha)[h(W_1, W_1)h(W_2, W_2) + h(W_1, W_2)h(W_2, W_1)].$$

Macchi's derivation of (7.54) by the method of conditioning proceeds as follows. By incorporating (7.12) into (7.45), we need to evaluate the expectation in

$$p[\{N(\sigma): t_0 \leq \sigma < T\}]$$

$$= E\left[\exp\left(-\alpha \sum_{i=1}^{\infty} |X_i|^2 + \int_{t_0}^{T} \ln \alpha |X(\sigma)|^2 dN(\sigma)\right)\right]. \tag{7.56}$$

If $N(T) = 0$, the counting integral is zero, and using the independence of the X_i, we have

$$p[\{N(\sigma): t_0 \leq \sigma < T\}] = \prod_{i=1}^{\infty} E[\exp(-\alpha |X_i|^2)].$$

But $u_i = |X_i|^2$ has an exponential distribution $p_{u_i}(U) = \lambda_i^{-1}\exp(-\lambda_i^{-1}U)$, $U \geq 0$. Consequently, $E[\exp(-\alpha |X_i|^2)] = (1 + \alpha\lambda_i)^{-1}$. This proves (7.54) for $N(T) = n = 0$. Now assume that $N(T) = n \geq 1$. Then, (7.56) becomes

$$p[\{N(\sigma): t_0 \leq \sigma < T\}]$$

$$= \alpha^n E\left\{|X(W_1)|^2 \cdots |X(W_n)|^2 \exp\left(-\alpha \sum_{i=1}^{\infty} |X_i|^2\right)\right\}$$

$$= \alpha^n \sum_{\substack{j_1,\cdots,j_n = 1 \\ k_1,\cdots,k_n = 1}}^{\infty} a(\mathbf{j}, \mathbf{k}) \prod_{l=1}^{n} \phi_{j_l}(W_l)\phi_{k_l}^*(W_l), \tag{7.57}$$

where we have used (7.51) to get the second equality and where

$$a(\mathbf{j}, \mathbf{k}) = E\left[X_{j_1} \cdots X_{j_n} X_{k_1}^* \cdots X_{k_n}^* \exp\left(-\alpha \sum_{i=1}^{\infty} |X_i|^2\right)\right]. \tag{7.58}$$

Let p_m and q_m be the number of j_i's and k_i's equal to m, respectively. There holds

$$\sum_{m=1}^{\infty} p_m = \sum_{m=1}^{\infty} q_m = n.$$

Because of the independence of the coefficients $\{X_i\}$, we have

$$a(\mathbf{j},\mathbf{k}) = \prod_{m=1}^{\infty} E\left[X_m^{p_m} X_m^{*q_m} e^{-\alpha|X_m|^2}\right]. \tag{7.59}$$

The moment factorization property of Gaussian random variables and the fact that $E(X_i^2) = 0$ imply that $a(\mathbf{j},\mathbf{k}) = 0$ if $p_m \neq q_m$. This and the exponential distribution for $|X_i|^2$ yield

$$a(\mathbf{j},\mathbf{k}) = \begin{cases} 0, & p_m \neq q_m \\ \prod_{m=1}^{\infty} \dfrac{m!}{1+\alpha\lambda_m}\left(\dfrac{m}{1+\alpha\lambda_m}\right)^{p_m}, & p_m = q_m. \end{cases} \tag{7.60}$$

The condition $p_m = q_m$ requires that the set of integers $k_1, \cdots, k_n$ be a permutation of the set of integers $j_1, \cdots, j_n$; denote this restriction on $k_1, \cdots, k_n$ by $\mathbf{k} \in \mathcal{P}(\mathbf{j})$. Then (7.57) becomes

$$p[\{N(\sigma): t_0 \leq \sigma < T\}] \tag{7.61}$$

$$= \alpha^n D_{\mathcal{F}}(\alpha)^{-1} \sum_{j_1, \cdots, j_n = 1}^{n} \left[\prod_{m=1}^{\infty}\left(\frac{\lambda_m}{1+\alpha\lambda_m}\right)^{p_m} \prod_{i=1}^{\infty} \phi_{j_i}\left(W_{j_i}\right)\right] f(j_1, \cdots, j_n),$$

where

$$f(j_1, \cdots, j_n) \tag{7.62}$$

$$\overset{\Delta}{=} \prod_{m=1}^{\infty} p_m! \sum_{\substack{k_1, \cdots, k_n = 1 \\ \mathbf{k} \in \mathcal{P}(\mathbf{j})}}^{\infty} \left[\phi_{k_1}^*(W_1) \cdots \phi_{k_n}^*(W_n)\right] = \sum_{\eta \in \mathcal{P}_n} \phi_{j_1}^*\left(W_{\eta_1}\right) \cdots \phi_{j_n}^*\left(W_{\eta_n}\right),$$

where the last equality can be verified by comparison with the defined expression for $f(j_1, \cdots, j_n)$. Thus,

$$p[\{N(\sigma): t_0 \le \sigma < T\}]$$

$$= \alpha D_{\mathcal{F}}^{-1}(\alpha) \sum_{\eta \in \mathcal{P}_n} \sum_{j_1,\cdots,j_n=1}^{\infty} [\prod_{m=1}^{\infty} \left(\frac{\lambda_m}{1+\alpha\lambda_m} \right)^{p_m} \prod_{l=1}^{n} \phi_{j_l}(W_l)\phi_{j_l}^*\left(W_{\eta_{l_i}}\right)$$

$$= \alpha D_{\mathcal{F}}^{-1}(\alpha) \sum_{\eta \in \mathcal{P}_n} h\left(W_1, W_{\eta_{1}}\right)\cdots h\left(W_n, W_{\eta_{n}}\right). \tag{7.63}$$

The last equality is obtained using (7.52). This establishes (7.54), the sample-function density for Gaussian light, in terms of the Fredholm determinant $D_{\mathcal{F}}(\alpha)$ and the function $h(t,u)$. These quantities are determined by the eigenfunctions and eigenvalues of the covariance function $K(t,u)$. Indicated below are four situations where $D_{\mathcal{F}}(\alpha)$ and $h(t,u)$ can be evaluated or closely approximated.

a. *Stationary Process, Long Observation Interval*. It is often reasonable to assume Gaussian light to be a stationary process. Then, $K(t,u) = K(t-u,0)$. If, in addition, the observation interval $[t_0, T)$ is long in comparison to the time over which the Gaussian process has significant correlation, then $D_{\mathcal{F}}(\alpha)$ and $h(t,u)$ can be specified in terms of the power-density spectrum $S(f) = \int_{-\infty}^{\infty} K(\tau,0)e^{-j2\pi f\tau}d\tau$. As discussed by H. L. Van Trees [47, p. 101, 150], there holds

$$\ln D_{\mathcal{F}}(\alpha) \approx (T-t_0)\int_{-\infty}^{\infty} \ln[1+\alpha S(f)]df$$

and

$$h(t,u) \approx \int_{-\infty}^{\infty} H(f)e^{j2\pi f(t-u)}df,$$

where $H(f) = S(f)[1+\alpha S(f)]^{-1}$.

b. *Separable Covariance*. The covariance function $K(t,u)$ is termed separable when it can be expressed as a finite sum

$$K(t,u) = \sum_{i=1}^{M} \lambda_i \phi_i(t)\phi_i^*(u),$$

where $\{\phi_i(t)\}$ is an orthonormal set. If follows readily that

$$D_{\mathcal{F}}(\alpha) = \prod_{i=1}^{M}(1+\alpha\lambda_i)$$

and

$$h(t,u) = \sum_{i=1}^{M}\frac{\lambda_i}{1+\alpha\lambda_i}\phi_i(t)\phi_i^*(u).$$

Separable covariance functions arise in optical communication and radar systems that employ time or frequency diversity. H. L. Van Trees [47, pp. 122-130] gives several specific examples.

c. *Finite-Dimensional Covariance*. The covariance function $K(t,u)$ is termed finite dimensional when it can be expressed in the form

$$K(t,u) = \begin{cases} \mathbf{a}'(t)\mathbf{b}^*(u), & t \geq u \\ \mathbf{b}'(t)\mathbf{a}^*(u), & t \leq u \end{cases}$$

for some finite-dimensional vector-valued functions $\mathbf{a}(\cdot)$ and $\mathbf{b}(\cdot)$. A Gaussian process with a finite-dimensional covariance can be realized with a linear dynamical system (see Ex. 5.3.2) excited by a broadband Gaussian process. This realization problem is discussed by B. Anderson, J. Moore, and S. Loo [1]. A. Baggeroer [3, Ch. 6], and H. L. Van Trees [47, p. 371] discuss the determination of $D_{\mathcal{F}}(\alpha)$ and $h(t,u)$ for finite-dimensional covariance functions.

d. *Low-Energy Coherence*. The expected energy in the ith mode of $X(t)$ is from (7.51)

$$E\left[\int_{t_0}^{T}|X_i\phi_i(t)|^2 dt\right] = \lambda_i. \tag{7.64}$$

Thus, the condition $\alpha\lambda_i \ll 1$ can be interpreted as requiring the expected energy in the ith mode to be much smaller than α^{-1}. Following H. L. Van Trees [47, p. 131], we use the term low-energy coherence for the condition that $\alpha\lambda_i \ll 1$ for all i. This does not require the total expected energy $\sum_{i=1}^{\infty}\lambda_i$ to be small. Under the low energy coherence condition, $D_{\mathcal{F}}(\alpha) \approx 1$ and $h(t,u) \approx K(t,u)$. ∎

From its definition in (2.30), the sample-function density can be written as

$$p[\{N(\sigma):t_0 \leq \sigma < T\}]$$

$$= \begin{cases} \Pr[N(T) = 0], & N(T) = n = 0 \\[2ex] p_w(\mathbf{W}\,|\,N(T) = n)\Pr[N(T) = n], & N(T) = n \geq 1. \end{cases} \tag{7.65}$$

Thus, the counting statistics can be determined from the integration

$$\Pr[N(T) = n] = \int\limits_{t_0 \leq W_1 < W_2 \cdots < W_n < T}\!\!\!\!\!\int \cdots \int p[\{N(\sigma): t_0 \leq \sigma < T\}]dW_1 dW_2 \cdots dW_n.$$

$$\tag{7.66}$$

For example, the photocount statistics $\Pr[N(T) = n]$ for zero-mean Gaussian light are seen from (7.66) and (7.54) to be

$$\Pr[N(T) = n] = \alpha^n D_{\mathscr{F}}^{-1}(\alpha) \sum_{\eta \in \mathscr{P}_n} \int\limits_{t_0 \leq W_1 < \cdots < W_n < T}\!\!\!\!\!\int \cdots \int \prod_{i=1}^n h\!\left(W_i, W_{\eta_i}\right)dW_1 dW_2 \cdots dW_n.$$

$$\tag{7.67}$$

7.3.2 Time Statistics by the Method of Self-Exciting Point Processes

According to Theorem 7.2.1, a doubly stochastic Poisson-process $\{N(t): t \geq t_0\}$ with intensity process $\{\lambda(t, \mathbf{x}(t)): t \geq t_0\}$ is a self-exciting process with intensity process $\{\hat{\lambda}(t): t \geq t_0\}$ defined in (7.18). Consequently, the time statistics for a doubly stochastic Poisson-process can be expressed in terms of those developed in Sec 7.3 for a self-exciting process. By combining (6.15) and (7.18), we have that the *conditional survival probability* for the (n+1)*st* point of a doubly stochastic Poisson-process is given for $t \geq W_n$ and $n \geq 1$ by

$$\mathscr{P}_{w_{n+1}|w_1, \cdots, w_n}(t \mid W_1, \cdots, W_n) = \exp\!\left(-\int_{W_n}^t \hat{\lambda}(\sigma)d\sigma\right), \tag{7.68}$$

where $\hat{\lambda}(\sigma) = E[\lambda(\sigma, \mathbf{x}(\sigma)) \mid N(\sigma) = n; W_1, \cdots, W_n]$. By combining (6.22), (6.23), and (7.18), we have that the *joint occurrence density* for the first n occurrence times of a doubly stochastic Poisson-process is given by

$$p_{\mathbf{w}}(\mathbf{W}) = \exp\left(-\int_{t_0}^{W_n} \hat{\lambda}(\sigma)\,d\sigma + \int_{t_0}^{W_n} \ln\hat{\lambda}(\sigma)\,dN(\sigma)\right), \qquad (7.69)$$

where

$$\hat{\lambda}(\sigma) = \begin{cases} E[\lambda(\sigma,\mathbf{x}(\sigma)) \mid N(\sigma) = 0], & N(\sigma) = 0 \\[2ex] E[\lambda(\sigma,\mathbf{x}(\sigma)) \mid N(\sigma) = n; W_1, \cdots, W_n], & N(\sigma) = n \geq 1. \end{cases} \qquad (7.70)$$

Finally, by combining (6.29) and (7.18), the *sample-function density* for a doubly stochastic Poisson-process is given by

$$p[\{N(\sigma): t_0 \leq \sigma < T\}] = \exp\left[-\int_{t_0}^{T} \hat{\lambda}(\sigma)\,d\sigma + \int_{t_0}^{T} \ln\hat{\lambda}(\sigma)\,dN(\sigma)\right], \qquad (7.71)$$

where $\hat{\lambda}(\cdot)$ is given in (7.70).

The estimate $\hat{\lambda}(\cdot)$ appearing in (7.68), (7.69) and (7.71) is in general difficult, if not impossible, to determine. Consequently, these expressions for the time statistics of a doubly stochastic Poisson-process must be viewed as representations rather than explicit formulas. The problem of determining $\hat{\lambda}(\cdot)$ is addressed in the next section.

7.4 Filtering

Estimation problems for observed doubly stochastic Poisson-processes are of basic theoretical and practical importance. As we have already seen in Sections 7.2 and 7.3, estimates of the intensity process appear as a natural component in expressions for the counting and time statistics. The estimation problem also occurs whenever it is desired to determine the information process that influences the intensity.

Our emphasis in this section is on a particular for of estimation problem commonly called the *filtering problem*. This problem is described as follows. Let $\{N(t): t \geq t_0\}$ be a doubly stochastic Poisson-process with intensity process $\{\lambda(t,\mathbf{x}(t)): t \geq t_0\}$, where $\{\mathbf{x}(t): t \geq t_0\}$ is the underlying information process. We suppose that $N(\cdot)$ is observed on the interval $[t_0, t)$ and that the entire counting path $\{N(\sigma): t_0 \leq \sigma < t\}$ is available for processing to form estimates. The end-point time t is allowed to increase from t_0, corresponding to a real-time parameter as additional data are accumulated. We desire to estimate the value at time t of some specified function of the information process; for example, this function could be the intensity process $\lambda(t,\mathbf{x}(t))$ or the information process $\mathbf{x}(t)$ itself. An estimate is sought that evolves in time as new data become available; all the data accumulated up to any time can be used to form the estimate at that time. This causal dependence of the estimate on the observed data implies that there is a potentiality for generating the estimate in real time as data arrive. It is important to note that in the filtering problem, we are not allowing the possibility of returning to update or improve any previous estimate after new data are obtained.

There are three main topics in this section. The first, in Sec. 7.4.1, is the linear filtering problem. In this, the estimate is constrained to be a linear functional of the data. The motivation for this constraint is that it leads to estimators that are relatively easy to realize with either hardware or numerical algorithms. The disadvantage is that for many applications, linearly constrained estimators are not accurate enough. Thus, the second topic, in Sec. 7.4.2, is the filtering problem without the linearity constraint. Here, the major difficulty is the analytic intractability of the estimates, but we do manage to make substantial progress in specifying suboptimal estimators that are found to work well in certain situations even at low count rates. The final topic, in Sec. 7.4.3, is the mean-square performance of estimators.

7.4.1 Linear Minimum Mean Square-Error Filtering

Let $\{N(t): t \geq t_0\}$ be a doubly stochastic Poisson-process with intensity process $\{\lambda(t): t \geq t_0\}$. We seek an estimate of the intensity at time t in terms of an observed counting path on the interval $[t_0, t)$. The estimator is constrained to be a linear functional of the data; that is, the estimate $\lambda_{\text{lin}}^*(t)$ of $\lambda(t)$ must have the form

$$\lambda_{\text{lin}}^*(t) = a(t) + \int_{t_0}^{t} h(t, u) \, dN(u) \tag{7.72}$$

for some deterministic function $a(t)$ and impulse response $h(t, u)$. As an example, a possible linear estimator occurs for the selection

$$a(t) = 0 \quad \text{and} \quad h(t, u) = \begin{cases} \dfrac{1}{\tau}, & t - \tau \leq u < t \\ 0, & \text{otherwise.} \end{cases}$$

For this choice,

$$\lambda_{\text{lin}}^*(t) = \frac{1}{\tau} \int_{t-\tau}^{t} dN(u) = \frac{1}{\tau}[N(t) - N(t - \tau)]. \tag{7.73}$$

This is a moving-average, histogram estimate of the intensity process. Such an estimate is widely used because it has the advantage of requiring almost no knowledge about the intensity process beyond that needed to select the averaging time τ. It is common practice to sample the moving-average estimate at discrete times, for instance at integral multiples of τ, and then to least-squares curve fit an assumed time function to the sampled values.

An estimator having a performance that is superior to the moving average estimator can be designed if more knowledge about the statistics of the intensity process is available. Such statistical knowledge might be available in practice because of having measured it experimentally or because of having a sufficient understanding of the physical mechanism involved to postulate it. The estimator we consider next requires a knowledge of the first- and second-order statistics of the intensity process for its implementation.

Let $\hat{\lambda}_{\text{lin}}(t)$ denote the linear estimate $\lambda^*_{\text{lin}}(t)$ that results in (7.72) when both $a(t)$ and $h(t,u)$ are selected to minimize the mean square-error $E[\{\lambda(t)-\lambda^*_{\text{lin}}(t)\}^2]$. This linear, minimum mean square-error estimate is given in the following theorem due to J. Grandell [15, 16], O. Macchi [29, 30, 31], and J. Clark [6]. We use Grandell's proof.

Theorem 7.4.1 (*Optimum Linear-Estimator*). Let $\lambda^*_{\text{lin}}(t)$ be a linear estimate of $\lambda(t)$ and $E[\{\lambda(t)-\lambda^*_{\text{lin}}(t)\}^2]$ the corresponding mean square-error. Denote the estimate that minimizes this error by $\hat{\lambda}_{\text{lin}}(t)$. Then,

$$\hat{\lambda}_{\text{lin}}(t) = E[\lambda(t)] + \int_{t_0}^{t} h_o(t,u)\{dN(u)-E[\lambda(u)]du\}, \qquad (7.74)$$

where the optimum impulse-response $h_o(t,u)$ is the solution to the integral equation

$$h_o(t,u)E[\lambda(u)] + \int_{t_0}^{t} h_o(t,\sigma)K_\lambda(\sigma,u)d\sigma = K_\lambda(t,u), \quad t_0 \le u \le t. \qquad (7.75)$$

Here, $K_\lambda(t,u)$ is the covariance function of the intensity process. Furthermore, the resulting minimum value of the mean square-error is

$$E\{[\lambda(t)-\hat{\lambda}_{\text{lin}}(t)]^2\} = h_o(t,t)E[\lambda(t)]. \qquad (7.76)$$

Proof. Let $\lambda^*_{\text{lin}}(t)$ be an arbitrary linear estimate, and suppose that $h_o(t,u)$ satisfies (7.75). Write $\lambda^*_{\text{lin}}(t)$ as

$$\lambda^*_{\text{lin}}(t) = a(t) + \int_{t_0}^{t} \{h_o(t,u)+g(t,u)\} \{dN(u)-E[\lambda(u)]du\}. \qquad (7.77)$$

By using the assertions in Prob. 7.2.6, we deduce that

$$E\left[\{\lambda(t)-\lambda^*_{\text{lin}}(t)\}^2\right] = E[\{(\lambda(t)-\overline{\lambda}(t))-(a(t)-\overline{\lambda}(t))$$

$$-\int_{t_0}^{t}[h_o(t,u)+g(t,u)][dN(u)-\overline{\lambda}(u)\,du\,]\}^2].$$

Thus,

$$E\left[\{\lambda(t)-\lambda^*_{\text{lin}}(t)\}^2\right]$$

$$= K_\lambda(t,t)+[a(t)-\overline{\lambda}(t)]^2+\int_{t_0}^{t}\overline{\lambda}(u)[h_o(t,u)+g(t,u)]^2du$$

$$+\int_{t_0}^{t}\int_{t_0}^{t}[h_o(t,u)+g(t,u)]K_\lambda(u,v)[h_o(t,v)+g(t,v)]du\,dv$$

$$-2\int_{t_0}^{t}[h_o(t,u)+g(t,u)]K_\lambda(t,u)\,du,$$

where $\overline{\lambda}(t)=E[\lambda(t)]$. Manipulation and use of (7.75) then yields

$$E\left[\{\lambda(t)-\lambda^*_{\text{lin}}(t)\}^2\right] = K_\lambda(t,t)+[a(t)-\overline{\lambda}(t)]^2-\int_{t_0}^{t}h_o(t,u)K_\lambda(t,u)\,du$$

$$+\int_{t_0}^{t}\int_{t_0}^{t}g(t,u)K_\lambda(u,v)g(t,v)\,du\,dv$$

$$+\int_{t_0}^{t}\overline{\lambda}(u)g^2(t,u)\,du. \tag{7.78}$$

Because $\overline{\lambda}(t)\geq 0$ and $K_\lambda(t,u)$ is nonnegative definite, there holds $\int_{t_0}^{t}\overline{\lambda}(u)g^2(t,u)\,du\geq 0$ and $\int_{t_0}^{t}\int_{t_0}^{t}g(t,u)K_\lambda(u,v)g(t,v)\,du\,dv\geq 0$. Consequently, $E\{[\lambda(t)-\lambda^*_{\text{lin}}(t)]^2\}$ is minimized by setting $a(t)=\overline{\lambda}(t)$ and $g(t,u)=0$. It follows from (7.77) that $\hat{\lambda}_{\text{lin}}(t)$ is given by (7.74). We have also that

$$E\{[\lambda(t)-\hat{\lambda}_{\text{lin}}(t)]^2\} = K_\lambda(t,t)-\int_{t_0}^{t}h_o(t,u)K_\lambda(t,u)\,du. \tag{7.79}$$

Equation (7.76) follows from this and (7.75). $\square$

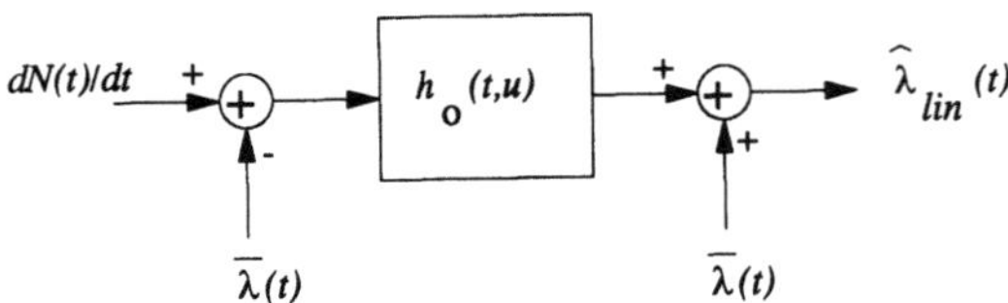

Figure 7.1 Optimum linear intensity estimator.

The linear estimator that minimizes the mean square-error is shown schematically in Fig. 7.1. The impulse response of the optimum linear filter $h_o(t,u)$ satisfies the integral equation (7.75). The mean square-error performance of this estimator is given in terms of the optimum impulse response in (7.76). The mean-value and covariance functions of the intensity process must be known in order to determine this estimator and predict its performance.

It is perhaps worth emphasizing that only the mean and covariance functions are needed for this linearly constrained estimator. Any additional statistical or structural information that may be available about the intensity process is not used and can be disregarded. Thus, for instance, while we have required that the mean of the intensity process be nonnegative for (7.78), we have not made use of the property that the intensity process itself is nonnegative. Two intensity processes with strikingly different sample paths will lead to the same linear intensity estimator if their mean and covariance functions are equal.

The integral equation (7.75) has been studied extensively as it arises with the linear filtering problem for observations that contain additive noise; see, for example, H. L. Van Trees [46, Ch. 6]. The additive model that leads to the same integral equation is the following. Let $\{\lambda(t): t \geq t_0\}$ be the intensity process defined above, and suppose that observations of it of the additive form

$$r(t) = \lambda(t) + \sqrt{\lambda(t)}\, w(t), \quad t \geq t_0 \tag{7.80}$$

are available, where $\{w(t): t \geq t_0\}$ is a zero-mean, white noise process with unit intensity and independent of $\lambda(\cdot)$. Then, the linear estimator that minimizes the mean square-error in estimating $\lambda(t)$ in terms of $\{r(\sigma): t_0 \leq \sigma < t\}$ is that shown in Fig. 7.1 with $h_o(t,u)$ specified by (7.75). This remark is very important. It implies that all of the highly developed, linear filtering theory for the additive observation model can be applied directly to the doubly stochastic Poisson-process observation model.

Perhaps the most notable aspect of this theory because of its computational advantage is the Kalman-Bucy formulation, about which extensive treatments are readily available; see, for example, E. Wong and B. Hajek [49, Ch. 3], H. L. Van Trees [46, Ch. 6], and A. Jazwinski [20, Ch. 7]. Rather than rederive this formulation here, we shall only indicate its use for doubly stochastic Poisson-process observations. We first need to postulate a linear dynamical model for the intensity process. This might be accomplished by knowing physical mechanisms generating the intensity process, but this is unnecessary because, as we have already emphasized, the model needs only to match the mean and covariance functions of the intensity process, so even a Gaussian model can be used as long as these functions are correct; systematic techniques are given by B. Anderson, J. Moore, and S. Loo for identifying a linear dynamical realization for a specified covariance function [1].

Suppose that the following linear dynamical model is available for the intensity process:

$$d\mathbf{x}(t) = \mathbf{A}(t)\mathbf{x}(t)\,dt + \mathbf{b}(t)\,du(t), \quad \mathbf{x}(t_0) = \mathbf{x}_0$$

$$\lambda(t) = E[\lambda(t)] + \mathbf{c}'(t)\mathbf{x}(t), \tag{7.81}$$

where: $\mathbf{x}(\cdot)$ is an n-dimensional state vector; $\mathbf{A}(\cdot)$ is a known $n \times n$ matrix, and $\mathbf{b}(\cdot)$ and $\mathbf{c}(\cdot)$ are known n-dimensional vectors; $\mathbf{x}_0$ is an initial random variable with zero mean and known covariance Σ_0; and $\{u(t): t \geq t_0\}$ is a zero-mean random process with stationary orthogonal increments. We normalize $u(\cdot)$ so that $E[u(s)u(t)] = \min(s,t)$. It now follows from (7.80), (7.81), and the Kalman-Bucy algorithm that $\hat{\lambda}_{\text{lin}}(t)$ is given by the following equations:

$$d\hat{\mathbf{x}}(t) = \mathbf{A}(t)\hat{\mathbf{x}}(t)\,dt + \Sigma(t)\mathbf{c}(t)\frac{1}{\overline{\lambda}(t)}[dN(t) - \overline{\lambda}(t)\,dt - \mathbf{c}'(t)\hat{\mathbf{x}}(t)\,dt], \tag{7.82a}$$

$$\hat{\mathbf{x}}(t_0) = \mathbf{0},$$

$$\frac{d\Sigma(t)}{dt} = \mathbf{A}(t)\Sigma(t) + \Sigma(t)\mathbf{A}'(t) + \mathbf{b}(t)\mathbf{b}'(t) - \frac{1}{\overline{\lambda}(t)}\Sigma(t)\mathbf{c}(t)\mathbf{c}'(t)\Sigma(t), \tag{7.82b}$$

$$\Sigma(t_0) = \Sigma_0,$$

$$\hat{\lambda}_{\text{lin}}(t) = \overline{\lambda}(t) + \mathbf{c}'(t)\hat{\mathbf{x}}(t). \tag{7.82c}$$

Furthermore, the mean square-error in estimating $\lambda(t)$ is given by

$$E\{[\lambda(t) - \hat{\lambda}_{\text{lin}}(t)]^2\} = \mathbf{c}'(t)\Sigma(t)\mathbf{c}(t). \tag{7.83}$$

Equation (7.82b) is a matrix Riccati equation and has been studied extensively. In some few examples, it is possible to solve (7.82a,b) analytically to get an explicit expression for the estimate (7.82c). More often, however, (7.82a,b) are used to define a numerical updating algorithm by which the estimate $\hat{\lambda}_{\text{lin}}(t)$ is propagated forward in time as data are accumulated. We now give some examples that illustrate the use of these equations.

Example 7.4.1 *Bayesian Regression* ─────────────────────────
Suppose that the intensity process has the form

$$\lambda(t) = \sum_{i=1}^{M} a_i c_i(t) = \mathbf{c}'(t)\mathbf{a} \tag{7.84}$$

for $t \geq t_0$, where each regression coefficient a_i is a random variable with known mean $E(a_i)$, and each regression function $c_i(t)$ is known. Denote the covariance matrix for the coefficient vector $\mathbf{a}$ by Σ_0; we assume that this matrix is known and positive definite. The regression functions could be a sequence of exponentials with known decay rates, a sequence of orthonormal functions, a sequence of polynomial spline functions, and so on. The problem is to observe the doubly stochastic Poisson-process with the intensity in (7.84) and then to estimate the coefficients $\mathbf{a}$. The linear minimum mean square-error estimator can be determined by modeling the intensity in the form of (7.81). For this purpose, define $\mathbf{x}(t)$ by $\mathbf{x}(t) = \mathbf{a} - E(\mathbf{a})$. Then, (7.81) becomes

$$d\mathbf{x}(t) = \mathbf{0}, \quad \mathbf{x}(t_0) = \mathbf{x}_0, \quad \lambda(t) = \mathbf{c}'(t)E(\mathbf{a}) + \mathbf{c}'(t)\mathbf{x}(t), \tag{7.85}$$

where $\mathbf{x}_0$ is a zero mean random variable with covariance Σ_0. The linear estimator of the coefficients is obtained by specializing (7.82) for the model of (7.85). The result is

$$d\hat{\mathbf{x}}(t) = \frac{1}{\bar{\lambda}(t)}\Sigma(t)\mathbf{c}(t)[dN(t) - \bar{\lambda}(t)dt - \mathbf{c}'(t)\hat{\mathbf{x}}(t)dt], \quad \hat{\mathbf{x}}(t_0) = \mathbf{0} \tag{7.86a}$$

$$\frac{d\Sigma(t)}{dt} = -\frac{1}{\bar{\lambda}(t)}\Sigma(t)\mathbf{c}(t)\mathbf{c}'(t)\Sigma(t), \quad \Sigma(t_0) = \Sigma_0, \tag{7.86b}$$

where $\bar{\lambda}(t) = E[\lambda(t)] = \mathbf{c}'(t)E(\mathbf{a})$. The solution to (7.86b) is

$$\Sigma(t) = \left[\mathbf{I} + \Sigma_0 \int_{t_0}^{t} \mathbf{c}(\sigma)\frac{1}{\bar{\lambda}(\sigma)}\mathbf{c}'(\sigma)d\sigma\right]^{-1}\Sigma_0. \tag{7.87}$$

The estimate of the regression coefficients is given by $\hat{\mathbf{a}}(t) = \hat{\mathbf{x}}(t) + E(\mathbf{a})$ and of the intensity by $\hat{\lambda}_{\text{lin}}(t) = \mathbf{c}'(t)\hat{\mathbf{a}}(t)$. ∎

Example 7.4.2 *Subcarrier Amplitude Modulation* ———————————
Suppose than an optical communication system is used to relay an analog information process $y(t)$ to a remote user. For this purpose, $y(t)$ amplitude modulates a subcarrier $\cos(\omega t)$ of frequency ω. This modulated subcarrier is then used to modulate the light intensity of an optical source. According to W. Pratt [34, p. 178], the term $|S(t,\vec{r})|^2$ appearing in (5.22) can be taken to be

$$|S(t,\vec{r})|^2 = P[1 + m\,y(t)\cos(\omega t)], \tag{7.88}$$

where P is a nonnegative constant, and m is the modulation index. The response of an ideal photodetector can be modeled as a doubly stochastic Poisson-process with intensity process $\{\lambda(t): t \geq t_0\}$, where

$$\lambda(t) = a[1 + \mu(t)\cos(\omega t)] + \lambda_0, \tag{7.89}$$

in which a is a nonnegative constant that depends on the detector area and quantum efficiency, and λ_0 is the thermoelectron rate due to dark current. We assume that the information process has a finite-dimensional realization as

$$d\mathbf{x}(t) = \mathbf{A}(t)\mathbf{x}(t)\,dt + \mathbf{b}(t)\,du(t), \quad \mathbf{x}(t_0) = \mathbf{x}_0$$

$$y(t) = \mathbf{h}'(t)\mathbf{x}(t), \tag{7.90}$$

where $\{u(t): t \geq t_0\}$ is a zero mean random process with orthogonal increments and normalized so that $E[u(s)u(t)] = \min(s,t)$. The initial state $\mathbf{x}_0$ has zero mean and covariance Σ_0. It follows that the intensity process has the representation in (7.81) with $\mathbf{c}(t) = am\cos(\omega t)\mathbf{h}(t)$ and $E[\lambda(t)] = a + \lambda_0$. Equations (7.82a,b) become

$$d\hat{\mathbf{x}}(t) = \mathbf{A}(t)\hat{\mathbf{x}}(t)\,dt$$

$$+ \frac{1}{a + \lambda_0}am\cos(\omega t)\Sigma(t)\mathbf{h}(t)\{dN(t) - [a(1 + m\mathbf{h}'(t)\hat{\mathbf{x}}(t)\cos(\omega t) + \lambda_0]dt\}$$

$$\tag{7.91}$$

and

$$\frac{d\Sigma(t)}{dt} = \mathbf{A}(t)\Sigma(t) + \Sigma(t)\mathbf{A}'(t) - \frac{1}{a+\lambda_0}[am\cos(\omega t)]^2\Sigma(t)\mathbf{h}(t)\mathbf{h}'(t)\Sigma(t).$$

$$(7.92)$$

The initial conditions are $\hat{\mathbf{x}}(t_0) = \mathbf{0}$ and $\Sigma(t_0) = \Sigma_0$. The linear minimum mean square-error estimate of the information process is $\hat{y}_{\text{lin}}(t) = \mathbf{h}'(t)\hat{\mathbf{x}}(t)$. The information process normally has a bandwidth that is much smaller than the subcarrier frequency ω. If this condition holds, then terms in (7.91) and (7.92) that vary at twice the subcarrier frequency can be eliminated. The corresponding minimum mean square-error is then

$$E\{[y(t)-\hat{y}_{\text{lin}}(t)]^2\} = \mathbf{h}'(t)\Sigma(t)\mathbf{h}(t).$$

$$(7.93)\blacksquare$$

Example 7.4.3 *On-Off Modulated Light* ——————————————

Suppose that a light source is turned on and off by a random telegraph wave. The rate $\mu(t)$ at which photoelectrons are generated in a photodetector will then follow the random telegraph wave; a typical sample function is shown in Fig. 7.2. This photocount intensity has the following structure. At time t_0, the intensity has the value 0 or a with equal probability. For $t \geq t_0$, the intensity alternates between the values 0 and a, switching at the occurrence times of a homogeneous Poisson process with constant intensity ν. Thus, for $t \geq t_0$, $\mu(t)$ can be expressed as

$$\mu(t) = \frac{a}{2}[1 + z(-1)^{M(t)}], \qquad (7.94)$$

Figure 7.2 Photocount intensity for on-off modulated light.

where z is a discrete random variable having the value -1 or +1 with equal probability, and where $\{M(t): t \geq t_0\}$ is a homogeneous Poisson process with intensity ν and independent of z. Let us suppose that the photodetector also generates thermoelectrons at a rate λ_0. Then, the detector output is a doubly stochastic Poisson-process with intensity process $\{\lambda(t): t \geq t_0\}$, where

$$\lambda(t) = \mu(t) + \lambda_0 = \left(\frac{a}{2} + \lambda_0\right) + z\left(\frac{a}{2}\right)(-1)^{M(t)}. \tag{7.95}$$

We need the mean-value and covariance functions for $\lambda(\cdot)$. It is easily seen that $E[\lambda(t)] = (a/2) + \lambda_0$. The covariance function is

$$K_\lambda(t, u) = E(z^2)\left(\frac{a}{2}\right)^2 E[(-1)^{M(t)+M(u)}] = \left(\frac{a}{2}\right)^2 e^{-2\nu(u-t)}, \quad t_0 \leq t < u.$$

In evaluating the expectation, we have used

$$E[(-1)^{M(t)+M(u)}] = E[(-1)^{M(u)-M(t)}(-1)^{2M(t)}] = E[(-1)^{M(u)-M(t)}].$$

As $K_\lambda(t, u) = K_\lambda(u, t)$, we have for all $t \geq 0$ and $u \geq 0$ that

$$K_\lambda(t, u) = \left(\frac{a}{2}\right)^2 e^{-2\nu|t-u|}. \tag{7.96}$$

We next determine the linear minimum mean square-error estimate of the signal intensity $\mu(t)$ in terms of the photodetector output. There are two ways to proceed. The first is to solve the integral equation (7.75) for the optimum impulse response, and the second is to formulate a linear dynamical model for $\lambda(\cdot)$ and use (7.82). We illustrate the latter approach. The model does not need to reproduce the sample function structure of the intensity process in (7.95). Rather, it only has to replicate the mean-value and covariance functions of $\lambda(\cdot)$. A one-dimensional dynamical model that accomplishes this is

$$dx(t) = -2\nu x(t)\,dt + du(t), \quad x(t_0) = x_0$$

$$\mu(t) = \frac{a}{2} + a\sqrt{\nu}\,x(t), \quad \lambda(t) = \mu(t) + \lambda_0, \tag{7.97}$$

where x_0 is a zero mean random variable with variance $1/4v$, and where $\{u(t): t \geq t_0\}$ is a zero mean random process with stationary, orthogonal increments, independent of x_0, and normalized so that $E[u(s)u(t)] = \min(s,t)$. The linearly constrained estimator is now seen from (7.82) to be specified by the equations

$$d\hat{x}(t) = -2v\hat{x}(t)dt + \frac{a\sqrt{v}}{(a/2)+\lambda_0}\Sigma(t)[dN(t)-\hat{\lambda}_{\text{lin}}(t)dt], \quad \hat{x}(t_0)=0$$

$$(7.98a)$$

$$\frac{d\Sigma(t)}{dt} = -4v\Sigma(t)+1-\frac{a^2v}{(a/2)+\lambda_0}\Sigma^2(t), \quad \Sigma(t_0)=\frac{1}{4v} \qquad (7.98b)$$

$$\hat{\lambda}_{\text{lin}}(t) = \hat{\mu}_{\text{lin}}(t)+\lambda_0 = \frac{a}{2}+a\sqrt{v}\hat{x}(t)+\lambda_0. \qquad (7.98c)$$

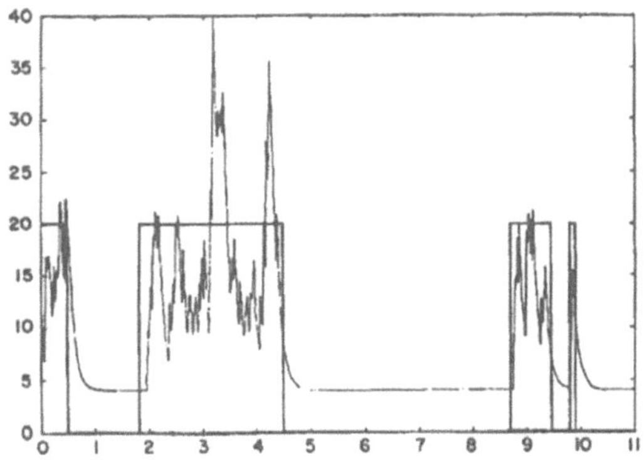

Figure 7.3 Linear estimation of on-off modulated light.

We performed a computer simulation of the processes in this example for the parameters $v=2.0$, $a=20.0$, and $\lambda_0=0.1$. The results are shown in Fig. 7.3. The process $\{\mu(t): t \geq 0\}$ had an initial value of 20.0, and transitions between the on and off states occurred at times 0.49, 1.81, 4.49, 8.70, 9.45, 9.80, and 9.85 seconds. During the ten second interval shown, a total of 72 points occurred in the observed doubly stochastic Poisson-process. The linearly constrained estimate $\{\hat{\mu}_{\text{lin}}(t): t \geq 0\}$ obtained by implementing (7.98) is shown. ∎

7.4.2 Nonlinear Minimum Mean Square-Error Filtering

We now examine the filtering problem for doubly stochastic Poisson-processes when the linearity constraint imposed in the previous section is removed. There are two important reasons for removing this constraint. The first is that for many applications, the linearly constrained estimators do not perform as well as alternatives designed without this constraint. The second reason is that linearly constrained, minimum mean square-error estimators do not hold the fundamental theoretical position of minimum mean square-error estimators without the constraint. This is evident from our investigations in earlier sections of the counting and time statistics for doubly stochastic Poisson-processes; the conditional mean estimators required in the representations developed are only infrequently linear.

We adopt the following notations throughout this subsection.

i.   The inner, or dot, product of two n-dimensional vectors v_1 and v_2 will be denoted by $< v_1, v_2 >$, where

$$< v_1, v_2 > = \sum_{i=1}^{n} v_{1i} v_{2i}. \tag{7.99}$$

ii.   The length of a vector will be denoted by $|v|$, where

$$|v| = \sqrt{< v, v >}. \tag{7.100}$$

iii. Let $\mathbf{h}(t, \mathbf{x}(t))$ be a possibly time-dependent vector-valued function of the information process such that $E[|\mathbf{h}(t, \mathbf{x}(t))|^2] < \infty$. An arbitrary filtering estimate of this function in terms of an observed counting path $\{N(\sigma): t_0 \leq \sigma < t\}$ will be denoted by $\mathbf{h}^*(t)$. The conditional expectation of this function given the counting path will be denoted by $\hat{\mathbf{h}}(t)$,

$$\hat{\mathbf{h}}(t) = E[\mathbf{h}(t, \mathbf{x}(t)) \mid N(\sigma): t_0 \leq \sigma < t]. \tag{7.101}$$

A well known relationship between $\mathbf{h}^*(t)$ and $\hat{\mathbf{h}}(t)$ is given in the following lemma.

Lemma 7.4.1 (*The conditional mean is optimum*). Let $\{N(t): t \geq t_0\}$ be a doubly stochastic Poisson-process with intensity process $\{\lambda(t, \mathbf{x}(t)): t \geq t_0\}$, where $\{\mathbf{x}(t): t \geq t_0\}$ is an information process. The estimate $\mathbf{h}^*(t)$ that minimizes the mean square-error $E[|\mathbf{h}(t, \mathbf{x}(t)) - \mathbf{h}^*(t)|^2]$ in terms of a given counting path $\{N(\sigma): t_0 \leq \sigma < t\}$ is the conditional mean $\hat{\mathbf{h}}(t)$, and the smallest value of the mean square error is

$$\min_{\mathbf{h}^*} E(|\,\mathbf{h}(t,\mathbf{x}(t)) - \mathbf{h}^*(t)|^2) = E(|\,\mathbf{h}(t,\mathbf{x}(t))|^2) - E(|\,\hat{\mathbf{h}}(t)|^2). \qquad (7.102)$$

proof. This lemma is easily established as follows. By expanding the mean square-error using the relation

$$E(<\mathbf{h}(t,\mathbf{x}(t)),\mathbf{h}^*(t)>) = E[E(<\mathbf{h}(t,\mathbf{x}(t)),\mathbf{h}^*(t)>|\,N(\sigma): t_0 \le \sigma < t)]$$

$$= E[<\hat{\mathbf{h}}(t),\mathbf{h}^*(t)>],$$

we have for an arbitrary estimate $\mathbf{h}^*(\cdot)$ that

$$E(|\,\mathbf{h}(t,\mathbf{x}(t)) - \mathbf{h}^*(t)|^2)$$

$$= E(|\,\mathbf{h}(t,\mathbf{x}(t))|^2) - 2E(<\mathbf{h}(t,\mathbf{x}(t)),\mathbf{h}^*(t)>) + E(|\,\mathbf{h}^*(t)|^2)$$

$$= E(|\,\mathbf{h}(t,\mathbf{x}(t))|^2) - 2E(<\hat{\mathbf{h}}(t),\mathbf{h}^*(t)>) + E(|\,\mathbf{h}^*(t)|^2)$$

$$= E(|\,\mathbf{h}(t,\mathbf{x}(t))|^2) + E(|\,\hat{\mathbf{h}}(t) - \mathbf{h}^*(t)|^2) - E(|\,\hat{\mathbf{h}}(t)|^2).$$

Only the second term on the right is influenced by the selection of $\mathbf{h}^*(t)$, and this term is nonnegative, so the mean square-error is smallest when the term is zero. $\square$

The nonlinear filtering problem is this. A doubly stochastic Poisson-process $\{N(t): t \ge t_0\}$ having intensity process $\{\lambda(t,\mathbf{x}(t)): t \ge t_0\}$ is observed on the interval $[t_0, t)$. Based on these observations we want to estimate at time t a specified function $\mathbf{h}(t,\mathbf{x}(t))$ of the information process. For instance, this function might be the intensity $\lambda(t,\mathbf{x}(t))$ or the information process $\mathbf{x}(t)$. There is no linearity constraint on the estimator. The problem is to determine the structure of the estimator for some performance criterion. The performance criterion we adopt is minimum mean square-error, so we know from Lemma 7.4.1 that the estimator must implement the conditional mean of $\mathbf{h}(t,\mathbf{x}(t))$ given the observed points. This conditional mean can be evaluated and the nonlinear filtering problem thus solved if we know the posterior statistics of the information process at any time given observations of the points up to that time.

Alternative performance criteria are also useful in some applications even though they do not seem to hold the same fundamental position as the minimum mean square-error criterion. An estimate that minimizes the expectation of the magnitude of the error is sometimes adopted. With this criterion, large errors are weighted less heavily than with the minimum mean square-error criterion. It can be shown that the estimate $\mathbf{h}^*(t)$ that minimizes $E(|\,\mathbf{h}(t,\mathbf{x}(t))-\mathbf{h}^*(t)|\,)$ in terms of $\{N(\sigma):t_0\le\sigma<t\}$ is the *conditional median* of $\mathbf{h}(t,\mathbf{x}(t))$ given $\{N(\sigma):t_0\le\sigma<t\}$; see Prob. 7.4.4. A uniform error weighting is also used. With this choice, it is desired to minimize $E[C_\varepsilon(|\,\mathbf{h}(t,\mathbf{x}(t))-\mathbf{h}^*(t)|\,)]$, where

$$C_\varepsilon(\alpha)=\begin{cases}0, & 0\le\alpha<\varepsilon\\ 1, & \varepsilon\le\alpha.\end{cases}$$

It can be shown that in the limit as $\varepsilon\to 0$, the estimate $\mathbf{h}^*(t)$ that minimizes $E[C_\varepsilon(|\,\mathbf{h}(t,\mathbf{x}(t))-\mathbf{h}^*(t)|\,)]$ in terms of $\{N(\sigma):t_0\le\sigma<t\}$ is the *conditional mode* of $\mathbf{h}(,\mathbf{x}(t))$ given $\{N(\sigma):t_0\le\sigma<t\}$. This estimate is termed the *maximum a posteriori probability estimate*; see H. L. Van Trees [46, Ch. 2]. The maximum a posteriori probability estimate is a natural extension of the maximum likelihood estimates used in previous chapters when the quantity to be estimated has a known prior probability distribution.

As with the conditional-mean estimate, both the conditional-median and conditional-mode estimates are determined by the posterior statistics of the information process at any time given observations of the point process up to that time. With this motivation, we now study these posterior statistics.

Posterior Statistics by the Characteristic Function Method

We shall assume in what follows that the intensity $\{\lambda(t,\mathbf{x}(t)):t\ge t_0\}$ is a positive function of an n-dimensional Markov process $\{\mathbf{x}(t):t\ge t_0\}$. These processes are defined in Ch. 5. In this subsection, we seek a characterization of the posterior statistics of $\mathbf{x}(t)$ given $\{N(\sigma):t_0\le\sigma<t\}$ for each $t\ge t_0$. We begin by characterizing the prior statistics of $\mathbf{x}(t)$. The characterization of the posterior statistics will be a natural generalization of this.

The time evolution of the prior statistics of a Markovian information process can be described by a differential equation for the characteristic function $M_t(j\mathbf{v})$ for $\mathbf{x}(t)$, where

$$M_t(j\mathbf{v})=E(e^{j<\mathbf{v},\mathbf{x}(t)>}). \tag{7.103}$$

For the purpose of deriving this differential equation, we need to assume that there exists a nonnegative function $g_t(\mathbf{v},\mathbf{x}(t))$ such that $E\{|\,g_t(\mathbf{v},\mathbf{x}(t))|\,\}<\infty$ and such that for all $\Delta t>0$ there holds

$$\frac{1}{\Delta t}|\,E\{e^{j<\mathbf{v},\Delta\mathbf{x}(t)>}-1|\,\mathbf{x}(t)\}\,|\le g_t(\mathbf{v},\mathbf{x}(t)), \tag{7.104}$$

where $\Delta x(t) = x(t + \Delta t) - x(t)$. Denote by $\Psi_t(v \mid x(t))$ the limit

$$\Psi_t(v \mid x(t)) = \lim_{\Delta t \downarrow 0} \frac{1}{\Delta t} E\{e^{j <v, x(t)>} - 1 \mid x(t)\}. \tag{7.105}$$

This function is termed the *characteristic form for the differential generator* of the Markov process $\{x(t): t \geq t_0\}$. It and the distribution of the initial condition $x(t_0)$ characterize $\{x(t): t \geq t_0\}$ statistically.

From these assumptions, it can be shown that the time evolution of the unconditional characteristic function of $x(t)$ is described for $t \geq t_0$ by

$$\frac{\partial M_t(jv)}{\partial t} = E\{e^{j <v, x(t)>}\Psi_t(v \mid x(t))\}, \quad M_{t_0}(jv) = E\{e^{j <v, x(t_0)>}\}.$$

$$\tag{7.106}$$

To see this, note first that

$$M_{t+\Delta t}(jv) = E\{e^{j <v, x(t+\Delta t)>}\} = E\{e^{j <v, x(t)>} e^{j <v, \Delta x(t)>}\}$$

$$= E\{e^{j <v, x(t)>} E(e^{j <v, \Delta x(t)>} \mid x(t))\}.$$

Thus,

$$\frac{1}{\Delta t}[M_{t+\Delta t}(jv) - M_t(jv)] = E\left\{ e^{j <v, x(t)>} \frac{1}{\Delta t} E(e^{j <v, \Delta x(t)>} - 1 \mid x(t)) \right\}.$$

Taking the limit as $\Delta t \downarrow 0$ results in the desired expression (7.106). The interchange of limit and expectation is justified by assumption (7.104) and the bounded convergence theorem, Theorem 1.4.1.

Four examples that illustrate the variety of information processes accommodated by the model are as follows.

Random Variables. Suppose that $x(t) = x(t_0)$ for $t \geq t_0$. It is easily seen from (7.105) that the function $\Psi_t(v \mid x(t))$ is zero. Substitution of this into (7.106) shows that for all $t \geq t_0$, $M_t(jv) = M_{t_0}(jv)$.

Poisson Driven Markov Processes. These processes are the subject of Ch. 5. The function $\Psi_t(v \mid x(t))$ is given in (5.100).

<u>*Markov Diffusion Processes*</u>. These processes are the continuous analogs of Poisson driven Markov processes. They differ in the sense that the excitation of the differential equation (5.39) is a continuous, independent-increment process rather than a discontinuous independent-increment counting-process. Here, we suppose that $\{\mathbf{x}(t): t \geq t_0\}$ is the solution to a stochastic differential equation of the form

$$d\mathbf{x}(t) = \mathbf{a}(t, \mathbf{x}(t))\,dt + \mathbf{B}(t, \mathbf{x}(t))\,d\mathbf{w}(t), \quad \mathbf{x}(t_0) = \mathbf{x}_0 \qquad (7.107)$$

in which $\{\mathbf{w}(t): t \geq t_0\}$ is a vector of independent Wiener processes such that $E[\mathbf{w}(t)\mathbf{w}'(u)] = \mathbf{I}\min(t, u)$, and where the initial condition $\mathbf{x}_0$ is a random variable independent of $\{\mathbf{w}(t): t \geq t_0\}$. Under conditions on $\mathbf{a}(t, \mathbf{x}(t))$ and $\mathbf{B}(t, \mathbf{x}(t))$ similar to those of Theorem 5.3.3, the solution to (7.107) exists and is unique; see E. Wong and B. Hajek [49, Ch. 4]. Furthermore, the solution forms a Markov process, which has as a special case stationary Gaussian processes having rational power-density spectra. The characteristic form for the differential generator is given by (see Prob. 7.4.5):

$$\Psi_t(\mathbf{v} \mid \mathbf{x}(t)) = j < \mathbf{v}, \mathbf{a}(t, \mathbf{x}(t)) > -\frac{1}{2} < \mathbf{v}, \mathbf{B}(t, \mathbf{x}(t))\mathbf{B}'(t, \mathbf{x}(t))\mathbf{v} > .$$

$$(7.108)$$

Substitution of this expression into (7.106) and inverse Fourier transformation shows that the probability density function $p_{\mathbf{x}(t)}(\mathbf{X}) \equiv p_t(\mathbf{X})$ for $\mathbf{x}(t)$ satisfies the following partial-differential equation

$$\frac{\partial p_t(\mathbf{X})}{\partial t} = -\sum_{i=1}^{n} \frac{\partial[a_i(t, \mathbf{X})p_t(\mathbf{X})]}{\partial X_i} + \frac{1}{2}\sum_{i=1}^{n}\sum_{j=1}^{n} \frac{\partial^2[(\mathbf{B}(t, \mathbf{x}(t))\mathbf{B}'(t, \mathbf{x}(t)))_{i,j} p_t(\mathbf{X})]}{\partial X_i \partial X_j},$$

$$p_{t_0}(\mathbf{X}) = p_{\mathbf{x}_0}(\mathbf{X}), \qquad (7.109)$$

where n is the dimension of $\mathbf{x}(\cdot)$. This equation is called the *Fokker-Planck* or *forward Kolmogorov equation*. It describes the time evolution of the prior or unconditional statistics of $\mathbf{x}(t)$. The differential operator

$$L(\cdot) = -\sum_{i=1}^{n} \frac{[a_i(t, \mathbf{X})(\cdot)]}{\partial X_i} + \frac{1}{2}\sum_{i=1}^{n}\sum_{j=1}^{n} \frac{\partial^2[(\mathbf{B}(t, \mathbf{X})\mathbf{B}'(t, \mathbf{X}))_{i,j}(\cdot)]}{\partial X_i \partial X_j}. \qquad (7.110)$$

Continuous-Time Markov Jump Processes. Denote by Z the denumerable state space associated with a continuous-time Markov jump process $\{x(t): t \geq t_0\}$. The sample paths of this process are piecewise constant with randomly occurring jumps between states $\zeta_i \in Z$. Define $a_{ik}(t)$ by

$$a_{ik}(t) = \begin{cases} \lim_{\Delta t \downarrow 0} \dfrac{1}{\Delta t} \Pr[x(t+\Delta t) = \zeta_k \mid x(t) = \zeta_i], & i \neq k \\ -\sum_{j(j \neq i)} a_{ij}(t), & i = k. \end{cases} \tag{7.111}$$

The function $a_{ik}(t)$ for $i \neq k$ is the instantaneous rate of jumps form state ζ_i to state ζ_k. The sum of $a_{ij}(t)$ on j for $j \neq i$ is the total instantaneous rate of jumps from state ζ_i to any other state. These Markov jump processes include as special cases the following: (i), Poisson counting processes when

$$a_{ik}(t) = \begin{cases} \mu(t), & k = i+1 \\ -\mu(t), & k = i \\ 0, & \text{otherwise,} \end{cases} \tag{7.112}$$

and Z is the set of nonnegative integers; (ii), Markov birth processes when

$$a_{ik}(t) = \begin{cases} \mu(t,i), & k = i+1 \\ -\mu(t,i), & k = i \\ 0, & \text{otherwise} \end{cases} \tag{7.113}$$

and Z is the set of nonnegative integers; and (iii), Markov birth-death processes when

$$a_{ij}(t) = \begin{cases} \mu(t,i), & k = i+1 \\ v(t,i), & k = i-1 \\ -\mu(t,i) - v(t,i), & k = i \\ 0, & \text{otherwise} \end{cases} \tag{7.114}$$

and Z is the set of integers. The characteristic form for the differential generator is given by (see Prob. 7.4.6)

$$\Psi_t(v \mid x(t) = \zeta_i) = e^{-jv\zeta_i} \sum_k a_{ik}(t) e^{jv\zeta_k}. \tag{7.115}$$

Let $\pi_i(t) = \Pr[x(t) = \zeta_i]$ be the probability that $x(t)$ is in state ζ_i at time t. Then, substitution of (7.115) into (7.106) and inverse Fourier transformation shows that these unconditional state probabilities for $x(\cdot)$ evolve according to the differential equation

$$\frac{d\pi_i(t)}{dt} = \sum_k a_{ki}(t)\pi_k(t), \quad \pi_i(t_0) = \Pr[x_0 = \zeta_i]. \tag{7.116}$$

This is known as the *forward Kolmogorov equation.* ∎

These examples illustrate the variety of information processes accommodated by the Markov model. The time evolution of the prior statistics of the information process is described by the differential equation (7.106) for the unconditional characteristic function. We next obtain an analogous description for the posterior statistics of $\mathbf{x}(t)$ given $\{N(\sigma): t_0 \le \sigma < t\}$. Denote the conditional characteristic function by

$$\hat{M}_t(j\mathbf{v}) = E\{e^{j<\mathbf{v},\mathbf{x}(t)>} \mid N(\sigma): t_0 \le \sigma < t\}. \tag{7.117}$$

The conditional expectation can be interpreted as the minimum mean square-error estimate of $\exp(j < \mathbf{v}, \mathbf{x}(t))$, which accounts for our notation. The main result is in the following theorem.

Theorem 7.4.2 (*Evolution of Posterior Statistics*). Let $\{N(t): t \ge t_0\}$ be a doubly stochastic Poisson-process with a positive intensity process $\{\lambda(t, \mathbf{x}(t)): t \ge t_0\}$. Assume that the information process $\{\mathbf{x}(t): t \ge t_0\}$ is a stochastically continuous Markov process and that there exists a function $g_t(\mathbf{v}, \mathbf{x}(t))$ such that $E\{|g_t(\mathbf{v}, \mathbf{x}(t))|\} < \infty$ and (7.104) holds. Denote the characteristic form for the differential generator by $\Psi_t(\mathbf{v} \mid \mathbf{x}(t))$, where this function is given in (7.105). Then the conditional characteristic function of $\mathbf{x}(t)$ given $\{N(\sigma): t_0 \le \sigma < t\}$ satisfies

$$d\hat{M}_t(j\mathbf{v}) = \hat{E}\{e^{j<\mathbf{v},\mathbf{x}(t)>}\Psi_t(\mathbf{v} \mid \mathbf{x}(t))\}dt$$

$$+ \hat{E}\{e^{j<\mathbf{v},\mathbf{x}(t)>}[\lambda(t,\mathbf{x}(t)) - \hat{\lambda}(t)]\}\frac{1}{\hat{\lambda}(t)}[dN(t) - \hat{\lambda}(t)dt]$$

$$\hat{M}_{t_0}(j\mathbf{v}) = E\left\{ e^{j<\mathbf{v},\mathbf{x}(t_0)>} \right\}, \tag{7.118}$$

where the circumflex "^" denotes the conditional expectation given $\{N(\sigma): t_0 \le \sigma < t\}$.

Proof. In the proof, we use the notation $\mathcal{N}_t$ to denote $\{N(\sigma): t_0 \leq \sigma < t\}$. Corresponding to an infinitesimal time increment $[t, t + \Delta t)$, there is an increment $\Delta N(t) = N(t + \Delta t) - N(t)$ in the counting path and an increment $\Delta \hat{M}_t(j\mathbf{v}) = \hat{M}_{t+\Delta t}(j\mathbf{v}) - \hat{M}_t(j\mathbf{v})$ in the conditional characteristic function for $\mathbf{x}(t)$, where

$$\hat{M}_{t+\Delta t}(j\mathbf{v}) = E\{e^{j<\mathbf{v},\mathbf{x}(t)>} e^{j<\mathbf{v},\Delta\mathbf{x}(t)>} | \mathcal{N}_t\} = \int e^{j<\mathbf{v},\mathbf{X}>} e^{j<\mathbf{v},\mathbf{Y}>} dP(\mathbf{X}, \mathbf{Y} | \mathcal{N}_{t+\Delta t}),$$

$$(7.119)$$

where $\Delta\mathbf{x}(t) = \mathbf{x}(t + \Delta t) - \mathbf{x}(t)$ is the increment in the information process and $P(\mathbf{X}, \mathbf{Y} | \mathcal{N}_{t+\Delta t})$ is the joint conditional distribution function for $\mathbf{x}(t)$ and $\Delta\mathbf{x}(t)$ given the path $\mathcal{N}_{t+\Delta t}$. To evaluate $\hat{M}_{t+\Delta t}(j\mathbf{v})$, define

$$\hat{M}_{t+\Delta t}^{\Delta}(j\mathbf{v}) = \int e^{j<\mathbf{v},\mathbf{X}>} e^{j<\mathbf{v},\mathbf{Y}>} dP(\mathbf{X}, \mathbf{Y} | \mathcal{N}_t, \Delta N(t)), \qquad (7.120)$$

where $P(\mathbf{X}, \mathbf{Y} | \mathcal{N}_t, \Delta N(t))$ is the joint distribution function for $\mathbf{x}(t)$ and $\Delta\mathbf{x}(t)$ given the path $\mathcal{N}_t$ and the increment $\Delta N(t)$. We first evaluate $\hat{M}_{t+\Delta t}^{\Delta}(j\mathbf{v})$ and then get $\hat{M}_{t+\Delta t}(j\mathbf{v})$ from this by using the fact that $\hat{M}_{t+\Delta t}^{\Delta}(j\mathbf{v})$ approaches $\hat{M}_{t+\Delta t}(j\mathbf{v})$ with probability one as Δt tends to zero; this holds because $(\mathcal{N}_t, \Delta N(t)) \to \mathcal{N}_{t+\Delta t}$ as $\Delta t \downarrow 0$, which implies by the Martingale convergence theorem that $P(\mathbf{X}, \mathbf{Y} | \mathcal{N}_t, \Delta N(t))$ approaches $P(\mathbf{X}, \mathbf{Y} | \mathcal{N}_{t+\Delta t})$ with probability one as $\Delta t \downarrow 0$; see W. Feller [10, p. 236]. We evaluate $\hat{M}_{t+\Delta t}^{\Delta}(j\mathbf{v})$ for fixed values of $\Delta N(t)$; namely, for $\Delta N(t) = 0$, $\Delta N(t) = 1$, and $\Delta N(t) = m$ for $m > 1$. For $\Delta N(t) = 0$,

$$\hat{M}_{t+\Delta t}^{\Delta}(j\mathbf{v}) = \int e^{j<\mathbf{v},\mathbf{X}>} e^{j<\mathbf{v},\mathbf{Y}>} dP(\mathbf{X}, \mathbf{Y} | \mathcal{N}_t, \Delta N(t) = 0).$$

The finite difference

$$\delta P(\mathbf{X}, \mathbf{Y} | \mathcal{N}_t, \Delta N(t) = 0)$$

$$= \Pr(\mathbf{x}(t) \in (\mathbf{X} - \delta\mathbf{X}, \mathbf{X}], \Delta\mathbf{x}(t) \in (\mathbf{Y} - \delta\mathbf{Y}, \mathbf{Y}] | \mathcal{N}_t, \Delta N(t) = 0)$$

can be evaluated by using Bayes' rule as

$$\delta P(\mathbf{X}, \mathbf{Y} | \mathcal{N}_t, \Delta N(t) = 0)$$

$$= \Pr(\Delta N(t) = 0 | \mathbf{x}(t) \in (\mathbf{X} - \delta \mathbf{X}, \mathbf{X}], \Delta \mathbf{x}(t) \in (\mathbf{Y} - \delta \mathbf{Y}, \mathbf{Y}], \mathcal{N}_t)$$

$$\times \frac{\delta P(\mathbf{X}, \mathbf{Y} | \mathcal{N}_t)}{\Pr(\Delta N(t) = 0 | \mathcal{N}_t)},$$

where, as $\delta \mathbf{X} \to \mathbf{0}$ and $\delta \mathbf{Y} \to \mathbf{0}$,

$$\Pr(\Delta N(t) = 0 | \mathbf{x}(t) \in (\mathbf{X} - \delta \mathbf{X}, \mathbf{X}], \Delta \mathbf{x}(t) \in (\mathbf{Y} - \delta \mathbf{Y}, \mathbf{Y}], \mathcal{N}_t)$$

$$= 1 - \lambda(t, \mathbf{X}) \Delta t + o(\Delta t),$$

and where

$$\Pr(\Delta N(t) = 0 | \mathcal{N}_t) \tag{7.121}$$

$$= E\{\Pr(\Delta N(t) = 0 | \mathbf{x}(t), \mathcal{N}_t) | \mathcal{N}_t\} = 1 - \hat{\lambda}(t) \Delta t + o(\Delta t),$$

where $o(\Delta t)/\Delta t \to 0$ as $\Delta t \downarrow 0$. By collecting these results, we obtain for $\Delta N(t) = 0$ that

$$\hat{M}^{\Delta}_{t + \Delta t}(j\mathbf{v}) \tag{7.122}$$

$$= E\{e^{j<\mathbf{v}, \mathbf{x}(t)>} e^{j<\mathbf{v}, \Delta \mathbf{x}(t)>}(1 - \lambda(t, \mathbf{x}(t)) \Delta t) | \mathcal{N}_t\} \frac{1}{1 - \hat{\lambda}(t)} + o(\Delta t)$$

$$= E\{e^{j<\mathbf{v}, \mathbf{x}(t)>} e^{j<\mathbf{v}, \Delta \mathbf{x}(t)>}(1 - \lambda(t, \mathbf{x}(t)) \Delta t + \hat{\lambda}(t) \Delta t) | \mathcal{N}_t\} + o(\Delta t).$$

Similarly, by using this same procedure, we obtain for $\Delta N(t) = 1$ that

$$\hat{M}^{\Delta}_{t + \Delta t}(j\mathbf{v})$$

$$= E\{e^{j<\mathbf{v}, \mathbf{x}(t)>} e^{j<\mathbf{v}, \Delta \mathbf{x}(t)>} \lambda(t, \mathbf{x}(t)) | \mathcal{N}_t\} \frac{1}{\hat{\lambda}(t)} + o(\Delta t) \tag{7.123}$$

and for $\Delta N(t) = m \geq 1$ that

$$\hat{M}^{\Delta}_{t + \Delta t}(j\mathbf{v}) = o(\Delta t). \tag{7.124}$$

It then follows that for $\Delta N(t)$ variable,

$$\hat{M}^{\Delta}_{t+\Delta t}(j\mathbf{v}) \tag{7.125}$$

$$= E\{e^{j<\mathbf{v},\mathbf{x}(t)>}e^{j<\mathbf{v},\Delta\mathbf{x}(t)>}(1-\lambda(t,\mathbf{x}(t))\Delta t+\hat{\lambda}(t)\Delta t)\,|\,\mathcal{N}_t\}\delta_{0,\Delta N(t)}$$

$$+E\{e^{j<\mathbf{v},\mathbf{x}(t)>}e^{j<\mathbf{v},\Delta\mathbf{x}(t)>}\lambda(t,\mathbf{x}(t))\,|\,\mathcal{N}_t\}\frac{1}{\hat{\lambda}(t)}\delta_{1,\Delta N(t)}+o(\Delta t).$$

Here, $\delta_{i,j}$ is the Kronecker delta function. Because $\delta_{0,\Delta N(t)}$ approaches $1-\Delta N(t)$ and $\delta_{1,\Delta N(t)}$ approaches $\Delta N(t)$ with probability one as $\Delta t\downarrow 0$, we can write

$$\hat{M}^{\Delta}_{t+\Delta t}(j\mathbf{v})=E\{e^{j<\mathbf{v},\mathbf{x}(t)>}e^{j<\mathbf{v},\Delta\mathbf{x}(t)>}\,|\,\mathcal{N}_t\} \tag{7.126}$$

$$+E\{e^{j<\mathbf{v},\mathbf{x}(t)>}e^{j<\mathbf{v},\Delta\mathbf{x}(t)>}(\lambda(t,\mathbf{x}(t))-\hat{\lambda}(t))\,|\,\mathcal{N}_t\}$$

$$\frac{1}{\hat{\lambda}(t)}(\Delta N(t)-\hat{\lambda}(t)\Delta t)+o(\Delta t).$$

By subtracting the conditional characteristic function for $\mathbf{x}(t)$ given $\mathcal{N}_t$, we therefore have

$$\hat{M}^{\Delta}_{t+\Delta t}(j\mathbf{v})-\hat{M}_t(j\mathbf{v})$$

$$=E\left\{e^{j<\mathbf{v},\mathbf{x}(t)>}\frac{1}{\Delta t}E(e^{j<\mathbf{v},\Delta\mathbf{x}(t)>}-1\,|\,\mathbf{x}(t),\mathcal{N}_t)\,|\,\mathcal{N}_t\right\}\Delta t$$

$$+E\{e^{j<\mathbf{v},\mathbf{x}(t)>}e^{j<\mathbf{v},\Delta\mathbf{x}(t)>}(\lambda(t,\mathbf{x}(t))-\hat{\lambda}(t))\,|\,\mathcal{N}_t\}\frac{1}{\hat{\lambda}(t)}(\Delta N(t)-\hat{\lambda}(t)\Delta t)$$

$$+o(\Delta t). \tag{1.127}$$

However, $E\{e^{j<\mathbf{v},\Delta\mathbf{x}(t)>}-1\,|\,\mathbf{x}(t),\mathcal{N}_t\}=E\{e^{j<\mathbf{v},\Delta\mathbf{x}(t)>}-1\,|\,\mathbf{x}(t)\}$ because $\mathbf{x}(\cdot)$ is a Markov process. Thus, we can now establish the theorem as follows:

$$d\hat{M}_t(j\mathbf{v})=\lim_{\Delta t\downarrow 0}[\hat{M}_{t+\Delta t}(j\mathbf{v})-\hat{M}_t(j\mathbf{v})]=\lim_{\Delta t\downarrow 0}[\hat{M}^{\Delta}_{t+\Delta t}(j\mathbf{v})-\hat{M}_t(j\mathbf{v})]$$

$$=E\{e^{j<\mathbf{v},\mathbf{x}(t)>}\Psi_t(\mathbf{v}\,|\,\mathbf{x}(t))\,|\,\mathcal{N}_t\}dt \tag{7.128}$$

$$+E\{e^{j<\mathbf{v},\mathbf{x}(t)>}(\lambda(t,\mathbf{x}(t))-\hat{\lambda}(t))\,|\,\mathcal{N}_t\}\frac{1}{\hat{\lambda}(t)}(dN(t)-\hat{\lambda}(t)dt),$$

where the interchange of limit and expectation required for the first term in the last equality is justified by assumption (7.104) and the bounded convergence theorem, Theorem 1.4.1. The second term follows from the stochastic continuity of $\mathbf{x}(\cdot)$. ❑

Equation (7.118) is one of our basic results for solving the nonlinear filtering problem for observed doubly stochastic Poisson-processes. The procedure for implementing an estimate of some specified function $\mathbf{h}(t, \mathbf{x}(t))$ of $\mathbf{x}(t)$ is clearly very complicated in general. First, it is necessary to solve (7.118) for $\hat{M}_t(j\mathbf{v})$ for a given observation $\{N(\sigma): t_0 \le \sigma < t\}$. Then, this characteristic function must be inverse Fourier transformed to determine the posterior distribution for $\mathbf{x}(t)$ given $\{N(\sigma): t_0 \le \sigma < t\}$. Finally, the desired estimate must be determined by evaluating the expectation $\hat{\mathbf{h}}(t) = E\{\mathbf{h}(t, \mathbf{x}(t)) | N(\sigma): t_0 \le \sigma < t\}$. These steps are generally intractable analytically so that approximations and numerical techniques will be required. As we shall see, the equations are well suited for this.

We now give several examples that illustrate the use of (7.118) in solving the filtering problem for Markovian information processes.

Example 7.4.4 *Estimation of Random Variables* —————————

Suppose that the information process is simply a time-independent vector $\mathbf{x}$ of random variables with prior probability density $p_{\mathbf{x}}(\mathbf{X})$. In this case, $\Psi_t(\mathbf{v} | \mathbf{x})$ of (7.105) is zero. If we let $p_t(\mathbf{X} | N(\sigma): t_0 \le \sigma < t)$ denote the conditional probability density function for $\mathbf{x}$ given $\{N(\sigma): t_0 \le \sigma < t\}$, and if we inverse Fourier transform (7.118), then we obtain

$$dp_t(\mathbf{X} | N(\sigma): t_0 \le \sigma < t)$$

$$= p_t(\mathbf{X} | N(\sigma): t_0 \le \sigma < t)[\lambda(t, \mathbf{X}) - \hat{\lambda}(t)]\frac{1}{\hat{\lambda}(t)}(dN(t) - \hat{\lambda}(t)dt),$$

$$p_{t_0}(\mathbf{X} | N(t_0)) = p_{\mathbf{x}}(\mathbf{X}), \tag{7.129}$$

where

$$\hat{\lambda}(t) = E\{\lambda(t, \mathbf{x}) | N(\sigma): t_0 \le \sigma < t\}$$

$$= \int_{\mathcal{R}^n} \lambda(t, \mathbf{X})p_t(\mathbf{X} | N(\sigma): t_0 \le \sigma < t)d\mathbf{X}. \tag{7.130}$$

This equation provides three lines of departure for solving the problem of estimating a specified function of $\mathbf{x}$ given the observed data. The first is that of solving the equation for the posterior density. This can be accomplished, as will be seen shortly. However, in more general situations where the information process is time dependent, we will be unable to pursue this line because of its analytic intractability. The second line is that of numerical solutions using (7.129) as an updating algorithm for the posterior density. This idea will be described in more detail after we examine the analytic solution of the equation. The third line is to use (7.129) to obtain estimates that are feasible to generate in practice but which are only approximations to the minimum mean square-error estimate. This is developed in Sec. 7.4.3.

The procedure we describe for solving (7.129) analytically makes use of the differential rule of Theorem 5.3.2., a rule that will be used frequently in the following sections. Define $\phi_t(\mathbf{X}) = \ln p_t(\mathbf{X} \mid N(\sigma): t_0 \le \sigma < t)$. Then, from (7.129), and the differential rule, we have

$$d\phi_t(\mathbf{X}) = -(\lambda(t, \mathbf{X}) - \hat{\lambda}(t))dt + \ln\left(\frac{\lambda(t, \mathbf{X})}{\hat{\lambda}(t)}\right)dN(t)$$

$$\phi_{t_0}(\mathbf{X}) = \ln p_{\mathbf{x}}(\mathbf{X}). \tag{7.131}$$

Consequently, (7.129) is equivalent to

$$p_t(\mathbf{X} \mid N(\sigma): t_0 \le \sigma < t) \tag{7.132}$$

$$= p_{\mathbf{x}}(\mathbf{X}) \exp\left[\int_{t_0}^{t}(\lambda(\sigma, \mathbf{X}) - \hat{\lambda}(\sigma))d\sigma + \int_{t_0}^{t}\ln\left(\frac{\lambda(\sigma, \mathbf{X})}{\hat{\lambda}(t)}\right)dN(\sigma)\right].$$

As $\hat{\lambda}(t)$ requires an integration with respect to the posterior density $p_t(\mathbf{X} \mid N(\sigma): t_0 \le \sigma < t)$, equation (7.132) is a nonlinear integral equation for the posterior density. We have traded the nonlinear differential-integral equation (7.129) for the nonlinear integral equation (7.132). The integral equation can be solved easily using the expressions developed in Sec. 7.3 for the sample-function density. In particular, by equating (7.45) and (7.71), we have the relation

$$E\left\{\exp\left[-\int_{t_0}^{t}\lambda(\sigma, \mathbf{X})d\sigma + \int_{t_0}^{t}\ln\lambda(\sigma, \mathbf{X})dN(\sigma)\right]\right\}$$

$$= \exp\left[-\int_{t_0}^{t}\hat{\lambda}(\sigma)d\sigma + \int_{t_0}^{t}\ln\hat{\lambda}(\sigma)dN(\sigma)\right]. \tag{7.133}$$

Substitution of this expression into (7.132) yields the desired solution for the posterior density:

$$p_t(\mathbf{X} \mid N(\sigma): t_0 \leq \sigma < t) = \frac{e^{H(t,\mathbf{X})}p_\mathbf{x}(\mathbf{X})}{E\{e^{H(t,\mathbf{x})}\}}, \tag{7.134}$$

where

$$H(t,\mathbf{x}) = -\int_{t_0}^{t} \lambda(\sigma,\mathbf{x})d\sigma + \int_{t_0}^{t} \ln\lambda(\sigma,\mathbf{x})dN(\sigma). \tag{7.135}$$

Equation (7.134) in principle solves the nonlinear filtering problem for information processes that are just time-independent random variables. From Lemma 7.4.1 and (7.134), we have that the minimum mean square-error estimate of a function $h(t,\mathbf{x})$ of $\mathbf{x}$ is given by

$$\hat{h}(t) = E\{h(t,\mathbf{x}) \mid N(\sigma): t_0 \leq \sigma < t\} = \frac{E\{h(t,\mathbf{x})e^{H(t,\mathbf{x})}\}}{E\{e^{H(t,\mathbf{x})}\}}. \tag{7.136}$$

In a limited number of situations, the required integrations with respect to the prior statistics of $\mathbf{x}$ can be performed analytically to give a closed-form expression for $\hat{h}(t)$. Such an example is given in Prob. 7.4.10. More generally, however, numerical evaluation of the integrals is required. The procedure for determining $\hat{h}(t)$ numerically is first to form $H(t,\mathbf{X})$ from the observed data $\{N(\sigma): t_0 \leq \sigma < t\}$ according to (7.135). Then, the two integrals in (7.136) must be evaluated. These integrations can, at least in principle, be done by multidimensional Gaussian quadratures or multi-dimensional Monte Carlo techniques; see S. Haber [18]. The advantage of this approach is that the resulting estimate is exact, the only source of error being in the numerical implementation. However, for the dimension of $\mathbf{x}$ greater than one or two, as is often the case in applications, the numerical integrations may require a substantial computational capability. More-over, a large amount of computer memory will be required for storage of the posterior density.

Equation (7.129) can also be viewed as defining an updating algo-rithm according to which the prior density $p_\mathbf{x}(\mathbf{X})$ is propagated forward in time to form the posterior density as data are accumulated. For this interpretation, it is convenient to rewrite (7.129) in the finite-difference form

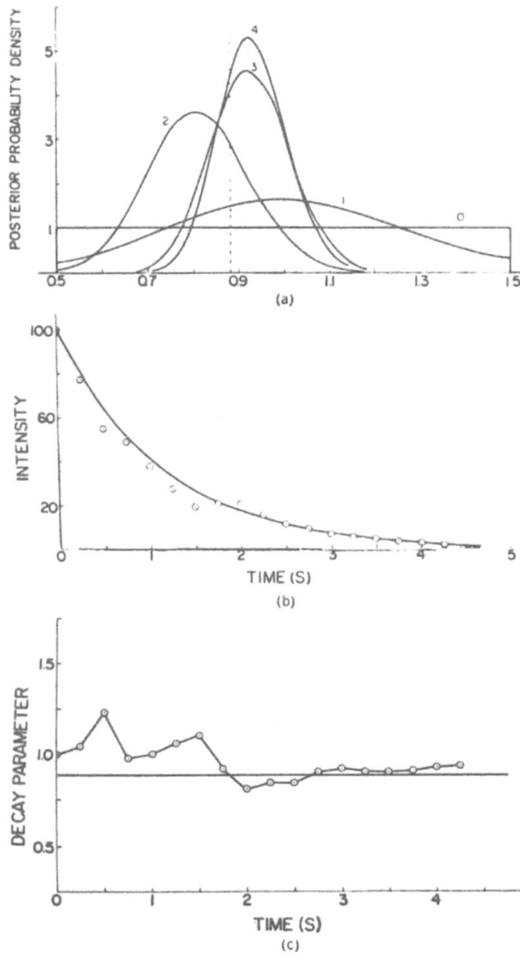

Figure 7.4 (a) Time evolution for the posterior probability density of x when $\lambda(t,x) = 100\exp(-xt)$. The true value of x equals 0.882. The density is shown for $t = 0, 1, 2, 3,$ and 4 seconds. All densities are zero outside [0.5,1.0]. (b) Time evolution for the minimum mean square-error estimate of the intensity. The solid curve is the intensity and the data points its estimate at 0.25 second intervals. (c) Time evolution for the minimum mean square-error estimate of x. The solid curve is the true value, and the data points are estimates at 0.25 second intervals.

$$p_{t+\Delta t}(\mathbf{X} \mid N(\sigma): t_0 \leq \sigma < t) \tag{7.137}$$

$$= p_t(\mathbf{X} \mid N(\sigma): t_0 \leq \sigma < t)\left\{1 + (\lambda(t,\mathbf{X}) - \hat{\lambda}(t))\frac{1}{\hat{\lambda}(t)}(\Delta N(t) - \hat{\lambda}(t)\Delta t)\right\} + o(\Delta t).$$

For Δt sufficiently small, the term $o(\Delta t)$ can be disregarded, and $\Delta N(t) = N(t + \Delta t) - N(t)$ will be either zero or one according to the nonoccurrence or occurrence of a point in $[t, t + \Delta t)$. The posterior density at time $t + \Delta t$ can be evaluated by knowing the density at time t and whether or not a point occurs in $[t, t + \Delta t)$; this evaluation requires an integration with respect to the density at t in order to determine $\hat{\lambda}(t)$. The estimate $\hat{h}(t)$ of a function $h(t,x)$ of x can be determined at each stage of the updating process by performing the integration

$$\hat{h}(t) = \int_{\mathcal{R}^n} h(t,X)p_t(X \mid N(\sigma): t_0 \le \sigma < t)dX$$

with respect to the updated density. Some simulation results illustrating this update method are summarized in Fig. 7.4. The intensity used in the simulation is $\lambda(t,x) = 100\exp(-xt)$ for $t \ge 0$, where x is a uniformly distributed random variable between 0.5 and 1.0. The randomly selected value of x used in the simulation equals 0.882. The numerically generated time evolution for $p_t(x \mid N(\sigma): t_0 \le \sigma < t)$ is shown as is $\hat{\lambda}(t)$ and $\hat{x}(t)$. ∎

Example 7.4.5 *Estimation of a Poisson-Driven Markov Process* ——————
Suppose that the information process $\{x(t): t \ge t_0\}$ is a Poisson driven Markov process satisfying (5.38) with $b(t,x(t),u) = b(t,u)$ not a function of $x(t)$. The function $\Psi_t(v \mid x(t))$ required in (7.118) is given in (5.100). Substitution of this function into (7.118) and inverse Fourier transformation shows that the posterior density of $x(t)$ given $\{N(\sigma): t_0 \le \sigma < t\}$ satisfies

$$dp_t(X \mid N(\sigma): t_0 \le \sigma < t)$$

$$= -\sum_{i=1}^{n} \frac{\partial[a_i(t,X)p_t(X \mid N(\sigma): t_0 \le \sigma < t)]}{\partial X_i}dt$$

$$+\lambda(t,X)\int_{\mathcal{U}} [p_t(X - b(t,U) \mid N(\sigma): t_0 \le \sigma < t) - p_t(X \mid N(\sigma): t_0 \le \sigma < t)dP_u(U)dt$$

$$+p_t(X \mid N(\sigma): t_0 \le \sigma < t)[\lambda(t,X) - \hat{\lambda}(t)]\frac{1}{\hat{\lambda}(t)}(dN(t) - \hat{\lambda}(t)dt),$$

with the initial condition being the probability density of the initial state. While this equation is difficult to solve analytically, it does provide an updating algorithm, in parallel with (7.137), by which numerical solutions can be produced. ∎

Example 7.4.6 *Estimation of a Markov-Diffusion Process* ——————
If $\{x(t): t \ge t_0\}$ is a Markov diffusion process satisfying (7.107), the function $\Psi_t(v \mid x(t))$ required in (7.118) is given in (7.108). Substitution of this function into (7.118) and inverse Fourier transformation shows that the posterior probability density for $x(t)$ given $\{N(\sigma): t_0 \le \sigma < t\}$ satisfies

$$dp_t(\mathbf{X} \mid N(\sigma): t_0 \le \sigma < t) = L[p_t(\mathbf{X} \mid N(\sigma): t_0 \le \sigma < t)]dt \qquad (7.138)$$

$$+ p_t(\mathbf{X} \mid N(\sigma): t_0 \le \sigma < t)[\lambda(t, \mathbf{X}) - \hat{\lambda}(t)]\frac{1}{\hat{\lambda}(t)}(dN(t) - \hat{\lambda}(t)dt),$$

where $L(\cdot)$ is the forward Kolmogorov differential operator defined in (7.110). This equation is the counterpart for observed doubly stochastic Poisson-processes of the Kushner-Stratonovich equation encountered in the additive white Gaussian noise observation model; see A. Jazwinski [20, p. 178]. Again, the equation is difficult to solve analytically, but it provides an updating algorithm analogous to (7.137) by which numerical solutions can be produced. ∎

Example 7.4.7 *Conditional Moments for a Markov Diffusion-Process* ⸺
The results of this example will be important in Sec. 7.4.3. Evolution equations for the conditional moments of the information process can be obtained using (7.118) and the moment generating properties of the conditional characteristic function. In particular, the first two conditional moments of $\mathbf{x}(t)$ given $\{N(\sigma): t_0 \le \sigma < t\}$ are given in terms of the conditional characteristic function by

$$\hat{\mathbf{x}}(t) = \left. \frac{\partial \hat{M}_t(j\mathbf{v})}{j\partial \mathbf{v}} \right|_{\mathbf{v}=0} \qquad (7.139)$$

and

$$\hat{E}\{\mathbf{x}(t)\mathbf{x}'(t)\} = \left. \frac{\partial^2 \hat{M}(j\mathbf{v})}{j^2 \partial \mathbf{v}^2} \right|_{\mathbf{v}=0}. \qquad (7.140)$$

Here, $\partial(\cdot)/\partial \mathbf{v}$ denotes the gradient vector of $(\cdot)$ with respect to $\mathbf{v}$, a vector in which the ith element is $\partial(\cdot)/\partial v_i$. And $\partial^2(\cdot)/\partial \mathbf{v}^2$ denotes a symmetric matrix having i,j-element $\partial^2(\cdot)/\partial v_i \partial v_j$. This idea can be illustrated when the information process is a Markov diffusion. Using (7.108), we obtain

$$\frac{\partial[e^{j<\mathbf{v},\mathbf{x}(t)>}\Psi_t(\mathbf{v} \mid \mathbf{x}(t))]}{\partial \mathbf{v}} \qquad (7.141)$$

$$= [j\mathbf{x}(t)\Psi_t(\mathbf{v} \mid \mathbf{x}(t)) + j\mathbf{a}(t, \mathbf{x}(t)) - \mathbf{B}(t, \mathbf{x}(t))\mathbf{B}'(t, \mathbf{x}(t))\mathbf{v}]e^{j<\mathbf{v},\mathbf{x}(t)>}$$

and

$$\frac{\partial^2[e^{j<\mathbf{v},\mathbf{x}(t)>}\Psi_t(\mathbf{v}\mid\mathbf{x}(t))]}{\partial\mathbf{v}^2}$$

$$= [j^2\mathbf{x}(t)\mathbf{x}'(t)\Psi_t(\mathbf{v}\mid\mathbf{x}(t))+j^2\mathbf{a}(t,\mathbf{x}(t))\mathbf{x}'(t)$$

$$+j^2\mathbf{x}(t)\mathbf{a}'(t,\mathbf{x}(t))-\mathbf{B}(t,\mathbf{x}(t))\mathbf{B}'(t,\mathbf{x}(t))$$

$$-j\mathbf{B}(t,\mathbf{x}(t))\mathbf{B}'(t,\mathbf{x}(t))\mathbf{v}\mathbf{x}'(t)$$

$$-j\mathbf{x}(t)\mathbf{v}'\mathbf{B}(t,\mathbf{x}(t))\mathbf{B}'(t,\mathbf{x}(t))]e^{j<\mathbf{v},\mathbf{x}(t)>}. \tag{7.142}$$

By taking the derivative of both sides of (7.118) with respect to $\mathbf{v}$, using (7.141), setting $\mathbf{v}=\mathbf{0}$, and using (7.139), the equation we obtain for the conditional mean is

$$d\hat{\mathbf{x}}(t)=\hat{\mathbf{a}}(t,\mathbf{x}(t))dt+\hat{E}\{\mathbf{x}(t)[\lambda(t,\mathbf{x}(t))-\hat{\lambda}(t)]\}\frac{1}{\hat{\lambda}(t)}(dN(t)-\hat{\lambda}(t)dt),$$

$$\hat{\mathbf{x}}(t_0)=E(\mathbf{x}_0). \tag{7.143}$$

By taking the derivative of both sides of (7.118) with respect to $\mathbf{v}$, using (7.141), setting $\mathbf{v}=\mathbf{0}$, and using (7.140), the equation we obtain for the conditional second-moment matrix is

$$d\hat{E}\{\mathbf{x}(t)\mathbf{x}'(t)\}$$

$$=\hat{E}\{\mathbf{a}(t,\mathbf{x}(t))\mathbf{x}'(t)+\mathbf{x}(t)\mathbf{a}'(t,\mathbf{x}(t))+\mathbf{B}(t,\mathbf{x}(t))\mathbf{B}'(t,\mathbf{x}(t))\}dt$$

$$+\hat{E}\{\mathbf{x}(t)\mathbf{x}'(t)[\lambda(t,\mathbf{x}(t))-\hat{\lambda}(t)]\frac{1}{\hat{\lambda}(t)}(dN(t)-\hat{\lambda}(t)dt),$$

$$\hat{E}\{\mathbf{x}(t_0)\mathbf{x}'(t_0)\}=E\{\mathbf{x}_0\mathbf{x}'_0\}. \tag{7.144}\blacksquare$$

The equations in Ex. 7.4.7 for the conditional first and second moments of a Markov diffusion process may not appear to be particularly useful because of the unknown and difficult to determine conditional expectations required on the right-hand sides. Nevertheless, we shall find them to be a natural point of departure in Sec. 7.4.3 for formulating approximations to the minimum mean square-error estimate of $\mathbf{x}(t)$. For this later purpose, it is convenient to carry (7.144) somewhat further by determining an equation for the conditional error-covariance matrix $\hat{\Sigma}(t)$ defined by

$$\hat{\Sigma}(t) = E\{[\mathbf{x}(t) - \hat{\mathbf{x}}(t)][\mathbf{x}(t) - \hat{\mathbf{x}}(t)]'\} = \hat{E}\{\mathbf{x}(t)\mathbf{x}'(t)\} - \hat{\mathbf{x}}(t)\hat{\mathbf{x}}'(t). \tag{7.145}$$

An expression for $d\hat{E}\{\mathbf{x}(t)\mathbf{x}'(t)\}$ is in (7.144). To get an expression for $d\hat{\Sigma}(t)$, we need to investigate the differential $d[\hat{\mathbf{x}}(t)\hat{\mathbf{x}}'(t)]$. We do this by using (7.143) and the differential rule in Theorem 5.3.2. Let $\mathbf{e}_i$ denote a unit vector with $i th$ element equal to one and all other elements equal to zero. Also, let $\phi_{ij}(\hat{\mathbf{x}}(t)) = \mathbf{e}_i'\hat{\mathbf{x}}(t)\hat{\mathbf{x}}'(t)\mathbf{e}_j$. Then,

$$\frac{\partial\phi_{ij}(\hat{\mathbf{x}}(t))}{\partial\hat{\mathbf{x}}(t)} = (\mathbf{e}_i\mathbf{e}_j' + \mathbf{e}_j\mathbf{e}_i')\hat{\mathbf{x}}(t).$$

From (7.143) and the differential rule, we conclude that

$$d\phi_{ij}(\hat{\mathbf{x}}(t)) \tag{7.146}$$

$$= <(\mathbf{e}_i\mathbf{e}_j' + \mathbf{e}_j\mathbf{e}_i')\hat{\mathbf{x}}(t), \mathbf{a}(t) - \hat{E}\{\mathbf{x}(t)[\lambda(t,\mathbf{x}(t)) - \hat{\lambda}(t)]\} > dt$$

$$+\mathbf{e}_i'\hat{E}\{[\mathbf{x}(t)\hat{\mathbf{x}}'(t) + \hat{\mathbf{x}}(t)\mathbf{x}'(t)][\lambda(t,\mathbf{x}(t)) - \hat{\lambda}(t)]\}\mathbf{e}_j\frac{1}{\hat{\lambda}(t)}dN(t)$$

$$+\mathbf{e}_i'\hat{E}\{\mathbf{x}(t)[\lambda(t,\mathbf{x}(t)) - \hat{\lambda}(t)]\}\hat{E}\{\mathbf{x}'(t)[\lambda(t,\mathbf{x}(t)) - \hat{\lambda}(t)]\}\mathbf{e}_j\frac{1}{\hat{\lambda}^2(t)}dN(t).$$

Consequently,

$$d[\hat{\mathbf{x}}(t)\hat{\mathbf{x}}'(t)] = [\hat{\mathbf{a}}(t)\hat{\mathbf{x}}'(t) + \hat{\mathbf{x}}(t)\hat{\mathbf{a}}'(t)]dt \tag{7.147}$$

$$+\hat{E}\{[\mathbf{x}(t)\hat{\mathbf{x}}'(t) + \hat{\mathbf{x}}(t)\mathbf{x}'(t)][\lambda(t,\mathbf{x}(t)) - \hat{\lambda}(t)]\frac{1}{\hat{\lambda}(t)}(dN(t) - \hat{\lambda}(t)dt)$$

$$+\hat{E}\{\mathbf{x}(t)[\lambda(t,\mathbf{x}(t)) - \hat{\lambda}(t)]\}\hat{E}\{\mathbf{x}'(t)[\lambda(t,\mathbf{x}(t)) - \hat{\lambda}(t)]\}\frac{1}{\hat{\lambda}^2(t)}dN(t).$$

The desired equation for the conditional error-covariance matrix is now obtained from (7.144), (7.147), and the relation $d\hat{\Sigma}(t) = d\hat{E}[\mathbf{x}(t)\mathbf{x}'(t)] - d[\hat{\mathbf{x}}(t)d\hat{\mathbf{x}}'(t)]$. The result is

$$d\hat{\Sigma}(t)$$

$$= \hat{E}\{[\mathbf{a}(t,\mathbf{x}(t)) - \hat{\mathbf{a}}(t)]\hat{\mathbf{x}}'(t) + \hat{\mathbf{x}}(t)[\mathbf{a}'(t,\mathbf{x}(t)) - \hat{\mathbf{a}}'(t)] + \mathbf{B}(t,\mathbf{x}(t))\mathbf{B}'(t,\mathbf{x}(t))\}dt$$

$$+ \hat{E}\{[\mathbf{x}(t)\mathbf{x}'(t) - \mathbf{x}(t)\hat{\mathbf{x}}'(t) - \hat{\mathbf{x}}(t)\mathbf{x}'(t)][[\lambda(t,\mathbf{x}(t)) - \hat{\lambda}(t)]]\frac{1}{\hat{\lambda}(t)}(dN(t) - \hat{\lambda}(t)dt)$$

$$- \hat{E}\{\mathbf{x}(t)[\lambda(t,\mathbf{x}(t)) - \hat{\lambda}(t)]\}\hat{E}\{\mathbf{x}'(t)[\lambda(t,\mathbf{x}(t)) - \hat{\lambda}(t)]\}\frac{1}{\hat{\lambda}^2(t)}dN(t),$$

$$\hat{\Sigma}(t_0) = E(\mathbf{x}_0\mathbf{x}_0') - E(\mathbf{x}_0)e(\mathbf{x}_0'). \tag{7.148}$$

These results are summarized in the following theorem. Equations (7.143) and (7.148) are modified slightly in the theorem for later use in Sec. 7.4.3.

Theorem 7.4.3 (*Conditional Mean and Error Covariance for a Markov Diffusion Process*).   Let $\{N(t): t \geq t_0\}$ be a doubly stochastic Poisson-process with a positive intensity process $\{\lambda(t,\mathbf{x}(t)): t \geq t_0\}$.   Suppose that the information process $\{\mathbf{x}(t): t \geq t_0\}$ is a Markov diffusion satisfying (7.107).   Suppose, further, that the conditional first and second moments for $\mathbf{x}(t)$ given $\{N(\sigma): t_0 \leq \sigma < t\}$ exist.   Then, the conditional mean $\hat{\mathbf{x}}(t)$, which is the minimum mean square-error estimate of $\mathbf{x}(t)$, and the conditional error-covariance matrix $\hat{\Sigma}(t)$ satisfy

$$d\hat{\mathbf{x}}(t) = \hat{\mathbf{a}}(t)dt + \hat{E}\{[\mathbf{x}(t) - \hat{\mathbf{x}}(t)][\lambda(t,\mathbf{x}(t)) - \hat{\lambda}(t)]\}\frac{1}{\hat{\lambda}(t)}(dN(t) - \hat{\lambda}(t)dt),$$

$$\hat{\mathbf{x}}(t_0) = E(\mathbf{x}_0) \tag{7.149}$$

and

$$d\hat{\Sigma}(t)$$

$$= \hat{E}\{[\mathbf{a}(t,\mathbf{x}(t)) - \hat{\mathbf{a}}(t)][\mathbf{x}(t) - \hat{\mathbf{x}}(t)]' + [\mathbf{x}(t) - \hat{\mathbf{x}}(t)][\mathbf{a}(t,\mathbf{x}(t)) - \hat{\mathbf{a}}(t)]'$$

$$+ \mathbf{B}(t,\mathbf{x}(t))\mathbf{B}'(t,\mathbf{x}(t))\}dt$$

$$+ \hat{E}\{[\mathbf{x}(t) - \hat{\mathbf{x}}(t)][\mathbf{x}(t) - \hat{\mathbf{x}}(t)]'[\lambda(t,\mathbf{x}(t)) - \hat{\lambda}(t)]\}\frac{1}{\hat{\lambda}(t)}(dN(t) - \hat{\lambda}(t)dt)$$

$$- \hat{E}\{[\mathbf{x}(t) - \hat{\mathbf{x}}(t)][\lambda(t,\mathbf{x}(t)) - \hat{\lambda}(t)]\}\hat{E}\{[\mathbf{x}(t) - \hat{\mathbf{x}}(t)]'[\lambda(t,\mathbf{x}(t)) - \hat{\lambda}(t)]\}\frac{1}{\hat{\lambda}^2(t)}dN(t),$$

$$\hat{\Sigma}(t_0) = E(\mathbf{x}_0\mathbf{x}_0') - E(\mathbf{x}_0)E(\mathbf{x}_0'). \tag{7.150}$$

Example 7.4.8 *Estimation of a Markov Jump-Process* ——————

Suppose that the information process is a continuous-time Markov jump process with states $\{\zeta_i\}$. Then, the function $\Psi_t(\mathbf{v}\mid\mathbf{x}(t))$ required in (7.118) is given in (7.115). Substitution of this function into (7.118) demonstrates that the conditional state occupancy probabilities $\hat{\pi}_i(t) = \Pr[x(t)=\zeta_i \mid N(\sigma): t_0 \le \sigma < t]$ satisfy

$$d\hat{\pi}_i(t) = \sum_k a_{ki}(t)\hat{\pi}_k(t)\,dt + \hat{\pi}_i(t)[\lambda(t,\zeta_i)-\hat{\lambda}(t)]\frac{1}{\hat{\lambda}(t)}(dN(t)-\hat{\lambda}(t)\,dt),$$

$$\hat{\pi}_i(t_0) = \Pr(x_0 = \zeta_i), \qquad\qquad\qquad (7.151)$$

where

$$\hat{\lambda}(t) = \sum_i \lambda(t,\zeta_i)\hat{\pi}_i(t), \qquad\qquad\qquad (7.152)$$

and where the instantaneous transition rates $\{a_{ij}(t)\}$ are defined in (7.111). Equation (7.151) provides an updating algorithm by which the conditional state-occupancy probabilities can be generated numerically. This procedure has the following recursive interpretation. Suppose that a point is observed to occur at time s, and no points occur during (s,t). Then, from (7.151), we have for $t > s$

$$\frac{d\hat{\pi}_i(t)}{dt} = \sum_k a_{ki}(t)\hat{\pi}_k(t) - \hat{\pi}_i(t)[\lambda(t,\zeta_i)-\hat{\lambda}(t)], \qquad\qquad (7.153)$$

$$\hat{\pi}_i(s^+) = \hat{\pi}_i(s^-)\lambda(s^-,\zeta_i)\frac{1}{\hat{\lambda}(s^-)}. \qquad\qquad (7.154)$$

Equations (7.153) and (7.154) can be used recursively to determine the state-occupancy probabilities. Thus, starting at time t_0, (7.153) is used to determine $\hat{\pi}_i(t)$ for t up to the first occurrence time w_1. Then (7.154) is used to determine $\hat{\pi}_i(w_i^+)$ from $\hat{\pi}_i(w_i^-)$, and (7.153) is used again for t up to the second occurrence time w_2. Then (7.154) is used again, and so forth. M. Rudemo shows that the solution to (7.153) can be obtained by solving an appropriate set of linear differential equations [38]. We now use (7.153) and (7.154) to reexamine the model of Ex. 7.4.3. ∎

Example 7.4.9 *On-off Modulated Light, continued* ——————————
The information process $\{\mu(t): t \geq t_0\}$ of Ex. 7.4.3 is a Markov jump process with two states $\{\zeta_0 = 0, \zeta_1 = a\}$. The transition rates for state ζ_0 to state ζ_1 and vice versa both equal ν. Thus, the instantaneous intensities of (7.111) are given by

$$\begin{bmatrix} a_{00}(t) & a_{01}(t) \\ a_{10}(t) & a_{11}(t) \end{bmatrix} = \begin{bmatrix} -\nu & \nu \\ \nu & -\nu \end{bmatrix}. \tag{7.155}$$

The problem of estimating the information process from an observed realization of the point process it influences can be solved using (7.153) and (7.154). For this purpose, we note that $\hat{\pi}_0(t) = 1 - \hat{\pi}_1(t)$ and that

$$\hat{\lambda}(t) = \lambda(t,0)\hat{\pi}_0(t) + \lambda(t,a)\hat{\pi}_1(t)$$

$$= \lambda_0[1 - \hat{\pi}_1(t)] + [a + \lambda_0]\hat{\pi}_1(t) = \lambda_0 + a\hat{\pi}_1(t).$$

From (7.153),

$$\frac{d\hat{\pi}_1(t)}{dt} = \nu - (2\nu + a)\hat{\pi}_1(t) + a\hat{\pi}_1^2(t). \tag{7.156}$$

The solution to this equation is

$$\hat{\pi}_1(t) = \frac{(\beta - \gamma) - (\beta + \gamma)\xi(s)e^{-2a\gamma(t-s)}}{1 - \xi(s)e^{-2a\gamma(t-s)}}, \tag{7.157}$$

where

$$\beta = \frac{1}{2}\left(1 + \frac{2\nu}{a}\right), \qquad \gamma = \frac{1}{2}\left(1 + \left(\frac{2\nu}{a}\right)^2\right)^{1/2},$$

and

$$\xi(s) = \frac{\hat{\pi}_1(s^+) - (\beta - \gamma)}{\hat{\pi}_1(s^+) - (\beta + \gamma)}.$$

From (7.154), the update of $\hat{\pi}_1(t)$ at an occurrence time is

$$\hat{\pi}_1(s^+) = \frac{\hat{\pi}_1(s^-)(\lambda_0 + a)}{\lambda_0 + a\hat{\pi}_1(s^-)}. \tag{7.158}$$

We processed the same computer-simulated data as used in producing Fig. 7.3 to compare the estimates of the information process obtained with and without a linearity constraint. The result is shown in Fig. 7.5. ∎

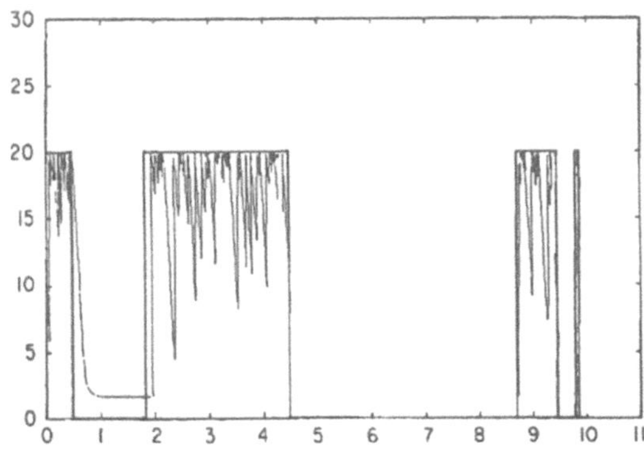

Figure 7.5 Nonlinear estimation of on-off modulated light.

Suboptimal Nonlinear Filtering

In Sec. 7.4.2, we developed a general framework for causally estimating information processes that influence the observed points of a doubly stochastic Poisson-process. The equations we have for the minimum mean square-error estimates are, in general, too complicated for analytic solution, and they even require a substantial computational capability for numerical solution. For this reason, we are motivated to develop suboptimal estimators that not only approximate the minimum mean square-error estimators but also have the advantage of a less extravagant computational requirement for their implementation. The estimators we obtain are nearly optimal when the estimation error is sufficiently small.

We will assume in this subsection that the information process $\{\mathbf{x}(t): t \geq t_0\}$ is a Markov diffusion satisfying the stochastic differential-equation

$$d\mathbf{x}(t) = \mathbf{a}(t,\mathbf{x}(t))dt + \mathbf{B}(t)d\mathbf{w}(t), \quad \mathbf{x}(t_0) = \mathbf{x}_0, \tag{7.159}$$

in which $\{\mathbf{w}(t): t \geq t_0\}$ is a vector of independent Wiener processes such that $E[\mathbf{w}(t)\mathbf{w}'(u)] = \mathbf{I}\min(t,u)$, and where the initial condition $\mathbf{x}_0$ is a random variable that is independent of $\{\mathbf{w}(t): t \geq t_0\}$. This is the model of (7.107) but with the assumption that $\mathbf{B}(\cdot)$ is not a function of $\mathbf{x}(\cdot)$; this assumption can be relaxed by a straightforward modification of the procedure we describe. The choice of the information process as a Markov diffusion includes as special cases time-independent random variables, Gaussian processes that have rational power-density spectra, and Gaussian processes having finite-dimensional state representations.

The procedure we adopt in developing suboptimal estimators parallels that used with other models [20, Ch. 9]. It can be described briefly as follows. We start with equations (7.149) and (7.150) for the conditional mean $\hat{\mathbf{x}}(t)$ and conditional error-covariance $\hat{\Sigma}(t)$ for $\mathbf{x}(t)$ given $\{N(\sigma): t_0 \leq \sigma < t\}$. The nonlinear functions $\mathbf{a}(t, \mathbf{x}(t))$ and $\lambda(t, \mathbf{x}(t))$ are expanded in a Taylor series about the estimate $\hat{\mathbf{x}}(t)$. This results in an infinite series in moments of the error $\mathbf{x}(t) - \hat{\mathbf{x}}(t)$ on the right hand sides of (7.149) and (7.150). The Gaussian moment-factorization property is used on these error moments, writing higher-order moments in terms of second-order moments as if the error were normally distributed. The infinite series are then truncated at the "most significant" terms on the assumption that the error moments are sufficiently small. Just what these most significant terms are does depend on the particular functions $\mathbf{a}$ and λ involved; consequently, the usefulness of the particular truncation we adopt must be verified in individual applications. Our truncations will be shown to be reasonable for certain situations encountered in radiation physics and optical communications.

The result of the manipulations described above is a pair of coupled equations defining the suboptimal estimator, as demonstrated below. The first equation is

$$d\mathbf{x}^*(t) = \mathbf{a}(t, \mathbf{x}^*(t))\,dt + \frac{1}{2}\sum_i \sum_j \sigma_{ij}^*(t)\frac{\partial^2 \mathbf{a}(t, \mathbf{x}^*(t))}{\partial x_i^* \partial_j^*}\,dt$$

$$+\Sigma^*(t)\left[\frac{\partial \lambda(t, \mathbf{x}^*(t))}{\partial \mathbf{x}^*(t)}\right]\frac{1}{\lambda(t, \mathbf{x}^*(t))}[dN(t) - \lambda(t, \mathbf{x}^*(t))\,dt], \quad (7.160)$$

$$\mathbf{x}^*(t_0) = E(\mathbf{x}_0)$$

where $\Sigma^*(t)$ is a symmetric matrix that satisfies the second equation, and $\sigma_{ij}^*(t)$ is the i,j-element of this matrix. The second equation is

$$d\Sigma^*(t) = \left[\frac{\partial \mathbf{a}(t, \mathbf{x}^*(t))}{\partial \mathbf{x}^*(t)}\right]\Sigma^*(t)\,dt + \Sigma^*(t)\left[\frac{\partial \mathbf{a}(t, \mathbf{x}^*(t))}{\partial \mathbf{x}^*(t)}\right]dt + \mathbf{B}(t)\mathbf{B}'(t)\,dt$$

$$-\Sigma^*(t)\left[\frac{\partial^2 \lambda(t, \mathbf{x}^*(t))}{\partial \mathbf{x}^*(t)^2}\right]\Sigma^*(t)\,dt + \Sigma^*(t)\left[\frac{\partial^2 \ln \lambda(t, \mathbf{x}^*(t))}{\partial \mathbf{x}^*(t)^2}\right]\Sigma^*(t)\,dN(t),$$

$$\Sigma^*(t_0) = \mathrm{cov}(\mathbf{x}_0), \qquad\qquad\qquad (7.161)$$

where $\partial \mathbf{a}(\mathbf{x})/\partial \mathbf{x}$ denotes the Jacobian matrix having as its i-row, j-column element $\partial a_j(\mathbf{x})/\partial x_i$, and where $\partial^2 \lambda(\mathbf{x})/\partial \mathbf{x}^2 = \partial[\partial \lambda(\mathbf{x})/\partial \mathbf{x}]/\partial \mathbf{x}$.

We refer to (7.160) as the *processor equation* and to (6.161) as the *variance equation*. It is seen that in general the processor and variance equations are coupled, and both depend on the observed doubly stochastic Poisson-process. The processor equation can be viewed as defining an updating procedure according to which the suboptimal estimate $\mathbf{x}^*(t + dt)$ at time $t + dt$ following an incremental observation is expressed in terms of the estimate $\mathbf{x}^*(t)$ at time t and the data $dN(t)$ observed on $[t, t + dt)$. Here, $dN(t)$ is zero or one depending upon whether or not a point occurs on $[t, t + dt)$. The only unknown quantity in this updating is the matrix $\Sigma^*(t)$, and this is specified by the variance equation, which can likewise be viewed as defining an updating procedure for $\Sigma^*(t)$.

As indicated in Ex. 7.4.6, the evolution equation for the posterior density $p_t[\mathbf{X} \mid N(\sigma): t_0 \leq \sigma < t]$ in (7.138) can be viewed as defining an updating procedure for estimating $\mathbf{x}(t)$. This procedure has the advantage compared to (7.160) that no approximations are required beyond those needed for the numerical implementation, so the estimates obtained are optimal. On the other hand, the advantages of the suboptimal estimator defined by (7.610) in comparison to this exact procedure are: a, no integrations are required; b, a large memory for storing the up-to-date version of the conditional density is not required; and c, only the prior mean and covariance of $\mathbf{x}(\cdot)$ are required, as opposed to the prior density needed in the exact analysis.

Derivation of the Processor Equation (7.160). We assume that $\mathbf{a}(t, \mathbf{x}(t))$ and $\lambda(t, \mathbf{x}(t))$ have Taylor series about the estimate $\hat{\mathbf{x}}(t)$ as

$$\mathbf{a}(t, \mathbf{x}(t)) \tag{7.162}$$

$$= \mathbf{a}(t, \hat{\mathbf{x}}(t)) + \left[\frac{\partial \mathbf{a}(t, \hat{\mathbf{x}}(t))}{\partial \hat{\mathbf{x}}(t)}\right]'[\mathbf{x}(t) - \hat{\mathbf{x}}(t)] + \frac{1}{2}\sum_{i,j}(x_i - \hat{x}_i)(x_j - \hat{x}_j)\frac{\partial^2 \mathbf{a}(t, \hat{\mathbf{x}}(t))}{\partial \hat{x}_i \partial \hat{x}_j} + \cdots,$$

and

$$\lambda(t, \mathbf{x}(t)) \tag{7.163}$$

$$= \lambda(t, \hat{\mathbf{x}}(t)) + \left[\frac{\partial \lambda(t, \hat{\mathbf{x}}(t))}{\partial \hat{\mathbf{x}}(t)}\right]'[\mathbf{x}(t) - \hat{\mathbf{x}}(t)] + \frac{1}{2}\sum_{i,j}(x_i - \hat{x}_i)(x_j - \hat{x}_j)\frac{\partial^2 \lambda(t, \hat{\mathbf{x}}(t))}{\partial \hat{x}_i \partial \hat{x}_j} + \cdots.$$

Upon substituting these expansions into (7.149), we obtain

$$d\hat{\mathbf{x}}(t) = \mathbf{a}(t, \hat{\mathbf{x}}(t))\,dt + \frac{1}{2}\sum_{i,j} \hat{\sigma}_{ij} \frac{\partial^2 \mathbf{a}(t, \hat{\mathbf{x}}(t))}{\partial \hat{x}_i \partial \hat{x}_j}\,dt \tag{7.164}$$

$$+ \hat{\Sigma}(t)\left[\frac{\partial \lambda(t, \hat{\mathbf{x}}(t))}{\partial \hat{\mathbf{x}}(t)}\right]\frac{1}{\hat{\lambda}(t)}[dN(t) - \hat{\lambda}(t)\,dt] + o[\mathbf{x}(t) - \hat{\mathbf{x}}(t)],$$

in which $\hat{\sigma}_{ij}$ is the i,j-element of the conditional error-covariance matrix $\hat{\Sigma}(t)$, and where $o(\mathbf{x} - \hat{\mathbf{x}})$ denotes terms of order three or more in the error moments. We now introduce three approximations:

a. replace $\hat{\lambda}(t)$ by $\lambda(t, \hat{\mathbf{x}}(t))$;
b. disregard $o(\mathbf{x} - \hat{\mathbf{x}})$;
c. replace $\hat{\mathbf{x}}$ by $\mathbf{x}^*$; and
c. replace $\hat{\Sigma}(t)$ by $\Sigma^*(t)$, which is defined below.

It is evident from (7.162) that approximation (a) is consistent with disregarding second-order terms as in (b). These approximations are ad hoc, so their usefulness must be examined for the specific functions encountered in a given application, but they appear to be reasonable if the estimation error is small. The result of these approximations is (7.160)

Derivation of the Variance Equation (7.161). By using the series (7.162) and (7.163), we obtain

$$\hat{E}\{[\mathbf{a}(t, \mathbf{x}(t)) - \hat{\mathbf{a}}(t)][\mathbf{x}(t) - \hat{\mathbf{x}}(t)]'\} = \left[\frac{\mathbf{a}(t, \hat{\mathbf{x}}(t))}{\partial \hat{\mathbf{x}}(t)}\right]\hat{\Sigma}(t) + \cdots,$$

$$\hat{E}\{[\lambda(t, \mathbf{x}(t)) - \hat{\lambda}(t)][\mathbf{x}(t) - \hat{\mathbf{x}}(t)]'\} = \left[\frac{\lambda(t, \hat{\mathbf{x}}(t))}{\partial \hat{\mathbf{x}}(t)}\right]\hat{\Sigma}(t) + \cdots,$$

and

$$\hat{E}\{[\lambda(t, \mathbf{x}(t)) - \hat{\lambda}(t)][\mathbf{x}(t) - \hat{\mathbf{x}}(t)][\mathbf{x}(t) - \hat{\mathbf{x}}(t)]'\}$$

$$= \sum_i \hat{E}\{(x_i - \hat{x}_i)[\mathbf{x}(t) - \hat{\mathbf{x}}(t)][\mathbf{x}(t) - \hat{\mathbf{x}}(t)]'\}\frac{\partial \lambda(t, \hat{\mathbf{x}}(t))}{\partial \hat{x}_i}$$

$$+ \frac{1}{2}\sum_{i,j} \hat{E}\{(x_i - \hat{x}_i)(x_j - \hat{x}_j)[\mathbf{x}(t) - \hat{\mathbf{x}}(t)][\mathbf{x}(t) - \hat{\mathbf{x}}(t)]'\}\frac{\partial^2 \lambda(t, \hat{\mathbf{x}}(t))}{\partial \hat{x}_i \partial \hat{x}_j} + \cdots$$

We now substitute these expressions into the right side of (7.150) and make the following approximations:

a. replace $\hat{\lambda}(t)$ by $\lambda(t, \hat{x}(t))$;
b. retain only the leading terms indicated in the above expansions;
c. disregard third-order moments of the error $\mathbf{x} - \hat{\mathbf{x}}$;
d. factor fourth-order moments of the error into products of second-order moments as if the error were normally distributed; and
e. replace $\hat{x}(t)$ by $\mathbf{x}^*(t)$.

These approximations lead us to define $\Sigma^*(t)$ according to

$$
d\Sigma^*(t) = \left[\frac{\partial \mathbf{a}(t, \mathbf{x}^*(t))}{\partial \mathbf{x}^*(t)} \right] \Sigma^*(t)\,dt + \Sigma^*(t)\left[\frac{\partial \mathbf{a}(t, \mathbf{x}^*(t))}{\partial \mathbf{x}^*(t)} \right] dt + \mathbf{B}(t)\mathbf{B}'(t)\,dt
$$

$$
+ \Sigma^*(t)\left[\frac{\partial^2 \lambda(t, \mathbf{x}^*(t))}{\partial \mathbf{x}^*(t)^2} \right] \Sigma^*(t) \frac{1}{\lambda(t, \mathbf{x}^*(t))}[dN(t) - \lambda(t, \mathbf{x}^*(t))dt]
$$

$$
- \Sigma^*(t)\left[\frac{\partial \lambda(t, \mathbf{x}^*(t))}{\partial \mathbf{x}^*(t)} \right]\left[\frac{\partial \lambda(t, \mathbf{x}^*(t))}{\partial \mathbf{x}^*(t)} \right] \Sigma^*(t) \frac{1}{\lambda^2(t, \mathbf{x}^*(t))}\,dN(t).
$$

The variance equation (7.161) results from a straightforward rearrangement of this equation and the use of the relation

$$
\frac{\partial^2 \ln \lambda(t, \mathbf{x})}{\partial \mathbf{x}^2} = -\left[\frac{\partial \lambda(t, \mathbf{x})}{\partial \mathbf{x}} \right]\left[\frac{\partial \lambda(t, \mathbf{x})}{\partial \mathbf{x}} \right]' \frac{1}{\lambda^2(t, \mathbf{x})} + \left[\frac{\partial^2 \lambda(t, \mathbf{x})}{\partial \mathbf{x}^2} \right]\frac{1}{\lambda(t, \mathbf{x})}.
$$

We now give some examples in which the suboptimal estimators defined by (7.160) and (7.161) are used.

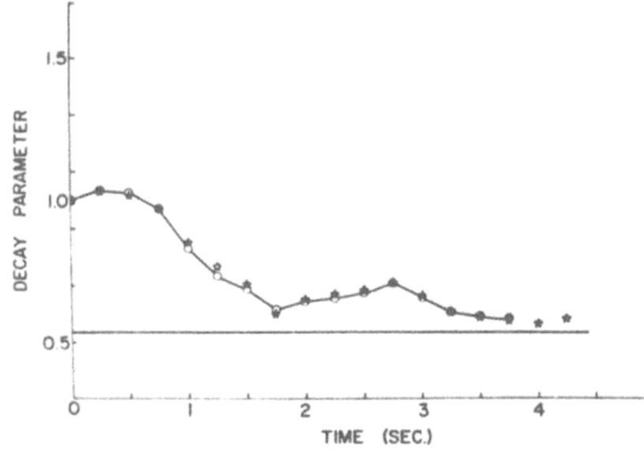

Figure 7.6 Simulation record for the optimal (circle) and suboptimal (star) estimates for the decay parameter x in the intensity $\lambda(t, x) = 100\exp(-xt)$. The true value of x is 0.538.

Example 7.4.10 *One-Parameter Exponential Intensity* —————————

To illustrate the use of the processor and variance equations as defining an updating algorithm, consider a one-parameter intensity process of the form $\lambda(t,x) = A \exp(-xt)$ for $t \geq 0$, where A is a known positive constant, and x is a nonnegative random variable with known mean $\overline{x}_0$ and variance σ_0^2. Then (7.160) and (7.161) become

$$dx^*(t) = At\Sigma^*(t)e^{-x^*(t)t} - t\Sigma^*(t)dN(t), \quad x^*(0) = \overline{x}_0$$

$$d\Sigma^*(t) = -At^2\Sigma^{*2}e^{-x^*(t)t}dt, \quad \Sigma^*(0) = \sigma_0^2 \tag{7.165}$$

We simulated a doubly stochastic Poisson-process with this intensity process. The value $A = 100$ and x uniformly distributed on the interval $[0.5,1.5]$ were used; then, $\overline{x}_0 = 1.0$ and $\sigma_0^2 = 1/12$. The optimal and suboptimal estimates of x are shown in Fig. 7.6 when the randomly selected value of x is 0.538. The procedure by which the data shown in this figure were obtained consisted of using a random number generator to select x, generating an inhomogeneous Poisson process with the appropriate exponential intensity, and then generating the estimates from the realization of the Poisson process. The optimal estimate was generated using the procedure described in Ex. 7.5.4; specifically, the estimates were obtained by numerically integrating $Xp_t(X \mid N(\sigma):0 \leq \sigma < t)$, where the posterior density was updated according to (7.137) with $\Delta t = 0.001$. This estimate is indicated by a circle. The suboptimal estimate was generated by the following finite difference version of (7.165):

$$x^*(t + \Delta t) = x^*(t) + 100t\Sigma^*(t)e^{-x^*(t)t}\Delta t + t\Sigma^*(t)\Delta N(t)$$

$$\Sigma^*(t + \Delta t) = \Sigma^*(t) - 100t^2\Sigma^{*2}(t)e^{-x^*(t)t}\Delta t,$$

with $\Delta t = 0.001$. This estimate is indicated by an asterisk. The agreement between the optimal and suboptimal estimates is striking in view of the small number of accumulated counts. ∎

Example 7.4.11 *Two-Parameter Exponential Intensity* —————————

For this example, consider a two-parameter intensity process of the form $\lambda(t,\mathbf{x}) = x_1\exp(-x_2t)$, $t \geq 0$, where x_1 and x_2 are uncorrelated, nonnegative random variables with known means $\overline{x}_1$ and $\overline{x}_2$ and variances σ_1^2 and σ_2^2. Then, (7.160) and (7.161) become

$$d\mathbf{x}^*(t) = -\Sigma^*(t)\begin{bmatrix} 1 \\ -tx_1^*(t) \end{bmatrix}e^{-x_2^*(t)t}\,dt + \Sigma^*(t)\begin{bmatrix} 1/x_1^*(t) \\ -t \end{bmatrix}dN(t), \quad \mathbf{x}^*(0) = \begin{bmatrix} \bar{x}_1 \\ \bar{x}_2 \end{bmatrix},$$

$$(7.166a)$$

and

$$d\Sigma^*(t) = -\Sigma^*(t)\begin{bmatrix} 0 & -t \\ -t & t^2 x_1^*(t) \end{bmatrix}\Sigma^*(t)e^{-x_2^*(t)t}\,dt - \Sigma^*(t)\begin{bmatrix} 1/x_1^{*2}(t) & 0 \\ 0 & 0 \end{bmatrix}\Sigma^*(t)\,dN(t),$$

$$\Sigma^*(0) = \begin{bmatrix} \sigma_0^2 & 0 \\ 0 & \sigma_1^2 \end{bmatrix}. \qquad (7.166b)$$

Radioactive decays from oxygen-15 were measured to illustrate the updating method we have described. Oxygen-15 has a radioactive half life of about two minutes. Physical principles suggest that the radioactivity disappears exponentially. With this in mind, we placed a quantity of oxygen-15 labeled water in the field of view of a scintillation detector and observed the times of occurrence of detected annihilation photons for about eight minutes.[1] We assumed that the detected events formed a doubly stochastic Poisson process with the two-parameter exponential intensity, where x_1 depends on the quantity of labeled water, and x_2 on the physical properties of the labeling isotope. Estimates of these parameters were generated according to (7.166) with the initial conditions $\mathbf{x}^*(0) = (1600, 0.7143 \times 10^{-2})$ and $\Sigma^*(0) = \mathrm{diag}(2500, 0.4 \times 10^{-5})$. The estimates are shown in Fig's. 7.7 and 7.8. Close agreement is seen between the estimated decay $x_2^*(t)$ and the known disappearance rate of oxygen-15 provided data spanning about one half-life of the isotope are processed. ∎

Example 7.4.12 *Subcarrier Angle-Modulation* ——————

Analog information can be transmitted at optical frequencies by modulating the intensity of a laser with an angle-modulated subcarrier, as discussed in Ex. 5.3.3. The output of an ideal photodetector at the receiver can be taken as a doubly stochastic Poisson-process with an intensity process given by

---

[1] Oxygen-15 is a cyclotron-produced isotope that can be used to form a variety of radioactive pharmaceutical tracers that are used in nuclear medicine and in biological and physiological research. The data described were collected in the Radiation Sciences Division of the Mallinckrodt Institute of Radiology, Washington University School of Medicine, St. Louis, MO.

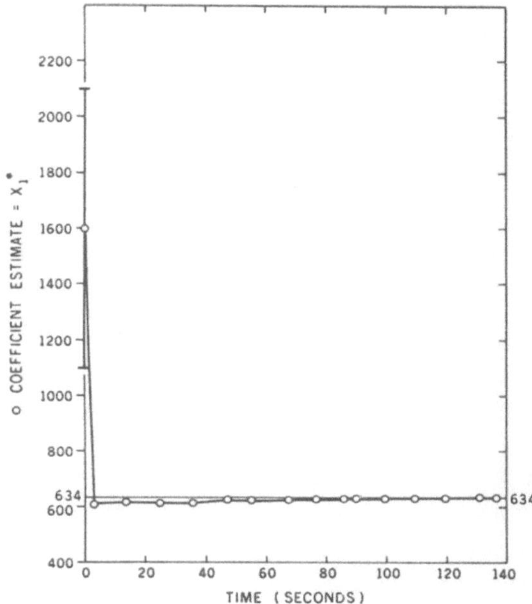

Figure 7.7 Time evolution of the suboptimal MMSE estimate of x_1 of the intensity $x_1 \exp(-x_2 t)$ for radiation from oxygen-15 labeled water.

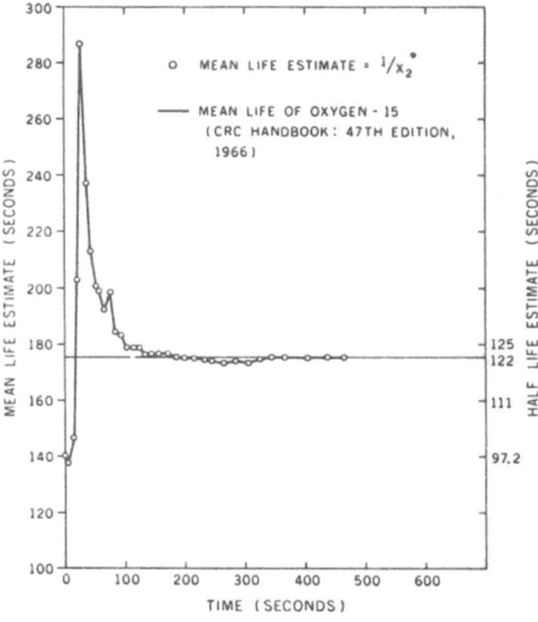

Figure 7.8 Time evolution of the suboptimal MMSE estimate of x_2 of the intensity $x_1 \exp(-x_2 t)$ for radiation from oxygen-15 labeled water.

$$\lambda(t,\mathbf{x}(t)) = \left(\frac{\eta \mathcal{A} P}{h\nu}\right)[1 + \alpha \cos(\omega t + \theta(t))] + \lambda_0, \qquad (7.167)$$

where η is the quantum efficiency of the detector, $\mathcal{A}$ is the detector area, P is the received signal power per unit area, h is Planck's constant, ν is the mean optical-frequency, α is the subcarrier modulation index and satisfies $|\alpha| \leq 1$, and λ_0 is the thermoelectron generation rate in the detector. The angle-modulation process $\{\theta(t): t \geq t_0\}$ is assumed to be a Gaussian process defined by $\theta(t) = \mathbf{c}'(t)\mathbf{x}(t)$, where $\{\mathbf{x}(t): t \geq t_0\}$ is a Gaussian diffusion given by (7.159) when $\mathbf{a}(t,\mathbf{x}(t)) = \mathbf{A}(t)\mathbf{x}(t)$, and $\mathbf{x}_0$ is a normal random variable with mean zero and covariance Σ_0. By selection of $\mathbf{c}(t)$ and $\mathbf{A}(t)$, this model includes the following useful angle-modulation formats:

a. *Phase Modulation.* For phase modulation, $\theta(t) = \beta y(t)$, where β is a constant called the modulation index, and $y(t)$ is the information process. Here, we take $y(t) = \mathbf{h}'(t)\mathbf{x}(t)$, where $\mathbf{x}(t)$ satisfies (7.159) with $\mathbf{a}(t,\mathbf{x}(t)) = \mathbf{A}(t)\mathbf{x}(t)$, and $\mathbf{c}(t) = \beta \mathbf{h}(t)$.

b. *Frequency Modulation.* For frequency modulation, $\theta(t) = d_f \int_{t_0}^{t} y(\sigma)d\sigma$, where d_f is a constant called the frequency-deviation factor, and $y(t)$ is the information process. Again, we take $y(t) = \mathbf{h}'(t)\mathbf{x}(t)$, where $\mathbf{x}(t)$ satisfies (7.159) with $\mathbf{a}(t,\mathbf{x}(t)) = \mathbf{A}(t)\mathbf{x}(t)$. To accommodate frequency modulation, we define an additional state $x_0(t) = \int_{t_0}^{t} y(\sigma)d\sigma$ and construct an expanded state vector $\tilde{\mathbf{x}}(t)$ of dimension $n+1$, where n is the dimension of $\mathbf{x}(t)$,

$$\tilde{\mathbf{x}}(t) = [x_0(t) \quad x_1(t) \quad \cdots \quad x_n(t)]'. \qquad (7.168)$$

Then, $\tilde{\mathbf{x}}(t)$ satisfies

$$d\tilde{\mathbf{x}}(t) = \tilde{\mathbf{A}}(t)\tilde{\mathbf{x}}(t)dt + \tilde{\mathbf{B}}(t)d\mathbf{w}(t), \qquad (7.169)$$

where

$$\tilde{\mathbf{A}}(t) = \begin{bmatrix} 0 & \mathbf{h}'(t) \\ 0 & \mathbf{A}(t) \end{bmatrix}, \quad \text{and} \quad \tilde{\mathbf{B}}(t) = \begin{bmatrix} \mathbf{0}' \\ \mathbf{B}(t) \end{bmatrix}.$$

In terms of $\tilde{\mathbf{x}}(t)$, the phase $\theta(t)$ can be written as $\theta(t) = \mathbf{c}'(t)\tilde{\mathbf{x}}(t)$, where $\mathbf{c}(t)$ is an $n+1$-dimensional vector of the form $\mathbf{c}(t) = [d_f \quad 0 \quad \cdots \quad 0]'$.

c. *Preemphasized Frequency Modulation.* For preemphasized frequency modulation, $\theta(t) = \int_{t_0}^{t} g(t,u)y(u)\,du$ for some specified preemphasis-filter response $g(t,u)$. If this response has a state-variable realization of dimension m, then preemphasized frequency modulation can be accommodated by defining an additional state variable $x_0(t)$ and an expanded state vector $\tilde{x}(t)$ of dimension $n+m+1$ having as elements $x_0(t)$, $\mathbf{x}(t)$, and the state vector associated with the preemphasis filter.

Simply to economize on the number of parameters in our estimation equations, we now rewrite the intensity of (7.167) in the form

$$\lambda(t,\mathbf{x}(t)) = G[1 + m\cos(\omega t + \theta(t))], \qquad (7.170)$$

where

$$G = \frac{\eta \mathcal{A}P}{h\nu} + \lambda_0 \quad \text{and} \quad m = \alpha\frac{\eta \mathcal{A}P}{h\nu G}.$$

As we have seen, this intensity includes useful angle-modulation formats, and the problem of estimating $\mathbf{x}(t)$ given $\{N(\sigma): t_0 \le \sigma < t\}$ is that of designing the receiver processing that should be performed on the output of the photodetector to recover the information processes that modulate the light incident on the detector. For the intensity (7.170), the processor equation (7.160) becomes

$$d\mathbf{x}^*(t) = \mathbf{A}(t)\mathbf{x}^*(t)$$

$$-\frac{m\sin(\omega t + \theta^*(t))}{1 + m\cos(\omega t + \theta^*(t))}\Sigma^*(t)\mathbf{c}(t)\{dN(t) - G[1 + m\cos(\omega t + \theta^*(t))]dt\},$$

$$\mathbf{x}^*(t_0) = E(\mathbf{x}_0) = \mathbf{0}, \qquad (7.171)$$

where $\theta^*(t) = \mathbf{c}'(t)\mathbf{x}^*(t)$. The variance equation (7.161) becomes

$$d\Sigma^*(t) = \mathbf{A}(t)\Sigma^*(t)dt + \Sigma^*(t)\mathbf{A}'(t)dt + \mathbf{B}(t)\mathbf{B}'(t)dt$$

$$+Gm\cos(\omega t + \theta^*(t))\Sigma^*(t)\mathbf{c}(t)\mathbf{c}'(t)\Sigma^*(t)dt$$

$$-\frac{m[m + \cos(\omega t + \theta^*(t))]}{[1 + m\cos(\omega t + \theta^*(t))]^2}\Sigma^*(t)\mathbf{c}(t)\mathbf{c}'(t)\Sigma^*(t)dN(t)$$

$$\Sigma^*(t_0) = \mathrm{cov}(\mathbf{x}_0) = \Sigma_0. \tag{7.172}$$

Equations (7.171) and (7.172) define the suboptimal estimate of $\mathbf{x}(t)$ and the phase $\theta(t)$. However, if ω is sufficiently large, as it is in practice, additional approximations can be made without degrading the performance of the estimates significantly. These result in estimation equations having the distinct advantage that the variance equation is uncoupled from both the suboptimal state estimate and the data process $N(t)$; the equation is a Riccati equation of the type occurring in Kalman-Bucy filtering, and which we encountered previously in (7.82) for the linear filtering problem. Furthermore, the estimation equations suggest the use of a tanlock loop operating on the output of the photodetector to estimate the phase process [35]. We assume for the additional approximations that ω is sufficiently large that the elements of $\mathbf{A}(t)$, $\mathbf{B}(t)$, $\mathbf{c}(t)$, $\mathbf{x}(t)$, $\mathbf{x}^*(t)$, and $\Sigma^*(t)$ are slowly varying in comparison to $\sin(\omega t + \theta^*(t))$ and $\cos(\omega t + \theta^*(t))$. It then appears reasonable to neglect the subtractive term $G[1 + m\cos(\omega t + \theta^*(t))]$ in (7.171) because of the implied time integration of the product of the slowly varying function $Gm\Sigma^*(t)\mathbf{c}(t)$ and $\sin(\omega t + \theta^*(t))$. We also replace $dN(t)$ in (7.162) by $G[1 + m\cos(\omega t + \theta^*(t))]dt$ with the following motivation

$$E(dN(t) \mid N(\sigma): t_0 \le \sigma < t)$$

$$= \hat{\lambda}(t)dt \approx \lambda(t, \mathbf{x}^*(t))dt = G[1 + m\cos(\omega t + \theta^*(t))]dt.$$

With this approximation, (7.172) becomes

$$d\Sigma^*(t) = \mathbf{A}(t)\Sigma^*(t)dt + \Sigma^*(t)\mathbf{A}'(t)dt + \mathbf{B}(t)\mathbf{B}'(t)dt$$

$$-\frac{Gm^2\sin^2(\omega t + \theta^*(t))}{1 + m\cos(\omega t + \theta^*(t))}\Sigma^*(t)\mathbf{c}(t)\mathbf{c}'(t)\Sigma^*(t)dt. \tag{7.173}$$

Now examine the fourth term on the right. The function

$$f(\phi) = \frac{\sin^2(\phi)}{1 + m\cos(\phi)} \tag{7.174}$$

is an even periodic function of ϕ. Consequently, it has a Fourier cosine series of the form

$$f(\phi) = a_0 + a_1 \cos(\phi) + a_2 \cos(2\phi) + \cdots,$$

where $a_0 = [1 - \sqrt{1 - m^2}]/m^2$. If we let $\phi = \omega + \theta^*(t)$ and substitute this series into the fourth term in (7.173), then it is consistent with our approximation to neglect all the harmonic components involving $\cos(n\phi)$ for $n \geq 1$ and retain only a_0 because of the implied time integration. The result of these approximations is the following pair of equations

$$d\mathbf{x}^*(t) = \mathbf{A}(t)\mathbf{x}^*(t)\,dt - \frac{m \sin(\omega t + \theta^*(t))}{1 + m \cos(\omega t + \theta^*(t))} \Sigma^*(t)\mathbf{c}(t)\,dN(t),$$

$$\mathbf{x}^*(t_0) = \mathbf{0} \tag{7.175}$$

and

$$\frac{d\Sigma^*(t)}{dt} = \mathbf{A}(t)\Sigma^*(t) + \Sigma^*(t)\mathbf{A}'(t) + \mathbf{B}(t)\mathbf{B}'(t) - G[1 - \sqrt{1 - m^2}]\Sigma^*(t)\mathbf{c}(t)\mathbf{c}'(t)\Sigma^*(t)$$

$$\Sigma^*(t_0) = \Sigma_0, \tag{7.176}$$

where $\theta^*(t) = \mathbf{c}'(t)\mathbf{x}^*(t)$. Equation (7.175) suggests the use of a form of tanlock loop to estimate $\theta(t)$ [35], and (7.176) is precomputable as it does not depend on the data. We will see in Sec. 7.4.4 that $\Sigma^*(t)$ lower bounds the performance in estimating $\mathbf{x}(t)$ and $\theta(t)$.

The computer simulation results shown in Fig. 7.9 were obtained for special case when $\theta(t) \equiv \theta$, a time independent random-variable uniformly distributed on $[-\pi, \pi]$. For this problem, $dx(t) = 0$, $E(x_0) = 0$, $\mathrm{cov}(x_0) = \pi^2/3$, and $\theta(t) = x(t)$. The processor and variance equations, (7.175) and (7.176), become

$$dx^*(t) = -\frac{m \sin(\omega t + \theta^*(t))}{1 + m \cos(\omega t + \theta^*(t))} \Sigma^*(t)\,dN(t), \quad x^*(0) = 0$$

$$\tag{7.177}$$

and

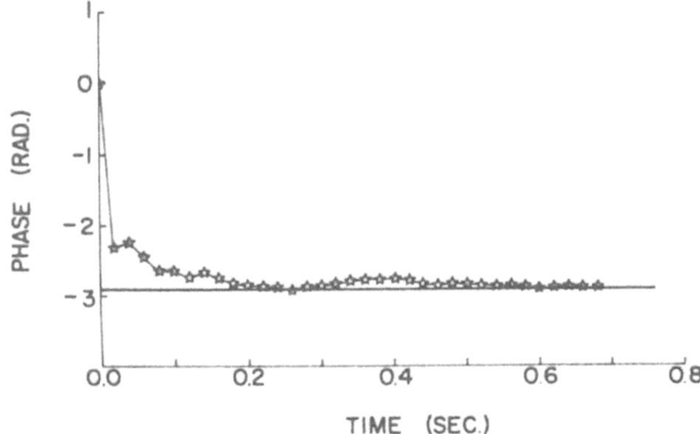

Figure 7.9 Simulation record for the suboptimal (star) estimate of θ for the intensity $\lambda(t,\theta) = G[1 + m\cos(\omega t + \theta)]$ with $G = 500$, $m = 0.75$, $\omega = 200\pi$, and $\theta = -2.905$ rad. The number of counts at $t = 0.6$ seconds is 267.

$$\frac{d\Sigma^*(t)}{dt} = -G[1 - \sqrt{1-m^2}]\Sigma^{*2}(t), \quad \Sigma^*(0) = \frac{\pi^2}{3}, \tag{7.178}$$

where $\theta^*(t) = x^*(t)$. The solution to (7.178) is

$$\Sigma^*(t) = \frac{1}{(3/\pi^2) + G[1 - \sqrt{1-m^2}]t}. \tag{7.179}$$

The procedure used in performing the simulation consisted of selecting a value of θ at random, generating a Poisson process with the intensity (7.170), and then generating the suboptimal estimate $\theta^*(t)$ by using (7.179).

The computer simulation results shown in Fig. 7.10 were obtained for the intensity $\lambda(t,\theta(t)) = G[1 + m\cos(\omega t + \theta(t))]$ in which $\{\theta(t): t \geq 0\}$ is a Wiener process with parameter $1/\tau_c$. The parameter τ_c can be regarded as the coherence time of an unstable oscillator; it is the time at which one radian r.m.s. of phase drift is accumulated. The problem of estimating $\theta(t)$ is that of establishing phase or time synchronization at a direct-detection receiver when the pilot-tone oscillator used to generate the subcarrier is unstable. The model for the phase is $\theta(t) = x(t)$, where $dx(t) = \tau_c^{-1/2} dw(t)$, $x(0) = 0$, where $\{w(t): t \geq 0\}$ is a unit-parameter Wiener process. The processor equation (7.175) is the same as when $\theta(t)$ is a constant independent of time, (7.177). The variance equation (7.176) becomes

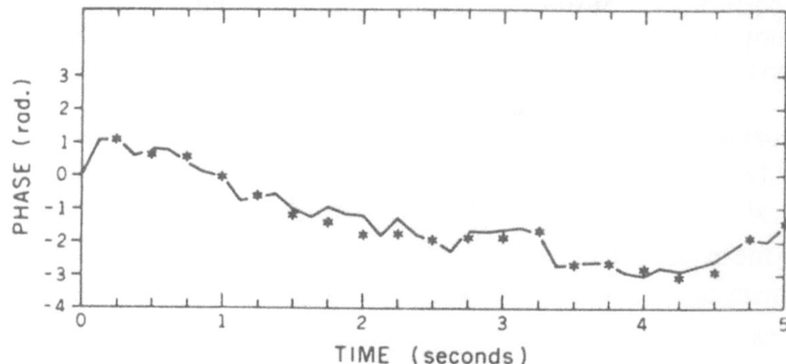

Figure 7.10 Simulation record for the suboptimal (star) estimate of $\theta(t)$ for the intensity $\lambda(t, \theta(t)) = G[1 + m\cos(\omega t + \theta(t))]$ with $G = 1000$, $m = 0.99$, $\omega = 200\pi$, and $\{\theta(t): t \geq 0\}$ is a Wiener process with coherence parameter $\tau_c = 1$.

$$\frac{d\Sigma^*(t)}{dt} = \frac{1}{\tau_c} - G[1 - \sqrt{1 - m^2}]\Sigma^{*2}(t), \quad \Sigma^*(0) = 0. \qquad (7.180)$$

The solution is

$$\Sigma^*(t) = \frac{\tanh\left[t\left(\frac{G}{\tau_c}(1 - \sqrt{1 - m^2})\right)^{1/2}\right]}{\sqrt{\tau_c G(1 - \sqrt{1 - m^2})}}. \qquad (7.181)$$

The procedure used in the simulation consisted of generating the phase process, then using this to generate a Poisson process, and then generating the suboptimal estimate of the phase process by using (7.177) as an updating algorithm. ∎

Nonlinear Filtering Performance for Gaussian Information Processes

The structure of optimal and suboptimal minimum mean square-error estimators is discussed in Sections 7.4.2 and 7.4.3. We now investigate the mean-square performance of these estimators. Unfortunately, only a limited amount can be said because in general the direct evaluation of the error presents insurmountable analytic difficulties. For this reason, we restrict our attention to Gaussian information processes and establish a lower bound on the mean square-error for any estimate casually derived from the observed doubly stochastic Poisson-process. We use the bound to study the performance of direct-detection optical communication systems that employ subcarrier angle modulation. The main result is stated in the following theorem [41].

Theorem 7.4.4 (*Performance Lower Bound*). Let $\{N(t): t_0 \le t \le T\}$ be a doubly stochastic Poisson-process with intensity process $\{\lambda(t, \mathbf{x}(t)): t_0 \le t \le T\}$, where $\{\mathbf{x}(t): t_0 \le t \le T\}$ is a Gaussian process with zero mean and autocovariance $\mathbf{K}(t, u) = E[\mathbf{x}(t)\mathbf{x}'(u)]$. Let $\mathbf{x}^*(t)$ denote any estimate of $\mathbf{x}(t)$ in terms of past data $\{N(\sigma): t_0 \le \sigma < t\}$. The particular estimate $E[\mathbf{x}(t) \mid N(\sigma) \le \sigma < t\}$ will be denoted by $\hat{\mathbf{x}}(t)$; this is the minimum mean square-error estimate. Then, for all $t \in [t_0, T]$ there holds

$$E\{[\mathbf{x}(t) - \mathbf{x}^*(t)][\mathbf{x}(t) - \mathbf{x}^*(t)]'\} \tag{7.182}$$

$$\ge E\{[\mathbf{x}(t) - \hat{\mathbf{x}}(t)][\mathbf{x}(t) - \hat{\mathbf{x}}(t)]'\} \ge \Sigma(t) \stackrel{\Delta}{=} \lim_{u \uparrow t} \Gamma(t, u),$$

where $\Gamma(t, u)$ satisfies the integral equation

$$\Gamma(t, u) + \int_{t_0}^{t} \Gamma(t, \tau) \mathbf{P}(\tau) \mathbf{K}(\tau, u) \, d\tau = \mathbf{K}(t, u), \quad t_0 \le u \le t \le T, \tag{7.183}$$

and where $\mathbf{P}(t)$ is a precomputable matrix defined by

$$\mathbf{P}(t) = E\left\{ \frac{1}{\lambda(t, \mathbf{x}(t))} \left[\frac{\partial \lambda(t, \mathbf{x}(t))}{\partial \mathbf{x}(t)} \right] \left[\frac{\partial \lambda(t, \mathbf{x}(t))}{\partial \mathbf{x}(t)} \right]' \right\}. \tag{7.184}$$

The matrix inequality $\mathbf{M} \ge \mathbf{N}$ appearing in (7.182) means that $\mathbf{M} - \mathbf{N}$ is nonnegative definite; for any vector $\mathbf{v}$, there holds $\mathbf{v}'\mathbf{M}\mathbf{v} \ge \mathbf{v}'\mathbf{N}\mathbf{v}$. Thus, for example, if $y(t) = \mathbf{c}'(t)\mathbf{x}(t)$, then

$$E\{[y(t) - y^*(t)]^2\} \ge E\{[y(t) - \hat{y}(t)]^2\} \ge \mathbf{c}'(t)\Sigma(t)\mathbf{c}(t), \tag{7.185}$$

where $y^*(t) = \mathbf{c}'(t)\mathbf{x}^*(t)$ and $\hat{y}(t) = \mathbf{c}'(t)\hat{\mathbf{x}}(t)$.

We make three important remarks before proving the theorem.

Remark 1. An interpretation of $\Sigma(t)$ can be given in terms of an appropriate linear filtering problem involving observations in additive Gaussian-noise. If $\mathbf{P}(t)$ is factored as $\mathbf{C}'(t)\mathbf{C}(t)$, then postmultiplication of (7.183) by $\mathbf{C}'(u)$ shows that $\Gamma(t, u)\mathbf{C}'(u)$ is the impulse response matrix of the causal linear, least-squares filter for estimating $\mathbf{x}(t)$ in terms of linear observations of the additive form $d\mathbf{x}(\sigma) = \mathbf{C}(\sigma)\mathbf{x}(\sigma)d\sigma + d\mathbf{w}(\sigma)$, $t_0 \le \sigma \le t$, where $\mathbf{w}(t)$ is a vector-valued Wiener process normalized so its covariance matrix is $\mathbf{I}\min(t,u)$. The lower-bound matrix $\Sigma(t)$ of (7.182) is the covariance of the corresponding estimation error.

Remark 2. When $\mathbf{x}(t)$ is a finite-dimensional Gauss-Markov process, the lower bound matrix $\Sigma(t)$ can be determined by solving a matrix Riccati equation. Suppose that $\mathbf{x}(t)$ has a finite-dimensional state representation as

$$d\mathbf{x}(t) = \mathbf{A}(t)\mathbf{x}(t)dt + \mathbf{B}(t)d\mathbf{w}(t), \quad \mathbf{x}(t_0) = \mathbf{x}_0, \tag{7.186}$$

where $\mathbf{w}(t)$ is a normalized Wiener process, and $\mathbf{x}_0$ is normally distributed with zero mean and covariance Σ_0 and independent of $\{\mathbf{w}(t): t \ge t_0\}$. Then, the lower-bound matrix $\Sigma(t)$ satisfies

$$\frac{d\Sigma(t)}{dt} = \mathbf{A}(t)\Sigma(t) + \Sigma(t)\mathbf{A}'(t) + \mathbf{B}(t)\mathbf{B}'(t) - \Sigma(t)\mathbf{P}(t)\Sigma(t),$$

$$\Sigma(t_0) = \Sigma_0. \tag{7.187}$$

Remark 3. An asymptotically efficient estimate is one for which equality in (7.182) is achieved asymptotically as some parameter, typically a measure of the signal-to-noise ratio, increases. Specifically, we will say that a causal estimate $\mathbf{x}^*(t)$ of $\mathbf{x}(t)$ is asymptotically efficient with a parameter α if for any $\varepsilon > 0$, there exists a real number a such that for all $\alpha \ge a$ there holds

$$\Sigma(t) \le E\{[\mathbf{x}(t) - \mathbf{x}^*(t)][\mathbf{x}(t) - \mathbf{x}^*(t)]'\} \le \Sigma(t) + \varepsilon\mathbf{I}. \tag{7.188}$$

We are unable to give conditions on the intensity that guarantee the existence of an asymptotically efficient estimate. An example for which the suboptimal estimators of Sec. 7.4.3 appear to be asymptotically efficient is given below for subcarrier angle-modulation.

Proof of Theorem 7.4.4. The first inequality in (7.182) is trivial because the conditional mean minimizes the mean square-error. To prove the second inequality, consider the approximation

$$\mathbf{x}_M(t) \stackrel{\Delta}{=} \sum_{m=1}^{M} \chi_m \psi_m(t), \quad t_0 \le t < T \tag{7.189}$$

to $x(t)$ over the interval $[t_0, T)$ motivated by the Karhunen-Loève representation in which $x_M(t)$ converges in the mean-square sense to $x(t)$ as $M \to \infty$ [46, p. 200]. The coefficients $\{\chi_m\}$ are the Fourier coefficients of the representation, and $\{\psi_m(t): t_0 \le t < T\}$ are the eigenvectors associated with the integral equation

$$\lambda_m \psi_m(t) = \int_{t_0}^{T} K(t, \tau) \psi_m(\tau) d\tau, \quad t_0 \le t < T \tag{7.190}$$

in which $K(t, u)$ is the autocovariance function for $x(\cdot)$. The coefficients are independent, zero-mean, Gaussian random variables with variances $\{\lambda_m\}$, where λ_i is the eigenvalue associated with the eigenvector $\psi_i(t)$.

Denote an observed counting path $\mathcal{N}_t = \{N(\sigma): t_0 \le \sigma < t\}$ by $\mathcal{N}_t^{(M)}$ when $x(\cdot)$ is replaced by its M-term approximation $x_M(\cdot)$. Let $p_\chi(X \mid \mathcal{N}_t^{(M)})$ be the posterior probability density of the coefficient vector χ given $\mathcal{N}_t^{(M)}$. Let $J_M(t)$ be the matrix with k,i-element

$$[J_M(t)]_{k,i} = -E\left\{ \frac{\partial^2 \ln p_\chi(X \mid \mathcal{N}_t^{(M)})}{\partial \chi_k \partial \chi_i} \right\}. \tag{7.191}$$

Finally, define $S_M(t) = J_M^{-1}(t)$. Then

$$E\{[\chi - \hat{\chi}(t)][\chi - \hat{\chi}(t)]'\} \ge S_M(t), \tag{7.192}$$

where $\hat{\chi}(t) = E[\chi \mid \mathcal{N}_t^{(M)}]$ is the causal minimum mean square-error estimate of χ in terms of $\mathcal{N}_t^{(M)}$. Eq. (7.192) is the Cramér-Rao inequality for random variables [46, p. 84].

We need next to evaluate the matrix $J_M(t)$ of (7.191). We have from the above definitions that $\{N_M(t): t_0 \le t < T\}$ is a doubly stochastic Poisson-process with an intensity process

$$\mu(t, X) = \lambda\left(t, \sum_{m=1}^{M} \chi_m \psi_m(t) \right) = \lambda(t, x_M(t)). \tag{7.193}$$

Thus, Bayes' rule and the normality of X imply that

$$p_{\chi}(X \mid \mathcal{N}_t^{(M)})$$

$$= C \exp\left\{ -\int_{t_0}^t \mu(\sigma, X)\, d\sigma + \int_{t_0}^t \ln \mu(\sigma, X)\, dN_M(\sigma) - \frac{1}{2} X' \Lambda^{-1} X \right\}, \quad (7.194)$$

where $\Lambda = \mathrm{diag}(\lambda_i)$ and C is a normalization constant. When this expression is used in (7.191), we obtain the following expression for the k,i-element after some manipulations

$$[\mathbf{J}_M(t)]_{k,i} = \frac{1}{\lambda_i}\delta_{ki} + \int_{t_0}^t E\left\{ \frac{1}{\mu(\sigma,X)}\left[\frac{\partial\mu(\sigma,X)}{\partial\chi_k}\right]\left[\frac{\partial\mu(\sigma,X)}{\partial\chi_i}\right]\right\} d\sigma. \quad (7.195)$$

From the chain rule and (7.193), there follows

$$\frac{\partial\mu(\sigma,X)}{\partial\chi_i} = \left[\frac{\partial\lambda(\sigma,\mathbf{x}_M(\sigma))}{\partial\mathbf{x}_M(\sigma)}\right]'\psi_i(\sigma).$$

Substitution of this expression into (7.195) yields the desired,

$$\mathbf{J}_M(t) = \Lambda^{-1} + \int_{t_0}^t \mathbf{\Psi}_M'(\sigma)\mathbf{P}_M(\sigma)\mathbf{\Psi}_M(\sigma)\,d\sigma, \quad (7.196)$$

where $\mathbf{\Psi}_M(t) = \mathbf{\Psi}_M(t) = [\psi_1(t) \quad \psi_2(t) \quad \cdots \quad \psi_M(t)]$ is the matrix of eigenvectors and $\mathbf{P}_M(t)$ is defined by

$$\mathbf{P}_M(t) = E\left\{ \frac{1}{\lambda(t,\mathbf{x}_M(t))}\left[\frac{\partial\lambda(t,\mathbf{x}_M(t))}{\partial\mathbf{x}_M(t)}\right]\left[\frac{\partial\lambda(t,\mathbf{x}_M(t))}{\partial\mathbf{x}_M(t)}\right]'\right\}. \quad (7.197)$$

We next use (7.192) and (7.196) to complete the proof. Let

$$\hat{\mathbf{x}}_M(t) = E\{\mathbf{x}_M(t) \mid \mathcal{N}_t^{(M)}\} = \sum_{m=1}^M \hat{\chi}_m(t)\psi_m(t) = \mathbf{\Psi}_M(t)\hat{X}(t) \quad (7.198)$$

be the causal minimum mean square-error estimate of $\mathbf{x}_M(t)$ in terms of $\mathcal{N}_t^{(M)}$. Using (7.192), we find

$$E\{[\mathbf{x}_M(t) - \hat{\mathbf{x}}_M(t)][\mathbf{x}_M(t) - \hat{\mathbf{x}}_M(t)]'\} = \mathbf{\Psi}_M(t)E\{[X - \hat{X}(t)][X - \hat{X}(t)]'\}\mathbf{\Psi}'_M(t)$$

$$\geq \mathbf{\Psi}_M(t)\mathbf{S}_M(t)\mathbf{\Psi}'_M(t) = \mathbf{\Sigma}_M(t). \quad (7.199)$$

Define

$$\Gamma_M(t,\tau) = \Psi_M(t)S_M(t)\Psi'_M(\tau), \quad t_0 \le t \le T. \tag{7.200}$$

Postmultiplying (7.200) by $\mathbf{P}_M(\tau)\psi_M(\tau)$, integrating over $[t_0, T]$, and using (7.196), and then using $\mathbf{S}_M(t)\mathbf{J}_M(t) = \mathbf{I}$, yields

$$\int_{t_0}^{t} \Gamma_M(t,\tau)\mathbf{P}_M(\tau)\psi_M(\tau)d\tau = \Psi_M(t)\mathbf{S}_M(t)[\mathbf{J}_M(t) - \Lambda^{-1}] = \Psi_M(t)[\mathbf{I} - \mathbf{S}_M(t)\Lambda^{-1}].$$

Now postmultiply by $\Lambda\Psi_M(u)$, use (7.200) and the definition $\mathbf{K}_M(t,u) = \Psi_M(t)\Lambda\Psi'_M(u)$ to obtain

$$\Gamma_M(t,u) + \int_{t_0}^{t} \Gamma_M(t,\tau)\mathbf{P}_M(\tau)\mathbf{K}_M(\tau,u)d\tau = \mathbf{K}_M(t,u), \tag{7.201}$$

for $t_0 \le u \le t \le T$. An interpretation of (7.201) in parallel to that of Remark 1 for (7.183) is that $\Gamma_M(t,u)\mathbf{C}'_M(u)$ is the impulse response of the optimum linear filter for causally estimating $\mathbf{x}_M(t)$ in terms of the additive observations $\{d\mathbf{z}(\sigma) = \mathbf{C}_M(\sigma)\mathbf{x}_M(\sigma) + d\mathbf{w}(\sigma): t_0 \le \sigma < t\}$, and $\Sigma_M(t)$ is the corresponding covariance of the error $\mathbf{x}_M(t) - \hat{\mathbf{x}}_M(t)$, where $\mathbf{P}_M(t) = \mathbf{C}'_M(t)\mathbf{C}_M(t)$. Now as $M \to \infty$, we have $\mathbf{P}_M(t) \to \mathbf{P}(t)$, $\mathbf{C}_M(t)\mathbf{K}_M(t,u)\mathbf{C}'_M(u) \to \mathbf{C}(t)\mathbf{K}(t,u)\mathbf{C}'(u)$, $\mathbf{x}_M(t) - \hat{\mathbf{x}}_M(t) \to \mathbf{x}(t) - \hat{\mathbf{x}}(t)$, and $\Gamma_M(t,u) \to \Gamma(t,u)$, where $\Gamma(t,u)$ satisfies (7.183). As (7.199) holds for all M, it holds in the limit as $M \to \infty$, and the proof of Theorem 7.4.4 is complete. $\square$

We now apply the performance lower bound of Theorem 7.4.4 to direct-detection optical communication systems.

Example 7.4.13 *Performance for Phase Modulation* ———————————
The model for subcarrier phase-modulation is given in Ex. 7.4.12. The intensity process is $\lambda(t,\mathbf{x}(t)) = G[1 + m\cos(\omega t + \theta(t))]$, where $\theta(t) = \mathbf{c}'(t)\mathbf{x}(t)$. Manipulations yield

$$\frac{1}{\lambda(t,\mathbf{x}(t))}\left[\frac{\partial\lambda(t,\mathbf{x}(t))}{\partial\mathbf{x}(t)}\right]\left[\frac{\partial\lambda(t,\mathbf{x}(t))}{\partial\mathbf{x}(t)}\right]' = Gm^2 f(\omega t + \theta(t))\mathbf{c}(t)\mathbf{c}'(t), \tag{7.202}$$

where $f(\cdot)$ is the function defined in (7.174); this function is even and periodic, so it can be written in a Fourier cosine series as

$$f(\phi) = a_0 + \sum_{n=1}^{\infty} a_n\cos(n\phi), \tag{7.203}$$

where $a_0 = [1 - \sqrt{1 - m^2}]m^{-2}$. For (7.183), we need the expectation of (7.202). By using the series for $f(\cdot)$ and assuming that $\theta(t)$ is a Gaussian random variable with zero mean and variance $\sigma^2(t)$, we obtain

$$\mathbf{P}(t) = \left\{ [1 - \sqrt{1 - m^2}]G + \sum_{n=1}^{\infty} a_n G e^{-n^2 \sigma^2(t)/2} \cos(n\omega t) \right\} \mathbf{c}(t)\mathbf{c}'(t). \qquad (7.204)$$

Because of the integration in (7.183) and the implied integration in (7.187), we retain only the first term in (7.204) on the assumption that the subcarrier frequency ω is sufficiently large that the rapid oscillations in the last term average to zero. Thus, for instance, if $\mathbf{x}(t)$ has a state representation as in (7.186), then we have that the mean square-error for any causal estimate $\theta^*(t)$ of $\theta(t) = \mathbf{c}'(t)\mathbf{x}(t)$ satisfies the inequality

$$E\{[\theta(t) - \theta^*(t)]^2\} \geq \mathbf{c}'(t)\Sigma(t)\mathbf{c}(t), \qquad (7.205)$$

where $\Sigma(t)$ satisfies the Riccati equation

$$\frac{d\Sigma(t)}{dt} = \mathbf{A}(t)\Sigma(t) + \Sigma(t)\mathbf{A}'(t) + \mathbf{B}(t)\mathbf{B}'(t) - G[1 - \sqrt{1 - m^2}]\Sigma(t)\mathbf{c}(t)\mathbf{c}'(t)\Sigma(t),$$

$$\Sigma(t_0) = \Sigma_0. \qquad (7.206)$$

In Ex. 7.4.12, we have shown that a tanlock loop can be used to generate an approximation to the causal, minimum mean square-error estimate of the phase process $\theta(t)$. The structure of this loop is specified in terms of the solution to a Riccati equation (7.176) that is identical to (7.206). Thus, the solution to (7.176) determines both the structure of this causal estimator as well as a lower bound on its performance.

The fact that (7.206) corresponds to the equation for the error covariance matrix of an appropriate Kalman-Bucy filtering problem implies that there are numerous specializations for which an explicit lower bound (7.205) can be given. One such example occurs when $t_0 = -\infty$ and $\theta(t)$ is a stationary Gaussian process with a rational power spectral density $S(f)$. Then, from (7.206) and the Yovits-Jackson equation (see M. Yovits and J. Jackson [50] or A. Viterbi [48, p. 143]), we deduce that

$$E\{[\theta(t) - \theta^*(t)]^2\} \geq \frac{1}{\alpha} \int_{-\infty}^{\infty} \ln[1 + \alpha S(f)]df, \qquad (7.207)$$

where $\alpha = G[1 - \sqrt{1 - m^2}]$.

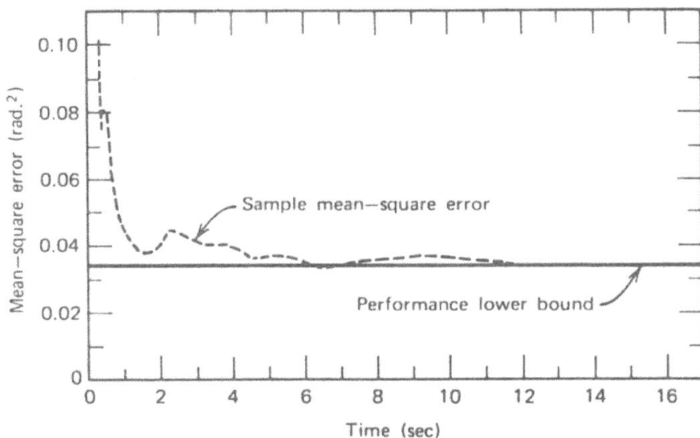

Figure 7.11 Sample mean square-error for the intensity $\lambda(t, \theta(t))$ $= G[1 + m \cos(\omega t + \theta(t))]$, where $G = 1000$, $m = 0.99$, and $\{\theta(t): t \geq 0\}$ is a Wiener process with coherence-time $\tau_c = 1$..

The simplest nonstationary example occurs when $\theta(\cdot)$ is a Wiener process with parameter $1/\tau_c$. As in Ex. 7.4.12, the problem of estimating the phase is then equivalent to that of establishing a reference subcarrier when the subcarrier oscillator is unstable and has a coherence time equal to τ_c. It follows from (7.206) and (7.181) that the steady-state mean square-error satisfies

$$\lim_{t \to \infty} E\{[\theta(t) - \overset{*}{\theta}(t)]^2\} \geq \frac{1}{\sqrt{\tau_c G[1 - \sqrt{1 - m^2}]}}. \tag{7.208}$$

Shown in Fig. 7.11 is the sample mean square-error obtained in the simulation of Fig. 7.10. It is seen that the sample mean square-error approaches a constant value with increasing time. The close proximity of this value to the performance lower bound (7.208) suggest that the suboptimal estimator defined by (7.177) and (7.178) is asymptotically efficient.

It is possible to analyze the performance somewhat more completely for this situation. By using the approximation procedure developed in Ex. 5.3.12, it is shown by R. Forrester and D. Snyder [12] that the steady state probability density of the phase estimation error is approximated by

$$p(\Phi) \approx C(1 + \Gamma \cos \Phi)^{\lambda \Gamma^{-1} - 1}, \quad |\Phi| \leq \pi, \tag{7.209}$$

where C is the normalization constant,

$$\Lambda = 2\frac{\sqrt{\tau_c G[1-\sqrt{1-m^2}]}}{1+\frac{1}{\sqrt{1-m^2}}},$$

and

$$\Gamma = \frac{\sqrt{1-m^2}-1}{\sqrt{1-m^2}+1}.$$

Shown in Fig. 7.12 is the sample mean square-error obtained in the same manner as that of Fig. 7.11 but for the parameter values $G = 300$, $m = 0.33$, and $\tau_c = 1$. The performance lower-bound (7.208) is shown as is the theoretical performance determined numerically form (7.209) according to $\int_{-\pi}^{\pi}\Phi^2 p_\infty(\Phi)d\Phi$. For the parameter values selected, the sample mean square-error does not appear to approach the lower bound, but it is accurately predicted from (7.209). ∎

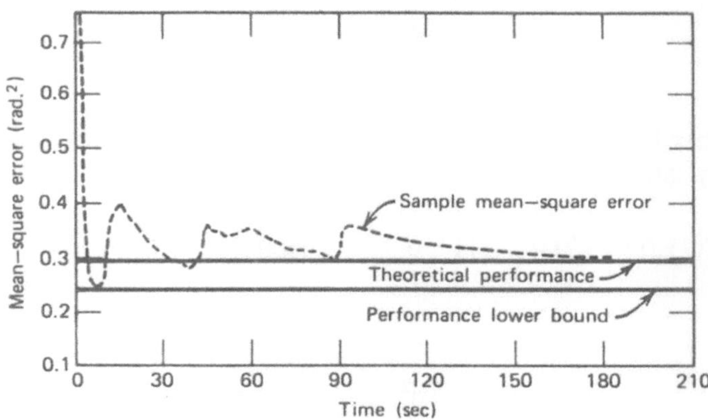

Figure 7.12 Sample mean square-error for the intensity $\lambda(t, \theta(t))$ = $G[1 + m\cos(\omega t + \theta(t))]$, where $G = 300$, $m = 0.33$, and $\{\theta(t): t \geq 0\}$ is a Wiener process with coherence-time $\tau_c = 1$.

Example 7.4.14 *Performance for Frequency Modulation* ────────

The model for subcarrier frequency modulation is given in Ex. 7.4.12. The intensity process is $\lambda(t, \tilde{x}(t)) = G[1 + m\cos(\omega t + \theta(t))]$, where $\theta(t) = c'(t)\tilde{x}(t) = d_f x_0(t)$. Here, $\tilde{x}(t)$ is an expanded state vector defined in (7.168), d_f is the frequency deviation factor, $x_0(t) = \int_{t_0}^{t} y(\sigma)d\sigma$ is the integrated message, and $y(t) = [0 \quad h'(t)]\tilde{x}(t)$. We have

$$\frac{1}{\lambda(t,\tilde{\mathbf{x}}(t))}\left[\frac{\partial\lambda(t,\tilde{\mathbf{x}}(t))}{\partial\tilde{\mathbf{x}}(t)}\right]\left[\frac{\partial\lambda(t,\tilde{\mathbf{x}}(t))}{\partial\tilde{\mathbf{x}}(t)}\right]' = Gm^2 f(\omega t + \theta(t))\mathbf{c}(t)\mathbf{c}'(t), \qquad (7.210)$$

where $f(\cdot)$ is defined in (7.174). Using the expansion for $f(\cdot)$ in (7.203) and dropping double-frequency terms, as before, we conclude from (7.182) that for any causal estimate $\tilde{\mathbf{x}}^*(t)$ of $\tilde{\mathbf{x}}(t)$ there holds

$$E\{[\tilde{\mathbf{x}}(t)-\tilde{\mathbf{x}}^*(t)][\tilde{\mathbf{x}}(t)-\tilde{\mathbf{x}}^*(t)]'\} \geq \tilde{\Sigma}(t), \qquad (7.211)$$

where $\tilde{\Sigma}(t)$ satisfies the Riccati equation

$$\frac{d\tilde{\Sigma}(t)}{dt} = \tilde{\mathbf{A}}(t)\tilde{\Sigma}(t) + \tilde{\Sigma}(t)\tilde{\mathbf{A}}'(t) + \tilde{\mathbf{B}}(t)\tilde{\mathbf{B}}'(t) - G[1-\sqrt{1-m^2}]\tilde{\Sigma}(t)\mathbf{c}(t)\mathbf{c}'(t)\tilde{\Sigma}(t),$$

$$(7.212)$$

with the initial condition

$$\tilde{\Sigma}(t_0) = \begin{bmatrix} 0 & \mathbf{0}' \\ \mathbf{0} & \Sigma_0 \end{bmatrix}.$$

It further follows that the mean square-errors in tracking phase and frequency satisfy

$$E\left\{[x_0(t)-x_0^*(t)]^2\right\} \geq [1 \quad 0 \quad \cdots \quad 0]\tilde{\Sigma}(t)[1 \quad 0 \quad \cdots \quad 0]', \qquad (7.213)$$

and

$$E\{[y(t)-y^*(t)]^2\} \geq [0 \quad \mathbf{h}'(t)]\tilde{\Sigma}(t)[0 \quad \mathbf{h}'(t)]', \qquad (7.214)$$

respectively, for any causal estimator. The fact that (7.212) corresponds to the error covariance matrix of an appropriate Kalman-Bucy filtering problem implies that there are numerous specializations for which explicit lower bounds (7.213) and (7.214) can be given. One such example occurs when $t_0 = -\infty$ and $y(t)$ is a stationary Gaussian process with a rational power-spectral density $S(f)$. Then, from (7.212) and the results in [42], we deduce for any causal estimator that

$$E\left\{[x_0(t)-x_0^*(t)]^2\right\} \geq \frac{1}{\alpha}I_1, \qquad (7.215)$$

and

$$E\{[y(t)-y^*(t)]^2\} \geq \frac{1}{3\alpha}I_1^3 + I_2, \tag{7.216}$$

where

$$\alpha = Gd_f^2[1-\sqrt{1-m^2}], \quad I_1 = \int_{-\infty}^{\infty} \ln[1+\alpha f^2 S(f)]df,$$

and

$$I_2 = \int_{-\infty}^{\infty} \frac{1}{\alpha}f^2 \ln[1+\alpha f^2 S(f)]df. \qquad \blacksquare$$

Summary. In this section, we have studied the filtering problem for observed doubly stochastic Poisson-processes. Four major topics were discussed: linear filtering, optimal nonlinear filtering, suboptimal nonlinear filtering, and filtering performance.

Estimation of the intensity process was emphasized for the linear filtering problem. Here, the intensity estimator is selected to minimize the mean square-error in estimating the intensity at time t subject to the constraint that the estimate must be a linear functional of the observed doubly stochastic Poisson-process up to time t. The optimum linear estimator is specified in Thm. 7.4.1. This estimator can be realized as a Kalman-Bucy filter for a wide class of intensity processes. Linear estimates of desired processes other than the intensity process are developed in Pbm. 7.4.2.

The main result when the linearity constraint is removed is contained in Thm. 7.4.2 of Sec. 7.4.2, where a representation is stated for the conditional characteristic function of the information process given the observed doubly stochastic Poisson-process. For alternative approaches to the filtering problem, we refer to P. Fishman and D. Snyder [11], P. Brémaud [Bre], R. Lipster and A. Shiryayev [27], and A. Karr [23].

Estimators without the linearity constraint are usually too complicated to implement. For this reason, suboptimal estimators were developed in Sec 7.4.3. While these estimators are more easily implemented and are nearly optimal for small estimation errors, they must be carefully examined in individual applications where alternative approximations may be more appropriate.

A precomputable lower bound on the mean-square performance in estimating Gaussian information processes was developed in Sec. 7.4.4. The lower bound applies to an arbitrary estimator including the linear, optimal, and suboptimal estimators we derived. It was shown that under favorable circumstances, the suboptimal estimators of Sec 7.4.3 can nearly achieve the lower bound. Furthermore, the lower bound was seen to be closely related to the design of the suboptimal estimators for certain applications arising in optical communication.

We now generalize these results to situations where more than one doubly stochastic Poisson-process is observed simultaneously.

7.5 Doubly Stochastic Multidimensional Poisson-Processes

The results in the preceding sections can be extended to accommodate multidimensional Poisson-processes. Doubly stochastic multichannel, time-space, and translated Poisson-processes are three versions of this extension that we will examine. They generalize the Poisson processes studied in Section 2.5 and Chapters 3 and 4 to include an information process that influences the intensity functions.

7.5.1 Doubly Stochastic Multichannel Poisson-Processes

A doubly stochastic Poisson-process defined on a time-space domain is termed a *doubly stochastic multichannel Poisson-process* when the spatial domain consists of discrete regions over each of which a temporal, doubly stochastic Poisson-process is observed while the spatial locations within the region are disregarded. An individual temporal point-process associated with one of the regions is termed a *component* of the multichannel process. A doubly stochastic multichannel Poisson-process may also be regarded as a marked doubly stochastic compound Poisson-process in which a mark on a point indicates the region in which the point occurs or some other discrete-valued random variable. Following are several applications that motivate the study of doubly stochastic multichannel Poisson-process point processes.

Example 7.5.1 *Nuclear Medicine* ──────────────────────────────
Multichannel data are routinely collected in nuclear medicine. This occurs in a variety of procedures and equipment configurations. Three examples are multi-isotope studies, the gamma-ray camera, and multi-probe detection systems.

> a. *Multi-Isotope Studies* (D. Ross and C. Harris [37, p. 179]). Multiple radioactive isotopes with different energy spectra are introduced simultaneously into an organ, and radioactivity is monitored externally with a scintillation detector. As in Fig. 4.5, point processes associated with different energy ranges are obtained by processing the pulses from the scintillation detector with energy discriminators. Ideally, each point process is influenced by only one isotope, but in practice, the energy spectra overlap so that this cannot be achieved perfectly.

b. *Gamma-Ray Camera* (H. Anger [2], N. Moody *et al.* [33]). The Anger gamma-ray camera is widely used in nuclear medicine. It consists of an array of photomultipliers that view a single scintillation crystal in the form of a large disk about 30 cm in diameter. The spatial location of a scintillation event is estimated with a circuit that weights the current pulses produced in the photomultipliers following the event. If the surface of the crystal is considered to be partitioned into discrete regions or pixels, then radioactivity monitored with the camera results in a multichannel point process with a number of components equal to the number of pixels.

c. *Multiprobe Detection Systems* (M. Ter-Pogossian *et al.* [45]). Multiprobe detection systems consist of an array of several independent scintillation detectors placed over an organ to monitor radioactivity in it. Each component of a multichannel point process is associated with one of the scintillation detectors. ∎

Example 7.5.2 *Intensity Interferometry* ─────────────────────────

In intensity interferometry, light incident on two spatially separated apertures is first detected and then processed to determine desired properties of the optical source, which is typically a stellar object (J. Goodman [14, p. 19]). The outputs of two photomultipliers used to detect the light can be modeled as components of a two-dimensional multichannel point process. ∎

Example 7.5.3 *Direct-Detection Optical Communication Systems* ──────────

Atmospheric turbulence causes the received light field in an optical communication system to be spatially incoherent [19]. This effect can be compensated for to some extent by using an array of photodetectors at the receiver. A component of a multichannel point process can be associated with the output one of the photodetectors in the array. ∎

Example 7.5.4 *Optical Position Sensing, continued* ────────────────

Consider Examples 4.4.1 and 4.4.2 again, but now suppose that the beam of light that is incident on the four-quadrant photodetector wanders about in a random walk due to temporal fluctuations in the propagation medium between the transmitter and the receiver. Thus, as indicated in Fig. 7.13, the location parameters α and β of the center of the light beam will now be random processes $\{\alpha(t), \beta(t) : t \geq t_0\}$, and we have for the i*th* quadrant that

$$\lambda_i(t, \alpha(t), \beta(t)) = \int_{\mathcal{A}_i} \int I(a, b \mid \alpha(t), \beta(t)) \, da \, db, \qquad (7.217)$$

where $\mathcal{A}_i$ is the region of the i*th* quadrant, and from (4.39)

$$I(a,b \mid \alpha(t),\beta(t)) = I_0 \exp\left\{-\frac{1}{2\rho^2}[(a-\alpha(t))^2 + (b-\beta(t))^2]\right\}. \quad (7.218)$$

This is an example of a doubly stochastic multichannel Poisson-process in which there are four component processes that are coupled by the motion processes $\alpha(\cdot)$ and $\beta(\cdot)$. The ith component process records the times of photoconversions that occur anywhere in the ith quadrant. ∎

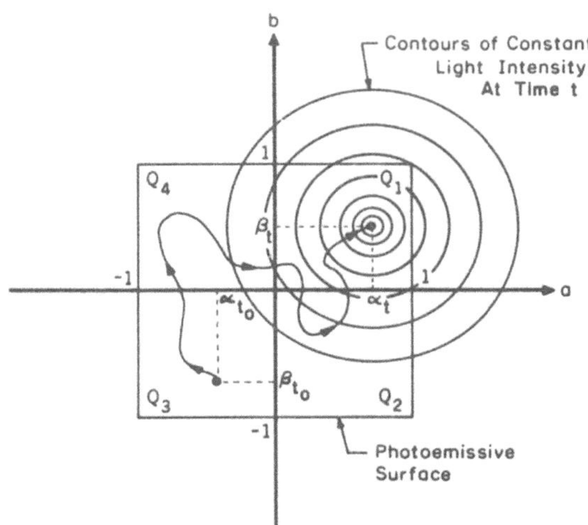

Figure 7.13 Four-quadrant Position Sensor.

Our model for a doubly stochastic multichannel Poisson-process is comprised of K single-channel doubly stochastic Poisson-processes that are conditionally mutually independent Poisson processes given the information process. Thus, let $\{N_k(t): t \geq t_0\}$ for $k = 1, 2, \cdots, K$ be doubly stochastic Poisson-processes with corresponding intensity processes $\{\lambda_k(t, \mathbf{x}(t)): t \geq t_0\}$, where $\{\mathbf{x}(t): t \geq t_0\}$ is an information process influencing them all. We assume that $N_1(\cdot)$, $N_2(\cdot)$, $\cdots$, and $N_K(\cdot)$ are mutually independent given $\{\mathbf{x}(t): t \geq t_0\}$. We term the vector $\mathbf{N}(\cdot)$, with components $N_1(\cdot)$, $N_2(\cdot)$, $\cdots$, and $N_K(\cdot)$, a doubly stochastic multichannel Poisson-process with intensity process

$$\boldsymbol{\lambda}(t, \mathbf{x}(t)) = [\lambda_1(t, \mathbf{x}(t)) \quad \lambda_2(t, \mathbf{x}(t)) \quad \cdots \quad \lambda_K(t, \mathbf{x}(t))]'.$$

In the next paragraphs, we indicate how the important results of the preceding sections are modified for doubly stochastic multichannel Poisson-processes. Most of these extensions are given without proof because the proofs are straightforward modifications of those for the single-channel model.

**Multichannel Differential Rule** (Extension of Theorem 5.3.2).   Let
$\{\zeta(t): t \geq t_0, \zeta(t) \in \mathcal{R}^n\}$ satisfy

$$d\zeta(t) = \mathbf{a}(t, \zeta(t))dt + \mathbf{B}(t, \zeta(t))d\mathbf{N}(t), \tag{7.219}$$

where $\{\mathbf{N}(t): t \geq t_0\}$ is a doubly stochastic multichannel Poisson-process.
Here, $\mathbf{B}(\cdot)$ is an $n \times K$-dimensional matrix. Suppose that $\phi(t, \mathbf{X})$ is a real
valued function of $\mathbf{X} \in \mathcal{R}^n$ with continuous partial derivatives in t and the
components of $\mathbf{X}$. Then

$$d\phi(t, \zeta(t)) = \left[\frac{\partial \phi(t, \zeta(t))}{\partial t} + < \frac{\partial \phi(t, \zeta(t))}{\partial \zeta(t)}, \mathbf{a}(t, \zeta(t)) > \right] dt$$

$$+ \sum_{k=1}^{K} \{\phi[t, \zeta(t) + \mathbf{B}(t, \zeta(t)\mathbf{e}_k] - \phi(t, \zeta(t))\}dN_k(t), \tag{7.220}$$

where $\mathbf{e}_k$ is a K-dimensional unit vector with a one as its _kth_ component
and zero for all others. Eq. (7.220) can be proven by paralleling the proof
of Theorem 5.3.2; see Problem 7.5.1.

**Multichannel Sample-Function Density** (Extension of (7.45) and (7.71)).
Let $\{\mathbf{N}(t): t \geq t_0\}$ be a doubly stochastic multichannel Poisson-process with
intensity process $\{\lambda(t, \mathbf{x}(t)): t \geq t_0\}$, where $\{\mathbf{x}(t): t \geq t_0\}$ is an information
process. Denote the joint sample-function density of the components of
$\mathbf{N}(\cdot)$ on the interval $[t_0, t)$ by $p[\{\mathbf{N}(\sigma): t_0 \leq \sigma < t\}]$. By using the method of
conditioning, the conditional independence of the components, and the
sample-function density for a doubly stochastic Poisson-process, we have

$$p[\{\mathbf{N}(\sigma): t_0 \leq \sigma < t\}] \tag{7.221}$$

$$= E\left\{ \exp\left[< \mathbf{1}, -\int_{t_0}^{t} \lambda(\sigma, \mathbf{x}(\sigma))d\sigma + \int_{t_0}^{t} \mathrm{diag}[\ln \lambda_k(\sigma, \mathbf{x}(\sigma))]d\mathbf{N}(\sigma) > \right] \right\},$$

where $\mathbf{1}$ is a K-dimensional vector of ones. This is the multichannel extension of (7.45). From it and the multichannel differential rule, we obtain

$$p[\{\mathbf{N}(\sigma): t_0 \leq \sigma < t\}] = \exp\left[< \mathbf{1}, -\int_{t_0}^{t} \hat{\lambda}(\sigma)d\sigma + \int_{t_0}^{t} \mathrm{diag}[\ln \hat{\lambda}_k(\sigma)]d\mathbf{N}(\sigma) > \right],$$

$$\tag{7.222}$$

where $\hat{\lambda}(t) = E[\lambda(t, \mathbf{x}(t)) \mid \mathbf{N}(\sigma): t_0 \leq \sigma < t]$ is the minimum mean square-error estimate of $\lambda(t, \mathbf{x}(t))$ in terms of $\{\mathbf{N}(\sigma): t_0 \leq \sigma < t\}$. This is the multichannel extension of (7.71). This equation is derived as follows. For fixed $\{\mathbf{N}(\sigma): t_0 \leq \sigma < t\}$, define

$$\mathbf{H}(t, \mathbf{x}(\sigma): t_0 \leq \sigma < t) \tag{7.223}$$

$$= -\int_{t_0}^{t} \lambda(\sigma, \mathbf{x}(\sigma)) d\sigma + \int_{t_0}^{t} \mathrm{diag}[\ln \lambda_k(\sigma, \mathbf{x}(\sigma))] d\mathbf{N}(\sigma).$$

Then

$$d\mathbf{H}(t, \mathbf{x}(\sigma): t_0 \leq \sigma < t) = -\lambda(t, \mathbf{x}(t)) dt + \mathrm{diag}[\ln \lambda_k(t, \mathbf{x}(t))] d\mathbf{N}(t).$$

$$\tag{7.224}$$

If $\phi(t, \mathbf{x}(\sigma): t_0 \leq \sigma < t) = \exp[< \mathbf{1}, \mathbf{H}(t, \mathbf{x}(\sigma): t_0 \leq \sigma < t) >]$, then (7.224) and the multichannel differential rule imply that

$$d\phi(t, \mathbf{x}(\sigma): t_0 \leq \sigma < t)$$

$$= - < \mathbf{1}, \lambda(t, \mathbf{x}(t)) \phi(t, \mathbf{x}(\sigma): t_0 \leq \sigma < t) > dt$$

$$+ < [\lambda(t, \mathbf{x}(t)) - \mathbf{1}] \phi(t, \mathbf{x}(\sigma): t_0 \leq \sigma < t), d\mathbf{N}(t) > . \tag{7.225}$$

By taking the expectation of both sides of this expression, recognizing from (7.221) that

$$E[\phi(t, \mathbf{x}(\sigma): t_0 \leq \sigma < t)] = p[\{\mathbf{N}(\sigma): t_0 \leq \sigma < t\}], \tag{7.226}$$

and using the following relation to be established

$$E[\lambda(t, \mathbf{x}(t)) \phi(t, \mathbf{x}(\sigma): t_0 \leq \sigma < t)] = \hat{\lambda}(t) E[\phi(t, \mathbf{x}(\sigma): t_0 \leq \sigma < t)], \tag{7.227}$$

we obtain

$$dp[\{\mathbf{N}(\sigma): t_0 \leq \sigma < t\}] \tag{7.228}$$

$$= -p[\{\mathbf{N}(\sigma): t_0 \leq \sigma < t\}](< \mathbf{1}, \hat{\lambda}(t) > dt + < \hat{\lambda}(t) - \mathbf{1}, d\mathbf{N}(t) >).$$

By applying the differential rule again, we have

$$d \ln p[\{\mathbf{N}(\sigma): t_0 \leq \sigma < t\}] \tag{7.229}$$

$$= - <\mathbf{1}, \hat{\lambda}(t)> dt + <\mathbf{1}, \mathrm{diag}[\ln \hat{\lambda}_k(t)]d\mathbf{N}(t)>.$$

Integration of this equation establishes (7.222). Equation (7.227) can be established using Bayes' rule, the sample-function density for a Poisson process, and (7.226) as follows. According to Bayes' rule, there holds

$$p[\{\mathbf{N}(\sigma): t_0 \leq \sigma < t\}] = \frac{p(\{\mathbf{N}(\sigma): t_0 \leq \sigma < t\} \mid \mathbf{x}(t) = \mathbf{X})p_t(\mathbf{X})}{E[p(\{\mathbf{N}(\sigma): t_0 \leq \sigma < t\} \mid \mathbf{x}(t))]}, \tag{7.230}$$

where $p_t(\mathbf{X})$ and $p_t(\mathbf{x} \mid \mathbf{N}(\sigma): t_0 \leq \sigma < t)$ are the prior and posterior probability densities for $\mathbf{x}(t)$, $p(\{\mathbf{N}(\sigma): t_0 \leq \sigma < t\} \mid \mathbf{x}(t) = \mathbf{X})$ is the conditional sample-function density for $\mathbf{N}(\cdot)$ given that $\mathbf{x}(t) = \mathbf{X}$, and the denominator is the required normalization factor. For the first term in the numerator, we use iterated expectations to obtain

$$p(\{\mathbf{N}(\sigma): t_0 \leq \sigma < t\} \mid \mathbf{x}(t) = \mathbf{X})$$

$$= E[p(\{\mathbf{N}(\sigma): t_0 \leq \sigma < t\} \mid \mathbf{x}(\sigma): t_0 \leq \sigma < t) \mid \mathbf{x}(t) = \mathbf{X})]$$

$$= E[\phi(t, \mathbf{x}(\sigma): t_0 \leq \sigma < t) \mid \mathbf{x}(t) = \mathbf{X}]. \tag{7.231}$$

For the denominator of (7.230), we have from (7.224) that

$$E[p(\{\mathbf{N}(\sigma): t_0 \leq \sigma < t\} \mid \mathbf{x}(t) = \mathbf{X})] \tag{7.232}$$

$$= p(\{\mathbf{N}(\sigma): t_0 \leq \sigma < t\}) = E[\phi(t, \mathbf{x}(\sigma): t_0 \leq \sigma < t)].$$

Hence, (7.230) can be rewritten using (7.231) and (7.232) as

$$p_t(\mathbf{X} \mid \mathbf{N}(\sigma): t_0 \leq \sigma < t)E[\phi(t, \mathbf{x}(\sigma): t_0 \leq \sigma < t)]$$

$$= p_t(\mathbf{X})E[\phi(t, \mathbf{x}(\sigma): t_0 \leq \sigma < t) \mid \mathbf{x}(t) = \mathbf{X}]. \tag{7.233}$$

Multiplying both sides of this equation by $\lambda(t, \mathbf{X})$ and integrating with respect to $\mathbf{X}$ then establishes (7.227) and, thus, the sample-function density (7.222).

Multichannel Linear Filtering (Extension of (7.74)). Let $\lambda_{\mathrm{lin}}^*(t)$ be a linear estimate of $\lambda(t)$ in terms of $\{\mathbf{N}(\sigma): t_0 \leq \sigma < t\}$; that is,

$$\lambda_{\mathrm{lin}}^{*}(t) = \mathbf{a}(t) + \int_{t_0}^{t} \mathbf{H}(t,u)\, d\mathbf{N}(u) \qquad (7.234)$$

for some selection of $\mathbf{a}(t)$ and $\mathbf{H}(t,u)$. Denote the linear estimate that minimizes the error covariance $E\{[\lambda(t) - \lambda_{\mathrm{lin}}^{*}(t)][\lambda(t) - \lambda_{\mathrm{lin}}^{*}(t)]'\}$ by $\hat{\lambda}_{\mathrm{lin}}(t)$; here, the minimization means that

$$E\{[\lambda(t) - \lambda_{\mathrm{lin}}^{*}(t)][\lambda(t) - \lambda_{\mathrm{lin}}^{*}(t)]'\} - E\{[\lambda(t) - \hat{\lambda}_{\mathrm{lin}}(t)][\lambda(t) - \hat{\lambda}_{\mathrm{lin}}(t)]'\}$$

is nonnegative definite for all choices of $\lambda_{\mathrm{lin}}^{*}(\cdot)$ defined by (7.234). Then, the multichannel extensions of (7.74) and (7.76) are

$$\hat{\lambda}_{\mathrm{lin}}(t) = E\{\lambda(t)\} + \int_{t_0}^{t} \mathbf{H}_{o}(t,u)[d\mathbf{N}(u) - E\{\lambda(u)\}\,du], \qquad (7.235)$$

where the optimum impulse-response matrix satisfies

$$\mathbf{H}_{o}(t,u)\mathrm{diag}[E\{\lambda_{k}(u)\}] + \int_{t_0}^{t} \mathbf{H}_{o}(t,\sigma)\mathbf{K}_{\lambda}(\sigma,u)\,d\sigma = \mathbf{K}_{\lambda}(t,u),$$

for $t_0 \le \sigma < t$, and

$$E\{[\lambda(t) - \hat{\lambda}_{\mathrm{lin}}(t)][\lambda(t) - \hat{\lambda}_{\mathrm{lin}}(t)]'\} = \mathbf{H}_{o}(t,t)\mathrm{diag}[E\{\lambda_{k}(t)\}]. \quad (7.236)$$

The derivation of these extensions is left as an exercise; see Problem 7.5.3.

In (7.82) and (7.83), we indicated how the Kalman-Bucy algorithm can be applied for linear filtering of single-channel doubly stochastic Poisson-processes. This algorithm can also be applied for multichannel observations. Thus, suppose that $\{\lambda(t): t \ge t_0\}$ has a state-variable realization as

$$d\mathbf{x}(t) = \mathbf{A}(t)\mathbf{x}(t)\,dt + \mathbf{B}(t)\,d\mathbf{u}(t), \quad \mathbf{x}(t_0) = \mathbf{x}_0$$

$$\lambda(t) = E\{\lambda(t)\} + \mathbf{C}(t)\mathbf{x}(t), \qquad (7.237)$$

where: $\mathbf{x}$ is an n-dimensional state vector; $\mathbf{A}$, $\mathbf{B}$, and $\mathbf{C}$ are matrices of dimensions $n \times n$, $n \times m$, and $K \times n$, respectively; $\mathbf{x}_0$ is an initial random variable with zero mean and covariance Σ_0; and $\{\mathbf{u}(t): t \ge t_0\}$ is a vector of independent, zero-mean random processes with stationary, orthogonal increments and normalized so that $E\{\mathbf{u}(t)\mathbf{u}'(\tau)\} = \mathbf{I}\min(t,\tau)$, where $\mathbf{I}$ is an $m \times m$ identity matrix. Then, $\hat{\lambda}_{\mathrm{lin}}(t)$ is given by the following equations:

$$d\hat{\mathbf{x}}(t) = \mathbf{A}(t)\hat{\mathbf{x}}(t)\,dt$$

$$+\Sigma(t)\mathbf{C}'(t)\,\mathrm{diag}^{-1}[E\{\lambda_k(t)\}][d\mathbf{N}(t) - E\{\lambda(t)\}\,dt - \mathbf{C}\hat{\mathbf{x}}(t)\,dt],$$

$$\hat{\mathbf{x}}(t_0) = \mathbf{0} \tag{7.238a}$$

$$\frac{d\Sigma(t)}{dt} = \mathbf{A}(t)\Sigma(t)$$

$$+\Sigma(t)\mathbf{A}'(t) + \mathbf{B}(t)\mathbf{B}'(t) - \Sigma(t)\mathbf{C}(t)\,\mathrm{diag}^{-1}[E\{\lambda_k(t)\}]\mathbf{C}'(t)\Sigma(t),$$

$$\Sigma(t_0) = \Sigma_0 \tag{7.238b}$$

$$\hat{\lambda}_{\mathrm{lin}}(t) = E\{\lambda(t)\} + \mathbf{C}(t)\hat{\mathbf{x}}(t). \tag{7.238c}$$

Multichannel Characteristic Functional (Extension of (7.118)). Let $\{\mathbf{N}(t): t \geq t_0\}$ be a doubly stochastic multichannel Poisson-process with intensity process $\{\lambda(t,\mathbf{x}(t)): t \geq t_0\}$, where $\{\mathbf{x}(t): t \geq t_0\}$ is a Markovian information process characterized by the differential generator $\psi_t(\mathbf{v}\mid\mathbf{x}(t))$. Our main result of nonlinear for single-channel doubly stochastic Poisson-processes, expressed in (7.118), has the following multichannel extension:

$$d\hat{M}_t(j\mathbf{v}) = \hat{E}\{e^{j<\mathbf{v},\mathbf{x}(t)>}\psi_t(\mathbf{v}\mid\mathbf{x}(t))\}\,dt$$

$$+\hat{E}\{e^{j<\mathbf{v},\mathbf{x}(t)>}[\lambda(t,\mathbf{x}(t)) - \hat{\lambda}(t)]'\}\,\mathrm{diag}^{-1}[\hat{\lambda}_k(t)][d\mathbf{N}(t) - \hat{\lambda}(t)\,dt],$$

$$\hat{M}_{t_0}(j\mathbf{v}) = E\left\{ e^{j<\mathbf{v},\mathbf{x}(t_0)>} \right\}. \tag{7.239}$$

The derivation and use of this equation are analogous to the derivation and use of (7.118).

Multichannel Suboptimal Filtering (Extension of (7.160) and (7.161)). Let $\{\mathbf{N}(t): t \geq t_0\}$ be a doubly stochastic multichannel Poisson-process with intensity process $\{\lambda(t,\mathbf{x}(t)): t \geq t_0\}$, where $\{\mathbf{x}(t): t \geq t_0\}$ is a Markov diffusion satisfying

$$d\mathbf{x}(t) = \mathbf{a}(t,\mathbf{x}(t))\,dt + \mathbf{B}(t)\,d\mathbf{w}(t), \quad \mathbf{x}(t_0) = \mathbf{x}_0 \tag{7.240}$$

in which $\{\mathbf{w}(t): t \geq t_0\}$ is a vector of independent Wiener processes normalized so that $E[\mathbf{w}(t)\mathbf{w}'(u)] = \mathbf{I}\min(t, u)$, and where $\mathbf{x}_0$ is independent of $\{\mathbf{w}(t): t \geq t_0\}$. The multichannel extension of the processor equation (7.160) is

$$d\mathbf{x}^*(t) = \mathbf{a}(t, \mathbf{x}^*(t))\,dt + \frac{1}{2}\sum_{i,j} \sigma_{ij}^* \frac{\partial^2 \mathbf{a}(t, \mathbf{x}^*(t))}{\partial x_i^* \partial x_j^*}\,dt$$

$$+\,\Sigma^*(t)\left[\frac{\partial \lambda(t, \mathbf{x}^*(t))}{\partial \mathbf{x}^*(t)}\right]\operatorname{diag}^{-1}[\lambda_k(t, \mathbf{x}^*(t))][d\mathbf{N}(t) - \lambda(t, \mathbf{x}^*(t))\,dt],$$

$$\mathbf{x}^*(t_0) = E(\mathbf{x}_0), \tag{7.241}$$

where $\Sigma^*(t)$ is a symmetric matrix that satisfies the multichannel variance equation, σ_{ij}^* is the i,j-element of $\Sigma^*(t)$, and $\partial \lambda(\mathbf{x})/\partial \mathbf{x}$ denotes the Jacobian matrix having as its i,j-element $\partial \lambda_j(\mathbf{x})/\partial x_i$. The multichannel extension of the variance equation is:

$$d\Sigma^*(t) = \left[\frac{\partial \mathbf{a}(t, \mathbf{x}^*(t))}{\partial \mathbf{x}^*(t)}\right]'\Sigma^*(t)\,dt + \Sigma^*(t)\left[\frac{\partial \mathbf{a}(t, \mathbf{x}^*(t))}{\partial \mathbf{x}^*(t)}\right]dt + \mathbf{B}(t)\mathbf{B}'(t)\,dt$$

$$-\,\Sigma^*(t)\frac{\partial}{\partial \mathbf{x}^*(t)}\left\{\left[\frac{\partial \lambda(t, \mathbf{x}^*(t))}{\partial \mathbf{x}^*(t)}\right]\operatorname{diag}^{-1}[\lambda_k(t, \mathbf{x}^*(t))][d\mathbf{N}(t) - \lambda(t, \mathbf{x}^*(t))\,dt]\right\}\Sigma^*(t)$$

$$\Sigma^*(t_0) = \operatorname{cov}(\mathbf{x}_0), \tag{7.242}$$

where $\partial \mathbf{v}(\mathbf{x})/\partial \mathbf{x}$ denotes the Jacobian matrix of the vector $\mathbf{v}(\mathbf{x})$ having as its i,j-element $\partial v_j(\mathbf{x})/\partial x_i$. The derivation of these equations closely parallels that for (7.160) and (7.161).

Example 7.5.5 *Optical Position Sensing, Continued* ———————————

Consider Ex. 7.5.4 again. There are four component doubly stochastic Poisson-processes having the intensities given in (7.217), and we wish to estimate the position processes $\alpha(t)$ and $\beta(t)$ at time t given observations of the component point processes on $[t_0, t)$. Assume that $\{\alpha(t): t \geq t_0\}$ and $\{\beta(t): t \geq t_0\}$ are independent, identically distributed Wiener processes such that $E[\alpha^2(t)] = E[\beta^2(t)] = t(2\tau_d)^{-1}$. The parameter τ_d characterizes the beam's stability; the beam undergoes a one unit (r.m.s.) displacement from the origin in τ_d units of time. We may then take the state vector $\mathbf{x}(t)$ in (7.237) to be

$$\mathbf{x}(t) = \begin{bmatrix} \alpha(t) \\ \beta(t) \end{bmatrix} = \frac{1}{\sqrt{2\tau_d}} \mathbf{w}(t),$$

and for this choice, $\mathbf{a}(t, \mathbf{x}(t)) = \mathbf{0}$ and $\mathbf{B}(t) = (1/\sqrt{2\tau_d})\mathbf{I}$, where $\mathbf{I}$ is the 2×2 identity matrix. Eq. (7.241) becomes

$$d\begin{bmatrix} \alpha^*(t) \\ \beta^*(t) \end{bmatrix} = \Sigma^*(t)\left\{ \sum_{i=1}^{4} \begin{bmatrix} \dfrac{\partial \ln \lambda_i(t, \alpha^*(t), \beta^*(t))}{\partial \alpha^*(t)} \\ \dfrac{\partial \ln \lambda_i(t, \alpha^*(t), \beta^*(t))}{\partial \beta^*(t)} \end{bmatrix} [N_i(dt) - \lambda_i(t, \alpha^*(t), \beta^*(t))dt] \right\}.$$

$$(7.243)$$

The process $\Sigma^*(t)$ is specified by (7.242). $\blacksquare$

Multichannel Filtering Performance (Extension of (7.182)).   Let $\{N(t) : t \geq t_0\}$ be a doubly stochastic multichannel Poisson-process with intensity process $\{\lambda(t, \mathbf{x}(t)) : t \geq t_0\}$, where $\{\mathbf{x}(t) : t \geq t_0\}$ is a Gaussian process with zero mean and autocovariance matrix $\mathbf{K}(t, u) = E[\mathbf{x}(t)\mathbf{x}'(u)]$. Let $\mathbf{x}^*(t)$ denote an arbitrary causal estimate of $\mathbf{x}(t)$ in terms of past multichannel data $\{N(\sigma) : t_0 \leq \sigma < t\}$. Denote the conditional mean estimate by $\hat{\mathbf{x}}(t)$. Then, there holds

$$E\{[\mathbf{x}(t) - \mathbf{x}^*(t)][\mathbf{x}(t) - \mathbf{x}^*(t)]'\} \tag{7.244}$$

$$\geq E\{[\mathbf{x}(t) - \hat{\mathbf{x}}(t)][\mathbf{x}(t) - \hat{\mathbf{x}}(t)]'\} \geq \Sigma(t) = \lim_{u \uparrow t} \Gamma(t, u),$$

where $\Gamma(\cdot)$ satisfies the integral equation

$$\Gamma(t, u) + \int_{t_0}^{t} \Gamma(t, \tau)\mathbf{P}(\tau)\mathbf{K}(\tau, u)d\tau = \mathbf{K}(t, u), \quad t_0 \leq u \leq t \tag{7.245}$$

in which $\mathbf{P}(t)$ is a precomputable matrix defined by

$$\mathbf{P}(t) = E\left\{ \left[\frac{\partial \lambda(t, \mathbf{x}(t))}{\partial \mathbf{x}(t)} \right] \mathrm{diag}^{-1}[\lambda_k(t, \mathbf{x}(t))] \left[\frac{\partial \lambda(t, \mathbf{x}(t))}{\partial \mathbf{x}(t)} \right]' \right\}, \tag{7.246}$$

where $\partial\lambda(\mathbf{x})/\partial\mathbf{x}$ denotes the Jacobian matrix for $\lambda(\mathbf{x})$. This is the multichannel extension of (7.182); the only modification is to incorporate the matrix $\mathbf{P}(t)$ into (7.183). This same modification carries over to (7.187) when $\mathbf{x}$ has a finite-dimensional state-variable realization.

7.5.2 Doubly Stochastic Time-Space Poisson-Processes

There are many and diverse applications in which points occur randomly in space and time. For instance, in nuclear medicine where the gamma-ray camera (see Examples 3.1.3 and 7.5.1) is used to monitor radioactivity, each detected photon has both a temporal and a spatial coordinate corresponding to the instant and location of its absorption in the detector. Seismic events occur randomly in time and space on the earth's surface, as do lightning discharges, storms and other phenomena. Photoevents in the two-dimensional photodetector of Examples 4.5.2 and 7.5.4 have both a temporal and a spatial coordinate. A point process in which points have both a temporal and a spatial coordinate can also be regarded as a marked point process in time, where the mark on a point is continuously distributed and indicates the spatial position of the point.

A doubly stochastic time-space Poisson-process can be regarded as the limit of a doubly stochastic multichannel Poisson-process as the number of component processes increases towards infinity. This idea is illustrated in the following example.

Example 7.5.6 *Optical Position Sensing, continued* ————————

Consider Ex. 7.5.4 again but suppose, now, that the surface of the photodetector is partitioned into more than four regions. The number of components in the multichannel point process that characterizes the sensing of photons is equal to the number of regions into which the detector is partitioned. As the partition is refined while keeping the detector the same size, the number of components grows towards infinity and, in the limit, there results a time-space point process in which each point has a time of occurrence in an observation interval and a spatial location in the detector. Let the intensity of this point process be denoted by $\{\lambda(t,\mathbf{u}): t \geq t_0, \mathbf{u} \in \mathcal{U}\}$, where $\mathcal{U}$ is the spatial domain of the photodetector. Then, from Examples 4.5.2 and 7.5.4,

$$\lambda(t,\mathbf{u},\mathbf{x}(t)) = I(\mathbf{u} \mid \mathbf{x}(t)) = I_0 \exp\left\{-\frac{1}{2\rho^2}\|\mathbf{u} - \mathbf{x}(t)\|^2\right\}, \qquad (7.247)$$

where $\mathbf{u} = [a \quad b]'$ and $\mathbf{x}(t) = [\alpha(t) \quad \beta(t)]'$. As in Ex. 7.5.4, the intensity $I(\cdot)$ of the spot of light moves in a random walk over the surface of the detector, with its position at time t being $\mathbf{x}(t)$. The problem of optical position sensing is that of estimating $\mathbf{x}(t)$ given the spatial location and occurrence times of photoconversions up to time t. ∎

Our model for a doubly stochastic time-space Poisson-process is the following. Let $N(T \times A)$ denote the number of points in $T \times A$, where $T \times A$ is a measurable set in the time-space domain $\mathcal{T} \times \mathcal{U}$ of the process. When $T = [t_0, t)$, we also use the notation $N(t, A) \equiv N([t_0, t) \times A)$ for brevity. Denote the intensity of this process by $\{\lambda(t, \mathbf{u}, \mathbf{x}(t)) : t \in \mathcal{T}, \mathbf{u} \in \mathcal{U}\}$, where $\{\mathbf{x}(t) : t \in \mathcal{T}\}$ is an information process. By definition, $N(\cdot)$ is a doubly stochastic time-space Poisson-process on $\mathcal{T} \times \mathcal{U}$ if it is conditionally a time-space Poisson process given the information process.

All of the results for doubly stochastic multichannel Poisson-processes have a natural counterpart for doubly stochastic time-space Poisson-processes. The modifications generally only require replacing discrete summations by continuous summations in the form of integration. We state these extensions without proof. Their use is precisely the same as in Sec. 7.5.1 and earlier chapters, so the discussion is brief.

Time-Space Sample-Function Density (Extension of (7.222)). The counterpart of (7.222) for a doubly stochastic time-space Poisson-process is the following. Let $\omega = \{N(t, \mathcal{U}), t_1, \cdots, t_{N(t, u)}, u_1, \cdots, u_{N(t, u)}\}$ denote a sample function of the time-space doubly stochastic Poisson-process on $[t_0, t) \times \mathcal{U}$. Then, the sample-function density is

$$p(\omega) = \exp\left[-\int_{t_0}^{t} \int_{\mathcal{U}} \hat{\lambda}(\sigma, \mathbf{u}) d\sigma d\mathbf{u} + \int_{t_0}^{t} \int_{\mathcal{U}} \ln \hat{\lambda}(\sigma, \mathbf{u}) N(d\sigma \times d\mathbf{u}) \right],$$

$$(7.248)$$

where

$$\hat{\lambda}(t, \mathbf{u}) = E[\lambda(t, \mathbf{u}, \mathbf{x}(t)) \mid N(t, \mathcal{U}), t_1, \cdots, t_{N(t, u)}, u_1, \cdots, u_{N(t, u)}]$$

is the minimum mean square-error estimate of $\lambda(t, \mathbf{u}, \mathbf{x}(t))$ in terms of the doubly stochastic time-space Poisson-process observed on $[t_0, t) \times \mathcal{U}$.

Time-Space Characteristic Functional (Extension of (7.239). The counterpart of (7.239) for a doubly stochastic time-space Poisson-process is:

$$d\hat{M}_t(j\mathbf{v}) = \hat{E}\{e^{j <\mathbf{v}, \mathbf{x}(t)>} \Psi_t(\mathbf{v} \mid \mathbf{x}(t))\} dt \qquad (7.249)$$

$$+ \int_{\mathcal{U}} \hat{E}\{e^{j <\mathbf{v}, \mathbf{x}(t)>} [\lambda(t, \mathbf{u}, \mathbf{x}(t)) - \hat{\lambda}(t, \mathbf{u})]\} \frac{1}{\hat{\lambda}(t, \mathbf{u})} [N(dt \times d\mathbf{u}) - \hat{\lambda}(t, \mathbf{u}) dt d\mathbf{u}]$$

$$\hat{M}_{t_0}(j\mathbf{v}) = E\left(e^{j <\mathbf{v}, \mathbf{x}(t_0)>}\right),$$

where "^" and $\hat{E}(\cdot)$ denote a conditional expectation given $\{N(t, \mathcal{U}), t_1, \cdots, t_{N(t,\mathcal{U})}, u_1, \cdots, u_{N(t,\mathcal{U})}\}$.

Time-Space Suboptimal Filtering (Extension of (7.241) and (7.242)). The counterparts of (7.241) and (7.242) doubly stochastic time-space Poisson-processes are the following.

$$d\mathbf{x}^*(t) = \mathbf{a}(t, \mathbf{x}^*(t))dt + \frac{1}{2}\sum_{i,j}\sigma_{ij}^*\frac{\partial^2\mathbf{a}(t, \mathbf{x}^*(t))}{\partial x_i^*\partial x_j^*}dt$$

$$+\Sigma^*(t)\int_{\mathcal{U}}\left[\frac{\partial\lambda(t, \mathbf{u}, \mathbf{x}^*(t))}{\partial\mathbf{x}^*(t)}\right]\frac{1}{\lambda(t, \mathbf{u}, \mathbf{x}^*(t))}[N(dt\times d\mathbf{u}) - \lambda(t, \mathbf{u}, \mathbf{x}^*(t))dtd\mathbf{u}],$$

$$\mathbf{x}^*(t_0) = E(\mathbf{x}_0), \tag{7.250}$$

$$d\Sigma^*(t) = \left[\frac{\partial\mathbf{a}(t, \mathbf{x}^*(t))}{\partial\mathbf{x}^*(t)}\right]\Sigma^*(t)dt + \Sigma^*(t)\left[\frac{\partial\mathbf{a}(t, \mathbf{x}^*(t))}{\partial\mathbf{x}^*(t)}\right]dt + \mathbf{B}(t)\mathbf{B}'(t)dt$$

$$+\Sigma^*(t)\left\{\int_{\mathcal{U}}\left[\frac{\partial^2\ln\lambda(t, \mathbf{u}, \mathbf{x}^*(t))}{\partial\mathbf{x}^{*2}(t)}\right]N(dt\times d\mathbf{u})\right.$$

$$\left.-\int_{\mathcal{U}}\left[\frac{\partial^2\lambda(t, \mathbf{u}, \mathbf{x}^*(t))}{\partial\mathbf{x}^{*2}(t)}\right]dtd\mathbf{u}\right\}\Sigma^*(t),$$

$$\Sigma^*(t_0) = \text{cov}(\mathbf{x}_0), \tag{7.251}$$

where all terms are as defined for (7.241) and (7.242)

Example 7.5.7 *Optical Position Sensing, continued* ─────────
Consider Ex. 7.5.6 again. Let the detector region be large compared to the beam width and the motions during data collection so that edge effects can be disregarded. We may then take $\mathcal{U} = \mathcal{R}^2$, the entire plane. The intensity process is given in (7.247). Suppose that $\{\mathbf{x}(t): t \geq t_0\}$ is a two-dimensional Wiener process with covariance $E[\mathbf{x}(t)\mathbf{x}'(u)] = (2\tau_d)^{-1}\mathbf{I}\min(t, u)$. Then, $\mathbf{a}(t, \mathbf{x}(t)) = 0$, and $\mathbf{B}(t) = (1/\sqrt{2\tau_d})\mathbf{I}$, and (7.250) and (7.251) become:

$$d\mathbf{x}^*(t) = \frac{1}{\rho^2}\Sigma^*(t)\int_{\mathcal{R}^2}[\mathbf{u} - \mathbf{x}^*(t)]N(dt\times d\mathbf{u}), \quad \mathbf{x}^*(0) = 0 \tag{7.252}$$

and

$$d\Sigma^*(t) = \frac{1}{2\tau_d}\mathbf{I}\,dt - \frac{1}{\rho^2}\Sigma^*(t)\Sigma^*(t)N(dt \times \mathcal{R}^2), \quad \Sigma^*(0) = \mathbf{0}. \qquad (7.253)$$

Note from (7.253) that $\Sigma^*(t)$ is a diagonal matrix. The evolution of position estimate $\mathbf{x}^*(t)$ proceeds as follows. At $t = 0$, $\mathbf{x}^*(0) = 0$ because the beam is known to be at the origin. The position estimate remains at the origin until the first photoevent is observed, say at time t_1 and location $\mathbf{u}_1$. During the interval $0 \le t < t_1$, $\Sigma^*(t) = (t/\sqrt{2\tau_d})\mathbf{I}$, which is the uncertainty in $\mathbf{x}(t)$. At time t_1^+, the estimated position is

$$\mathbf{x}^*(t_1^+) = \frac{t_1}{2\tau_d\rho^2}\mathbf{u}_1.$$

Thus, the estimate moves in the direction of the observed photoevent by an amount proportional to the uncertainty in $\mathbf{x}(t_1)$, as represented by $t_1/2\tau_d$, and inversely proportional to the beam width, as represented by $1/\rho^2$. The position estimate remains fixed at $\mathbf{x}^*(t_1^+)$ until the second photoevent is observed, say at time t_2 and location $\mathbf{u}_2$. During the interval $t_1 \le t < t_2$,

$$\Sigma^*(t) = \frac{t - t_1}{2\tau_d}\mathbf{I} + \frac{t_1}{2\tau_d}\left[1 - \frac{t_1}{2\tau_d\rho^2}\right]\mathbf{I}.$$

At time t_2^+, the estimated position is

$$\mathbf{x}^*(t_2^+) = \mathbf{x}^*(t_1) + \frac{1}{\rho^2}\Sigma^*(t_2)[\mathbf{u}_2 - \mathbf{x}^*(t_1^+)].$$

Thus, the estimate moves in the direction of the observed photoevent by an amount proportional to the conditional uncertainty in $\mathbf{x}(t)$ given the first photoevent, as represented by $\Sigma^*(t_2)$, and inversely proportional to the beam width. The estimated position $\mathbf{x}^*(t)$ continues in this fashion, remaining fixed between photoevents and instantaneously moving towards each photoevent as it occurs, the displacement being proportional to the conditional uncertainty in $\mathbf{x}(t)$ at that instant and inversely proportional to the beam width.

We have illustrated with this example how the suboptimal estimator equations (7.250) and (7.251) are used. The position-sensing problem in this example is solved without any approximations (that is, for the optimal estimator) by D. Snyder and P. Fishman [43]. ∎

7.5.3 Doubly Stochastic Translated Poisson-Processes

These processes generalize the translated Poisson-processes of Ch. 3 to include an information process that influences the underlying point process whose points are observed after random translations. A representative problem for which these processes provide a model is that of forming an image of a faint object that appears to be moving in a random walk when viewed with an imager having a finite point-spread function, as illustrated in Fig. 7.14. We envision an object in an input space $X = \mathcal{R}^2$ whose intensity function is $\{\lambda(x): x \in X\}$. This object is viewed with an imager whose output has intensity $\{\mu_0(u): u \in \mathcal{U}\}$, where $\mathcal{U} = \mathcal{R}^2$ is the output space, and

$$\mu_0(u) = \int_X h(u - x)\lambda(x)\,dx, \qquad (7.254)$$

in which $h(\cdot)$ is the nonnegative, intensity point-spread function of the imager, normalized as a probability-density.

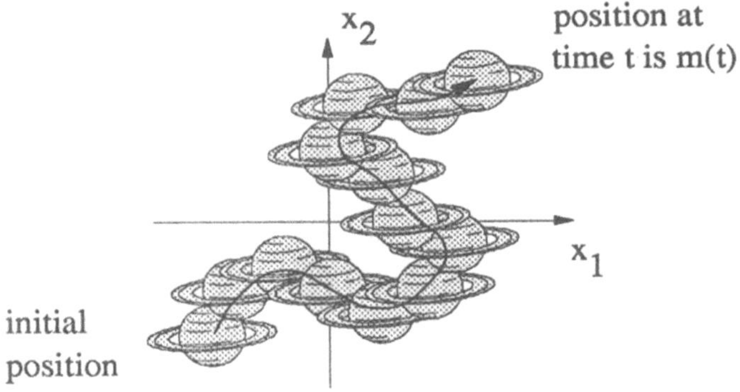

Figure 7.14 A randomly moving object whose spatially varying intensity-distribution is to be estimated from measured points.

In Ch. 3, we demonstrated that if the object produces a Poisson process on X with intensity $\lambda(\cdot)$, then the imager produces a translated Poisson-process on $\mathcal{U}$ with intensity $\mu_0(\cdot)$, and we developed a method for estimating the object intensity $\lambda(\cdot)$ based upon regularized maximum-likelihood estimation implemented via the expectation-maximization algorithm. Suppose, now, that the medium through which radiation passes is turbulent, causing the intensity $\mu_0(\cdot)$ to move in a random walk, resulting in a time-space intensity process given by:

$$\mu(t,u) = \mu_0(u - m(t)) = \int_X h(u - m(t) - x)\lambda(x)dx, \qquad (7.255)$$

where $\{m(t): t \geq t_0\}$ is a random process that describes the motions. Based upon measurements of the occurrence times and locations of the resulting doubly stochastic translated Poisson-process, we wish to form an estimate of the object's intensity function $\lambda(\cdot)$ using regularized maximum-likelihood. We assume that a complete statistical characterization of $\{m(t): t \geq t_0\}$ is known in the form of the joint density of $m(t_1)$, $m(t_2)$, $\cdots$, and $m(t_n)$ given any countable collection of times t_1, t_2, $\cdots$, t_n in the observation interval $\mathcal{T}$. This problem has been addressed by D. Snyder and T. Schulz and extends the results in Examples 3.1.4, 3.3.3, and 3.3.6 to accommodate a statistical description of the motions when it is available [44].

We introduce the following notation to summarize the results in [44] and to specify the incomplete-data loglikelihood. Let $N(\mathcal{T} \times \mathcal{U}) \equiv N$ denote the number of points that are observed, $\mathbf{t} = [t_1 \ \ t_2 \ \ \cdots \ \ t_N]'$ their occurrence times, and $\mathbf{u} = [u_1 \ \ u_2 \ \ \cdots \ \ u_N]'$ their locations in $\mathcal{U}$. Also, let $\mathbf{m} = [m(t_1) \ \ m(t_2) \ \ \cdots \ \ m(t_N)]'$ be samples of the motion process at the instants when points on $\mathcal{U}$ are observed. The locations $\mathbf{u}$ of points in the output space $\mathcal{U}$ can be viewed as random translations to the output space of points at positions $\mathbf{x} = [x_1 \ \ x_2 \ \ \cdots \ \ x_N]'$ in the object space X according to

$$\mathbf{u} = \mathbf{x} + \mathbf{b} + \mathbf{m}. \qquad (7.256)$$

Here, $\mathbf{b} = [b_1 \ \ b_2 \ \ \cdots \ \ b_N]'$ represents the random translation of points from X to $\mathcal{U}$ in the absence of motion, $\mathbf{m} = \mathbf{0}$. As in Ch. 3, these occur with a joint density

$$p_{\mathbf{b}|N}(\mathbf{B} \mid n) = \prod_{i=1}^{n} h(B_i) \qquad (7.257)$$

that is governed by the point spread function of the imager. Thus, the effect of blurring by the imager is to translate independently each point in X by a random variable with a probability density equal to the imager's point-spread function. In the presence of motion, the overall translation, or displacement, vector is $\mathbf{d} = \mathbf{b} + \mathbf{m}$. If the effects of blurring by the imager and motions are assumed to be independent, the conditional probability density of $\mathbf{d}$, given N and the N occurrence times, is

$$p_{\mathbf{d}|t,N}(\mathbf{D} \mid \mathbf{T},n) = \int_{X^n} p_{\mathbf{m}|t,N}(\mathbf{D} - \mathbf{B} \mid \mathbf{T},n)p_{\mathbf{b}|N}(\mathbf{B} \mid n)d\mathbf{B}, \quad (7.258)$$

where $p_{\mathbf{m}|t,N}(\mathbf{M} \mid \mathbf{T},n)$ is the joint probability density of the samples of the motion process given the instants at which points are observed. The conditional probability density of the spatial locations $\mathbf{u}$ in the output space $\mathcal{U}$ is given in terms of the conditional density of $\mathbf{d}$ and the conditional density of the locations $\mathbf{x}$ of points in the input space X according to

$$p_{\mathbf{u}|t,N}(\mathbf{U} \mid \mathbf{T},n) = \int_{X^n} p_{\mathbf{d}|t,N}(\mathbf{U} - \mathbf{X} \mid \mathbf{T},n)p_{\mathbf{x}|N}(\mathbf{X} \mid n)d\mathbf{X}$$

$$= \int_{X^n} p_{\mathbf{d}|t,N}(\mathbf{U} - \mathbf{X} \mid \mathbf{T},n)\left\{ \frac{1}{\Lambda^n} \prod_{i=1}^{n} \lambda(X_i) \right\} d\mathbf{X}, \quad (7.259)$$

where

$$\Lambda = \int_X \lambda(x)dx. \quad (7.260)$$

The final quantity needed to specify the incomplete-data loglikelihood is the conditional density $p_{t|N}(\mathbf{T} \mid n)$ of the occurrence times given the total number of points observed. This is readily specified when it is noted that points in time, with the spatial coordinate disregarded, form a Poisson process $N(t) = \int_{\mathcal{U}} N(t,du)$. This is a homogeneous Poisson process because its intensity $\mu(t) = \int_{\mathcal{U}} \mu(t,u)du = \int_{\mathcal{U}} \mu_0(u)du = \Lambda$ is a constant. Consequently, from Sec. 2.3.3,

$$p_{t|N}(\mathbf{T} \mid n) = \frac{1}{\|\mathcal{T}\|^n} \quad T_i \in \mathcal{T}, \quad i = 1,2,\cdots,n. \quad (7.261)$$

where $\|\mathcal{T}\| = \int_{\mathcal{T}} dt$ is the duration of the measurement.

The incomplete-data are $\mathcal{D}_I = \{N(\mathcal{T} \times \mathcal{U}),t_1,\cdots,t_N,u_1,\cdots,u_N\}$, so the incomplete-data likelihood is $\Pr[N = 0]$ if $N \equiv N(\mathcal{T} \times \mathcal{U}) = 0$ and

$$p_{\mathbf{u}|\mathbf{t},N}(\mathbf{U}\mid\mathbf{T},n)p_{\mathbf{t},N}(\mathbf{T}\mid n)\Pr(N=n)$$

if $N\geq 1$, where $p_{\mathbf{u}|\mathbf{t},N}(\mathbf{U}\mid\mathbf{T},n)$ is given in (7.259), $p_{\mathbf{t},N}(\mathbf{T}\mid n)$ is given in (7.261), and N is Poisson distributed with parameter $\Lambda\|\,\mathcal{T}\|$. Thus, the incomplete-data loglikelihood for $N(\mathcal{T}\times\mathcal{U})\geq 1$ is

$$l(\mathcal{D}_I)=-\|\,\mathcal{T}\|\int_X\lambda(X)\,dX+\ln\!\left[\int_{X^n}p_{\mathbf{d}|\mathbf{t},N}(\mathbf{U}-\mathbf{X}\mid\mathbf{T},n)\left\{\prod_{i=1}^{n}\lambda(X_i)\right\}d\mathbf{X}\right],$$

$$(7.262)$$

where all terms that do not depend on the object's intensity $\lambda(\cdot)$ have been suppressed.

If it exists, the maximum-likelihood estimate of $\{\hat{\lambda}(x):x\in X\}$ is a function $\{\lambda(x):x\in X\}$ that maximizes the incomplete-data loglikelihood (7.262) given the data $\mathcal{D}_I$. A necessary condition that this estimate must satisfy is obtained by applying variational calculus to the incomplete-data loglikelihood, which results in

$$\sum_{j=1}^{n}\left[\frac{\displaystyle\int_{X^{n-1}}p_{\mathbf{d}|\mathbf{t},N}(\mathbf{U}-\mathbf{X}\mid\mathbf{T},n)\left\{\prod_{i=1,i\neq j}\hat{\lambda}(X_i)\,dX_i\right\}}{\displaystyle\|\,\mathcal{T}\|\int_{X^n}p_{\mathbf{d}|\mathbf{t},N}(\mathbf{U}-\mathbf{X}\mid\mathbf{T},n)\left\{\prod_{i=1}\hat{\lambda}(X_i)\,dX_i\right\}}\right]=1.\qquad(7.263)$$

This nonlinear integral-equation has no closed-form solution in general, and remains difficult to solve even under simpler conditions. For example, if the components of the displacement vector $\mathbf{d}$ are conditionally independent of one another and of the occurrence times $\mathbf{t}$ given N, then (7.263) becomes

$$\sum_{j=1}^{n}\left[\frac{p_{d_j|N}(U_j-X\mid n)}{\displaystyle\|\,\mathcal{T}\|\int_X p_{d_j|N}(U_j-X\mid n)\hat{\lambda}(X)\,dX}\right]\qquad(7.264)$$

$$=\int_{\mathcal{U}}\left[\frac{p_{d_j|N}(U-X\mid n)}{\displaystyle\|\,\mathcal{T}\|\int_X p_{d_j|N}(U-X\mid n)\hat{\lambda}(X)\,dX}\right]N(dU)=1.$$

This is the same as the equation (3.22) for estimating the intensity of a translated Poisson-process. Even in this simpler situation, a numerical procedure, such as the expectation-maximization algorithm developed in Sec. 3.3, is needed to produce a solution. Moreover, unconstrained estimates are unstable and need to be regularized, as in Sec. 3.4.

The use of regularization constraints and the expectation-maximization algorithm for producing stable maximum-likelihood estimates of the intensity function of a doubly stochastic translated Poisson-process is developed by D. Snyder and T. Schulz [44]. Generalizing (3.72), a desired function $\{d(w): w \in \mathcal{W} = \mathcal{R}^2\}$ is introduced to suppress edge artifacts

$$d(w) = \int_X r(w \mid x)\lambda(x)\,dx, \qquad (7.265)$$

where $r(\cdot)$ is a given, nonnegative resolution-kernel that is normalized as a conditional-probability density. The maximum-likelihood estimate of this desired function is sought, subject to a sieve constraint that the estimate must be a member of the smooth set of nonnegative functions defined by

$$S = \left\{ d(w): d(w) = \int_Z s(w \mid z)\xi(z)\,dz \right\}, \qquad (7.266)$$

where $Z = \mathcal{R}^2$, and $s(\cdot)$ is a given, nonnegative kernel that is normalized as a conditional-probability density. The sieve and resolution kernels must be selected so that there exists a density function $q(U \mid W)$ satisfying the following generalization of (3.73):

$$\int_{W^n} \left[p_{m \mid t, N}(U - W \mid T, n)\left\{ \prod_{i=1}^{n} h(W_i - X_i) \right\} - q(U \mid W)\left\{ \prod_{i=1}^{n} r(W_i \mid X_i) \right\} \right] dW$$

$$= 0. \qquad (7.267)$$

One choice of resolution kernel where this equation can be solved easily is $r(w \mid x) = h(w - x)$. Then $d(\cdot)$ is the blurred version of $\lambda(\cdot)$ that appears at the output of the imager in the absence of motion effects, and the goal of the estimation procedure is to compensate as well as possible for additional blurring due to motion. For this selection, a solution is $q(U \mid W)$ equal to the joint density of the samples of the motion process, $p_{m \mid t, N}(U - W \mid T, n)$.

A sequence of estimates converging towards the constrained maximum-likelihood estimate of $d(\cdot)$ is produced by the following expectation-maximization algorithm, which generalizes (3.77) and (3.78):

$$\hat{\xi}^{(k)}(Z) = \hat{\xi}^{(k-1)}(Z) \sum_{i=1}^{n} \left[\frac{\int_{Z^{n-1}} k(\mathbf{U} \mid \mathbf{Z}) \left\{ \prod_{j=1, j \neq i} \hat{\xi}^{(k-1)}(Z_j) dZ_j \right\}}{\|T\| \int_{Z^n} k(\mathbf{U} \mid \mathbf{Z}) \left\{ \prod_{j=1} \hat{\xi}^{(k-1)}(Z_j) dZ_j \right\}} \right], \tag{7.268}$$

and

$$\hat{d}^{(k)}(w) = \int_Z r(w \mid z) \hat{\xi}^{(k)}(z) dz, \tag{7.269}$$

where the kernel $k(\cdot)$ in (7.268) is defined by

$$k(\mathbf{U} \mid \mathbf{Z}) = \int_{W^n} q(\mathbf{U} \mid \mathbf{W}) \left\{ \prod_{i=1}^{n} s(W_i \mid Z_i) \right\} d\mathbf{W}, \tag{7.270}$$

which is the generalization of (3.79). The detailed derivation of (2.268) and (2.269) is in [44].

The specialization of (2.268) for particular choices of the statistics of the motion process are examined in the following examples. These include a rapidly varying motion, a piecewise constant motion, and a very slow motion. We assume for these examples that the resolution kernel is selected as $r(w \mid x) = h(w - x)$. Then, the function $q(\mathbf{U} \mid \mathbf{W})$ in (7.267) may be taken as equal to the joint density of the samples of the motion process, $p_{m \mid t, N}(\mathbf{U} - \mathbf{W} \mid \mathbf{T}, n)$.

Example 7.5.8 *Rapid Motions* ————————————————————————

Suppose that $\{m(t) : t \in \mathcal{T}\}$ is such that at the observed occurrence times $\mathbf{t}$, the samples $m(t_1), m(t_2), \cdots, m(t_n)$ are statistically independent and identically distributed random variables, with their common probability density being $p_m(M)$. Then,

$$p_{m \mid t, N}(\mathbf{U} - \mathbf{W} \mid \mathbf{T}, n) = \prod_{j=1}^{n} p_m(U_j - W_j), \tag{7.271}$$

and

$$k(\mathbf{U} \mid \mathbf{Z}) = \prod_{j=1}^{n} p(U_j \mid Z_j), \tag{7.272}$$

where

$$p(U_j \mid Z_j) = \int_W p_m(U_j - W_j)s(W_j \mid Z_j)\,dW_j. \qquad (7.273)$$

Substitution into (2.268) then yields a sequence of estimates defined by

$$\hat{\xi}^{(k)}(Z) = \hat{\xi}^{(k-1)}(Z) \int_U \left[\frac{p(U \mid Z)}{\|T\| \int_Z p(U \mid Z)\hat{\xi}^{(k-1)}(Z)\,dZ} \right] dU. \qquad (7.274)$$

The corresponding sequence of estimates of the desired function $d(\cdot)$ that would appear at the output of the imager in the absence of motion is obtained using (7.269). These equations are entirely analogous to those in (3.77) and (3.78). Thus, a motion process that varies so rapidly that samples of it are statistically independent results in a relatively minor modification of the situation when there are no motions. ∎

Example 7.5.9 *Slow Motions* ───────────────────────────────
     Suppose that $m(t) = m$ for $t \in T$, where m is a random variable with probability density $p_m(M)$. Then

$$p_{\mathbf{m}\mid t,N}(\mathbf{M} \mid T, n) = p_m(M_i) \prod_{j=1, j \neq i}^{n} \delta(M_j - M_i). \qquad (7.275)$$

For this choice, (7.268) can be written as

$$\hat{\xi}^{(k)}(Z) = \hat{\xi}^{(k-1)}(Z) \sum_{i=1}^{n} \left[\frac{\int_X p_m(X)s(U_i - X \mid Z)\left\{ \prod_{j=1, j \neq i}^{n} \hat{d}^{(k-1)}(U_j - X) \right\} dX}{\|T\| \int_X p_m(X)\left\{ \prod_{j=1}^{n} \hat{d}^{(k-1)}(U_j - X) \right\} dX} \right],$$

$$\qquad (7.276)$$

where the required estimates of $d(\cdot)$ are obtained using (7.269). There is no appreciable simplification of this expression, which holds when the motions vary so slowly that they can be regarded as constant over the measurement interval. ∎

Example 7.5.10 *Piecewise-Constant Motions* ──────────────────
     Suppose that $m(t) = m_l$ for $t \in T_l$, $l = 1, 2, \cdots, L$. Here, the motion process is constant within subintervals that together comprise the total observation interval $T = \cup_{l=1}^{L} T_l$. The subintervals might correspond to short exposures during which snapshots are taken of the moving object, with the motion being frozen during the snapshot. If the motions are independent from snapshot to snapshot, then

$$p_{\mathbf{m}|\mathbf{t},N}(\mathbf{M}\,|\,\mathbf{T},n) = \prod_{\ell=1}^{L} p_{\mathbf{m}_\ell|\mathbf{t}_\ell,N}(\mathbf{M}_\ell\,|\,\mathbf{T}_\ell,n), \qquad (7.277)$$

where N_ℓ is the number of points observed during $\mathcal{T}_\ell$, $\mathbf{t}_\ell$ is the vector of N_ℓ occurrence times in $\mathcal{T}_\ell$, and $\mathbf{m}_\ell$ is the vector of sample values of the motion process at the times in $\mathcal{T}_\ell$. Within $\mathcal{T}_\ell$, there holds

$$p_{\mathbf{m}_\ell|\mathbf{T}_\ell,N_\ell}(\mathbf{M}_\ell\,|\,\mathbf{T}_\ell,n_\ell) = p_m(M_{\ell i}) \prod_{j=1,\,j\neq i}^{n_\ell} \delta(M_{\ell i}-M_{\ell j}). \qquad (7.278)$$

Then, (7.268) can be written as

$$\hat{\xi}^{(k)}(Z) = \hat{\xi}^{(k-1)}(Z) \sum_{\ell=1}^{L} \sum_{i=1}^{n_\ell} \left[\frac{\int_X p_m(X)\,s(U_{\ell i}-X\,|\,Z)\left\{\prod_{j=1,\,j\neq i}^{n_\ell} \hat{d}^{(k-1)}(U_{\ell j}-X)\right\} dX}{\|\,\mathcal{T}\,\| \int_X p_m(X)\left\{\prod_{j=1}^{n_\ell} \hat{d}^{(k-1)}(U_{\ell j}-X)\right\} dX} \right],$$

$$(7.279)$$

where the required estimates of $d(\cdot)$ are obtained using (7.269). There is no appreciable simplification of this expression, which holds when the motions are piecewise constant and vary independently between positions.

A computer simulation was performed to provide a comparison between an image estimate produced with a single long exposure and one produced a series of short exposures [44]. The object intensity used in the simulation corresponds to a one-dimensional binary star that is similar to that used by L. de Freitas, M. Northcott, B. Brames, and C. Dainty [13]; this is shown in Fig. 7.15. Data were simulated for 166 exposure intervals, with each exposure having an average of three points. The data in each exposure were created by simulating point locations determined by the object's intensity and then translating each point by a Gaussian random variable m with a mean of zero and a standard deviation of three pixel units, with the same random variable being used for all the points in an exposure and independent random variables used between exposures. For the *histogram estimate* shown in Fig. 7.15, each point contributes unity to the pixel in which it occurs. The resulting intensity estimate is the same as that obtained with a single long exposure without further processing. The constrained maximum-likelihood estimate in Fig 7.15 was produced using (7.279) and (7.269) for 100 iterations. Other estimates are compared in [44]. ∎

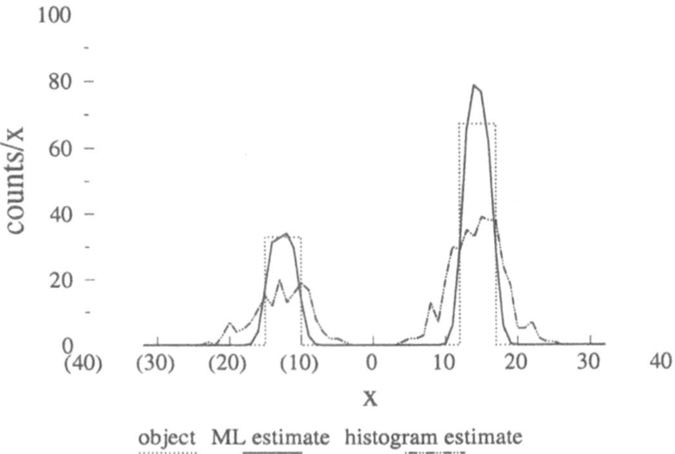

Figure 7.15 Comparison of estimates of a randomly translating object using a single long exposure (histogram estimate) and a series of short exposures (ML estimate).

7.6 References

1.   B. D. O. Anderson, J. B. Moore, and S. G. Loo, "Spectral Factorization of Time-Varying Covariance Functions," *IEEE Transactions on Information Theory*, Vol. IT-15, pp. 550-557, September 1969.

2.   H. O. Anger, "the Scintillation Camera for Radioisotope Localization," Proc. Conf. on Nuclear Medicine, Heidelberg, October 1966.

3.   A. B. Baggeroer, *State Variables and Communication Theory*, Research Monograph No. 61, The M.I.T. Press, Cambridge, MA, 1970.

4.   P. Brémaud, *Point Processes and Queues: Martingale Dynamics*, Springer-Verlag, Berlin, 1981.

5.   J. Clark and E. V. Hoversten, "Poisson Process as a Statistical Model for Photodetectors Excited by Gaussian Light," Quarterly Progress Report No. 98, Research Laboratory of Electronics, M. I. T., Cambridge, MA, pp. 95-101, July 1970.

6.   J. R. Clark, "Estimation for Poisson Processes with Applications in Optical Communication," Ph.D. Thesis, Dep't. of Electrical Engineering, M. I. T., Cambridge, MA, September 1971.

7.   D. R. Cox, "Some Statistical Methods Connected with Series of Events," *J. Royal Statistical Society B*, Vol. 17, pp. 129-164, 1955.

8.   D. R. Cox and P. A. W. Lewis, *The Statistical Analysis of Events*, Metheun, London, 1966.

9. D. J. Daley and D. Vere-Jones, "A Summary of the Theory of Point Processes," in: *Stochastic Point Processes: Statistical Analysis, Theory, and Applications* (P. A. W. Lewis, Ed.), Wiley, New York, 1972

10. W. Feller, *An Introduction to Probability Theory and Its Applications: Volume II*, Wiley, New York, 1966.

11. P. M. Fishman and D. L. Snyder, "Statistical Inference for Time Space Point Processes," *IEEE Transactions on Information Theory*, IT-22, No. 3, May 1976.

12. R. H. Forrester, Jr. and D. L. Snyder, "Phase-Tracking Performance of Direct-Detection Optical Receivers," *IEEE Transactions on Communications*, Vol. COM-23, pp. 1037-1039, September 1973.

13. L. C. de Freitas, M. J. Northcott, B. J. Brames, and J. C. Dainty, "Object Reconstruction from Photon-Limited Centroided Data of Randomly Translating Images," *Proc. SPIE Conf. on Digital Image Recovery and Synthesis*, Vol. 828, pp. 62-72, August 1987.

14. J. W. Goodman, "Synthetic-Aperture Optics," in *Progress in Optics: Vol. 111*, E. E. Wolf (Ed.), North-Holland, Amsterdam, 1970.

15. J. Grandell, "On Stochastic Processes Generated by a Stochastic Intensity Function," *Skndinavisk Aktuarietidskrift*, pp. 204-240, 1971.

16. J. Grandell, "Statistical Inference for Doubly Stochastic Poisson Processes," in *Stochastic Point Processes, Statistical Analysis, Theory, and Applications*, P. A. W. Lewis (Ed.), Wiley, New York, 1972.

17. J. Grandell, *Doubly Stochastic Poisson Processes*, Research Report 102, Inst. of Actuarial Mathematics and Mathematical Statistics, Univ. of Stockholm, 1975.

18. S. Haber, "Numerical Evaluation of Multiple Integrals," *SIAM Review*, Vol. 12, pp. 481-526, October 1970.

19. E. V. Hoversten, R. O. Harger, and S. J. Halme, "Communication Theory for the Turbulent Atmosphere," *Proc. IEEE*, Vol. 58, pp. 1626-1650, October 1970.

20. A. H. Jazwinski, *Stochastic Processes and Filtering Theory*, Academic Press, New York, 1970.

21. S. Karp and J. R. Clark, "Photon Counting: A Problem in Classical Noise Theory," *IEEE Transactions on Information Theory*, Vol. IT-16, pp. 672-680, November 1970.

22. S. Karp, E. O'Neill, and R. Gagliardi, "Communication Theory for the Free-Space Optical Channel," *Proc. IEEE*, Vol. 58, pp. 1611-1626, October 1970.

23. A. F. Karr, *Point Processes and Their Statistical Inference*, Marcel-Dekker, New York, 1986.

24. R. S. Kennedy, "Communication Through Optical Scattering Channels: An Introduction," *Proc. IEEE*, Vol. 58, pp. 1651-1665, October 1970.

25. T. Kailath, "Fredholm Resolvents, Wiener-Hopf Equations, and Riccati Differential Equations," *IEEE Transactions on Information Theory*, Vol. IT-15, pp. 665-672, November 1969.

26. G. Lachs, "Theoretical Aspects of Mixtures of Thermal and Coherent Radiation," *Physical Review*, Vol. 138, pp. B1012-B116, 1965.

27. R. S. Lipster and A. N. Shiryayev, *Statistics of Random Processes*, Springer-Verlag, NY, 1977.

28. O. Lundberg, *On Random Processes and Their Application to Sickness and Accident Statistics*, Almqvist & Wiksells Boktryckeri A. B., Uppsala, 1964.

29. O. Macchi, "Distribution Statistique des Instants D'émission des Photoélectrons d'une Lumiére Thermique," *C. R. Acad. Sci. Paris, Ser. A*, Vol. 272, pp. 437-440, February 1971.

30. O. Macchi and B. Picinbono, "Estimation and Detection of Weak Optical Signals," *IEEE Transactions on Information Theory*, Vol. IT-18, pp. 562-573, September 1972.

31. O. Macchi, "Processus Ponctuels et Coincidences," D. Sc. These, Centre D'Orsay, Univ. de Paris Sud, January 1972.

32. C. L. Mehta, "Theory of Photoelectron Counting," in *Progress in Optics*, Vol. VIII, E. Wolf (Ed.), North-Holland, Amsterdam, 1970.

33. N. F. Moody, W. Paul, and M. L. G. Joy, "A Survey of Medical Gamma-Ray Cameras," *Proc. IEEE*, Vol. 58, pp. 217-242, February 1970.

34. W. K. Pratt, *Laser Communication Systems*, Wiley, New York, 1969.

35. L. M. Robinson, "Tanlock: A Phase-Lock Loop of Extended Tracking Capability," Proc. National Conf. on Military Electronics, 1962.

36. M. Rosenblatt, *Random Processes*, Oxford Univ. Press, New York, 1962.

37. D. A. Ross and C. C. Wagner, Jr., "Measurement of Radioactivity," in *Principles of Nuclear Medicine*, H. N. Wagner, Jr. (Ed.), Saunders, Philadelphia, 1968.

38. M. Rudemo, "Doubly Stochastic Poisson Processes and Process Control," *Advan. Appl. Prob.*, Vol. 4. pp. 318-338, August 1972.

39. B. E. A. Saleh, "Probability Distribution of Time of Arrival of Photon Counts for a Stationary Optical Field," *IEEE Transactions on Information Theory*, Vol. IT-20, pp. 262-263, March 1974.

40. R. F. Serfozo, "Conditional Poisson Processes," *J. Appl. Prob.*, Vol. 9, pp. 288-302, September 1972.

41. D. L. Snyder and I. B. Rhodes, "Filtering and Control Performance Bounds with Implications on Asymptotic Separation," *Automatica*, Vol. 8, pp. 747-753, November 1972.

42. D. L. Snyder, *The State Variable Approach to Continuous Estimation with Applications to Analog Communication Theory*, M. I. T. Press, Cambridge, MA, 1969.

43. D. L. Snyder and P. M. Fishman, "How to Track a Swarm of Fireflies by Observing Their Flashes," *IEEE Transactions on Information Theory*, November 1975.

44. D. L. Snyder and T. J. Schulz, "High Resolution Imaging at Low Light-Levels Through Weak Turbulence," *Journal of the Optical Society of America, A*, July 1990.

45. M. M. Ter-Pogossian, J. O. Eichling, D. O. Davis, and M. J. Welch, " the Measure in Vivo of Regional Cerebral Oxygen Utilization by Means of Oxyhemoglobin Labeled with Radioactive Oxygen-15," *J. Clin. Invest.*, Vol. 49, pp. 381-391, 1970.

46. H. L. Van Trees, *Detection, Estimation, and Modulation Theory: Part I*, Wiley, New York, 1968.

47. H. L. Van Trees, *Detection, Estimation, and Modulation Theory: Part III*, Wiley, New York, 1971.

48. A. J. Viterbi, *Principles of Coherent Communication*, McGraw-Hill, New York, 1966.

49. E. Wong and B. Hajek, *Stochastic Processes in Engineering Systems*, Springer-Verlag, New York, 1985.

50. M. Yovits and J. Jackson, "Linear Filter Optimization with Game Theory Considerations," 1955 IRE National Convention Record, Part 3, pp. 193-199, 1955.

7.7 Problems

7.2.1 Let $\{N(t): t \geq t_0\}$ be a doubly stochastic Poisson-process with constant intensity x.

a. Show that

$$\Pr[N(t)=n] = \frac{1}{n!}(-t)^n \frac{d^n \Pr[N(t)=0]}{dt^n}.$$

b. Suppose that x has the negative binomial distribution

$$\Pr(x=k) = \frac{\Gamma(k+\beta+1)}{\Gamma(k+1)\Gamma(\beta+1)}\left(\frac{\alpha}{\alpha+t}\right)^{\beta+1}\left(\frac{t}{\alpha+t}\right)^k$$

for $k = 0, 1, 2, \cdots$. Show that

$$\Pr[N(t)=0] = \left[\frac{1-\xi}{1-\xi e^{-t}}\right]^{\beta+1},$$

where $\xi = t(\alpha+t)^{-1}$. Deduce from (a) that

$$\Pr[N(t)=n] = \frac{t^n(1-\xi)^{\beta+1}}{n!\Gamma(\beta+1)} \sum_{j=0}^{\infty} \frac{j^n}{j!}\Gamma(\beta+1+j)\xi^j e^{-jt}.$$

7.2.2 Let $\{N(t): t \geq t_0\}$ be a doubly stochastic Poisson-process with intensity process $\{\lambda(t, \mathbf{x}(t)): t \geq t_0\}$. Let

$$P_t(\Lambda) = \Pr\left[\frac{1}{t}\int_{t_0}^{t} \lambda(\sigma, \mathbf{x}(\sigma))d\sigma \leq \Lambda\right].$$

Show that

$$\Pr[N(t) = n] = \int_0^{\infty} \frac{1}{n!}(\Lambda t)^n e^{-\Lambda t} dP_t(\Lambda).$$

7.2.3 Let $\{N(t): t \geq t_0\}$ be a doubly stochastic Poisson-process with intensity process $\{x\nu(t): t \geq t_0\}$, where x is a nonnegative, integer-valued random variable, and $\nu(\cdot)$ is a deterministic function. Suppose that x is Poisson distributed with parameter μ. Show that

$$\Pr[N(t) = n] = \frac{1}{n!}\xi^n(t)\exp[-\mu(1 - e^{-\xi(t)})]m_n(\mu e^{-\xi(t)}),$$

where $\xi(t) = \int_{t_0}^{t} \nu(\sigma)d\sigma$, and $m_n(\lambda)$ denotes the nth central moment of a Poisson distribution with parameter λ.

7.2.4 Let $\{N(t): t \geq t_0\}$ be a doubly stochastic Poisson-process with intensity process $\{\lambda(t): t \geq t_0\}$. Define $\Lambda(t) = \int_{t_0}^{t} \lambda(\sigma)d\sigma$.

a. Establish the following relation between the covariance functions for N and Λ:

$$\mathrm{cov}[N(t), N(u)] = \mathrm{cov}[\Lambda(t), \Lambda(u)] + \min\{E[\Lambda(t)], E[\Lambda(u)]\}.$$

b. The *index of dispersion* of a random variable z is defined by $\mathrm{ID}(z) = \mathrm{var}(z)/\mathrm{mean}(z)$. Show for the processes of part (a) that

$$\mathrm{ID}[N(t)] = \mathrm{ID}[\Lambda(t)] + 1 \geq 1.$$

Conclude, therefore, that the index of dispersion for a doubly stochastic Poisson-process is never smaller than that for a Poisson process with the same mean-value function.

7.2.5 Let $\{N(t): t \geq t_0\}$ be a doubly stochastic Poisson-process with intensity process $\{\lambda(t): t \geq t_0\}$. Define $\Lambda(t) = \int_{t_0}^{t} \lambda(\sigma)d\sigma$. Let $M_{N(t)}(s) = E\{\exp[sN(t)]\}$ and $M_{\Lambda(t)}(s) = E\{\exp[s\Lambda(t)]\}$ be the moment generating functions for $N(t)$ and $\Lambda(t)$, respectively. Similarly, let $\Gamma_{N(t)}(s) = \ln M_{N(t)}(s)$ and $\Gamma_{\Lambda(t)}(s) = \ln M_{\Lambda(t)}(s)$ be the corresponding cumulant generating functions; the cumulants are defined in Prob. 2.2.3.

a. Show that

$$M_{N(t)}(s) = M_{\Lambda(t)}(e^s - 1) \quad \text{and} \quad \Gamma_{N(t)}(s) = \Gamma_{\Lambda(t)}(e^s - 1).$$

b. Use the moment generating properties of $M_{N(t)}(s)$ and $M_{\Lambda(t)}(s)$ to establish the following relations between the moments of $N(t)$ and $\Lambda(t)$:

$$E[N(t)] = E[\Lambda(t)]$$

$$E[N^2(t)] = E[\Lambda^2(t)] + E[\Lambda(t)]$$

$$E[N^3(t)] = E[\Lambda^3(t)] + 3E[\Lambda^2(t)] + E[\Lambda(t)]$$

$$E[N^4(t)] = E[\Lambda^4(t)] + 6E[\Lambda^3(t)] + 7E[\Lambda^2(t)] + E[\Lambda(t)].$$

c. Use the cumulant generating properties of $\Gamma_{N(t)}(s)$ and $\Gamma_{\Lambda(t)}(s)$ to establish the following relations between the cumulants $\gamma_k[N(t)]$ and $\gamma_k[\Lambda(t)]$ of $N(t)$ and $\Lambda(t)$, respectively:

$$\gamma_1[N(t)] = \gamma_1[\Lambda(t)]$$

$$\gamma_2[N(t)] = \gamma_2[\Lambda(t)] + \gamma_1[\Lambda(t)]$$

$$\gamma_3[N(t)] = \gamma_3[\Lambda(t)] + 3\gamma_2[\Lambda(t)] + \gamma_1[\Lambda(t)]$$

$$\gamma_4[N(t)] = \gamma_4[\Lambda(t)] + 6\gamma_3[\Lambda(t)] + 7\gamma_2[\Lambda(t)] + \gamma_1[\Lambda(t)].$$

d. Use the result of (c) to show for a Poisson process with mean value function $\Lambda(t)$ that for all integers $k \geq 1$ there holds $\gamma_k[N(t)] = \Lambda(t)$.

7.2.6 Let $\{N(t): t \geq t_0\}$ be a doubly stochastic Poisson-process with intensity process $\{\lambda(t): t \geq t_0\}$. Let $\{y(t): t \geq t_0\}$ be a filtered doubly stochastic Poisson-process,

$$y(t) = \int_{t_0}^{t} h(t,u)\,dN(u).$$

Use the method of conditioning and the results of Sec. 5.2.1 to establish the following second-moment properties for $y(\cdot)$.

a.   Show that

$$E[y(t)] = \int_{t_0}^{t} h(t,u)E[\lambda(u)]\,du.$$

b.   Show that

$$\mathrm{var}[y(t)] = \int_{t_0}^{t} h^2(t,u)E[\lambda(u)]\,du + \int_{t_0}^{t}\int_{t_0}^{t} h(t,u)K_\lambda(u,v)h(v,u)\,du\,dv,$$

where K_λ is the covariance function for $\lambda(\cdot)$.

c.   Let

$$\mathrm{cov}[\lambda(s), y(t)] = E[\lambda(s)y(t)] - E[\lambda(s)]E[y(t)].$$

Show that

$$\mathrm{cov}[\lambda(s), y(t)] = \int_{t_0}^{t} h(t,u)K_\lambda(s,u)\,du.$$

7.2.7   Suppose that the covariance function $K(\sigma,u)$ for the process $\{X(t): t \geq t_0\}$ in Ex. 7.2.2 has the separable form

$$K(\sigma,u) = \sum_{i=1}^{m} \lambda_i \phi_i(\sigma)\phi_i^*(u),$$

where m is finite. In this case, K has a finite number, m, of nonzero eigenvalues. The situation when all these eigenvalues $\lambda_1, \lambda_2, \cdots, \lambda_m$ are equal to the same constant, say σ^2, is of some importance in practice. For instance, if the background radiation $B(t) = X(t) - M(t)$ is a white, band-limited process of bandwidth W, the constant σ^2 is the noise intensity (i.e., $\sigma^2 W$ is the total noise power), and $m \approx 2Wt + 1$ (see H. Van Trees [46, p. 193] and S. Karp and J. Clark [21]). For another example, $X(t)$ in an optical radar-ranging system can correspond to the optical field resulting from reflections off m rough, equal-strength scatterers; $M(t)$ corresponds to a specular component in the received optical radar pulse.

a. Show that the characteristic function for $N(t)$ is given by

$$M_{N(t)}(ju) = \frac{\exp\left\{\alpha(e^{ju}-1)\int_{t_0}^{t}|M(\tau)|^2 d\tau[1-\alpha\sigma^2(e^{ju}-1)]^{-1}\right\}}{[1-\alpha\sigma^2(e^{ju}-1)]^m}.$$

b. Show that when $M(t)=0$, the count $N(t)$ has the negative binomial distribution

$$\Pr[N(t)=n] = \frac{\Gamma(n+m)}{\Gamma(n+1)\Gamma(m)}\left[\frac{1}{1+\alpha\sigma^2}\right]^m\left[\frac{\alpha\sigma^2}{1+\alpha\sigma^2}\right]^n.$$

c. The situation $\sigma^2=0$ corresponds to the absence of background radiation or to rough scatterers. In this situation, $N(t)$ clearly has a Poisson distribution with parameter $\int_{t_0}^{t}|M(\tau)|^2 d\tau$. Suppose that

$m \gg 1$ and $\alpha\sigma^2 \ll 1$, corresponding to a large number of weak background-noise modes or to a large number of weak scatterers. Show that $N(t)$ is approximately Poisson distributed with parameter $\alpha\left[\sigma^2 m + \int_{t_0}^{t}|M(\tau)|^2 d\tau\right]$; here, $\sigma^2 m$ is the total expected energy in $B(\cdot)$.

7.2.8 (O. Lundberg [28]). Let $\{N(t): t \geq t_0\}$ be a homogeneous Poisson process given its intensity parameter x, where x is a nonnegative random variable. It is shown in Sec 7.2.2 that the counting statistics for $N(\cdot)$ are the same as those of a pure Markov birth process with birthrate $\mathscr{l}(t,n) = E(x \mid N(t)=n)$. Here, you are to establish several properties of $\mathscr{l}(t,n)$.

a. Demonstrate that $\mathscr{l}(t,n)$ satisfies the following recursive equation

$$\mathscr{l}(t,n+1) = \mathscr{l}(t,n) - \frac{d[\ln \mathscr{l}(t,n)]}{dt}.$$

b. Show for fixed n that $\mathscr{l}(t,n)$ is a nonincreasing function of t. Hint: Establish and use the expression

$$\mathfrak{L}(t,n) = \frac{E(x^{n+1}e^{-xt})}{E(x^n e^{-xt})}.$$

The Schwarz inequality can be used with this to show that $\mathfrak{L}(t,n+1) \geq \mathfrak{L}(t,n)$.

c.   Show for fixed t that $\mathfrak{L}(t,n)$ is a nondecreasing sequence in n.

d.   Suppose for fixed t that $\mathfrak{L}(t,n)$ is linear in n; that is, for some nonnegative functions $a(t)$ and $b(t)$, there holds $\mathfrak{L}(t,n) = a(t) + nb(t)$. Show that $\mathfrak{L}(t,n)$ must then have the form $\mathfrak{L}(t,n) = (n+c_1)/(t+c_2)$ for some constants c_1 and c_2. Conclude, therefore, that $\{N(t): t \geq t_0\}$ must be a homogeneous Polya process.

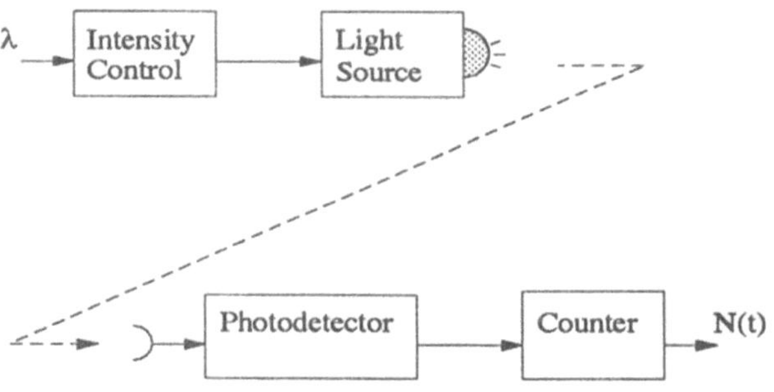

Figure 7.2.9

7.2.9 The optical communication system shown diagrammatically in Fig. P.7.2.9 is used to transmit a random variable λ to a remote user. The variable λ modulates the intensity of the transmitted light beam. The receiver consists of a photodetector and a counter that registers the total number of photoelectrons generated in any observation interval. Denote the number of photoelectrons registered by this counter in a T second interval by $N(T)$. Make the following assumptions:

i. $N(T)$ is conditionally a Poisson distributed random variable with parameter λT, so

$$\Pr[N(T) = n] = \frac{1}{n!}(\lambda T)^n e^{-\lambda T}$$

for $n = 0, 1, 2, \cdots$.

ii.   The probability density function for λ is

$$
p_\lambda(\Lambda) = \begin{cases} \dfrac{\alpha^{\beta+1}}{\Gamma(\beta+1)} \Lambda^\beta e^{-\alpha\Lambda}, & \Lambda \geq 0 \\ 0, & \Lambda < 0 \end{cases}
$$

where $\alpha > 0$ and $\beta < 0$ are known constants, and where $\Gamma(\cdot)$ is a gamma function.

   a.   Determine the mean of λ.

   b.   Determine the variance of λ.

   c.   Determine the mean of $N(T)$.

   d.   Determine the variance of $N(T)$.

   e.   Suppose that the number of registered photoelectrons in a T-second interval is observed to be n and based on this observation we want to estimate the value of λ. To do this, we first propose to use for simplicity a *linear* estimate $\hat{\lambda}_{\mathrm{lin}}(n)$ of the form

$$
\hat{\lambda}_{\mathrm{lin}}(n) = n c_1 + c_2,
$$

where c_1 and c_2 are constants. Determine these constants so that the mean square error $E[(\lambda - \hat{\lambda}_{\mathrm{lin}}(n))^2]$ is minimized.

   f.   Suppose now that the linearity restriction is removed in part (e). Determine an expression for the estimate $\hat{\lambda}(n)$ of λ in terms of n that minimizes $E[(\lambda - \hat{\lambda}(n))^2]$. Discuss your results.

7.2.10 Let $\{N(t): t \geq t_0\}$ be a doubly stochastic Poisson-process with intensity process $\{x\nu(t): t \geq t_0\}$, where x is a nonnegative random variable with mean m_x and variance σ_x^2, and where $\nu(\cdot)$ is a deterministic function. Show that if $\nu(\cdot)$ has infinite area, then $N_{\mathrm{cn}}(t) = [N(t) - m_N(t)]\sigma_N^{-1}(t)$ converges in distribution to a zero-mean, unit-variance Gaussian random variable as t tends to infinity.

7.2.11 Let $N_{cn}(t)$ and $\Lambda_{cn}(t)$ be defined as in (7.30). Show that if $\Lambda_{cn}(t)$ converges in distribution to a zero-mean, unit-variance Gaussian random variable as t tends to infinity, then so does $N_{cn}(t)$.

7.2.12 Let $\{N(t): t \geq t_0\}$ be the doubly stochastic Poisson-process of Example 7.2.5.

a. Show that the cumulant generating function for $N(t)$ is given by

$$\mu(s) = \ln E\{e^{sN(t)}\} = \left[\frac{\alpha}{1+\alpha-e^s}\right]^{\beta t+1}.$$

b. By using a Chernoff bound (see H. Van Trees [46]), show that

$$\Pr[N(t) < \eta] \leq \eta^{-\eta} \left[\frac{\beta t + 1 + \eta}{1+\alpha}\right]^{\beta t + 1 + \eta} \left[\frac{\alpha}{\beta t + 1}\right]^{\beta t + 1}.$$

7.3.1 Let $\mathbf{x}$ be a vector of k independent, identically distributed Gaussian random variables, each with mean zero and variance σ^2. Suppose that $\{N(t): t \geq t_0\}$ is a doubly stochastic Poisson-process with intensity process $\{\lambda(t, \mathbf{x}): t \geq t_0\}$, where $\lambda(t, \mathbf{x}) = \mathbf{x}'\mathbf{x}\xi v(t) = (x_1^2 + x_2^2 + \cdots + x_k^2)\xi v(t)$, where ξ is a nonnegative parameter, and $v(t)$ is a known, nonnegative function.

a. Suppose that at time T, it is observed that $N(T) = n$. What is the maximum likelihood estimate of ξ in terms of this observed count?

b. Suppose that a path of N is observed on the interval $[0, T)$, and for this path, $N(T) = n$ and jumps occur at times $W_1, W_2, \cdots,$ and W_n. What is the maximum likelihood estimate of ξ in terms of this counting path?

c. Suppose now that $\xi = 1$ and that the variance σ^2 of the variables x_i is the unknown parameter. Indicate from your results in (a) and (b) what the maximum likelihood estimate of σ^2 is for both the count and count-record data.

7.3.2 Consider an orderly point process on the interval $[0, T)$. Let $\{u_i\}$ denote the *unordered* occurrence times on $[0, T)$. These times and the total count $N(T)$ are equivalent to the entire counting record $\{N(\sigma): 0 \leq \sigma < T\}$. Define the entropy of the counting paths by

$$H(0,T) \overset{\Delta}{=} -E(\ln p[\{N(\sigma): t_0 \le \sigma < T\}]),$$

where $p(\cdot)$ is the sample-function density. Observe that H can be written in terms of the statistics of the unordered occurrence times as

$$H(0,T) = -\sum_{n=0}^{\infty} \Pr[N(T)=0]\ln\Pr[N(T)=0]$$

$$-\sum_{n=1}^{\infty} \Pr[N(T)=n] \int_0^T \cdots \int_0^T p_u(U \mid N(T)=n)\ln p_u(U \mid N(T)=n)dU_1 \cdots dU_n.$$

a. Show that subject to the constraint that $E[N(T)] = \bar{n} \overset{\Delta}{=} \lambda T$, the entropy $H(0,T)$ is maximized by the selection

$$p_u(U \mid N(T)=n)\Pr[N(T)=n] = \frac{1}{T^n(1+\bar{n})}\left[\frac{\bar{n}}{1+\bar{n}}\right]^n$$

for $n \ge 1$. Thus, conclude that the entropy is maximized when the unordered occurrence times are independently and uniformly distributed on $[0,T)$ and the counting statistics are Bose-Einstein. Hint: maximize the entropy subject to the constraints:

i. $\displaystyle\sum_{n=0}^{\infty} \Pr[N(T)=n] = 1,$

ii. $\displaystyle\sum_{n=0}^{\infty} n\Pr[N(T)=n] = \bar{n},$

iii. $\displaystyle\int_0^T \cdots \int_0^T p_u(U \mid N(T)=n)dU_1 \cdots dU_n = 1.$

b. Show that

$$H_{\text{max}}(0,T) = (1+\bar{n})\ln(1+\bar{n}) - \bar{n}\ln\left(\frac{\bar{n}}{T}\right).$$

c. Let $\{N(t): 0 \le t < T\}$ be a doubly stochastic Poisson-process with a time independent intensity process x. We know from Ch. 2 that for any given value of x, the unordered occurrence times of N are independently and uniformly distributed given that $N(T)=n \ge 1$. It is natural to inquire if there is a distribution for x for which N is the maximum entropy counting process; i.e., is there a density $p_x(X)$ such that

$$\int_0^\infty \frac{1}{n!}(XT)^n e^{-XT} p_x(X)\,dX = \frac{1}{1+\overline{\lambda}T}\left[\frac{\overline{\lambda}T}{1+\overline{\lambda}T}\right]^n.$$

Show that this holds when x has the gamma distribution (7.3) with $\beta = 0$ and $\alpha = 1/\overline{\lambda}$; that is, when x is exponentially distributed.

7.3.3 Let $\{N(t): t \geq t_0\}$ be an inhomogeneous Polya process. Use the sample-function density for this process in (7.48) to determine the probability that $N(T) = n$ for $n \geq 1$.

7.3.4 Let $\{N(t): t \geq t_0\}$ be a doubly stochastic Poisson-process with intensity process $\{\lambda(t, x(t)): t \geq t_0\}$. Show that the sample-function density $p_t = p[\{N(\sigma): t_0 \leq \sigma < t\}]$ satisfies the stochastic differential equation

$$dp_t = -p_t\hat{\lambda}(t)\,dt + p_t(\hat{\lambda}(t) - 1)\,dN(t), \quad p_{t_0} = 1.$$

Hint: Use the differential rule in Thm. 5.3.2 and (7.71)

7.3.5 (*Interarrival-Time Distribution*). Let $\{N(t): t \geq t_0\}$ be a doubly stochastic Poisson-process with intensity process $\{\lambda(t, \mathbf{x}(t)): t \geq t_0\}$. Define $\Lambda(t) = \int_{t_0}^t \lambda(\sigma, \mathbf{x}(\sigma))\,d\sigma$, and denote the interarrival-time sequence by $\{t_n\}$.

a. The forward-occurrence density is given (2.47) for a Poisson process. Show for $\{N(t): t \geq t_0\}$ that

$$p_{t_n|t_{n-1},\cdots,t_1}(T \mid T_{n-1}, \cdots, T_1)$$

$$= -\frac{\partial E[\exp\{-\Lambda(t_0 + T_1 + \cdots + T_{n-1} + T) + \Lambda(t_0 + T_1 + \cdots + T_{n-1} + T)\}]}{\partial T}.$$

b. Conclude from (a) that if $\{\Lambda(t): t \geq t_0\}$ is a stationary process, then

$$p_{t_n}(T) = -\frac{\partial M_\Lambda(1;T)}{\partial T},$$

where $p_{t_n}(T)$ is the unconditional probability density function for the n*th* interarrival and where

$$M_\Lambda(s;T) = E[e^{s\Lambda(T)}]$$

is the moment-generating function for $\Lambda(T)$. This property was noted by B. Saleh [39].

7.4.1 In the linear filtering problem discussed in Sec. 7.4.1, it is assumed that count-record data are available. This problem is examined for histogram data in this problem.

Let $\{N(t): t \ge t_0\}$ be a doubly stochastic Poisson-process with intensity process $\{\lambda(t): t \ge t_0\}$. Assume that the mean $E[\lambda(t)]$ and covariance function $K_\lambda(t,u)$ for the intensity process are known. An observation interval $[t_0, t)$ is partitioned into m disjoint intervals according to the times $t_0 < t_1 < t_2 < \cdots < t_m = T$, and the number of points occurring in each subinterval is observed. Denote these observations by $N_1, N_2, \cdots, N_m$, where $N_k = N(t_k) - N(t_{k-1})$. Define v_k by

$$v_k = \frac{1}{t_k - t_{k-1}} \int_{t_{k-1}}^{t_k} \lambda(\sigma) d\sigma.$$

a. Show that

$$\mathrm{cov}(v_k, v_l) = \frac{1}{(t_k - t_{k-1})(t_l - t_{l-1})} \int_{t_{k-1}}^{t_k} \int_{t_{l-1}}^{t_l} K_\lambda(u,v) du dv.$$

b. Show that

$$E(N_k) = (t_k - t_{k-1})E(v_k)$$

and

$$\mathrm{cov}(N_k, N_l) = (t_k - t_{k-1})(t_l - t_{l-1})\mathrm{cov}(v_k, v_l) + (t_k - t_{k-1})E(v_k)\delta_{kl},$$

where δ_{kl} is the Kronecker delta function.

c. Define the m-vectors **N** and v by $\mathbf{N} = [N_1 \quad N_2 \quad \cdots \quad N_m]'$ and $v = [v_1 \quad v_2 \quad \cdots \quad v_m]'$ and the $m \times m$ diagonal matrix **T** of subinterval durations $\mathbf{T} = \mathrm{diag}(t_k - t_{k-1})$. Furthermore, let $\mathbf{K}_N$ and $\mathbf{K}_v$ be the covariance matrices for **N** and **v**. Notice from part (b) that $E(N) = TE(v)$, and $\mathbf{K_N} = \mathbf{TK_vT} + \mathbf{T}\text{diag}[E(v_k)]$. Suppose that we observe **N** and want to estimate **v**. Denote the estimate by $\mathbf{v}^* = \mathbf{v}^*(\mathbf{N})$, and suppose it is constrained to be a linear function of **N**. Thus, $\mathbf{v}^*$ can be written in the form

$$\mathbf{v}^* = \mathbf{a} + \mathbf{H}[\mathbf{N} - \mathbf{T}E(v)]$$

for some selection of **a** and **H**, where **a** is an m-vector and **H** is an $m \times m$ matrix. Show that the selection of **a** and **H** that minimizes the mean square-error $E(|\mathbf{v} - \mathbf{v}^*|^2)$ is $\mathbf{a} = E(v)$ and $\mathbf{H} = \mathbf{\Sigma T K_N^{-1}}$, where

$$\mathbf{\Sigma}^{-1} = \mathbf{TK_N^{-1}T} + \mathbf{K_v^{-1}}.$$

d. Show that the minimum mean-square error obtained in part (c) is

$$\min_{\mathbf{v}^*} E(|\mathbf{v} - \mathbf{v}^*|^2) = \text{trace}(\mathbf{\Sigma}).$$

7.4.2 Let $\{N(t): t \geq t_0\}$ be a doubly stochastic Poisson-process with intensity process $\{\lambda(t): t \geq t_0\}$. Suppose that a path of $N(\cdot)$ is observed on the interval $[t_0, T)$ and based on these observations it is desired to estimate some function of the intensity $\lambda(t)$ at time t. Denote this desired estimate by $d(t)$; $d(t)$ could possibly equal $\lambda(t)$, $\int_{t_0}^t \lambda(\sigma)d\sigma$, $d\lambda(t)/dt$, or some nonlinear function of $\lambda(t)$. Assume that the mean-value functions, $E[d(t)]$ and $E[\lambda(t)]$, and the cross-covariance function, $K_{d\lambda}(t,u) = E[d(t)\lambda(u)] - E[d(t)]E[\lambda(u)]$, are known. The time t may or may not equal T. The situation $t = T$ is the filtering problem of Sec. 7.4. The *prediction problem* occurs for $t > T$ and the *smoothing problem* for $t < T$. If $t = T - \tau$ and T increases as a real-time parameter, the smoothing problem is termed *fixed lag* with lag time τ. If t is fixed and T increases, the smoothing problem is termed *fixed point*. And if T is fixed and t increases in $[t_0, T)$, the smoothing problem is termed a *fixed interval sweep*. Suppose that the estimator $d^*(t)$ of $d(t)$ is constrained to be of the linear form

$$d^*(t) = a(t) + \int_{t_0}^T h(t,u)dN(u)$$

for some deterministic function $a(t)$ and (possibly noncausal) impulse response $h(t,u)$. Denote by $\hat{d}(t)$ the linear estimate that results when $a(t)$ and $h(t,u)$ are selected to minimize the mean square-error $E\{[d(t)-d^{*}(t)]^{2}\}$ subject to the linearity constraint.

a.   Show that

$$\hat{d}(t)=E[d(t)]+\int_{t_0}^{T} h_o(t,u)\{dN(u)-E[\lambda(u)]du\},$$

where the optimum impulse response $h_o(t,u)$ is the solution to the following integral equation

$$h_o(t,u)E[\lambda(u)]+\int_{t_0}^{T} h_o(t,\sigma)K_\lambda(\sigma,u)d\sigma=K_{d\lambda}(t,u), \quad t_0 \leq u < t$$

where $E[\lambda(u)]$ and $K_\lambda(\sigma,u)$ are the mean and covariance functions for the intensity process.

b.   Show that the minimum mean square-error is given by

$$E\{[d(t)-\hat{d}(t)]^{2}\}=K_d(t,t)-\int_{t_0}^{T} h_o(t,\sigma)K_{d\lambda}(t,\sigma)d\sigma.$$

7.4.3 Let $\{N(t): t \geq t_0\}$ be a doubly stochastic Poisson-process with intensity process $\{\lambda(t,\mathbf{x}(t)): t \geq t_0\}$. Suppose that $\mathbf{h}^{*}(t)$ is an arbitrary filtering estimate of $\mathbf{h}(t,\mathbf{x}(t))$ in terms of an observed record of N on $[t_0, t)$ and that $\hat{\mathbf{h}}(t) = E\{\mathbf{h}(t,\mathbf{x}(t))\,|\,N(\sigma): t_0 \leq \sigma < t\}$. Show for all choices of $\mathbf{h}^{*}(t)$ and for any nonnegative definite matrix $\mathbf{M}(t)$ that

$$E[\{\mathbf{h}(t,\mathbf{x}(t))-\mathbf{h}^{*}(t)\}'\mathbf{M}(t)\{\mathbf{h}(t,\mathbf{x}(t))-\mathbf{h}^{*}(t)\}]$$

$$\geq E[\{\mathbf{h}(t,\mathbf{x}(t))-\hat{\mathbf{h}}(t)\}'\mathbf{M}(t)\{\mathbf{h}(t,\mathbf{x}(t))-\hat{\mathbf{h}}(t)\}].$$

Conclude, therefore, that the minimum mean square-error estimate of $\mathbf{h}(t,\mathbf{x}(t))$ also minimizes the weighted mean square-error. Also, use this result to argue that the minimum mean square-error estimate of a linear combination of the components of $\mathbf{h}(t,\mathbf{x}(t))$ equals the corresponding linear combination of the components of $\hat{\mathbf{h}}(t)$.

7.4.4 Let $\{N(t): t \geq t_0\}$ be a doubly stochastic Poisson-process with intensity process $\{\lambda(t, x(t)): t \geq t_0\}$. Suppose that $x^*(t)$ is an arbitrary filtering estimate of $x(t)$ in terms of an observed record of N on the interval $[t_0, t)$ and that $\hat{x}_{abs}(t)$ is the value of X for which the posterior distribution $P_{x(t)}(X \mid N(\sigma): t_0 \leq \sigma < t)$ for $x(t)$ is equal to 0.5. Thus, $\hat{x}_{abs}(t)$ is the conditional median of $x(t)$ given $\{N(\sigma): t_0 \leq \sigma < t\}$. Show that

$$E\{|x(t) - x^*(t)|\} \geq E\{|x(t) - \hat{x}_{abs}(t)|\}.$$

7.4.5 Consider the finite-difference version of (7.107),

$$\Delta \mathbf{x}(t) = \mathbf{a}(t, \mathbf{x}(t))\Delta t + \mathbf{B}(t, \mathbf{x}(t))\Delta \mathbf{w}(t),$$

where $\Delta \mathbf{x}(t) = \mathbf{x}(t + \Delta t) - \mathbf{x}(t)$, $\Delta \mathbf{w}(t) = \mathbf{w}(t + \Delta t) - \mathbf{w}(t)$, and $\{\mathbf{w}(t): t \geq t_0\}$ is a vector of independent Wiener processes such that $E[\mathbf{w}(t)\mathbf{w}'(u)] = \mathbf{I}\min(t, u)$. In this case,

$$E[e^{j<\mathbf{v}, \Delta \mathbf{x}(t)>} - 1 \mid \mathbf{x}(t)] = e^{j<\mathbf{v}, \mathbf{a}(t,\mathbf{x}(t))>\Delta t} E[e^{j<\mathbf{v}, \mathbf{B}(t,\mathbf{x}(t))\Delta \mathbf{w}(t)>} \mid \mathbf{x}(t)] - 1.$$

Use the property that $\Delta \mathbf{w}(t)$ is normally distributed with zero mean and covariance $\mathbf{I}\Delta t$ to deduce that

$$E[e^{j<\mathbf{v}, \mathbf{B}(t,\mathbf{x}(t))\Delta \mathbf{w}(t)>} \mid \mathbf{x}(t)] = e^{<\mathbf{v}, \mathbf{B}(t,\mathbf{x}(t))\mathbf{B}'(t,\mathbf{x}(t))\mathbf{v}>\Delta t/2}.$$

Conclude, therefore, that

$$\Psi_t(\mathbf{v} \mid \mathbf{x}(t)) = \lim_{\Delta t \downarrow 0} \frac{1}{\Delta t} E[e^{j<\mathbf{v}, \Delta \mathbf{x}(t)>} - 1 \mid \mathbf{x}(t)]$$

$$= j < \mathbf{v}, \mathbf{a}(t, \mathbf{x}(t)) > - \frac{1}{2} < \mathbf{v}, \mathbf{B}(t, \mathbf{x}(t))\mathbf{B}'(t, \mathbf{x}(t))\mathbf{v} > .$$

7.4.6 Derive equation (7.115).

7.4.7 Define

$$f(t, \mathbf{X}) = \frac{e^{H(t, \mathbf{X})} p_{\mathbf{x}}(\mathbf{X})}{E[e^{H(t, \mathbf{x})}]},$$

where

$$H(t,\mathbf{x}) = -\int_{t_0}^{t} \lambda(\sigma,\mathbf{x})\,d\sigma + \int_{t_0}^{t} \ln\lambda(\sigma,\mathbf{x})\,dN(\sigma),$$

and the expectation in the denominator is with respect to the probability density $p_{\mathbf{x}}(\mathbf{X})$ of $\mathbf{x}$. According to (7.134), $f(t,\mathbf{X})$ is the posterior density of $\mathbf{x}$ given observations $\{N(\sigma): t_0 \le \sigma < t\}$ of a doubly stochastic Poisson-process with intensity process $\{\lambda(t,\mathbf{x}): t \ge t_0\}$. This can be verified directly by using the differential rule of Thm. 5.3.2 to show that $f(t,\mathbf{X})$ satisfies (7.129). Let $P(t,\mathbf{X})$ and $Q(t)$ be the numerator and denominator of $f(t,\mathbf{X})$, respectively.

a. Show that

$$dP(t,\mathbf{X}) = -\lambda(t)P(t,\mathbf{X})\,dt + [\lambda(t,\mathbf{X})-1]P(t,\mathbf{X})\,dN(t),$$

$$P(t_0,\mathbf{X}) = p_{\mathbf{x}}(\mathbf{X}).$$

b. Show that

$$dQ(t) = -\hat{\lambda}(t)Q(t)\,dt + [\hat{\lambda}(t)-1]Q(t)\,dN(t),$$

$$Q(t_0) = 1,$$

where $\hat{\lambda}(t) = E[\lambda(t,\mathbf{x})\,|\,N(\sigma): t_0 \le \sigma < t]$.

c. Use the results of (a) and (b) to show that

$$df(t,\mathbf{X}) = f(t,\mathbf{X})[\lambda(t,\mathbf{X}) - \hat{\lambda}(t)]\frac{1}{\hat{\lambda}(t)}[dN(t) - \hat{\lambda}(t)\,dt],$$

$$f(t_0,\mathbf{X}) = p_{\mathbf{x}}(\mathbf{X}).$$

7.4.8 Let $\{N(t): t \ge t_0\}$ be an inhomogeneous Poisson process having an intensity $\{v(t): t \ge t_0\}$ that is randomly scaled by a nonnegative random variable x. The result is a doubly stochastic Poisson-process with intensity process $\{xv(t): t \ge t_0\}$. Suppose that x is gamma distributed with parameters α and β, so

$$p_x(X) = \frac{\alpha^{\beta+1}}{\Gamma(\beta+1)} X^\beta e^{-\alpha X}.$$

a. Show that x is also conditionally gamma distributed given $\{N(\sigma): t_0 \leq \sigma < t\}$ and that in the conditional distribution α and β become $\alpha + \int_{t_0}^{t} v(\sigma) d\sigma$ and $\beta + N(t)$.

b. Evaluate the estimate $\hat{x}(t) = E[x|N(\sigma): t_0 \leq \sigma < t]$.

c. Show that $\{N(t): t \geq t_0\}$ is a Markov birth process, and determine the birth rate $\mu(t,n)$, where

$$\mu(t,n) = \lim_{\Delta t \downarrow 0} \frac{1}{\Delta t} \Pr[N(t, t+\Delta t) = 1 | N(t) = n].$$

Hint: Use Theorems 6.3.1 and 7.2.1

7.4.9 (*Continuation*). Let the model be that of Prob. 7.4.8 with $\hat{x}(t)$ as defined in part (b).

a. Evaluate the conditional mean square-error

$$E\{[x(t) - \hat{x}(t)]^2 | N(\sigma): t_0 \leq \sigma < t\}.$$

b. Evaluate the minimum mean square-error $E\{[x(t) - \hat{x}(t)]^2\}$

c. Evaluate the conditional bias $B(x) = E[\hat{x}(t) - x | x]$ of the minimum mean square-error estimate of x.

d. Evaluate the bias $B = E[B(x)]$ of the minimum mean square-error estimate of x.

7.4.10 Let $\{N(t): t \geq t_0\}$ be a doubly stochastic Poisson-process with intensity process $\{v(t)e^{-xt}: t \geq t_0\}$, where x is a nonnegative random variable with prior probability density $p_x(X)$. Show that the minimum mean square-error estimate $\hat{x}(t)$ of x given $\{N(\sigma): t_0 \leq \sigma < t\}$ is given by

$$\hat{x}(t) = f\left(\int_{t_0}^{t} \sigma dN(\sigma) \right),$$

where

$$f(t,z) = E\left[\exp\left(-zx - \int_{t_0}^{t} \nu(\sigma)e^{-x\sigma}d\sigma\right)\right]$$

in which $E(\cdot)$ indicates integration with respect to the prior density of x. Conclude, therefore, that the sum of the occurrence times is a sufficient statistic for estimating x.

7.4.11 Let $\{N(t): t \geq t_0\}$ be a homogeneous, doubly stochastic Poisson-process with a time independent intensity $\lambda(\mathbf{x}) = \sum_{i=1}^{k} x_i^2$, where x_1, x_2, $\cdots$, and x_k are independent, identically distributed Gaussian random variables with zero mean and variance σ^2. Suppose at time T that $N(T) = n$.

a. Determine the minimum mean square-error estimate $\hat{\lambda}_{\text{MMSE}}(n)$ of the intensity given $N(T) = n$. This estimate is the conditional mean of λ given the data.

b. Determine the maximum *a posteriori* probability estimate $\hat{\lambda}_{\text{MAP}}(n)$ of the intensity given $N(T) = n$. This estimate is the conditional mode of λ given the data.

c. Show that the maximum likelihood estimate $\hat{k}_{\text{ML}}(n)$ of the number of independent components in λ given $N(T) = n$ is the value of k that maximizes

$$\frac{\Gamma\left(n + \frac{k}{2}\right)}{\Gamma\left(\frac{k}{2}\right)\sqrt{1 + 2\sigma^2 T}^{k}}$$

where $\Gamma(\cdot)$ is the gamma function.

d. Determine the minimum men square-error estimate of λ given an entire counting path $\{N(\sigma): t_0 \leq \sigma < t\}$.

7.4.12 Suppose that the random telegraph wave of Examples 7.4.3 and 7.4.9 is modified so that the transition rates from state ζ_0 (off) to state ζ_1 (on) and from state ζ_1 to state ζ_0 are not equal; denote these rates by ν_{01} and ν_{10}, respectively. The model is unchanged otherwise.

a. Specify a differential equation for the conditional occupancy probability $\hat{\pi}_1(t)$ of state ζ_1 between observed photoelectron emission times.

b.   Solve the differential equation of part (a).

c.   Indicate how $\hat{\pi}_1(t)$ should be updated at the instant a photoelectron emission occurs.

7.4.13 Let $\{N(t): t \geq t_0\}$ be a doubly stochastic Poisson-process with intensity process $\{\lambda(t, x(t)): t \geq t_0\}$, where $\{x(t): t \geq t_0\}$ is a continuous time Markov jump process. Let $\{\zeta_i\}$ denote the states of $x(t)$ and $\pi_i(t) = \Pr[x(t) = \zeta_i]$ the state occupancy probabilities at time t.

a.   Show that the conditional state occupancy probabilities given an observed counting path $\{N(\sigma): t_0 \leq \sigma < t\}$ have the representation

$$\hat{\pi}_i(t) = \frac{E\{e^{H(t)} \mid x(t) = \zeta_i\}}{E\{e^{H(t)}\}} \pi_i(t),$$

where

$$H(t) = -\int_{t_0}^{t} \lambda(\sigma, x(\sigma)) d\sigma + \int_{t_0}^{t} \ln \lambda(\sigma, x(\sigma)) dN(\sigma).$$

b.   Use the result of (a) and the differential rule to obtain an alternative derivation of the equation (7.151).

7.5.1 Parallel the proof of Theorem 5.3.2 to establish the multichannel differential rule (7.220).

7.5.2 Let $\{N(t): t \geq t_0\}$ be a multichannel doubly stochastic Poisson-process with intensity process $\{\lambda(t): t \geq t_0\}$. Let $\{y(t): t \geq t_0\}$ be a filtered multichannel doubly stochastic Poisson-process,

$$y(t) = \int_{t_0}^{t} H(t, u) dN(u).$$

Establish the following extension of the results in Prob. 7.2.6.

a. Show that

$$E[y(t)] = \int_{t_0}^{t} H(t, u) E[\lambda(u)] du.$$

b.   Let $\mathbf{K}_y(t,u) = E[\mathbf{y}(t)\mathbf{y}'(u)] - E[\mathbf{y}(t)]E[\mathbf{y}'(u)]$ be the covariance matrix of $\mathbf{y}(\cdot)$. Show that

$$\mathbf{K}_y(t,t) = \int_{t_0}^{t} \mathbf{H}(t,u)\mathrm{diag}[E\{\lambda_k(u)\}]\mathbf{H}'(t,u)\,du$$

$$+ \int_{t_0}^{t}\int_{t_0}^{t} \mathbf{H}(t,u)\mathbf{K}_\lambda(u,v)\mathbf{H}'(t,v)\,du\,dv,$$

where $\mathbf{K}_\lambda$ is the covariance matrix for λ.

c.   Suppose that $\mathbf{H}$ is a square matrix so that the dimensions of $\mathbf{y}$ and $\mathbf{N}$ are the same. Show that

$$\mathrm{cov}[\mathbf{y}(t),\lambda(x)] = E[\mathbf{y}(t)\lambda'(s)] - E[\mathbf{y}(t)]E[\lambda'(s)] = \int_{t_0}^{t} \mathbf{H}(t,u)\mathbf{K}_\lambda(u,s)\,du.$$

7.5.3 Derive the equations (7.235) and (7.236) for linear multichannel linear filtering by using the results of Prob. 7.5.2 and paralleling the proof of Theorem 7.4.1.

7.5.4 Let $\{\mathbf{N}(t): t \geq t_0\}$ be a multichannel doubly stochastic Poisson-process with intensity process $\{\lambda(t,\mathbf{x}): t \geq t_0\}$, where $\mathbf{x}$ is a random variable with probability density $p_\mathbf{x}(\mathbf{X})$. Let $p_t(\mathbf{X}|\mathbf{N}(\sigma): t_0 \leq \sigma < t)$ be the posterior probability density of $\mathbf{x}$ given a counting path on $[t_0, t)$.

a. Show as an extension of (7.132) that

$$p_t(\mathbf{X}|\mathbf{N}(\sigma): t_0 \leq \sigma < t)$$

$$= p_\mathbf{x}(\mathbf{X})\exp\left[<1,-\int_{t_0}^{t}[\lambda(\sigma,\mathbf{X})-\hat{\lambda}(\sigma)]\,d\sigma + \int_{t_0}^{t}\mathrm{diag}\left\{\ln\left[\frac{\lambda_k(\sigma,\mathbf{X})}{\hat{\lambda}_k(\sigma)}\right]\right\}d\mathbf{N}(\sigma)>\right].$$

b.   What is the multichannel extension of (7.134)?

AUTHOR INDEX

EXAMPLES INDEX

SUBJECT INDEX

MIX
Papier aus verantwortungsvollen Quellen
Paper from responsible sources
FSC® C105338
www.fsc.org

If you have any concerns about our products,
you can contact us on
ProductSafety@springernature.com

In case Publisher is established outside the EU,
the EU authorized representative is:
**Springer Nature Customer Service Center GmbH
Europaplatz 3, 69115 Heidelberg, Germany**

Printed by Libri Plureos GmbH
in Hamburg, Germany